T0174756

ENVIRONMENTAL ILLNESS
Myth and Reality

ENVIRONMENTAL
SERIES

Herman Staudenmayer, Ph.D.

ENVIRONMENTAL ILLNESS

Myth and Reality

CRC Press
Taylor & Francis Group
Boca Raton London New York

CRC Press is an imprint of the
Taylor & Francis Group, an **informa** business

CRC Press
Taylor & Francis Group
6000 Broken Sound Parkway NW, Suite 300
Boca Raton, FL 33487-2742

© 1999 by Taylor & Francis Group, LLC
CRC Press is an imprint of Taylor & Francis Group, an Informa business

First issued in paperback 2019

No claim to original U.S. Government works

ISBN 13: 978-0-367-44769-4 (pbk)
ISBN 13: 978-1-56670-305-5 (hbk)

This book contains information obtained from authentic and highly regarded sources. Reasonable efforts have been made to publish reliable data and information, but the author and publisher cannot assume responsibility for the validity of all materials or the consequences of their use. The authors and publishers have attempted to trace the copyright holders of all material reproduced in this publication and apologize to copyright holders if permission to publish in this form has not been obtained. If any copyright material has not been acknowledged please write and let us know so we may rectify in any future reprint.

Except as permitted under U.S. Copyright Law, no part of this book may be reprinted, reproduced, transmitted, or utilized in any form by any electronic, mechanical, or other means, now known or hereafter invented, including photocopying, microfilming, and recording, or in any information storage or retrieval system, without written permission from the publishers.

For permission to photocopy or use material electronically from this work, please access www. copyright.com (http://www.copyright.com/) or contact the Copyright Clearance Center, Inc. (CCC), 222 Rosewood Drive, Danvers, MA 01923, 978-750-8400. CCC is a not-for-profit organization that provides licenses and registration for a variety of users. For organizations that have been granted a photocopy license by the CCC, a separate system of payment has been arranged.

Trademark Notice: Product or corporate names may be trademarks or registered trademarks, and are used only for identification and explanation without intent to infringe.

Visit the Taylor & Francis Web site at
http://www.taylorandfrancis.com

and the CRC Press Web site at
http://www.crcpress.com

Library of Congress Cataloging-in-Publication Data

Staudenmayer, Herman
        Environmental illness : myth and reality / by Herman Staudenmayer.
            p. cm.
        Includes bibliographical references and index.
        ISBN 1-56670-305-0 (alk. paper)
        1. Multiple chemical sensitivity.   2. Environmentally induced
diseases. 3. Medical misconceptions. 4. Mental illness — Environmental aspects.
        5. Chemicals — Health aspects. I. Title.
            [DNLM: 1. Environmental Illness — psychology.   2. Environmental
Illness — etiology.   3. Environmental Illness — therapy.
        4. Psychotherapy — methods.   WM 31S798e   1998]
        RB 152.6.S73
        615.9'02 — dc21
        DNLM/DLC
        for Library of Congress   98-8694
                                                                                              CIP

Library of Congress Card Number 98-8694

Foreword

This is an important book — a *very* important book. It concerns the set of phenomena called "environmental illness" (EI). Dr. Staudenmayer has studied patients who are supposedly suffering from exposure to low-level amounts of substances in their environments — foods, chemicals, toxins, biologics, and so forth. The patients feel bad and are frustrated. Dr. Staudenmayer describes his own frustration in trying to understand these patients, the sources of their discomfort, and the many practitioners who reinforce their ideation. The frustration is real because, as Dr. Staudenmayer discusses, there is no evidence that their perceived disabilities are in fact due to these exposures. It's a house of cards.

How could such situations arise and be perpetuated? That is the subject of this book. It encompasses a wide variety of discussions of various psychological mechanisms, some or all of which are deeply involved in "environmental illness".

In addition to his own experience with these patients over many years, psychologist Staudenmayer has read deeply, widely, and carefully. He evaluates the literature. I estimate that there are over 1000 references. Thus, this is not only a thorough exploration of many psychological pathways, but an encyclopedia, as well. The references are drawn from a vast array of medical sources, some of which would not be consulted by the usual practitioner.

There is no other place in the literature where the reader can find such an account of EI together with authoritative discussions of personality disorders, the placebo response, stress and reactions to stress, somatization disorders, phenomena such as "limbic kindling", and an evaluation of "clinical ecology".

This is a passionate book. Dr. Staudenmayer is deeply concerned for the welfare of people alleged to suffer from EI. If he weren't concerned, he never would have made the heroic effort this book represents.

He is a true advocate for his patients. His book is worth very serious consideration.

Henry N. Claman, M.D.
Distinguished Professor of Medicine and Immunology
University of Colorado School of Medicine

Preface

This book addresses the complex issues surrounding a phenomenon that goes by many names. It originated in the 1940s as ecological or environmental illness (EI) and now is most commonly called multiple chemical sensitivities (MCS). One might expect that books on environmental medicine would be written by physicians and toxicologists who are familiar with the disease processes induced by toxic environmental agents. In fact, medical and toxicological experts have published voluminous material about verifiable allergic, dermatologic, respiratory, neurologic, and other responses to toxic environmental agents. But, this book is not about environmentally induced illness. The environmental illness phenomenon has no generally accepted definition. It is not a medical diagnosis. It does not qualify as a disease or even as a recognized medical syndrome. Nonetheless, its advocates would have us believe that EI affects 15% of the population, with disastrous mortality and morbidity.

Several prestigious medical organizations have issued position papers criticizing the theories of environmental illness and the unsubstantiated practices of their proponents. Each of these organizations has also pointed out how the MCS phenomenon is really not within its purview and that the pathological mechanisms it recognizes cannot explain it. Thus, as if by default, I find myself writing a comprehensive book on environmental illness. Someone once said it is what we do not know that scares us. But what I *do* know about this subject frightens me.

I am informed in writing this book by many years of research and clinical experience with hundreds of patients who believed they were affected or disabled by food, chemical, or other environmental sensitivities. During the 1970s, while working at the National Jewish Hospital and Research Center, I observed the often dramatic and life-threatening allergic reactions of asthma patients and became familiar with the devastating psychosocial effects associated with chronic, life-threatening illness. In the late 1970s, allergist John C. Selner, M.D., and I began looking for a connection between low-level environmental exposures — to both foods and chemicals — and allergic mechanisms. Having encountered patients who claimed to be exquisitely sensitive to such exposure, we decided to investigate their claims. For 5 years, we assessed and treated inpatients in a specially constructed "environmental care unit" at Presbyterian Medical Center in Denver. This facility included a chamber for conducting double-blind, placebo-controlled provocation-challenge studies under controlled conditions. After 1985, we also saw outpatients in our respective private practices, under the organizational umbrella of the Allergy Respiratory Institute of Colorado, which also contained a challenge chamber.

At first, we were excited with the prospect of finding scientific evidence to corroborate the harmful effects that others had attributed to low-level exposures to foods and chemicals. We did find evidence of hypersensitivity in a few individuals, but these effects were limited to well-known IgE mediated immunologic responses or well-characterized toxicological or irritant effects. These responses were compatible with dose-duration parameters described for these chemicals. One of our first demonstrations of a chemically induced pulmonary reaction involved an aircraft painter with a long history of exposure to isocyanate who responded to minute exposure of the paint containing this accelerator. He was demonstrated to have a profound drop in baseline pulmonary function when exposed to paint at levels approximating his workplace exposure. Early on, this and many other similar experiences educated this psychologist to the reality of chemical intolerance and its specificity. However, we could not find a single case in which exposure to low levels of many different classes of chemicals and foods induced the gamut of symptoms attributed to EI. Our alternative explanations for the EI are presented in this book.

During the early 1980s, working with difficult patients who were largely shunned by the established medical community left me open to criticism. Advocates of toxicogenic explanations for EI referred to themselves as "clinical ecologists". Some colleagues mistakenly considered me guilty by association simply because I treated patients professing chemical sensitivities. Many conventional physicians saw me as a pariah. On one occasion I went toe-to-toe with the hospital's chief of medicine who wanted those environmental "kooks" out of the environmental care unit because they had no objective immunologic disease. A chief of psychiatry informed me that referrals to his ward were unwelcome because these patients were too difficult to manage. Psychiatrists for medication consults were hard to find — one case was typically enough for them. Several academic friends and psychology colleagues thought I had made a poor decision in working with these patients. The early years of working with EI were among the most interesting of my career. I saw the depths of human fear and despair, and experienced the satisfaction of helping some afflicted individuals find a path through a hostile medical world to wellness. Regrettably, I also came to learn the limits to which despair can be exploited.

My motivation for writing this book includes incredulity, disgust, sadness, and frustration upon repeatedly witnessing something terribly wrong. A sage once cautioned, "He is a fool who cannot get angry, but he is a wise man who will not." It was probably the same wise man who uttered, "Speak out when you are angry and you will produce the best argument you will ever regret." EI patients may be unsuspecting. They may be vulnerable, desperate, and gullible, and some may indeed be exploiting the system. But, too many are easy prey for charlatans, manipulators, and social parasites who offer diagnostic tests which have never been scientifically validated and phantom treatment practices couched in pseudoscientific rhetoric. I believe that many of their patients contract an iatrogenic (physician-induced) illness, which, at its worse, leaves them anxious, depressed, hopeless, and socially isolated. The "craziness" inherent in EI phenomena is not within the patients.

Traditional medicine too often rejects these patients. I acknowledge they do not meet the cost and time-efficient parameters imposed by the managed care mentality. But I still believe physicians take a public oath upon entering their profession and I do not recall any mention of cost or time sworn to. "Do no harm" is still a dictum that measures the medical profession. It seems to me that abandoning the EI patient to the charlatans is inconsistent with "do no harm." EI patients suffer greatly. They require and should expect appropriate treatment. Clinical experience has taught me that many have been ravaged by childhood trauma or adult misfortune and stress. For those so affected, the adult sequelae of trauma and stress include ineffective psychological defenses, disrupted personality structure, biological abnormalities from chronic activation of the stress response, and the inability to cope effectively with problems of living. Many of these individuals see the world and the environment as hostile, and not without justification. In severe cases, individuals live in isolation because they perceive everything around them as life threatening: "The world is an unsafe place to live." The psychologically detrimental consequences of such fears and the resulting assumed or imposed isolation should not be ignored. Even patients who are initially adamant in their rejection of obvious alternative explanations for their symptoms can be helped with proper comprehensive medical and psychological care.

If no harm were associated with the induction of unsubstantiated beliefs about environmental intolerances, if there were no harm in raising false hopes with unsubstantiated methods of treatment, if there were no despair resulting from unjustifiable withdrawal from society and isolation, if there were no harm to the families of these victims, then I would not have undertaken the writing of this book. To this day I vividly recall reading a postcard sent from Hawaii by a former patient thanking me for helping her live the last year of her life enjoyably until she succumbed to cancer. When I first saw her in our inpatient facility, she expressed acute xenophobia, experienced panic, and was emaciated to the point she was at risk of starvation because she did not eat for fear of food intolerances. She had come to believe she was allergic to everything after being treated by a clinical ecologist. Medical examination revealed a history of cancer which accounted for her symptoms but had never been evaluated by the clinical ecologist. Psychotherapy helped her overcome the iatrogenic beliefs about EI. She ate again and gained weight. After 10 days of hospitalization she could be reintroduced into the outside world without difficulty. At 6 months' follow-up, she had returned to her usual endeavors, was again enjoying the symphony and gourmet dining, and had gained an additional 15 pounds. She spent her last months on the beach at Maui.

The harm resulting from this phenomenon extends to all elements of society. It has ramifications for science, medicine, social attitudes about the environment, communities and schools compelled to have "chemically free zones," insurance coverage, regulatory policy, legislation, civil litigation, disability and entitlement programs, and politics. Harm is involved in the chemical sensitivity phenomenon, but it does not come from the physical environment. Rather, the "craziness" lies with the exploiters and proponents of a practice and belief system which resembles the conviction of metaphysics, not science. This "craziness" is tolerated by a society intimidated and immobilized by policies catering to victimization and invalidism. The motivation for this book is the unnecessary harm done to these suffering souls and the unnecessary burden perpetrated on society by what I believe is tantamount to organized crime.

Moreover, no previous comprehensive, science-based account of the biological, psychological, social, and political facets of this phenomenon has been published. This book aims to fill this void for those who are affected by EI as patients, as family members who lose loved ones, as targets of liability and disability claims, or as citizens asked to accommodate "environmentally sensitive" individuals.

The author

Herman Staudenmayer, Ph.D., is a psychologist in private practice in Denver, CO. During the past 25 years, he has conducted research on the relationship between medical and psychological disorders. He has collaborated with allergists, immunologists, internists, chest physicians, toxicologists, neurologists, chemists, pharmacists, and psychologists in both inpatient (National Jewish Center, Presbyterian Hospital) and outpatient settings. His research interests and publications have spanned basic cognitive science, language and reasoning, medical decision-making, antisocial behavior, psychophysiology and biofeedback, psychosomatic medicine and psychogenic illness, psychoneuroimmunology, personality disorders, and the methodology of provocation chamber challenges. He has also evaluated or treated several hundred patients with environmental illness. He had a 20-year collaborative relationship with John C. Selner, M.D., director of the Allergy Respiratory Institute of Colorado, during which they addressed many of the issues presented in this book. Dr. Staudenmayer has held academic positions at the Graduate Faculty of the New School for Social Research in New York City and the University of Colorado Health Sciences Center in Denver. He has also been Adjunct Professor of Psychology at the New Mexico State University. He has organized three international conferences on the topic of this book. In 1996, he participated in a panel convened in Berlin by the World Health Organization, the German Federation for the Environment, and the German Health Ministry to develop a consensus on environmental illness (MCS). His writings include six book chapters and 40 papers in the scientific literature.

Acknowledgments

The author wishes to thank the following individuals for their help during preparation of this manuscript.

The CRC/Lewis editorial staff:	Kenneth McCombs
	Susan Alfieri
Scientific consultants:	Henry Claman, M.D.
	John C. Selner, M.D.
Source materials:	E. Marcus Barnes, P.E.
	Timothy Capshandy, J.D.
	David T. Janigan, M.D.
	Eliot Jubelierer, Esq.
	Cindy Lynn Richard
	State-of-Science speakers of the Aspen Environmental Medicine Conferences, 1994, 1995, 1996.
Editorial assistance:	Stephen Barrett, M.D.
	Ginger Burrus, M.S.
	Sarah Nicely Fortener

Contents

Dedication

To John C. Selner, M.D.
my colleague and friend

Dedication

chapter one

What is *"environmental illness"*?

The subject of this book has many names. Whenever a phenomenon is called so many things one can be sure that none describes it very well. Some say it is so mysterious that its essence cannot be conceived, much less defined. Its effects are said to be powerfully harmful. There is consensus that it has significant impact on our society. It has ramifications for science, medicine, social attitudes, the media, insurance coverage, regulatory policy, legislation, civil litigation, disability and entitlement programs, politics, and, last but not least, the individuals said to suffer from it. Yet,

- We don't know what to call it.
- It can't be defined.
- It can't be diagnosed.
- Responsible scientific disciplines tells us it does not qualify as a disease or as a syndrome.
- It has no consistently identified biological mechanism of action.
- It is based on toxicologic and physiologic rationale that is implausible.
- It is inconsistent with the laws of chemistry and physics.

For these reasons, it is best to think of "environmental illness" as a phenomenon rather than a disease.

I begin this chapter by contrasting environmental illness with well-characterized, chemically induced diseases. Second, I define some of the concepts of environmental illness (EI) and the parties who advocate environmental illness as a toxicogenic theory, meaning that ill-health effects are caused by environmental exposures to chemicals at levels tolerated by most of the population. Third, I describe the characteristics of patient symptoms, all of which need to be explained. And, fourth, the remainder of the chapter is a brief outline of the two opposing theories — toxicogenic and psychogenic — highlighting the issues to be addressed in later chapters of the book.

What it is not

Historically, the effects of chemicals on disease and particularly the functioning of the nervous system are firmly rooted in antiquity and are universally accepted tenets of modern medical practice. The clearest example in medicine is the *Physician's Desk Reference* of drugs, which is an encyclopedia of the healing as well as the adverse side effects that can be caused by medicines prescribed by physicians. The mortality and associated costs related to adverse drug events in the U.S. are greater than that of motor vehicle injuries, assaults with firearms, and AIDS combined. According to Multum Information Services (Denver, CO), 19% of medical treatments are associated with adverse drug events, which cause 11% of all hospitalizations and account for 60,000 to 140,000 inpatient fatalities per year.

There is no question that illness related to chemicals in the environment can be produced by toxicologic as well as immunologic mechanisms. A variety of physical factors, infection, or direct toxic or pharmacologic effects of assorted natural and synthetic substances can do the same. Generally, these diseases present with characteristic symptoms and physical and laboratory findings, as well as a definable pathology. One example is the causal link between ingestion of alfalfa seeds and symptoms of systemic lupus erythematosus (SLE) (Selner and Condemi, 1988). A second example is the causal link between contaminated rapeseed oil and an epidemic of scleroderma-like syndrome that erupted in Spain in 1981, affecting close to 20,000 people (Lopez-Ibor et al., 1985). Death and respiratory disorders were sobering consequences in the methyl isocyanate chemical spill in Bhopal, India (Murthy and Isaac, 1987). And, possibly the worse technological disaster was the nuclear radiation accident at Chernobyl, Ukraine, which killed instantly, caused terminal cancer and chronic diseases, contaminated the region, and created an anxiety throughout Europe.

Specialists in the fields of allergy and immunology, endocrinology, occupational and environmental medicine, pulmonary medicine, toxicology and pharmacology, neurology, and neurotoxicology have demonstrated adverse effects associated with exposure to many chemicals in well-documented, critically reviewed scientific publications (Third Task Force for Research Planning in Environmental Health Science, 1984; Gammage and Kaye, 1985; Frank et al., 1985; Wallace et al. 1987; National Research Council, 1992b; U.S. Congress, Office of Technology Assessment, 1992; Spencer and Schaumburg, 1980, 1985; Schaumburg and Spencer, 1987; Dybing et al., 1996; Tsien and Spector, 1997). In our laboratory, we have identified individuals who have consistent adverse responses to provocation challenges with low levels of chemicals, illustrated by the following case.

> A 53-year-old male aircraft painter complained of respiratory symptoms characterized by bronchospasm on exposure to fumes generated by spray painting. Despite the use of a respirator, he would experience tightness of the chest and wheezing within 30 minutes of exposure to paint fumes. His symptoms cleared on weekends or prolonged holidays, only to recur when he was re-exposed at the workplace. Inhalation challenge testing with methacholine to rule out reversible airway disease was inconclusive, with a drop in FEV_1 of 15% (Figure 1.1) occurring only at high doses (25mg/ml). These results suggested that observable respiratory effects attributable to paints could be due to something more than the irritant effect. A simple provocation procedure employing aerosol polyurethane paint and isocyanate hardening accelerators used in the workplace resulted in bronchospasm, as measured by pulmonary functions, with a drop in FEV_1 of more than 50% soon after exposure. Because the exposure involved a suspected chemical from the workplace and the challenge demonstrated objective pulmonary changes corroborating symptom complaints, no further evaluation was required. Avoidance options were explored in a harmonious arbitration of the patient's disability status (Selner and Staudenmayer, 1985).

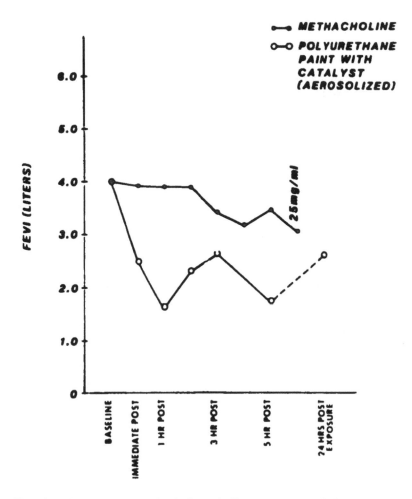

Figure 1.1 Closed circles indicate methacholine challenges at five inhalations each dose, with 3-minute intervals between doses. Dose sequence: placebo, 0.07, 0.31, 1.25, 5.0, 25.0 mg/mL. (From Selner, J.C. and Staudenmayer, H., *Ann. Allergy*, 55(5), 665–673, 1985. With permission.)

To someone unfamiliar with the current usage of the term "environmental illness", it could mean a host of chemically related diseases resulting from well-characterized, high-level environmental exposures that have toxicogenic ill-health effects and associated emotional sequelae, such as those just presented. This example is presented to emphasize what this book is *not* about. The subject of this book is much more difficult to define.

Examples of what it is

The 1973 Arab oil embargo impacted environmental illness. To conserve energy for heating and cooling, large buildings were designed to be air-tight, and the rates of air exchange were reduced through alterations in heating, ventilation, and air-conditioning (HVAC) systems. Because of these design and engineering changes, the quality of indoor air was affected and some people fell ill. These effects are called building-related illness, defined as being an objective disease or discomfort caused by identifiable agents in the building such as excessive carbon monoxide or carbon dioxide, inappropriate lighting, or allergens such as molds, pollens, and others which proliferate or are trapped in the accumulation of moisture or dust (Bardana and Montanaro, 1997). In contrast to building-related illness, the term "sick building syndrome" was coined as an opportunistic expression which did not require objective indices of disease, nor exposure to measurable levels of chemicals or allergens that exceed some toxicologic standard. Initial speculation focused on the quality of "stale air", but buildings with documented poor air circulation, as measured by comparative carbon dioxide concentrations, were found to be unrelated to symptom complaints.

One double-blind study in Canada tested the hypothesis that symptoms were associated with the accumulation of contaminants resulting from reduced supply of outdoor air (Menzies et al., 1993). The study involved 1546 workers in four buildings. Unknown to the workers, the ventilation systems were manipulated to deliver varying amounts of outdoor air (7 and 32%) for one-week periods, randomly counterbalanced over consecutive weeks. Each week the workers rated their symptoms. There were no differences in the symptom ratings associated with the amount of outdoor air flow, even when corrected for work-site manipulations of ventilation, temperature, humidity, and air velocity. The investigators concluded that the supply of outdoor air did not appear to affect workers' perception of their office environment or their reporting of symptoms considered to be typical of the "sick building syndrome".

Furthermore, "sick building syndrome" could occur in buildings that were built before 1973 and did not have the airtight design or decreased air-exchange rate. This led to alternative hypotheses about other chemicals being the toxic culprits (Kreiss, 1993); however, studies showed that symptoms did not correlate with measurements of indoor air levels of particulates, volatile organic compounds, and specific chemicals such as formaldehyde (Hodgson et al., 1991).

The purported causes of "sick building syndrome" include anything that can be found in buildings, even agents that go beyond chemicals. For example, electromagnetic force (EMF) fields are postulated to affect body polarity adversely. These EMF fields may emanate not only from inanimate electrical sources such as power lines, heating blankets, televisions, or computer visual display units, but also from another person. Each person is said to have unique EMF fields which may be incompatible with those of another person such that one individual can induce a reaction in the other if they come into close proximity.

It appears that there is more to "sick building syndrome" than the 1973 oil embargo. Several studies have shown that social and organizational factors in the workplace offer the most probable explanation of "sick building syndrome" (Baker, 1989; Hedge, 1995; Ryan and Morrow, 1992).

With apprehension, the next environmental manifestation of environmental illness was awaited. The wait was not long, and the most recent manifestation was ironically also related to the Mideastern oil politics resulting in the Persian Gulf War. Veterans who participated in Desert Shield, Desert Storm, and the peacekeeping afterwards began complaining of a mysterious illness that has been labeled "Gulf War Syndrome" and is officially referred to as Gulf War Veterans' Illnesses by the Presidential Advisory Committee. Not only were the veterans deployed in the Gulf affected, but also their family members with whom they had contact after returning to the U.S. To date, after extensive investigation, no evidence of a toxicologic effect has been demonstrated, while numerous toxicogenic theories have been disproved (Lashof, 1996).

Table 1.1   Some Labels Applied to the EI Phenomenon

Aircraft workers' syndrome	Gulf War syndrome
Allergic toxemia	Multiple chemical sensitivity
Cerebral allergy	Multiple chemical sensitivity syndrome
Chemical AIDS	Sick building syndrome
Chemical allergy/sensitivity	Total allergy syndrome
Chemical hypersensitivity syndrome	Toxic carpet syndrome
Chemically induced immune	Toxic-induced loss of tolerance
dysregulation syndrome	Total environmental allergy
Chemical intolerance	Total immune disorder syndrome
Ecologic illness	Toxic encephalopathy
Environmental hypersensitivity	Toxic response syndrome
Environmental illness	Toxicant-induced loss of tolerance
Environmental irritant syndrome	20th-century disease
Environmental maladaption syndrome	Universal allergy
Generalized immune deficiency	Universal reactivity, universal reactors

Definitions

Naming the indefinable

> If one conception does not differ practically from a second, then
> acceptance of it adds nothing to our acceptance of that second con-
> ception; and if it has no assignable practical effects whatsoever, then
> the works that express it are so much meaningless verbiage.
>
> C.S.Peirce's *Philosophy of Pragmatism* (see Gallie, 1952)

A seemingly endless list of synonyms for this phenomenon has come into our vocabulary
(Table 1.1). The most common labels are environmental illness, ecologic illness, chemical
sensitivities, and multiple chemical sensitivity (MCS).

The label "multiple chemical sensitivity" was coined by Cullen for the purpose of
identifying a population to study experimentally and is defined as (Cullen, 1987):

> An acquired disorder characterized by recurrent symptoms, refer-
> able to multiple organ systems, occurring in response to demon-
> strable exposure to many chemically unrelated compounds at doses
> far below those established in the general population to cause harm-
> ful effects. No single widely accepted test of physiologic function can
> be shown to correlate with symptoms.

Table 1.2 summarizes the seven criteria Cullen proposed to establish the diagnosis. The
MCS diagnostic criteria are inclusive, and they define sufficiency. MCS casts a broad net
which fails to exclude individuals whose symptoms are explained by alternative, well-
established diagnoses. In statistical terms, these criteria have sensitivity (minimizing the
likelihood of missing someone), but lack specificity (inclusion of non-cases). For example,
cases that are explained by psychiatric or psychological conditions need not be excluded.
Another limitation is the lack of operational definition of certain terms, leaving them open

Table 1.2   Cullen's Multiple Chemical Sensitivities Diagnostic Criteria

1.   The disorder is acquired in relation to some documentable environmental exposure(s), insult(s), or illness(es).
2.   Symptoms involve more than one organ system.
3.   Symptoms recur and abate in response to predictable stimuli.
4.   Symptoms are elicited by exposures to chemicals of diverse structural classes and toxicologic modes of action.
5.   Symptoms are elicited by exposures that are demonstrable (albeit of low level).
6.   Exposures that elicit symptoms must be very low, by which we mean many standard deviations below "average" exposures known to cause adverse human responses (generally lower than 1% of established threshold limit values, TLV).
7.   No single, widely available test of organ system function can explain symptoms.

Source: Adapted from Cullen, M.R., *Occup. Med.: State Art Rev.*, 2(4), 655–662, 1987.

to interpretation and theoretical speculation. What is a documentable exposure? What is the association between exposure and harm? Is self-report sufficient to establish either? What are predictable stimuli? When are exposures demonstrable? If several people testify to the presence of an offending agent, does it strengthen the likelihood of there having been an exposure event?

These questions address problems which have been carried over into pseudo-epidemiological attempts to estimate the prevalence of MCS and to establish unique subpopulations with implied different manifestations based solely on survey, self-report data (Kipen et al., 1995; Meggs et al., 1996). These surveys suggest that there are distinct manifestations of MCS, but self-report alone cannot differentiate toxicologic effects from beliefs about attribution.

Despite the colloquial usage of MCS, Cullen's criteria may not hold up under scientific scrutiny. To protect the toxicogenic theory from refutation one would expect EI advocates to dissociate themselves from MCS by proposing new labels and concepts. This has begun. Iris Bell, a long-standing EI advocate (Bell, 1975; 1982), has proposed a new definition of "cacosmia" to describe aversive responses (e.g., headache, nausea, dizziness) to common chemical odors (e.g., gasoline, perfume) that are neutral or mildly unpleasant to most people (Bell, 1994b; Bell et al., 1993, 1995). Cacosmia-related syndromes include MCS, sick building syndrome, chronic fatigue syndrome, Persian Gulf War syndrome, and all the other labels in Table 1.1, all of which are said to share a toxicogenic etiology mediated by mechanisms of neural sensitization (Bell et al., 1995).

It would also be expected that traditional scientists would offer labels to capture what is known about this phenomenon. Recently, Cohn (1994), a traditional allergist, proposed the term "multi-organ dysesthesia", which does not imply a cause for symptoms. While technically correct in terms of the appraisal of sensory experiences as unpleasant, this term is silent on the question of attribution. While acceptable in the medical community, this tongue-twisting label is less attractive to the general public who have adopted the MCS acronym. Another proposed neutral term is "hypersensitivity syndrome" (Jewett, 1992a), but it, too, has not caught on, possibly because the term "hypersensitivity" may be interpreted to imply an immune mechanism. Moreover, the term syndrome implies an as-yet-to-be-discovered mechanism whose existence is plausible. To date, there exists no such evidence to suggest a plausible toxicodynamic hypersensitivity mechanism, and therefore I prefer the term "phenomenon", which has no such tacit implications.

Another approach to labeling this phenomenon is via the path of social factors seen in group manifestations of it. In Sweden, Gothe et al. (1995) reviewed the history of mass psychogenic illness epidemics involving workers in Europe who believed that their symptoms were caused by exposure to tangible components of the external environment or by ergonomic stress or work. They described epidemics of unsubstantiated arsenic poisoning, chronic carbon monoxide poisoning after the gas was no longer generated, amalgam and oral galvanism, and electromagnetic fields or "electric allergy" allegedly induced by computer monitors. Their investigation found that mental contagiousness and the tendency to cluster — common features of epidemic hysteria — best explained these epidemics. Because of somatization and unsubstantiated beliefs they proposed the label "environmental somatization syndrome". Whether this label will be used in Europe remains to be seen.

Idiopathic environmental intolerances

In February 1996, I was invited to participate in a workshop on multiple chemical sensitivity jointly convened in Berlin by agencies of the German government and the International Programme on Chemical Safety (IPCS) of the World Health Organization (WHO). Although not acted on by the IPCS board to date and therefore not official policy of IPCS, the conclusions and recommendations of the workshop panel have been published and are germane to several topics in this book, especially naming this phenomenon (IPCS, 1996).

The IPCS panel recommended that the use of the label "multiple chemical sensitivity" (MCS) be discontinued, despite its familiarity in the literature and colloquial use. An appropriate way to identify the phenomenon is by description without any implication of chemical etiology or sensitivity mechanisms. The panel consensus was the descriptor "idiopathic environmental intolerances" (IEI), which emphasizes that the cause remains unknown. This term incorporates a number of disorders sharing similar symptomatologies, including what is described as MCS. The descriptor may be qualified by the putative origin of the disorder, e.g., IEI (chemical).

A working definition of this disorder is that it is

1. An acquired disorder with multiple recurrent symptoms.
2. Associated with diverse environmental factors tolerated by the majority of people.
3. Not explained by any known medical or psychiatric/psychological disorder.

IEI places the burden of proof on EI advocates. As expected, EI advocates launched a campaign of resistance on the Internet and obtain 81 signatures on a letter of opposition to the recommendations and conclusions of the IPCS workshop panel, arguing, in part, that the conclusions were influenced by some members of industry who were present and that the integrity of the IPCS process had been compromised. To my knowledge, only two of the signatories were present in Berlin, Claudia S. Miller and Howard Kipen. What I witnessed was not the influence of industry, but rather the lack of credibility of the arguments put forward by EI advocates. Kipen was the chairman of the meeting, and at no time during the workshop did I hear him express objection to the forum or conduct of the proceedings. The letter in question was published as a letter to the editor in the *Archives of Environmental Health* (1996), along with letters in response.

I contributed to the IPCS definition of IEI and endorse it. I was inclined to use idiopathic environmental intolerances, or simply IEI, to refer to the subject of this book.

After extensive discussion, however, editorial advisors convinced me not to use it because the term IEI is unfamiliar. Instead, I have chosen the label originally used by Theron Randolph, the founding father of this phenomenon — environmental illness or ecologic illness, or simply the acronym EI. To maintain consistency, this book uses the terms environmental illness, EI, or EI phenomenon for passages written in the author's voice but uses the original language when quoting original sources. Individuals who have been labeled to have or who profess to suffer from EI are referred to as EI patients. EI is no more than an identifying label for the synonyms listed in Table 1.1. EI does not imply the existence of an underlying disease process, toxicodynamic or otherwise. EI is not a diagnosis.

Clinical ecology and EI advocates

Most physicians who diagnose and treat EI identify themselves as "specialists in environmental medicine". In the past, they referred to themselves as "clinical ecologists", creating and/or following theories and practices of diagnosis and treatment which defined clinical ecology. Its distinguishing feature is the belief that exposure to extremely low levels of environmental chemicals (or other stressors) can wreak havoc within the human body. Environmental medicine is a component of the recognized specialty of preventive medicine (public health), but the theories and practices of clinical ecology are not. Clinical ecology is not a recognized medical specialty and is not included in standard medical textbooks. To avoid confusion, this book refers to practitioners who espouse the principles or practices of clinical ecology as clinical ecologists or EI advocates, even though they dissociate themselves from this label.

Clinical ecologists portray EI as a systemic, polysymptomatic disorder that can include complaints from every organ system and can be caused by exposure to virtually any environmental agent (Rea, 1994), even at levels not known to affect the general population. Its definition, as formulated by Theron Randolph, has appeared in each issue of the promotional magazine originally called *Clinical Ecology*:

> "Ecologic illness is a chronic multi-system disorder, usually poly-symptomatic, caused by adverse reactions to environmental incitants, modified by individual susceptibility and specific adaptation. The incitants are present in the air, water, food, drugs and our habitat."

According to the principles and practices of clinical ecology, there is no requirement of independent corroboration of subjective complaints and attribution to a putative cause, e.g., chemicals, electromagnetic forces. A core presupposition is that the patient always knows best. And if the patient does not know, the connection to EI can be readily made by the clinical ecologist; on occasion, even over the phone during an initial 10-minute consultation. To counteract the lack of objective medical or laboratory evidence to corroborate the symptoms, EI advocates may argue that there does not exist instrumentation sensitive enough to detect the subtle biological or physiological effects of EI. Or, they may argue that measurements do exist, but only they can interpret the subtleties of laboratory measurements and the ever-changing patterns observed on brain-imaging studies.

Prevalence

How many people are affected by EI? It is impossible to answer this question epidemiologically because there is no case definition. There are no medically accepted inclusion criteria. To make matters even more confusing, there are also no agreed upon exclusion criteria. For example, EI patients with alternative medical diagnoses or psychiatric disorders which can account for the symptoms are nonetheless still said to suffer from EI by advocates. Two scenarios are speculated — first, people with psychological problems acquire EI or, second, EI is reason to acquire psychological problems.

Estimates of prevalence have been based on surveys of people who self-report having EI. According to sound epidemiological principles (e.g., sampling), such surveys cannot be used to estimate prevalence. Nevertheless, these estimates have been reported to be as high as 15% by government panels that included clinical ecologists (National Research Council, 1987). When projected onto the entire population, this estimate comes to 37.5 million people in a nation of 250 million (such as the U.S.) and approximately 750 million cases worldwide.

Another estimate is based on the number of subscribers to EI patient support/advocacy newsletters. In 1994, one such newsletter, *National Ecology and Environmental Delivery System*, distributed out of Syracuse, NY, reported that 25 newsletters contacted had a combined readership of over 40,000 which was described to be ever growing.

By comparison, Cullen (1987) at the Yale Occupational and Environmental Medicine Clinics was able to identify only a very small number of patients who met criteria for his definition of MCS. During a 5-year period (1986 to 1991), 49 patients coded as MCS in a database were identified. This comes to about one case per month over the 53 months. No attempt at etiologic imprints or assessment of the strategy used for evaluation or diagnoses was intended (Cullen et al., 1992). Despite the pervasive prevalence projected by the EI advocates and even taking into account the restricted operational definitions such as MCS, created in attempt to study the phenomenon scientifically, the incidence of objectively verified cases of EI to date is still zero.

Philosophy of science

The fundamental presuppositions of a theory are called "hard-core postulates". They define the mechanisms and processes by which the theory explains the phenomena. They also determine how many phenomena the theory explains. The more phenomena it explains and the more difficult the problems it addresses, the greater the theory's value. The more precise the postulates and the more testable the hypotheses they generate, the greater the risk of refutation. For a theory to survive scientific scrutiny, hard-core postulates must be validated by hypotheses that are empirically testable and therefore refutable. The degree to which a hypothesis risks refutation of the postulates, the greater its value to the theory if the evidence fails to refute it. Although never provable, a theory gains scientific recognition as the accumulating evidence from rigorous tests of hypotheses fails to refute the postulates. The validity of the theory is judged by its ability to withstand rigorous attempts at refutation. Research programs which test the postulates aggressively and contribute to modification and refinement in the theory are considered progressive. Research programs which do not generate testable hypotheses or generate data which do not advance the refinement of the theory are considered degenerative.

One indication of a degenerative theory is the number of hard-core and auxiliary postulates it has. Another indication is the complexity of a hypothesis, the convoluted nature of conditions under which it is testable. A pseudoscientific theory is characterized by hard-core postulates which are difficult to translate into testable hypotheses. At its worse, pseudoscientific theories have an elaborate structure of auxiliary hypotheses to protect the hard-core postulates from refutation. When a hypothesis generated from an auxiliary hypotheses is refuted, its refuting implications stop at the auxiliary hypothesis, leaving hard-core postulates intact. For further details on the philosophy-of-science methods which distinguish progressive from degenerative research programs, the reader is referred to Appendix A.

Competing theories of EI

Recently, a review panel of physicians presented four views to summarize the various etiological theories for EI (MCS): (1) MCS is a physical or psychophysiologic reaction to multiple environmental chemicals. (2) MCS symptoms may be precipitated by low-level environmental chemical exposures, but the underlying increased sensitivity is initiated primarily by psychological stress. (3) MCS is a misdiagnosis, and chemical exposure is not the cause of the symptoms; in this case the symptoms may be due to misdiagnosed physical or psychological illness. (4) MCS is simply a belief system instilled by certain practitioners, the media, or others in society; MCS is therefore the manifestation of culturally shaped illness behavior (Sparks et al., 1994).

The three different views to explain EI (first, second, and fourth) fall into two categories — toxicogenic and psychogenic. The third view, misdiagnosis, excludes consideration of EI as a unique illness if an alternative medical or psychiatric/psychological explanation can account for the symptoms. Exclusion criteria include well-characterized, objective toxicologic illness such as chronic respiratory disease induced by exposure to high levels of a toxic chemical, as was the case in Bophal, India. In discussing these views in the book, I have defined them in terms of etiologic mechanisms and testable hypotheses.

Toxicogenic theory

The term "toxicogenic" as used in this book is meant to include the meaning of toxicokinetic (the movement of a chemical in the body, especially in the brain) and the meaning of toxicodynamic (the effect of a chemical on the body). As used, toxicogenic means an adverse effect on the body caused by a toxic agent. A toxic agent refers to all agents in the external physical environment (exogenous) or those in the body (endogenous) hypothesized to have a toxic or "poisonous" effect on human physiology. While most toxic agents are chemicals, including those in foods, some represent other forms of matter such as magnetic and electrical energy.

The first view presented by Sparks et al. (1994) represents two different explanations, one held by traditional toxicologists and the other by EI advocates. They share the presupposition that the symptoms and effects seen clinically are caused by low-level exposure to chemicals agents which initiates physical and psychophysiologic mechanisms (e.g., chemical-receptors). Pre-morbid psychological or psychophysiological effects are not considered relevant to the etiology. All psychological and psychophysiologic symptoms (including conditioning effects) are postulated to be toxicodynamic effects, that is, consequences of chemical activation of the underlying physiologic mechanisms. The two explanations differ in that EI advocates hold that basic postulates of toxicology and neurotoxicology (e.g., dose-duration-response and consistency of effect) (Schaumburg and Spencer, 1987; Waddell, 1993) need not apply. When referring to the advocates' version, I shall use the general term "toxicogenic theory" to distinguish it from accepted toxicology.

I shall refer to the second toxicogenic view presented by Sparks et al. as the interactive explanation, defined as toxicologic effects observable only when an EI patient is susceptible due to additional psychological vulnerabilities or psychophysiologic effects from stress. I believe the interaction explanation is an elaboration of the clinical ecology theory,

Table 1.3 Some Postulates of the Toxicogenic Theory

Hard-core postulates
Threshold of onset
  Total body load
Long-term effects
  Sensitization
  Super low threshold: one-molecule effect
  Heightened sensitization, subsensory threshold
  Hyperosmia and cacosmia
  Context specific sensitivity
Threshold variability
  Adaptation (masking) and de-adaptation (unmasking)
Sensitivity to multiple environmental agents
  Spreading of effects to multiple triggers
  Natural vs. synthetic chemicals
Time course of a hypersensitivity reaction
  Arousal and exhaustion
  Time-dependent sensitization
Variability of effects
  Bipolarity   hyperarousal and hypoarousal
  Switching of symptoms
Addicted to exposure
  Craving the offending agent

Auxiliary postulates
Temporal cohesiveness
Demographic diversity
Route of exposure
Dose dependence
Inter-individual variability: uniqueness
Rapid onset and cessation of symptoms
Self-regulation of symptoms

and I will also refer to it as the toxicogenic theory throughout the book except when discussing specific models.

History of EI

The history of EI is linked to 19th-century neurasthenia, an illness characterized by fatigue and general malaise as well as a psychological overlay which is presented in greater detail in Chapter 6. About 60 years ago, EI originated as an allergy theory for neurasthenia with Randolph's (1945) publication of "Fatigue and weakness of allergic origin to be differentiated from nervous fatigue or neurasthenia" in *Annals of Allergy*. According to Randolph's theory, minute levels of any chemical in the air or in food or in any form that could come into contact with humans could make susceptible people intolerant of environmental exposures for the rest of their lives. In time, the theory was elaborated with complex postulates and mechanistic hypotheses (Table 1.3), discussed at length in Chapter 2. EI could account for any symptom in any physiological system, e.g., immunologic, cardiovascular, central nervous. Once environmental illness was contracted, Randolph argued, there is a "spreading phenomenon" that could explain generalized sensitivity to any and all other chemicals and other non-chemical stressors. Reactivity to a chemical did not have to elicit symptoms consistently from exposure to exposure, called "switching". Different doses could have different effects at different times, often within minutes. Theoretical constructs such as "adaptation/de-adaptation", well recognized in physiology, were postulated to explain it. Randolph proposed a theory that could account for any event, or its absence, under any and all circumstance. In brief, EI could explain all of the ailments of mankind, physical and psychological, yet there were no substantiated methods for diagnosis or effective treatment.

EI progressed from unfounded speculation to a socio-political movement through a network of disciples devoted to Randolph. These were mostly physicians who called themselves clinical ecologists and formed the Society for Clinical Ecology, later to be renamed the American Academy of Environmental Medicine (AAEM). A network of patient education and support groups also evolved, such as the Human Ecology Action League (HEAL) in Atlanta, founded in 1977 by Randolph himself. With the death of Randolph, the torch has passed to William Rea, director of the Environmental Health Center in Dallas, TX, author of a four-volume text on the theory and practices of clinical ecology (Rea, 1992, 1994, 1995, 1997).

Causal agents

The environmental agents that have been targeted over the past 60 years are ever-changing. In the early years of the EI movement, the focus of putatively hazardous chemicals was foods (Rinkel, 1944; Rinkel et al., 1964), food additives and preservatives (Feingold, 1975), ambient air (Randolph, 1955), and drinking water (Rea, 1994). The list of environmental chemicals generally emphasizes synthetic products, implying that the "natural" form of the same chemical is harmless or even beneficial.

Specific chemical agents commonly implicated are petrochemicals, heavy metals, chlorinated compounds, hydrocarbons, phenol, ethanol, ammonia, isocyanates, and aldehydes such as formaldehyde. Exposure to these chemicals occurs through various vehicles, including combustion exhaust, household cleaners, industrial solvents, fragrance products, pesticides, herbicides, fungicides, paint, lacquers, electronic and plastic products, computers, copier machines, telephones, printed material and inks, synthetic clothing and fabrics, building materials, carpets and adhesives, window coverings, automobiles, pharmaceuticals, mercury amalgams, etc. Ironically, in a survey of EI in selected European countries, different putative causative agents were associated with EI in the various countries (Ashford et al., 1995). In order to be unique toxicological manifestations, one might expect that the exposure levels for a specific agent would be highest in the country most affected. Such data were not available in the Ashford et al. report. The alternative explanation suggests that the effects represent different beliefs about the attribution of EI in the different countries.

It is true that thousands of new chemical products are produced annually. The implication that sheer numbers prove that the effects are both harmful and potentially endless (Rea, 1994) is unfounded. Radical environmentalists are not beyond creating alarmist stories, and the media are often too quick to promote them. For example, in 1993, 64 cases of chronic pyrethroid intoxication were reported to the Federal Health Office in Germany. Shortly afterwards the media spoke of thousands of cases of pyrethroid intoxication in homes from carpets, moth killers, pesticide sprays, and wood preservatives. In a study of 23 of the 64 cases, eight presented with EI. All eight had normal physical and laboratory findings, and in six of the cases a link to pyrethroid exposure could be established. However, there was not a single case in which evidence for irreversible peripheral or central nervous system lesions could be found (Altenkirch et al., 1996). The media often neglect to report that exposure does not necessarily mean there was harm. In an attempt to discourage the media from irresponsible reporting, the 1996 IPCS Berlin workshop panel found that:

> "Scientific research on IEI will lead to an improved understanding of environmental hazards that should translate to improved risk communication. Public information should be based on established facts and not on speculation."

Once disproved, the latest chemical scare seems to disappear from public attention, only to be replaced by yet another environmental agent so the cycle of illusion can be repeated. One example is long-term, low-level exposure to trichloroethylene (TCE), an important industrial degreasing solvent for metal parts, also used in household cleaning products and commercial dry cleaning of clothes. Adverse psychological and behavioral abnormalities have been reported in industrial overexposure and include symptoms of headache, fatigue, lightheadedness, depression, insomnia, irritability, and confusion (Ellenhorn and Barceloux, 1988). In the U.S., a community adjacent to a toxic dump site reported ill-health effects and attributed them to traces of TCE in domestic well water. In 1996, the Agency for Toxic Substances and Disease Registry (ATSDR) published the results of self-reported information for general health effects obtained from individuals on the TCE subregistry who had documented environmental exposure to TCE (Burg et al., 1995). Several of the reported ill-health effects were different from a cohort-control database, including speech impairment, hearing impairment, stroke, liver disease, anemia and other blood disorders, diabetes, kidney disease, urinary tract disorders, and skin rashes. These problems were not substantiated by medical records, and the causes of the symptoms are unspecified. Moreover, the surveys were conducted after the individuals who registered were given documentation on the ill effects of TCE, as required by government policy. The report acknowledges that "the response rates of the TCE subregistry members might, in part, have been influenced by the participants' knowledge of and concern about TCE exposure." However, this disclaimer will unlikely be the headline in media reports, and the article will likely be introduced as plaintiff's evidence in civil tort cases.

Microorganisms are also included in EI, specifically *Candida albicans* (Truss, 1986) referred to as the "yeast connection" (Crook, 1986). Endogenous hormones are considered potentially harmful chemicals, specifically progesterone (Mabray et al., 1982). And even body odors from another person are said to induce a reaction, as evidenced by several female EI patients I treated who attributed dyspareunia to semen sensitivity.

The source of putatively toxic agents may be ubiquitous or specific, natural or synthetic, exogenous or endogenous, depending on the perceived context of their manifestation.

Biological mechanisms and susceptibility

Speculation about the effects of chemical sensitivities on the immune system began in the 1940s, when Randolph (1945, 1947, 1962) divined that allergy was the cause of fatigue, weakness, irritability and behavior problems, depression, confusion, and nervous tension and hyperactivity in children. Throughout the early course of the history of EI, chemically induced symptoms, including central nervous system effects, were postulated to be secondary to immune dysregulation; there was little said to suggest direct toxic effects on the central nervous system (Bell, 1982; Levin and Byers, 1987). In the past decade, theories shifted their focus to central nervous system mechanisms referred to as neural desensitization

(Bell et al., 1992), with immune abnormalities now seen as secondary to central nervous system irregularities rather than being directly induced by low-level exposure to chemicals. Commonly accepted neurobiological theories for endogenous neurotransmitter imbalances (e.g., dopamine, epinephrine, norepinephrine, serotonin) pioneered in the field of biological psychiatry are incorporated into these theories to account for the florid psychological disturbances seen in severely affected EI patients.

It is striking that most, if not all, of the symptoms associated with EI, like so many other manifestations of somatization in history, may be characterized as general malaise. It is also striking that the cluster of predominant symptoms and the explanations for them seem to follow what is socially and culturally acceptable at that point in history (Kleinman, 1982, 1986; Shorter, 1992; Zola, 1966). In the 1990s, in keeping with the spirit of the so-called "decade of neuroscience", toxicogenic theory focused on models of brain mechanisms such as limbic kindling, sensitization, conditioning (Bell et al., 1992), and time-dependent sensitization (Bell et al., 1993). Furthermore, there are also hypothesized host amplification mechanisms. Predisposition to susceptibility for ecologic illness is postulated to be heritable (Bell et al., 1992; Bell, 1994b) and associated with shyness, a personality trait identified independently in studies about personality development unrelated to EI (Kagan et al., 1988). Biological predisposition to acquire EI is also postulated to mediate depression and other affective disorders (Bell, 1994a). In toxicogenic theory, if a factor cannot be explained as an effect of EI, it is assumed to be a pre-existing condition for susceptibility.

Because not everyone is susceptible to contracting EI, individual differences require explanation. Host susceptibility as a biological construct is a truism. But is it reasonable to reframe known etiologic factors of illness as susceptibilities mediating toxicogenic mechanisms for which there is no evidence? For example, when research showed that EI patients had pre-morbid (before the alleged onset of EI) histories of psychiatric and stress disorders, this finding was readily incorporated into the theory by assuming that emotional and stress disorders make an individual more susceptible to acquiring chemical sensitivities. Even psychiatric disorders believed to be primarily of genetic or biological etiology (e.g., bipolar affective disorder) are incorporated by postulating that the neurophysiological and biochemical imbalances are really caused or exacerbated by chemical sensitivities. The interaction of chemicals with trauma and stress is summarized as (Bell, 1994a):

> "The original cause that sets an amplification process in motion may have been psychological or physical trauma, but then any number of environmental factors might later cross-sensitize with the original stressor, perhaps including new life stressors, common foods (e.g., opioid-releasing wheat, milk, sweets, fats), drugs (e.g., morphine, stimulants, ethanol), and chemicals (e.g., pesticides, solvents), at currently lower intensities otherwise tolerated by the majority of the population."

Each elaboration of speculation is meant to identify yet another unique subpopulation of chemically sensitive individuals. What is not acknowledged is that personality disorders and other psychiatric conditions have a similar idiopathic etiology involving complex interactions among inherited disposition and temperament, child rearing and nurturing, and traumatic experiences.

The counter-intuitive nature of individual differences to explain EI is reflected in an observation reported in the epidemiological data set obtained at the Yale Occupational Medicine and Environmental Clinics. It is the continued low rate of EI cases occurring in those industrial sectors that is responsible for most other chemical and physical injuries, namely, blue-collar males in an industrial setting (Mooser, 1987).

Is there any plausible rationale for a biological, toxicogenic, chemical-receptor mechanism that can honestly claim to account for the bizarre, hysterical, and sometimes frank delusional presentations of many severely affected EI patients?

To beg the question by postulating different susceptible populations is only a dodge to preserve the toxicogenic theory.

Explaining psychological symptoms

According to the toxicogenic theory, psychological symptoms are biological responses to low levels of chemicals (Ashford and Miller, 1989; Philpott and Kalita, 1980). EI advocates have gone so far as to argue that mood disorders cannot be reliably assessed in patients with *bona fide* medical disease (e.g., diabetes, cancer) because the somatic symptoms of depression overlap those of the disease. While there is some truth that symptoms overlap, that is not to say that somatic symptoms cannot be differentiated. But EI advocates postulate that all somatic symptoms associated with depression are caused by EI. This implies that toxic effects from exposure to low levels of chemicals or other environmental agents can never be ruled out as a cause of mood disorders and other psychological symptoms. They suggest that symptom-checklist indices of depression are invalid unless somatic items are removed or separated (Davidoff, 1992a). EI advocates contend that neurophysiologic effects common to EI patients and psychiatric patients reflect chemical sensitivities, the difference being that the psychiatric patients do not know it (Davidoff et al., 1991).

Another possible explanation raised by EI advocates is that psychological effects are independent of the symptoms that they attribute to chemical exposure. This postulate is based on the fact that depressed and otherwise psychologically impaired individuals, like psychologically healthy individuals, are subject to exposure. There is no argument about this. However, if this were the case in EI patients, then it could only explain the psychological symptoms elicited by a chemical reaction. Explanation of the occurrence of psychological symptoms during those periods when the individual is not exposed requires further elaborating speculation. For example, one may postulate that patients with premorbid depression are more susceptible to acquiring EI and once so afflicted have lower thresholds for their affective symptoms, which could be elicited or exacerbated by chemical exposure or some other form of stress, defined by the appraisal of the patient. And, if the etiology of the symptoms cannot be differentiated, it is safe to assume they must have been caused by exposure to some unidentifiable toxic environmental agent.

All of these arguments presume that psychological factors cannot be causally linked to the symptoms of EI. Based on this misconception, advocates of EI contend that it is premature to focus research in the psychological, psychiatric, or psychophysiologic arenas. Instead, they say, all effort should be directed toward identifying environmental agents that they believe cause EI symptoms (Davidoff, 1992a,b).

Unique principles of toxicology

EI advocates claim unique principles of toxicology in the sense that classic toxicology cannot explain EI (Ashford and Miller, 1989, 1991, 1998; Rea, 1992). Exposure to the same chemical agent may elicit debilitating symptoms on one occasion but none the next. But that does not mean there was not a toxicogenic effect. Rather, the effects were too subtle to be perceived. Despite being incapacitating, symptoms defy objective measurement. The

Table 1.4   Hill's Criteria for Toxicologic Disease

1. *Strength* — estimated by the rate of increase in the symptom/disorder in the exposed population, e.g., mortality from lung cancer is 30 times higher in heavy smokers.
2. *Consistency* — the association has been repeatedly observed by different investigators, in different circumstances and at different times.
3. *Specificity* — the association is limited to people with specific exposures and to particular physiological systems and types of disease.
4. *Temporality* — which is the cart and which the horse; does environmental exposure cause EI or does belief in EI lead to environmental "reactions"?
5. *Biological gradient* —there is a dose-response curve, e.g., death rate from lung cancer correlates with the number of cigarettes smoked daily.
6. *Plausibility* — the association is biologically plausible, and the criteria are based on the biological knowledge of the day.
7. *Coherence* — the cause-and-effect interpretation of data should not seriously conflict with the generally known facts of the natural history and biology of the disease.
8. *Experiment* — Some preventative action does, in fact, prevent the association, e.g., improvement of symptoms in non-fragrance zones.
9. *Analogy* — in some circumstances it would be fair to judge by analogy to a well-characterized disease, but how realistic is the analogy of EI to AIDS?

Source: Adapted from Hill, A.B., *Proc. Roy. Soc. Med.: Sect. Occup. Med.*, 58(5), 295–300, 1965.

time course of low-level exposure effects is indeterminate; they can come and go in a flash. A fundamental premise of toxicology, the dose-duration-response relation, does not apply; less can cause more, but more may have no effect at all. Medical toxicology has established criteria to assess the cause-and-effect association between the environment and disease, summarized by Hill (1965) (Table 1.4).

Politics

One of the most effective lobbying groups is the National Center for Environmental Health Strategies (NCEHS) in New Jersey, founded in 1986 by Mary Lamielle, who by her own account — "Vexations, scents, and sensitivity" (as reported by C. H. Crowley in the *Washington Post*, April 26, 1994) — claims hypersensitivity to multiple chemicals including detergents, air fresheners, and pesticides. She claims to having been homebound and on oxygen much of the time since 1979 because of inflammation of the joints attributed to reactions to lawn-care products and air pollutants from a nearby sewage plant. NCEHS has been influential in exerting political pressure in Congress to obtain U.S. federal moneys, channeled through ATSDR, to organize national meetings at which Mary Lamielle and other proponents of EI presented their speculations and testimonials (ATSDR, 1994a,b). The political agenda for the presentations in Baltimore and those open to the public in Berkeley was evident. The Baltimore conference commenced with the introduction of Theron Randolph, in honor of his being the founding father of clinical ecology. At the Berkeley conference, a speaker who presented himself as an EI sufferer mimicked bizarre symptoms attributed to chemical exposure, symptoms which psychiatrists call hysterical conversion reactions. William Rea, recognized as a leading EI advocate, made a presentation under the condition that no questions were to be asked of him. A tacit understanding at these conferences seems to have been that unsubstantiated theoretical speculation about the plausibility of the mechanisms and etiology of EI was not to be criticized; it is as if the medically unsubstantiated clinical manifestations of EI patients and the controversy resulting from unsubstantiated methods employed in diagnosis and treatment by clinical ecologists did not exist. I, like others in attendance at the Baltimore conference, was watching the emperor's parade in a kingdom of loyal, blinded followers, waiting for the exclamation "The emperor has no clothes!" which never came, out of deference to civility. Such restraint was overcome during the Berkeley conference on one occasion to which I shall return in Chapter 3. From these conferences and the published proceedings that followed, EI advocates claimed legitimacy for ideas which have not gained acceptance through the scientific process.

EI advocates hide behind the cloak of uncertainty that enshrouds science. They profess to be scientists and researchers with a "different opinion" and suggest an atmosphere of credibility. They present a facade of being dedicated researchers and clinicians excluded by the fraternity of the medical-industrial establishment. They view themselves as leaders of a populist movement. They do not discover scientific truths; they divine unfounded beliefs. When challenged about the lack of scientific evidence for their diagnostic and treatment practices, they ignore or trivialize such criticism and may even invoke conspiracy theories against them. Instead of providing empirical evidence, they respond with lawsuits for alleged restraint of trade or charges. Especially contemptible are charges of ethical misconduct against scientists who presented contrary evidence, filed through an advocacy organization in Maryland, MCS Resources & Referrals. EI advocates present themselves as the champions of victims, but turn out to be champions for themselves who practice victimization and foster invalidism.

EI patients

> "Many times it is more important to know what kind of patient has the disease than what kind of disease the patient has." (quoted by Rosch, 1979)

The heterogeneous and often complex psychiatric, psychological, and psychophysiological characteristics of EI patients are presented in detail in subsequent chapters. For an initial impression, I summarize a vignette presented in one of our earliest publications on this subject.

> The patient was 28 when she arrived at our environmental care unit in Denver one evening, wheeled in on a gurney draped with IV bottles, wearing a full-faced gas mask with an oxygen line. Her presenting condition was severe malnutrition, profound leukopenia, and a suspected monilial sepsis. She had been transported from another city because she required the sanctuary of an ECU due to acute phobia of exposure to the ambient environment. She had been living under the care of her parents, either in their home in a specially constructed, aluminum-foil-lined room, or a porcelain-lined trailer parked in the mountains. She was fed parenterally, in a manner other than through the digestive tract, via Hickman catheter which had been implanted by a clinical ecologist in a Texas facility. Self-administered intravenous bicarbonate salts were used to alleviate acute psychiatric distress. The patient claimed beneficial response to these salts within seconds of administration. Her history revealed

long-standing psychiatric treatment with a variety of disorders including schizophrenia. In the ECU, she was given oral foods despite her protests and was soon eating a general diet including pizza. She showed dissociative disorders and related childhood traumas of severe emotional, physical, and sexual abuse. Foods had specific associations to the abuse in the home. Her brother had also been affected and had entered Scientology to escape the home. The mother had a history of schizophrenia diagnosis. During 4 weeks of treatment in the ECU, the patient received intensive psychotherapy to overcome her beliefs in EI. She had projected her childhood traumas onto environmental illness, assisted by iatrogenic illness. Her family continued to control her life. They would bring "special water" to accommodate her environmental sensitivities. They reinforced her belief that her psychiatric problems were EI related. I recall spending an hour with the father during which time he said nothing when I asked him how his daughter came to be this way; he showed signs of acute anxiety. The mother was delusional and disoriented during my interview. The patient was moved to the hospital inpatient psychiatric ward for an additional 6 weeks, during which she received intensive psychotherapy and medication. In the ensuing years, she lived in halfway houses and enrolled in a professional school, but she had numerous relapses resulting in inpatient hospitalizations. Over time, though, she mastered the sequelae of her trauma, overcame her fears, and resolved the issues with her parents. Eighteen years later, she has earned a Ph.D. in the biological sciences and is pursuing a professional career.

Symptoms

The presenting complaints of EI patients can be from any physical system and in any pattern. Typically, they are symptoms of general malaise (Table 1.5), and any one of these symptoms can occur in diseases with specified organ pathology. During medical assessment, differential diagnosis unrelated to EI can account for some of the symptoms in certain EI patients (Selner and Staudenmayer, 1985). Also true is that any one or several of these symptoms may be caused by well-characterized toxic effects of numerous chemicals (Ellenhorn and Barceloux, 1988). Nevertheless, it seems unlikely that any one individual, EI patient or otherwise, could be experiencing the multitude of high-dose toxic exposures (or comorbid diseases) to explain such a myriad of symptoms.

Some EI patients manifest paralyzing depression with feelings of hopelessness and despair. While not unique to depression, some of the following symptoms may help the doctor formulate an impression: negative self-image, low self-esteem, lack of confidence, fears of incompetence, feeling inept in social or personal relations, difficulty in relaxing, feeling the world is hazardous or unsafe, general unhappiness, a resolve to accept chronic illness.

Strength of belief in EI

Some EI patients are candid about the psychological factors that may impact their condition. They do not hold overvalued, closed beliefs about the validity of EI. I have seen patients complaining of sensitivity to chemicals who say they have never heard of EI or

Table 1.5 Some of the Multi-System Symptoms of EI Patients

Head, eye, ear, nose, and throat
Eye burning, phoria, tinnitus, vertigo, rhinorrhea, nasal obstruction,
   nasal burning, glossodynia, pharyngeal irritation, dysphonia

Pulmonary
Dyspnea, cough, chest pain

Cardiovascular
Palpitations, irregular heart beat

Gastrointestinal
Dyspepsia, eructation, flatus, recurrent diarrhea, constipation,
   cramping pain

Genitourinary
Dysmenorrhea, vaginitis, dyspareunia, dysuria, urinary frequency,
   urinary retention

Musculoskeletal
Myalgia, weakness, muscle tension, arthralgia, dyskinesia

Skin
Irritation, pruritus without rash

Lymphatic
Fluctuation in size of lymph nodes

Neurologic, psychological
Headaches, neurasthenia, numbness, anesthesia, paresthesia,
   hyperaesthesia, fatigue, irritability, hyperactivity, sleep
   dysfunction, cognitive dysfunction, decreased attention span, loss
   of concentration, memory loss, anxiety, depression, mood swings

consulted a clinical ecologist. Some patients obtained information from a friend or a report in the media. For some, the influence of a misguided practitioner has reinforced their belief. When it is explained to them that what they have heard or read is unsubstantiated, some of these patients accept that they have been misinformed and respond well to education. Many of these patients have no motivation to maintain a toxicogenic belief about EI. More importantly, they recognize symptoms to be stress related. They show insight for the psychological aspects of their condition and are able to examine the irrationality of their symptom appraisals and environmental attributions without anxiety. A frank, rational discussion about both their symptoms and the pitfalls of the EI diagnosis are often all these patients require for management. They typically do not manifest psychiatric disorders, and if they do they seem to have no motivation to project them onto the physical-environment.

Such patients contrast with the more severely affected EI patients who believe that psychological factors could not possibly be responsible for their symptoms. Some harbor deep-seated antagonism and resentment toward medical, psychiatric, and psychological professions. They reject the possibility that the scientific method applies to their condition. In the extreme, they deny psychological symptoms on self-report inventories or during interviews, apparently motivated by a fear of being labeled "crazy". If they admit to having symptoms of depression or anxiety, they attribute them to either chemical sensitivity or the disruption of their life by their illness.

Table 1.6   Some Characteristics Identifying EI Patients

Somatization
Over-utilization of medical services
Irrational beliefs about illness
Overvalued ideas, a closed belief system
Illness worry (hypochondria)
Vigilance for environmental triggers
Symptom amplification
Illness as center of life
Non-adherence to psychotropic medications
Disability and entitlement
Inability to take responsibility
Embittered emotion
Anxiety, phobias, and panic attacks
Resolve to be ill, adopting the sick role
Victimization
Melancholy, *la belle indifference*

Personality and behavioral characteristics

EI patients can be difficult to live with and difficult to manage. Table 1.6 lists some clinical characteristics of severely incapacitated EI patients. Over-utilization of medical services may be apparent from the patient's history. The medical records are usually voluminous (20 pounds of records are not uncommon), with diagnostic evaluations from several medical specialists, invariably with negative laboratory tests and no objective findings. The patient's belief about his/her illness is unsubstantiated and often irrational. Explanations about the etiology and manifestation of the illness tend to reflect the prevailing models of unsubstantiated, fringe medical practices, including clinical ecology practice. The belief system may vary in terms of the strength of conviction about symptoms caused by chemical sensitivities. When the attribution is inflexible, the belief system is considered closed, and the patient often has the characteristics associated with overvalued ideas. Patients may show morbid absorption with bodily sensations and processes, with catastrophic predictions about their future. They also worry about aspects of daily life, and see life as a burden, characteristics associated with hypochondriasis

Victims can be very vocal about entitlement demands. While they demand extensive and costly medical assessment and care, they nonetheless may take poor precautions about health and self care (e.g., smoking, eating disorders, and obesity). Displaced aggression is evident in EI patients. Some displace their anger toward the traditional physicians and staff for not believing in EI, or they may disparage science and medicine for a lack of methods to make them well. These individuals are usually not shy.

Psychogenic theory

The core presupposition of psychogenic theory is that psychological factors are necessary and sufficient to account for the clinical presentations of EI patients. Psychogenic theory emphasizes belief, somatization, psychophysiologic stress and anxiety responses, and psychogenic etiology (Sparks et al., 1994). A defining characteristic of EI patients is somatization, defined as (Lipowski, 1988):

"A tendency to experience and communicate somatic distress and symptoms unaccounted for by pathological findings, to attribute them to physical illness, and to seek medical help for them. It is usually assumed that this tendency becomes manifest in response to psychosocial stress brought about by life events and situations that are personally stressful to the individual."

The psychogenic theory postulates that EI is a learned phenomenon, although two mechanisms of learned sensitivity are hypothesized. Classical Pavlovian conditioning necessitates a provoking exposure experience; cognitive mediated beliefs do not.

Belief

In my view, EI is a disorder of belief. Beliefs explain EI patients' appraisal of harmful effects (recurrent multi-system symptoms) attributed to agents in the physical-environment. Beliefs are acquired through learning. Classical Pavlovian conditioning may explain some cases with documented exposure events. EI patient beliefs are more likely to be acquired without identifiable exposures, where perceived exposure substitutes for actual exposure. In other cases, patients do not appraise exposure but only perceived symptoms. Regardless of exposure history, or the lack of it, belief in chemical sensitivity is acquired (1) iatrogenically from a clinical ecologist, (2) through the media, or (3) through contact with an EI victim network. These forms of suggestion and influence are not mutually exclusive, and the methods employed are discussed in Chapters 7 and 8.

Patient reasons for acquiring beliefs are also involved, and they are discussed in Chapter 14. Unsubstantiated beliefs serve as psychological defenses for vulnerable individuals, especially those with personality disorders. EI patients are motivated by resistance to facing anxieties stemming from vulnerability to rejection, abandonment, uncertainty about the future. They are buttressed by fear and terror and other psychodynamic issues. As a result EI patients displace the cause of these anxieties away from self by projecting them onto the physical environment.

Carroll Brodsky (1983) was the first psychiatrist to describe how EI patients organize their life around their illness. These patients spend inordinate energy and time immersed in the phenomenon. They read scholarly treatments on the potential hazards of chemicals, and they are even more engaged by what must be termed junk science literature. Unfortunately, they are unable or unwilling to distinguish the difference. Expert testimony that fails to acknowledge fundamental principles of chemistry and physics does little to relieve this dilemma. These patients attend support/advocacy groups and visit doctors who reinforce their beliefs, arranging visits to special clinics to undergo detoxification procedures and attending to workers' compensation claims or law suits. They may express a motivation to pursue work and life's interests were it not for the disability caused by EI. Rejection of responsibility is consistent with secondary gain motivation, such as avoiding work at home or in the workplace. Avoidance of responsibility may also serve as a psychological defense in service of primary gain, the motivation to avoid deep-seated personal anxieties, taken up in Chapter 13. Attribution of suffering and life's problems to EI preserves the illusion that social, marital, and job-related problems would vanish were it not for EI.

Stress-response

Multi-system symptoms are not unique to EI patients. They are mediated by the physiological systems involved in the human stress-response, presented in detail in Chapter 9.

Somatization, anxiety, and psychophysiological disorders are associated with several psychiatric disorders described in Chapters 10, 11, and 12. A consistent finding in psychiatric epidemiology is that risk of psychiatric disorder increases linearly with the number of patient complaints (Goldberg and Huxley, 1970; Howard and Wessely, 1993). The severity of both current and lifetime psychiatric disorders correlates highly with the number of somatization symptoms (Russo et al., 1995). Central nervous system complaints (e.g., mental confusion, difficulty concentrating, poor memory, irritability, and mood alterations) are universal among EI patients. Fatigue is also typical, especially in EI patients with comorbid chronic fatigue syndrome (CFS) or "sick building syndrome" (Witorsch and Schwartz, 1994). Headache is the most common pain symptom. General arthralgia (aching joints) and myalgia (muscle aches) are also common and can lead to a diagnosis of fibromyalgia.

The stressors underlying the pathophysiology of the stress-response are numerous and diverse. Most compelling among EI patients are the long-term sequelae of childhood stress and trauma which manifest as psychophysiological disorders and psychiatric disorders, particularly post-traumatic stress disorder (PTSD).

Psychiatric disorders

EI patients are heterogeneous with respect to psychiatric disorders. Anxiety seems universal among EI patients. Many view the world as an unsafe place, with good reason but unrelated to chemicals in the environment. Phobic avoidance is seen in many patients. So-called chemical reactions often have the cognitive and physiological characteristics of panic attacks. Some EI patients show a surprising lack of concern about their illness and appear resolved to be ill. I have heard many doctors remark about the lack of worry and anxiety some of these patients display about an illness they describe to be debilitating. Only the most histrionic of EI patients display the emotions of imminent death, more characteristic of some patients with verifiable chronic and terminal illness such as cancer or AIDS. The attitude of *la belle indifference*, a relative lack of concern about illness, describes some EI patients well. In the 19th century, Pierre Janet first described this in patients with melancholy, who often lack insight into the distressing symptoms of emotional illness that they display. If they deny symptoms of depression and anxiety, they somatize them. Some also lack knowledge or insight into the irrationality of their appraisals of environmental intolerances and usually become indignant if their perception is challenged. Vigilance paid to bodily sensation and amplification of symptoms distort perception and appraisal of harmful effects and reinforce hypochondriacal beliefs about illness.

Treatment

Most EI patients express low tolerance for psychotropic medication, often because they interpret the side effects of these medications as chemical intolerance. Antidepressant and anti-anxiety drugs are virtually impossible to mask in double-blind, placebo-controlled studies because of both their therapeutic effects and their sensory side effects. EI patients can easily detect them. It would be a mistake to conclude, based on observations from patients vested in their illness, that these medications could not be effective in managing symptoms. Some EI patients who are not motivated to maintain their illness respond well to antidepressant medication (Rix et al., 1984). However, only a small minority of EI patients try psychotropic medications, and many of these do not adhere to them. This has implications for treatment, where current health-care management practices encourage short-term intervention, with emphasis on psychotropic medication. These patients are difficult to manage, and their therapy may be complex and lengthy.

Conclusion

EI advocates believe chemical sensitivity accounts for any and all somatic or psychological symptoms, even in patients with well-documented medical diseases or psychiatric disorders. EI can be anything for anybody. The multi-system symptoms of EI patients are said to be caused by vanishing levels of virtually any environmental agent, levels that are tolerated by the general population. The basis for establishing the onset of the phenomenon and subsequent triggering of its effects are left to the attributions of the afflicted or their clinical ecology doctors. The toxicogenic theory is nebulous and ever-changing but never testable. The bewildering hard-core postulates are impermeable to refutation, protected by a set of flexible auxiliary postulates, from which EI advocates have opportunistically argued any one of several conflicting outcomes. There are no testable hypotheses, because anything can happen under any circumstance.

The psychogenic theory is the alternative explanation for EI. The multi-system symptoms are explained by the biological mechanisms of the stress-response. The personality characteristics and motivations of patients are explained by heterogeneous but well-known psychiatric disorders, many of which can be traced to childhood disruption of personality development induced by abuse and neglect. Psychology can explain the motivation and practices of clinical ecologists and why patients are attracted to these unsubstantiated beliefs and practices. The EI phenomenon illustrates the adversarial nature of science and pseudoscience, and how the search for truth may be diverted to political forums, a topic taken up in Chapter 16.

chapter two

Toxicogenic theory

A tale of misguided exploration

The following tale serves as an example of the endless search for verification of a theory for which no corroborative evidence exists. This tale is about an imaginary case of planetary misbehavior, told by Imre Lakatos, the esteemed philosopher of science. To begin the tale, a physicist of the pre-Einsteinium era takes Newton's mechanics and his law of gravitation, the accepted initial conditions, and with their help calculates the path of a newly discovered small planet, p.

But the planet deviates from the calculated path. Does our Newtonian physicist consider that the deviation was forbidden by Newton's theory and therefore that, once established, it refutes the theory? No.

He suggests that there must be a hitherto unknown planet p' which perturbs the path of p. He calculates the mass, orbit, etc. of this hypothetical planet and then asks an experimental astronomer to test his hypothesis.

The planet p' is so small that even the biggest available telescopes cannot possibly observe it. The experimental astronomer applies for a research grant to build yet a bigger one. In three years' time, the new telescope is ready.

Were the unknown planet p' to be discovered, it would be hailed as a new victory of Newtonian science. But it is not. Does our scientist abandon Newton's theory and his idea of the perturbing planet? No.

He suggests that a cloud of cosmic dust hides the planet from us. He calculates the location and properties of this cloud and asks for a research grant to send up a satellite to test his calculations.

Were the satellite's instruments (possibly new ones based on a little-tested theory) to record the existence of the conjectural cloud, the result would be hailed as an outstanding victory for Newtonian science. But the cloud is not found. Does our scientist abandon Newton's theory, together with the idea of the perturbing planet and the idea of the cloud which hides it? No.

He suggests that there is some magnetic field in that region of the universe which disturbed the instruments of the satellite. A new satellite is sent up. Were the magnetic field to be found, Newtonians would celebrate a sensational victory. But it is not. Is this regarded as a refutation of Newtonian science? No.

Either yet another ingenious auxiliary hypothesis is proposed or the whole story is buried in the dusty volumes of periodicals and never mentioned again (Lakatos, 1970; with permission).

Table 2.1 Some Hard-Core Postulates
of the Toxicogenic Theory

Threshold of onset
Total body load

Long-term effects
Sensitization
Super-low threshold — one-molecule effect
Heightened sensitization, subsensory threshold
Hyperosmia and cacosmia
Context specific sensitivity

Threshold variability
Adaptation (masking)
Deadaptation (unmasking)
Context specificity

Sensitivity to multiple environmental agents
Spreading of effects to multiple triggers
Natural vs. synthetic chemicals

Time course of a hypersensitivity reaction
Arousal and exhaustion
Time-dependent sensitization

Variability of effects
Bipolarity — hyperarousal and hypoarousal
Switching symptoms

Addicted to exposure
Craving the offending agent

Postulates

Clinical ecologists and EI advocates (Ashford and Miller, 1991, 1998; Bell, 1982; Dickey, 1976; Miller, 1997; Rea, 1977, 1992, 1994a, 1995, 1997) elaborated the toxicogenic theory of EI as originally formulated by Theron Randolph (Randolph 1945, 1947, 1959, 1962; Randolph and Moss, 1980). Toxicogenic theory employs numerous postulates which I have classified in terms of the clinical phenomena for which they are most likely invoked. I make no claim that this classification of postulates into hard-core (Table 2.1) and auxiliary (Table 2.2) is definitive nor that the postulates are exhaustive. Generally, EI advocates employ these postulates without any apparent priority and in any combination. I selected the hard-core postulates on the basis that they have been consistently present in the writings of EI advocates (Randolph, 1962; Rea, 1992, 1994a; Bell, 1982; Ashford and Miller, 1991, 1998). The auxiliary hypotheses are more recent (Miller, 1992, 1996, 1997).

In this chapter, I define each of the postulates of the toxicogenic theory. Each postulate is then criticized in terms of the rationality of the underlying logic, as well as the scientific constructs to which it alludes.

Threshold of onset: total body load

The straw that broke the camel's back, "total body load" is defined as the "mechanism" by which the body incorporates and assimilates all of the stressors that impinge on it in order to maintain homeostasis. The primary stressors are "pollutants" found in food, drinking

Table 2.2 Some Auxiliary Postulates
of the Toxicogenic Theory

Temporal cohesiveness
Demographic diversity
Route of exposure
Dose dependence
Inter-individual variability — uniqueness
Rapid onset and cessation of symptoms
Self-regulation of symptoms

water not obtained from specific wells, food additives and preservatives, outdoor and indoor air pollution — basically, all inorganic and organic chemicals ubiquitous in the environment. Nonchemical agents contribute to "total body load" including psychological stress, natural meteorological effects such as barometric pressure, and technological energies such as electromagnetic force fields (EMF).

EMF is an abbreviation meaning 60-Hertz (Hz; also cycles per second) power-frequency electric magnetic fields, those fields around powerlines and electric appliances. The North American standard power-frequency is mostly 60 Hz, but 50 Hz is usually found elsewhere in the world. EMF is as ubiquitous in our everyday lives as air and water. EMF attracted attention as a potential health risk in 1979 when an epidemiological study suggested certain kinds of cancer risks (Barnes, 1994). EI advocates immediately incorporated EMF into the toxicogenic theory. Some idea of this is reflected in the transcript of a question-and-answer session conducted by William Rea (October 2, 1993; recorded by and available through Share, Care & Prayer; Frazier Park, CA). Some of the more poignant comments were published in an advocacy newsletter (Table 2.3).

The adverse health effects result either from one major high-dose exposure or through chronic low-level exposure that most of the population can tolerate; level of exposure does not seem to matter. "Total body load" invokes the well-known physiological principle of homeostasis: living organisms survive by maintaining an immensely complex dynamic

Table 2.3 Dr. Rea's Answers to the Question, "Does EMF Sensitivity Exist?"

- A growing number of people are sensitive to items that use electricity like florescent lights, hair dryers, microwave ovens, computers, television sets, electric blankets, answering machines, water pipes, AM radio. As with all sensitivities, symptoms vary depending on which target organ is involved. Examples include aches, pains, fuzzy head, sinus problems, skin problems, colon problems, heart irregularities.
- Some people can clear up by living in a canyon where radio and TV signals cannot be picked up — usually out in west Texas or Idaho.
- You can also help yourself by grounding yourself. Run around outside in bare feet for at least 1/2 hour each day. You can also put a copper bracelet around your leg and attach it to a metal pipe while you sleep at night.
- Consider cutting off electricity to your bedroom at night. Most importantly, don't sleep with your head against a wall where electric wires are. Sleep with your head in the middle of the room.

Source: Adapted from the Allergy and Environmental Health Association, *AEHA Q.*, Winter/Spring, XVI(4), 5–6, 1994/95.

and harmonious equilibrium that is constantly challenged by intrinsic or extrinsic disturbing forces or stressors (Chrousos and Gold, 1992). In the 19th-century Claude Bernard assigned it a physiological mechanism. Walter Cannon extended the concept to psychophysiology to include emotional factors and coined the label used today, "homeostasis" (Cannon, 1927, 1929). Cannon helped initiate the scientific study of stress psychophysiology with the fight-or-flight model of the stress-response which today involves pioneering research in neuroscience. Some of this research is reviewed in later chapters, including neuroimaging of panic attacks and the many effects of neurotransmitters and neuropeptides on the hypothalamic-pituitary adrenal (HPA) axis, the immune system, and the autonomic nervous system (ANS). Stress and stress disorders, while restricted by careful operational definitions in scientific studies, are now accepted aspects of neurophysiology, neuroendocrinology, psychoneuroimmunology, and general medicine.

"Total body load" is identified with Hans Selye's General Adaptation Syndrome (GAS) (Selye, 1936, 1946). GAS employed concepts such as stress and strain, borrowed from mechanical physics, to describe forces that impact biological systems. As a precautionary note, theory based on analogy to an epistemology appropriate in a different context is always risky. Stressors are characterized as anything that puts a strain on a biological system, be it physical, biological, psychological, or psychosocial. Stressors' effects on the body were not specific but were general and largely left undefined.

EI advocates illustrate the concept of "total body load" and other principles of chemical sensitivity with a barrel of water that overflows (Figure 2.1). Symptoms result when the barrel overflows from too much water (stress), mostly caused by environmental agents. Deviations from homeostasis are in either direction, leading to symptoms of hyperarousal or hypoarousal. Deviation from homeostasis may be transitory or permanent.

Three criticisms directed to Selye's GAS are equally apropos to "total body load":

1. The causal agent, stress, is confounded with the stress-response.
2. Animal models have limited implications for humans.
3. Mechanical analogy is too simple for an organic, dynamic system.

Selye's research was restricted to animals, and the outcome measures were physiological pathology (e.g., adrenal glands, lymphatic structures, ulcers in the stomach and intestines) observed at necropsy. Selye postulated the cause of these effects to be nonspecific, resulting from a heterogeneous set of stressors used in animal research, including heat, infection, trauma, hemorrhage, nervous irritation, and many other aversive stimuli. Selye's concept of stress was consistent with Charles Darwin's formulation of the term "ecology", encompassing stressors in both the physical environment and the person environment. The main limitation of animal models is that they are restricted to assessing the effects of stressors external to the individual and thereby cannot incorporate factors that originate within the individual — the human elements of personality, thinking, and emotion.

Human models of stress most analogous to the GAS are those emphasizing coping. Stress is defined as a situation where demands exceed coping abilities (French et al., 1974; Lazarus, 1966; Sells, 1970). In coping models for non-specific stress, as in the GAS, there is a tautology in the definition and an experimental confounding between the stressors and the stress-response. If stressors are indistinguishable from symptoms, then there is no need to specify the stressor in stress research (Dohrenwend et al., 1984). By analogy, the identification of the alleged triggering chemical agent is superfluous to the toxicogenic theory. This limits empirical evidence to symptom report, because the chemical agent appraised to be the stressor is left to the perception of the patient or the clinical ecologist. This creates

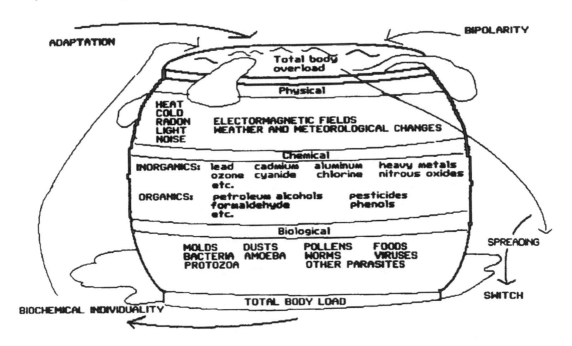

Figure 2.1 Illustration of principles of chemical sensitivity, "total body load". (From Rea, W.J,. *Chemical Sensitivity*. Vol. 1. *Principles and Mechanisms*, Lewis Publishers, Boca Raton, FL, 1992, p. 18. With permission.)

the untenable situation that attribution of cause is left to EI advocates without need for independent corroboration.

The concept of stress is very complex, and difficult to define in humans. Selye noted that emotions (e.g., love, hate, joy, anger, challenge, and fear) and thoughts elicit the stress-response. In discussing the human stress-response, Selye (1982) suggested that psychological arousal is one of its most frequent activators. Definitions of stress in humans emphasize intrinsic factors. According to Hobfoll (1989), when people must behave or experience themselves in a way dissonant with their basic view of the world or self, they are likely to experience psychological stress. According to Spielberger (1972), certain events are stressful if they are thought to be threats to the physical self or the phenomenological self.

One source of human stress is worry, defined as anticipation of events that have not yet occurred or may never occur (Caplan et al., 1985; Spacapan and Cohen, 1983). Another source is that created by an individual's own coping behavior — for example, substance abuse, eating disorders, anxiety-provoking activities, or false beliefs. Many kinds of individual differences are detailed in the psychological literature on stress including personality, cognitive resources, emotions, perception and appraisal of the stressor, the process by which stressful events are translated into health outcomes, and the context in which the stress response occurs (Appley and Trumbull, 1986; Lazarus and Folkman, 1984; Meichenbaum, 1977; Pearlin and Schooler, 1978; Sarason, 1975; Thoits, 1983).

There is considerable confusion and disagreement concerning the exact meaning of constructs such as stress, coping, and health; the causal relationships among these constructs; and the appropriate methodological approaches for the assessment of these constructs and their interrelationships (Kasl and Cooper, 1987). Selye's postulate that stress is a unitary phenomenon is no longer tenable (Weiner, 1977). It is equally untenable that "total body load" is a unitary phenomenon which explains any and all symptoms. Today, stress theory has addressed the three criticisms of GAS. Stress research separates the confounding of the causal stressors (independent variables) and the stress-response effects (dependent variables) (Chrousos et al., 1995). The same cannot be said for the research programs of the toxicogenic theory.

Long-term effects

Sensitization

The putative offending environmental exposure is said to create a priming effect called "sensitization". Once an individual is sensitized, a hypersensitivity reaction occurs to subsequent, low-level chemical exposure. The effect may be long-lasting. Behavioral and biological sensitization is one of the most fundamental forms of learning and reflects change in an organic process due to prior experience (Razran, 1971). Numerous studies have shown sensitization of behavioral and physiologic parameters, even at the cellular level.

A fascinating demonstration of sensitization at the cellular level was presented in an unpublished study reported by Djuric and Bienenstock (1993). In the immune system, mast cells degranulate in response to micromolar doses of substance P, a neuropeptide. In this electrophysiologic study, electrical current was recorded from the mast cell in response to lower, picomolar concentrations of substance P. Upon first stimulation with substance P, the cell responded with a brief, measurable current which was not accompanied by degranulation. When this subthreshold stimulation was repeated on the same cell, temporal summation of the current frequently led to the onset of degranulation. The investigators suggested an interaction between the neuronal and immune systems such that substance P-containing neurons, through this subthreshold stimulation of mast cells, potentiate their cellular responsiveness to subsequent neuronal or immunologic stimuli. Neural stimulation of mast cells thus not only can result in their immediate degranulation but can also lower their activation thresholds. These biological effects were reported to be consistently reliable over time and stimulation.

In contrast, the effects of low-level chemical sensitivities are said to be sporadic and intermittent and have as yet to be demonstrated at the biological level. Sensitization and desensitization are well-recognized processes in medicine (Hyman and Nestler, 1996), particularly in the field of allergy. However, the speculation that underlying sensitization mechanisms apply to the toxicogenic theory remain unsubstantiated.

One-molecule effect

Low-level dose has been defined both relatively and absolutely. A dose tolerated by most of the population is a normative definition. Threshold limit values (TLVs) are used in industrial hygiene and are established by regulatory agencies such as the Occupational Safety and Health Administration (OSHA). The EI advocates theorize that the dose defined as low level may be as low as one molecule of a chemical agent (Rea et al., 1978).

Gots (1993a) reviewed the criticism of the one-molecule postulate. Dr. Sidney Shindell called this the "receding zero". Dr. George Koelle, a chemist, put this amount of matter into perspective. He estimated that if one pint of water were spilled in the ocean, once it mixed fully and completely, 5000 molecules of that water would wind up in every pint of water taken anywhere in the world. The same analogy can be made for any chemical disposed of in the water or air.

One-molecule exposures to common household products such as formaldehyde are ubiquitous. EI advocates build chemically less contaminated, so-called ecologically safe environments, environmental care units (ECU) in hospitals, or "safe houses" for outpatients. These environments are built with "safe" building materials and purified by special air-filtration systems. Occupants use only "hypo-allergenic" toiletries and cosmetics, their foods are organic, and electronic devices are kept to a minimum. Whenever possible, these enclaves are located away from urban, industrial, or highly trafficked areas, preferably where geothermal and air-flow characteristics are said to move "pollutants" around the buildings. In the extreme, aluminum-lined rooms or porcelain-lined trailers have been prescribed by clinical ecologists as medical necessities. But no matter how or where "safe houses" are built, eventually they, too, are judged unfit for habitation by EI patients.

Heightened sensation: hyperosmia and cacosmia

Low-level exposure effects require a postulate to allow for heightened sensation. The postulate suggests that EI patients have lower sensory thresholds for odor perception, called "hyperosmia" (Ashford and Miller, 1989, 1991) and claimed by some EI patients (Morrow et al., 1989). Heightened sensation to non-perceivable stimuli requires supposition of an additional mechanism, yet to be defined. However, the postulate of lower sensory threshold generated a hypothesis testable by methods commonly used in odor-detection studies. Under controlled conditions, subjects sniff progressive doses of an odor and judge whether they can detect it or not. In a study using phenyl ethyl alcohol and methyl ethyl ketone, the sensory thresholds for EI patients were not lower compared to control subjects (Doty et al., 1988). Hummel (1997) reported findings of no difference between EI patients and controls when olfactory-evoked potentials (a statistical parameter derived from the EEG brain waves with repeated sensory stimulation) were recorded in response to varying doses of hydrogen sulfide (rotten egg smell). These studies refuted specific hypotheses that EI patients are hyperosmic. There are no scientific studies substantiating that EI patients are hyperosmic. The weight of the scientific evidence suggests that the theory of heightened sensitivity is refuted and incorrect.

In order to preserve the toxicogenic theory, mechanistic hypotheses moved from the sensory mechanisms of odors to their effects. "Cacosmia" was defined as hypersensitivity to aversive effects of odors (Bell, 1994b). The empirical foundation for postulating "cacosmia" comprised survey data from college students and geriatric subjects showing that a percentage of these subjects reported elevations in ratings of subjective discomfort from odors. In a companion study, brain-wave EEG data were interpreted to represent the brain correlates of "cacosmia" (Schwartz et al., 1994). Independent neurophysiologic studies by Lorig (1989) investigating the effects of odor found that food odor had demonstrable effects on the EEG, as would be expected. However, these same studies also showed that food imagery in the absence of odors had similar effects on the EEG. These results demonstrate that odor perception and associated central nervous system responses are cognitively mediated.

Additional evidence for the non-specificity of activation of cortical regions involved in olfactory perception comes from brain-activation studies which employed positron emission tomography (PET), reviewed by Roland (1994). The olfactory pathways start in the olfactory receptor cells located in the nasal mucosa. They are relatively broadly tuned and send the axons to the mitral cells in the olfactory bulb. The mitral cells in turn send the axons to several places: the amygdala, the hippocampus, and the piriform cortex in the anterior insula in the cerebral cortex. PET studies with normals have shown that olfactory cells, as well as other parts of the limbic and paralimbic cortex, are activated through a variety of vegetative, olfactory, cognitive, and emotional functions. In one unpublished Japanese study by Kawashima (described by Roland, 1994), normal subjects received an injection of vitamin B_1 which excites the olfactory cells so that they sense a sulfuric smell. Roland also summarized other studies with normals which had shown that different regions of the insular cortex were activated by various behavioral tasks and processes including rapid eye movement sleep, calculation of prime numbers and memorizing historical dates, tactile learning and tactile recognition of geometric objects, somatosensory discrimination of shapes, and subjects repeating words shown to them. Of particular relevance to EI patients who have a high incidence of panic-like reactions, the anterior part of the insular cortex has been activated in panic-disorder patients with lactate-induced panic attacks (Reiman et al., 1989).

Neurophysiologic responses do not reflect the sensory, molecular aspect of the odor alone, as has been assumed by EI advocates to support the olfactory-limbic kindling mechanism to explain "cacosmia" (Bell et al., 1992), discussed in detail in Chapter 12.

Threshold variability

Threshold variability is postulated to explain an EI patient's inconsistency of response to environmental agents upon repeated exposure (Ashford and Miller, 1989, 1991; Rea, 1992). The threshold of reactivity is said to vary with time, history of exposure, and host susceptibility. The threshold may be higher (less likely to react) such that the "set point" has been adjusted. The effect of a particular exposure on the body is difficult to detect, it is argued, because it could be masked by other agents, factors, or conditions elevating the threshold of reactivity for the agent in question. This idea is analogous to the signal-to-noise ratio for sensory perception. The signal cannot be detected unless the background noise level is reduced, an experience familiar to those who have tried to talk to someone at a rock concert. The effects may last, and it is difficult to hear because your ears may be ringing for several hours due to the deafening background noise. The signal is said to be masked. The effect of low-level sensitivity is said to be detectable only if it is "unmasked" after prolonged sequestration in an environmental care unit, possibly assisted by detoxification methods (Ashford and Miller, 1989, 1991, 1998), as discussed in Chapter 3. Clinical ecologists estimate the period required for "deadaptation" to be 4 to 7 days, analogous to the time course of alcohol withdrawal (Randolph and Moss, 1980). It is a common experience for normal subjects exposed to a constant odor level (perfume) to recognize a decreasing detection of the odor over time.

EI advocates propose *adaptation*, defined as an evolutionary survival mechanism which apparently allows an individual to get used to an acute toxic exposure (Rea, 1992). "Adaptation" is synonymous with other terms such as masking, habituation, acclimatization, inurement, or developing tolerance (Ashford and Miller, 1989, 1991; Randolph, 1962). Adaptation is a valid physiological concept and typically occurs with frequent or continuous exposure, as in the case of alcohol and drug use (Hyman and Nestler, 1996).

Deadaptation or *unmasking* is a postulate to counteract the effects of "adaptation", allowing for detection of symptoms again. Deadaptation may result from an "unusually high" dose exposure which lowers the threshold (more likely to react) to lower doses of any chemical to which the individual adapted and tolerated. Deadaptation to low-level chemical exposure after a period of non-reactivity is hypothesized to be similar to the principle of time-dependent sensitization in animal research (Antelman, 1988), discussed shortly. It is also thought to underlie the course of certain drug responses in the treatment of human affective disorders (Post et al., 1984). In these studies, the effects are reliable and predictable; according to the toxicogenic theory, they are not. The time course of swings between reaction and non-reaction after provocation is not specified *a priori* or by independent assessment. Rather, adaptation and deadaptation are invoked *post hoc* to explain any sequence of events.

The sporadic, unpredictable variability of low-level chemical reactions over short time periods (even over a few minutes) is not consistent with central nervous system mechanisms of toxicodynamic reactions (Spencer and Schaumburg, 1980). Some of these mechanisms have been hypothesized by the toxicogenic theory, including limbic kindling, partial limbic kindling, conditioning, sensitization, and time-dependent sensitization (Bell et al., 1992, 1993b). Scientific studies demonstrated that the effects of these mechanisms are detectable, consistent, long-term or permanent, and reproducible with repeated challenge (Adamec, 1978, 1990; Antelman, 1988; Gilbert, 1992a,b; Post et al., 1988a,b, 1992; Racine, 1978).

Another postulate is *context specificity*, defined as the sensitivity threshold varying depending on the context. For example, a chemical agent intolerable in the workplace may be tolerable at home or while on vacation. The analogy is to animal conditioning studies in which the effects of cocaine differed with respect to the cage the animal was tested in. Animals repeatedly injected with cocaine in the same cage environment were much more responsive to rechallenge when compared to animals injected with cocaine in a different cage environment and then challenged with cocaine in the test cage for the first time (Post et al., 1987). This effect is readily replicable and appears to represent the psychological aspects of animal pharmacology (Weiss et al., 1986).

Sensitivity to multiple environmental agents

After EI onset, sensitivity increases for a host of other chemicals, foods, and non-chemical environmental factors. One suggestion to explain this is *spreading*. It is different from the stimulus generalization in classical Pavlovian condition. Unlike stimulus generalization, which requires a stimulus gradient (e.g., color, odor, taste), spreading does not require similarity of molecular structure or sensory property among the multiple environmental agents, which do not even have to be perceivable as is the case with EMF.

Another postulate is *cross-sensitization*. In conditioning studies, animals are repeatedly subjected to a variety of stressors such that reactivity occurs in response to a lower magnitude of any one stressor when used alone. There are also interactive effects between pharmacological and non-pharmacological stress (foot shock) (Kalivas et al., 1986).

While low-level chemical sensitivity is said to spread from one agent to include any or all agents, it has the peculiarity of not spreading to the same chemical that derived from a different origin. EI patients report sensitivities to a synthetic manifestation of the chemical, but not its natural form. For example, low-levels of formaldehyde measured in parts per million (ppm) coming from building materials or office furniture are said to have acute debilitating effects which could last for weeks and sensitizing effects which could last a lifetime. At the same time, the 60 mg of formaldehyde in an orange, the 30 mg in a cup of coffee, the 300 mg or more from a cigarette, or the 58,000 mg generated daily by an adult male from intermediary metabolism are said to have no effect at all (Waddell, 1993).

The idea that a toxic response to one chemical can lead to sensitivity to all other chemicals — "spreading" — transgresses basic principles of toxicology and clinical science. (Gots et al., 1993).

Time course of a hypersensitivity reaction

Clinical ecology publications make constant reference to Selye's GAS model of stress, although there are problems with Selye's model, which have been previously discussed.

Yet another aspect of GAS has been misrepresented — the time course of a hypersensitivity reaction.

The General Adaptation Syndrome defines stress as a physiological defense reaction designed to protect the body from environmental (physical and social) stressors. Three sequential stages span life: the alerting or alarm response, the resisting response, and exhaustion. The occurrence of stress is determined by observation of any of the three stages (the circularity in the model). The alarm response is distinguished by a marked increase in sympathetic-adrenal activity resulting in behavioral arousal. Attentional hypervigilance and the near panic emotional state associated with it have also been identified with the alarm response in humans (Janis, 1982). The alerting response is followed by a stage of physiologic resistance during which time the subject copes with the stressor. Finally, when the coping resources are depleted, a stage of collapse or exhaustion sets in. Selye (1982) offered an analogy for the operation of GAS in humans:

> "These three stages are reminiscent of childhood, with its character-
> istic low resistance and excessive response to any kind of stimulus;
> adulthood, during which the body has adapted to most commonly
> encountered agents and resistance is increased; and senility, charac-
> terized by loss of adaptability and eventual exhaustion, ending with
> death."

Selye's stage model is consistent with contemporary neurophysiological and biochemical models of aging and stress in both animals and humans (Sapolsky, 1992).

Alluding to Selye's GAS model, Randolph postulated that the after-effects triggered by chemical exposure follow a linear time sequence (Randolph and Moss, 1980). But in sharp contrast to GAS, Randolph did not specify the absolute periodicity of the time course; it could be seconds, minutes, hours, or days. Short-term physiological effects are called time-limited state changes. Long-term, permanent changes such as those postulated in GAS over the course of a lifetime are called trait changes. State and trait effects are often mediated by different biological mechanisms and physiological systems. Full recovery is the end point in Randolph's explanation of a temporary effect, a state change. Selye described long-term trends characterized by permanent adjustments in physiological func-tioning relative to homeostasis, called tonic levels. Toxicogenic theory often does not distinguish between short-lived state effects and long-term, persistent effects.

Toxicogenic theory postulates that the sequence of state effects elicited after exposure begin with immediate pronounced stimulatory symptoms, followed by withdrawal symp-toms induced by adaptation, as illustrated in Figure 2.2. Behaviorally, the extreme at the stimulatory end of the continuum is mania or hysterical conversion symptoms (seizure-like reactions or convulsions, panic, agitation, etc.), with lesser symptoms including hypo-mania, anxiety, hyperactivity, and irritability. The extreme at the withdrawal end of the continuum is depression with or without altered consciousness (depersonalization, dere-alization), with lesser symptoms that include disturbed thinking, headache and myalgia, mild depression, and various gastrointestinal complaints (Randolph and Moss, 1980).

The theory suggests testable hypotheses with clear empirical predictions in double-blind, placebo-controlled provocation challenge studies. Only arousal symptoms are ex-pected immediately after challenge with an active agent. Delayed arousal symptoms or immediate withdrawal symptoms would refute the hypothesis. However, were this ex-periment to be done and the results were not confirming, additional postulates about variability of effects could be invoked *post hoc* to explain them away (discussed next).

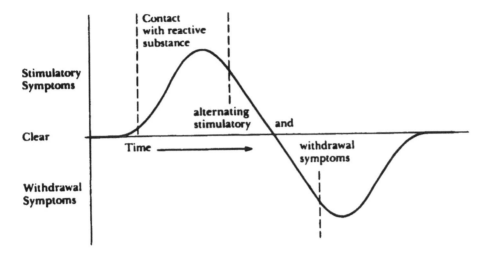

Figure 2.2 Time course of symptom progression after a hypersensitivity reaction. (From O'Banion, D.R., *Ecological and Nutritional Treatment of Health Disorders*, Charles C Thomas, Springfield, IL, 1981, p. 68. With permission.)

Variability of effects

Individual EI patients may have unique symptoms selected from the list of multiple system complaints presented in Chapter 1. The symptoms elicited by any chemical do not have to be similar among individuals, much less conform to any set recognized in medical toxicology. Within any individual, different chemicals may elicit the same symptoms or completely different ones. Finally, the same chemical and dose exposure may elicit different symptoms with repeated exposures in the same individual. Toxicogenic theory offers several ways to account for this total lack of specificity of symptoms (Rea, 1992).

According to the idea of *bipolarity*, the potential for reactivity may be at one extreme or another on a continuum of hyporeactivity to hyperreactivity. If an EI patient is in a low phase of reactivity, withdrawal symptoms are said to occur. If in a high phase of reactivity, arousal symptoms are said to occur. Bipolarity rationalizes why symptoms may not be consistently related to exposure and why the results of challenge tests depend on how recently exposure has occurred (Rea, 1992). For example, withdrawal symptoms may occur after a period of exposure to a chemical agent which did not elicit any excitatory symptoms. Such an event would violate the proposal for the time course of a hypersensitivity reaction just described above.

Switching symptoms accounts for different symptoms observed in the same individual to the same exposure at different times. The concepts of bipolarity and switching symptoms are not compatible with the contention that EI patients can identify subsensory threshold exposures based solely upon the unique pattern of symptoms they experience (Miller, 1992). These contentions are more likely made in civil litigation.

Addiction to exposure

EI patients report craving a "toxic" substance to avoid withdrawal effects, analogous to the process associated with alcohol and drug addiction. *Addiction to exposure* explains why EI patients succumb to the cravings which lead to the body eventually breaking down when the toxic substances exceed the "total body load". This rationale is used to explain why certain EI patients cannot adhere to avoidance of offending agents, most notable in eating disorders.

I believe this to be toxicogenic rationalization of psychological processes. Addiction to exposure is consistent with what Sigmund Freud called the "repetition compulsion", a compulsive tendency to repeat some aspect of a negative experience (Freud, 1914; Horowitz, 1986). In related historical developments, Janet noted that traumatized people become attached to the trauma and seem unable to go on with their lives (van der Kolk and van der Hart, 1989), what Freud called "fixation". For some EI patients, environmental intolerances seem to take over as the main purpose in life, as if the perceived illness itself becomes their reason for living. The inherent bias of the mind toward explanation of experience can lead to false appraisal and misattribution of cause-and-effect (Weiner, 1985). Recurrent obsessions can lead to overvalued ideas and closed belief systems, while the underlying psychological issues may be symbolically repeated.

Auxiliary postulates, the protective belt

When hard-core postulates cannot explain a phenomenon, auxiliary postulates are often invoked to protect them from refutation. "Epicyclical elaboration" is the poignant term used by the philosopher Polanyi to characterize this endless list of excuses to avoid refutation (Waddell, 1993).

Temporal cohesiveness

Temporal cohesiveness suggests that exposure events being appraised as harmful by many people concurrently in the same physical environment indicates that a real problem exists (Miller, 1992). Temporal cohesiveness assumes group testimonials are intrinsically valid without the need for corroborating scientific evidence. Patient histories must be taken seriously and are an essential aspect of medical diagnosis, especially in the case of documented toxic exposure, but it is another thing to suggest that patient histories are sufficient for diagnosis, especially in the case of EI. The following news story illustrates the problems with temporal cohesiveness (*Chemical & Engineering*, February 20, 1995):

> "George Tiers sent from St. Paul, MN, a report from a local newspaper, the *Pioneer Press*, about a painting job scheduled for the fourth floor of a building occupied at least in part by county employees. The maintenance people told employees of the scheduled painting date, as they do in the event of any work involving materials that emit fumes. As it happens, the painters fell behind on other jobs, and therefore the painting at issue didn't occur on the scheduled day. Nevertheless, about 20 employees of the county attorney's office complained about irritating fumes."

Epidemic hysteria, known by multiple synonymous labels, including mass hysteria, mass psychogenic illness, and mass sociogenic illness, has been defined as a constellation of symptoms suggestive of organic illness, but without an identifiable cause, that occurs between two or more people who share beliefs related to those symptoms. Epidemic hysteria is more common than recognized, and the literature on the topic has been reviewed by Boss (1997). Without verification, group testimonials have no more validity than individual testimonials, although they appear more convincing and can be more influential. People experience their environment mediated by conceptual categories, some of which may be fabricated in social interaction (Douglas, 1985). In 1982, Colligan et al. (1982) edited a social analysis of mass psychogenic illness (MPI) epidemics, and extracted two fundamental points. First, MPI must be ruled out for mass effects to be physical. Second, MPI cannot be ruled out just because a physical agent is present. Singer (1982) added an observation specifically germane to low-level exposure: it may very well be that physical agents that are below toxic levels have effects only because MPI or similar psychological processes are potentiating them.

Social information, defined as information provided by others in the individual's psychosocial environment, may induce stress behavior. Research in the area of job satisfaction indicates that social information may influence individual's environmental perceptions and responses (Blau and Katerberg, 1982). This research has implications for workplace dissatisfaction attributed to poor indoor air quality, confounding the reports of illness in "sick buildings" (Boxer, 1990; Guidotti et al., 1987). Suggestion can have a profound effect on illness behavior in a residential community near a toxic waste site, as was demonstrated in the Love Canal incident in upstate New York. The region was portrayed as a "disaster area" based solely on anecdotal testimonials which led people to abandon their homes and sell them below appraised value (Levine, 1982). A report by the State of New York Governor's panel concluded that scientific evidence supporting the health effects feared by residents of Love Canal never materialized (Thomas, 1980). Years later, people moved back into the same homes which resold at real estate market value, supporting the initial conclusions that the residents had overreacted.

Another example of EI being a case of MPI occurred in Frayser, a community in Memphis, TN (U.S. House Subcommittee on Oversight and Investigation, 1980). Rumors spread that houses in the community were built on what was once a toxic waste dump. The Environmental Protection Agency (EPA) and the Army Corps of Engineers made great efforts costing $1,000,000 to find large numbers of reported buried drums of contaminants but could find only a few. Newspapers published hundreds of articles on hazardous waste issues, and television reporters were often present at public meetings. Banner headlines emphasized the existence of environmental-induced health effects such as, "Sickness: way of life in Frayser neighborhood," and "Genetic damage blamed on chemicals." Residents visited their physicians and complained of various symptoms which they attributed to emanations from the dump beneath their community.

Behavioral scientists concluded that social contagion was the most plausible explanation for the effects (Schwartz et al., 1985). They also concluded that Frayser residents had good reason to be concerned. Neighbors had similar health complaints, harmful chemicals and buried steel drums had been found (though in small quantities), a former health department employee swore under oath that there had been dumping of hazardous materials, the news media and some outside experts called it "the next Love Canal only bigger," and the responses of local and state health departments could reasonably be interpreted as a cover-up. The one fact that remained indisputable was that the dump never existed.

Demographic diversity

Initially, EI advocates defined four subpopulations of EI patients based on locus of exposure: industrial workers, occupants of tight buildings, residents of communities whose air or water has been contaminated, and individuals who have personal or unique chemical exposure in domestic contexts (Ashford and Miller, 1991). A fifth subpopulation has been added — veterans from the 1991 war with Iraq, allegedly suffering from Gulf War Syndrome (Miller, 1994).

This categorization is descriptive, but unwarranted. Do geographical or building sources of exposure have different toxicologic effects? Are industrial workers affected differently from domestic workers by agents ubiquitous at low levels? Are we to accept the suggestion that every country in the world has its own unique form of EI? (Ashford et al., 1995). Demographic diversity implies that scientific evidence about any one subpopulation does not generalize to the others.

Route of exposure

Toxicants may enter the body by multiple exposure routes: inhalation, ingestion, mucosal contact, dermal absorption, or injection. The concept of *route of exposure* suggests that different routes of administration produce similar symptoms. The rationalization for this comes from EI patient reports of having reactions to a chemical whether they inhale it, eat it, or touch it. Different exposure routes is a fundamental toxicokinetic truth, but that similar symptoms are produced via these routes is not (Office of Technology Assessment, 1990). Route of exposure is incompatible with other hard-core postulates. It is also inconsistent with EI patients' reports that, faced with an unavoidable exposure (e.g., traffic exhaust or perfume in an elevator), inhaling via their mouths instead of their noses mitigates symptoms (Miller, 1992).

Dose dependence

Dose dependence proposes that the severity of symptoms is proportional to dose (duration and magnitude) of the exposure in the sensitized individual (Miller, 1992). This seems consistent with accepted toxicologic principles, yet it is blatantly inconsistent with the hard-core postulate of rapid changes in sensitization brought on by adaptation and de-adaptation. This auxiliary postulate, taken literally, appears to generate testable hypotheses. But there is a hedge in the nebulous phrase of in the "sensitized individual".

Individual variability, uniqueness

Uniqueness complements the variability-of-effect postulates. Under most considerations of statistical design, the assumption of uniqueness precludes the employment of group comparisons. The exception would be designs that have a very large number of distinct subjects in each group, where it is assumed that individual differences average out from main effects through random assignment. Given the absence of any accepted case definition or epidemiological basis of identifying patients other than by their own beliefs, EI advocates can question the validity of the assumption of random assignment to groups. However, this does not preclude testing. Statistical analyses for single-case designs may be employed in which each subject is a study (Kazdin, 1976). How many studies does it take to refute a hypothesis? Or, how many negative findings are necessary before one is confident there is no effect? Statistically, when is it reasonably probable that the null hypothesis is not rejected? Answers to these questions may be contemplated in the context of Lakatos' example of planetary explanation by the Newtonian scientists, presented above.

Rapid onset and cessation of symptoms

Within a few seconds of exposure, EI patients report immediate symptom onset, which lasts only a few seconds when they are neutralized by a "rescue medication" (e.g., intravenous vitamin C or alkali salts). Such responses lead to speculations about rapid shifts of ions across cell membranes, a neurologic or immunologic mechanism, or possibly conditioning (Miller, 1992).

A psychological explanation of "rapid onset and cessation of symptoms" involves more complex thought processes. Allow me to share a serendipitous revelation I experienced while working with a patient in our inpatient environmental care unit (ECU) at Presbyterian Hospital in Denver. To summarize the case briefly, the patient was admitted to the ECU because of severe malnutrition resulting from fear of food allergy, profound leukopenia, and a suspected monilial sepsis secondary to complications from an arterial Hickman catheter implanted by a clinical ecologist through which she was fed because of alleged intolerance of foods by ingestion. She could lapse into periods of florid psychosis. To control these reactions, the clinical ecologist told her to use intravenous bicarbonate salts. Much to the bewilderment of the staff, the patient demonstrated remarkable beneficial response to these salts within seconds of intravenous administration on several occasions. The pertinent event occurred one day while I was talking to her. Suddenly, she became emotionally distressed and asked me to turn on the intravenous delivery unit next to her. While fumbling with the mechanism, I asked her how she was doing. She showed profound improvement, speaking calmly and rationally, and said she could feel the healing effects of the bicarbonate salts. I had not been able to turn on the apparatus.

EI patients report that biofeedback, self-hypnosis, or other self-regulation techniques improve their ability to cope during exposures by lessening symptom severity. Self-regulation techniques may be legitimately employed to facilitate relaxation and reduction of arousal. However, such methods should not to be misconstrued to be effective in ameliorating the symptoms allegedly caused by exposure to low-level chemical exposure. Any effectiveness of these techniques in mitigating true toxicologic symptoms is limited to a placebo effect, very much like that of the sodium bicarbonate ritual in this case.

Hypothesized biologic mechanisms

Speculation about mechanisms can be endless. The less testable a mechanistic hypothesis and the less it resembles a well-characterized and empirically validated biological mechanism, the longer it is likely to survive scientific scrutiny. Besides, refutation of a mechanistic hypothesis does not directly impact the hard-core postulates. In pseudoscience, in particular, refutation generates new and even less testable hypotheses.

The mechanisms proposed to account for EI have, until recently, focused primarily on the immune system. For example, a hypersensitivity mechanism in which environmental chemicals act as allergens was hypothesized (Bell, 1982; Rea et al., 1978). An autoimmune model in which environmental chemicals induce the production of autoantibodies has also been proposed (Levin and Byers, 1987). Allergist and immunologist Terr (1992; 1993c) reviewed these theories and the spinoff from them and concluded that EI clinical symptoms have no resemblance to immune complex sensitivity diseases. The absence of clinical or laboratory evidence for autoimmune diseases, coupled with the absence of corroborating evidence that numerous common environmental chemicals can damage the immune system at low-level exposure, renders the toxicogenic theory highly implausible (Terr, 1993a,b,c).

Limbic kindling

Speculation about biological mechanisms has recently shifted its focus to the central nervous system (CNS). This shift was coupled with a variety of neurological diagnoses (e.g., toxic encephalopathy) which remain unsubstantiated for EI patients in the field of occupational and environmental neurology (Rosenberg, 1995).

Limbic kindling was the first of several CNS mechanisms incorporated by the toxicogenic theory. In the scientific literature, limbic kindling was defined by an animal model for seizure disorders induced by low levels of stimulation or high doses of a pharmacologic agent (Gilbert, 1992a,b; Wada, 1990). In 1969, Goddard (1969) reported that daily electrical stimulation of limbic and cortical brain structures sensitized the animals so that they would convulse in response to smaller stimuli. The sensitizing current did not damage the site where it was applied, so that the effect was not detectable by anatomical brain-imaging techniques. The lowered seizure threshold was observed to last as long as one year (Wada et al., 1974). The effect does not remain localized but gradually spreads to brain structures beyond those stimulated, typically the corresponding cortical site in the other hemisphere, thus the term "kindling" (Racine, 1978). The theory that limbic kindling contributes to non-experimentally-induced human seizure activity is controversial (Wada, 1990). EI advocates suggest that low-level chemicals serve as the inducing agents.

Because seizure activity by definition is the only measurable effect of limbic kindling, it is not a viable model for the subjective symptoms associated with EI. Although they use the term, EI advocates are not really talking about limbic kindling. They have redefined limbic kindling or partial limbic kindling so that behavioral and subjective symptoms substitute for seizure activity, and odorous chemicals replace cortical electrical stimulation and ingestion of a pharmacologic agent. The route of exposure was changed to the nose where odors are transported via the olfactory bulb into the limbic system, illustrated in Figure 2.3, either directly or by proximal excitation of neurons. From there, excitation initiated by odors is said to spread to other regions of the brain that evoke diverse cognitive and emotional symptoms. This limbic kindling hypothesis bears no resemblance to what neurologists call limbic kindling. Besides, EI advocates concede there is no evidence to support it (Bell et al., 1992).

Time-dependent sensitization

Time-dependent sensitization is another mechanistic hypothesis employed in support of the toxicogenic theory (Bell et al., 1993b). In the scientific literature, time-dependent sensitization is defined as the ability of a strong, potentially threatening stimulus to enhance the response when presented subsequently at the same or weaker dose, even when presented much later (Antelman, 1988). The scientifically established time-dependent aspects of sensitization are that (1) the effect is stronger if the interstimulus interval is longer, and (2) the effect does not occur if the stimulus is presented continually or immediately after the first sensitizing exposure. The effect occurs in all species, can manifest in any physiologic system, in response to all types of stressors, pharmacological and non-pharmacological or psychological, even to an injection of saline (Antelman et al., 1991, 1992). In essence, sensitization represents an unusual form of memory, enabling the organism to make an accelerated defense response to a previously experienced threat. Novelty of the stimulus rather than its stressful pharmacologic properties seems to be the essential factor for the organism to perceive potential harm (Antelman et al., 1987).

The obvious lack of specificity makes sensitization and time-dependent sensitization unlikely mechanisms to validate the toxicogenic theory. Antelman (1988) concluded that

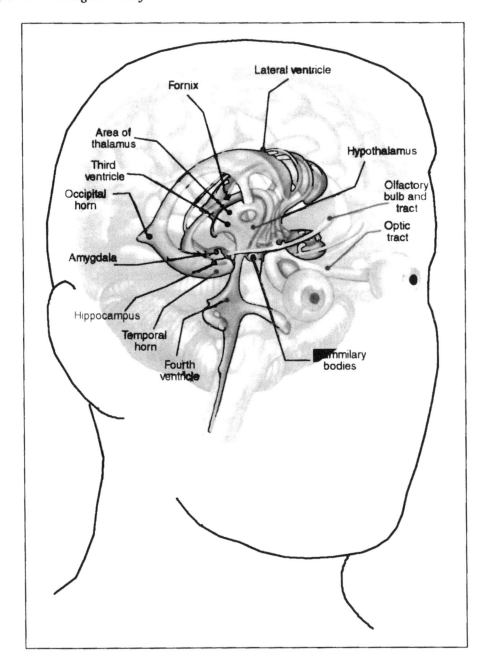

Figure 2.3 Anatomical sketch including the limbic system and the olfactory system. (Adapted from Fincher, J., *The Brain: Mystery of Matter and Mind*, Torstar Books, New York, 1984.)

sensitization and time-dependent sensitization are best viewed as cognitive processing effects involved in the formulation of memory because they hold equally for pharmacologic agents and saline, as well as psychological stressors perceived to be threatening.

Antelman's interpretation, as we shall see, is more consistent with psychogenic theory. Psychiatrists such as Breuer and Freud, and Janet independently, observed a latency period between the occurrence of a stressful event and the onset of symptoms (Horowitz, 1986; van der Kolk and van der Hart, 1989). Janet postulated a breakdown in stress/coping processes, such that the latency of expression is determined by the time required to perceive the inescapable reality of the event, the time necessary for the individual to expend useless efforts to fight the inescapable, and the occurrence of other stressful events that deplete the individual's last reserves. Once symptoms were formed, however, there was a remarkable tendency toward recurrence or persistence long after the event's termination and its immediate effects. Yehuda and Antelman (1993) proposed time-dependent sensitization as an animal model for the study of post-traumatic stress disorder (PTSD) .

Smell and taste

Smell and taste pervade our daily experiences and are integrally involved in health and disease (Getchell et al., 1991). Bell and colleagues hypothesize a link between odor perception and neurotoxic effects of odorous chemicals and called it "cacosmia" (Bell et al., 1993a, 1996). Ziem and Davidoff (1992) postulated that any alteration in olfactory function reflects a toxicogenic effect, be that a diminution or loss in perception (hyposmia, anosmia, olfactory fatigue or paralysis) or an enhancement (cacosmia, hyperosmia). Several diseases, physical trauma, or obstruction can contribute to olfactory dysfunction, limited to olfactory loss (Bartoshuk and Beauchamp, 1994). For, example, patients with Huntington's disease have higher thresholds for odor detection, intensity discrimination, quality discrimination, and identification (Nordin et al., 1995). Significant olfactory dysfunction has been identified in patients with Down's Syndrome (Murphy and Jinich, 1996), and Alzheimer's disease (Nordin et al., 1995). Odor detection sensitivity decreases with age as measured by olfactory-evoked potentials (Murphy et al., 1994) and psychophysical scaling methods, but not by self-report questionnaire which reflect normal smell sensitivity (Nordin et al., 1995). By comparison, the mechanisms of lowered odor threshold or hyperosmia are not well understood, and the toxicogenic theory of cacosmia is based on self-report questionnaires (Bell et al., 1993a).

The sense of smell is integrally involved in memory and emotion and involves the limbic system. As early as 1937, Papez (1937) demonstrated that several parts of the nose or smell brain (rhinencephalon) and particularly the olfactory cortex (archipallium) are not used for olfactory functions in the human being but for the experience of emotion. In spite of the diminished importance of olfaction in the evolution of species, these anatomical areas have expanded, rather than decreased, in humans and have become associated with vast neocortical areas which mediate abstract thinking.

Odorants also produce sensory irritation or perceived pungency, mediated through the trigeminal nerve. Olfaction, in contrast to irritation, has multiple dimensions, including intensity, qualitative odor identification, and an aesthetic dimension (Cain, 1988). Odors can initiate both pleasant and unpleasant memories (Engen et al., 1991), and flavor aversions are often learned in a variety of contexts (Bernstein, 1991). Odors elicit annoyance reactions and are viewed as nuisance phenomena (Shusterman, 1992). Odors also contribute to symptom report in the workplace or a hazardous waste site (Lees-Haley and Brown, 1992), with symptoms enhanced by worry about the harmful effects of perceived exposure (Shusterman et al., 1991). Shusterman's (1992) contention that reaction to odors is an annoyance response or a sensory irritation response rather than a toxicologic response has been dismissed by EI advocates as "a failure to understand the health significance of odors" (Ziem and Davidoff, 1992).

Epidemiological research has confirmed the large discrepancy between self-perception of allergy and sensitivities among the public and its detection by objective methods

(Howard and Wessely, 1993). For example, Young et al. (1987), in an epidemiology study, asked 30,000 people to complete a mailed questionnaire in High Wycombe, a municipality in Britain. Of the 18,582 (61.9%) responders, 7.4% stated that they had a problem (including behavioral/mood changes) attributed to food additives, foods, and aspirin. Of these, 132 whose symptoms were regarded as sufficiently suggestive to warrant investigation were approached to participate in double-blind, placebo-controlled studies with capsules. There were 81 responders, three of whom showed consistent responses to low and high doses of food additives. The first reactor was a 50-year-old atopic (meaning allergic by immunoglo-bulin parameters) male who reported a 5-year history of headaches after ingestion of colorants within 12 hours of ingestion. He reacted to annatto, but he also reacted to placebo on one occasion. The second was a 31-year-old non-atopic female who reported upper abdominal pain attributed to ingestion of preservatives and antioxidants. She reported symptoms in response to challenge with annatto. The third was a 5-year-old atopic child with eczema and a family history of hay fever who showed a change in mood 1 to 2 hours after azo dye challenge, but also after placebo on two occasions. The investigators expressed some doubt about this last case. The estimated prevalence based on 132 was 4.9, for a population rate of 0.026%. To put this percentage into perspective, the prevalence rate established under controlled studies was 285 times less than that estimated from self-perceived food allergy or sensitivity.

Conclusion

> "Man has such a predilection for systems and abstract deductions
> that he is ready to distort the truth intentionally, he is ready to deny
> the evidence of his senses in order to justify his logic." (Dostoyevsky,
> *Notes from the Underground*)

Toxicogenic theory relies on a web of postulates to protect the presupposition that harmful effects of low-level chemical exposure are unpredictable and not measurable and that demonstration of cause and effect is not necessary. The postulates explain unreliability of reactivity over time or repeated exposure events. The semantics are such that any event and any non-event can be rationalized. For example, once a susceptible individual is "chemically sensitized" there is no relief from reactivity, unless one is in a state of "adap-tation" in which there are no perceivable symptoms to exposure because they have been "masked" by the effects of other agents. Symptom "spread", triggered by any and all environmental agents at any dose depending on the "set point" of the sensitivity threshold, without regard to a known process of stimulus generalization, be that molecular structure, sensory qualities, toxicokinetic mechanisms, or physiologic pathways. Chemicals "cross-sensitize" with non-chemical stressors to contribute to the "total load" of stress on the body, judged in excess when the patient perceives symptoms. Any given chemical is deemed harmful if it is synthetic, but safe if it is natural (so-called organic).

Such speculations are unsubstantiated in the scientific fields of toxicology and phar-macology (Ames and Gold, 1995; Waddell, 1993). Furthermore, these extreme formulations were not endorsed in an opinion survey of physicians, including clinical ecologists, who had clinical experience in treating EI patients (Nethercott et al., 1993). When theoretical speculation is curtailed by knowledge, EI advocates deflect argument by the concept of "temporal cohesiveness", saying "so many EI victims can't be wrong." But linking symp-toms to putatively harmful environmental agents is scientifically meaningless because to date not one single substantiated case of EI has been reported in the credible scientific literature. Despite this, there have been more than 50 years of unsubstantiated theorizing, diagnosis, and treatment, issues addressed in the following chapters. That the absence of evidence does not preclude a possibility is logical. But, in the world of science, there is another truism: the lack of epidemiological evidence for the existence of a disease precludes establishing a diagnosis for it.

Toxicogenic theory suffers from the ahistorical fallacy, a neologism referring to the absence of a sense of history (Andreasen, 1994). The ahistorical fallacy is based on three faulty assumptions:

1. Proposition X must be true because it is what the "experts" are teaching.
2. Proposition X must be true because it is the most recent one to be put forth.
3. If information increases, knowledge increases as well.

chapter three

Unsubstantiated diagnoses and treatments

> "We should not be embarrassed to say 'so what?' when confronted with an observation that an isolated finding of 'statistical significance,' without independent confirmation or a theoretical framework, might have a high clinical correlation but make no common sense." (Brodie, 1996)

The terms "unproven" and "unsubstantiated" are synonymous and describe diagnostic and treatment methods and practices for which there is no scientific evidence. There is no clear boundary between the methods as applied to diagnosis and treatment, and the logic is often circular. The rationale for these methods is dubious at best. Unproven and unsubstantiated is not to be confused with the term "experimental" which defines diagnostic or treatment methods undergoing scientific assessment. I also discuss some inappropriate methods and practices defined as valid when employed properly and in the right context, but invalid as alluded to by the toxicogenic theory.

Unsubstantiated diagnostic and treatment practices

The unproven and unsubstantiated diagnostic and treatment methods practiced by clinical ecologists (Table 3.1) have been reviewed elsewhere (Condemi, 1991; Selner and Condemi, 1988; Selner and Staudenmayer 1991; Terr, 1993c). I have selected one of these to illustrate how illusive these practices are to scientific scrutiny and that they persist in the face of refuting evidence.

Provocation-neutralization testing

The so called provocation-neutralization technique is employed to diagnose and treat allergy to foods (Hansel, 1968) and chemical sensitivities (Galapeaux, 1976). Sublingual or subcutaneous provocation testing and neutralization therapy are difficult to distinguish in the procedure because one dose provokes a reaction, a weaker dose neutralizes the reaction, and a still weaker dose again provokes a reaction. Clinical ecologists claim that minute quantities of allergens can both induce and alleviate symptoms. The postulated mechanisms for the neutralization procedure is a "phasic" immune phenomenon capable

Table 3.1 Some Unsubstantiated Diagnostic
and Treatment Methods Practiced
by Clinical Ecologists

Diagnostic methods
Cytotoxic testing
Rinkel technique
Subcutaneous/sublingual provocation testing
Hair analysis
Blood and urine amino acid screens
Electrodermal testing (electro-acupuncture)
Applied kinesiology

Treatment methods
Subcutaneous/sublingual neutralization therapy
Purge therapy
Salt neutralization therapy
Orthomolecular therapy
Autogenous urine injection
Antioxidants: vitamins and supplements
Acupuncture
Homeopathy
Sauna depuration
Imprinted water

of influencing specific immune responses involving antigen processing, immune complex formation, and mediator release in non-IgE-mediated symptom induction (Rapp, 1986). The reference in this definition to non-IgE-mediated is important. IgE is the recognized immunoglobulin that mediates classical allergies. The rationale for subcutaneous and sublingual delivery are the same, with the sublingual method favored to make the testing and treatment procedure more acceptable to the patient.

The sublingual method used for diagnosis usually involves placing three drops of aqueous extract under the patient's tongue and waiting up to 10 minutes for symptoms to appear. Similarly, a subcutaneous dose is injected, with subjective symptoms being the primary dependent variable, while the cutaneous response at the site of injection may be considered important or unimportant depending on the subjective symptoms. If no symptoms are reported, higher doses are administered in a serial fashion until symptoms are provoked. There is no standardization of the nature or intensity of the symptoms to interpret a positive test. Symptoms are not specific, those reported by the patient and attributed to EI need not be the ones experienced during testing. Even the absence of symptoms may be interpreted as a positive test. The neutralization dose is determined by progressively administering a series of lower doses until no symptoms are reported, and that dose is used for subsequent treatments. There are many variations of this protocol, but subjectivity is the consistent characteristic.

In its review of provocation-neutralization procedures, the panel of the American College of Physicians (1989) reported 15 published investigations. The studies purporting to show a positive finding were methodologically flawed. Most challenges were conducted in an open, unblinded fashion. This and other critical reviews have not supported the validity of this technique (American Academy of Allergy and Immunology, 1981; Golbert, 1971, 1975). There were also two notable double-blind, placebo-controlled studies, conducted by physicians who believed in the efficacy of the procedures, which showed the effects were attributable to placebo.

In 1971, Kailin and Collier (1971) evaluated sublingual and subcutaneous provocation testing conducted by five experienced clinical ecologists, using selected patients from their practices who previously tested positive under open challenges. Confident in the reliability and validity of these reactions, the physicians conducted the procedures in a double-blind fashion, using saline as a placebo control. The results showed no difference between reactions to aqueous food extracts and placebo. Furthermore, in evaluating the neutralization dose, there was no difference between food extracts and saline in their ability to relieve symptoms. Clinical ecologists, including the investigators, deemed this and other studies with negative findings as methodologically flawed and continued to utilize the procedures.

The second study was conducted by Jewett, an orthopaedic surgeon who was diagnosed and treated for EI by Rea in Dallas. Jewett initially subscribed to the method and was motivated to conduct a double-blind, placebo-controlled, and methodologically flawless study which could definitively show the positive effects of the subcutaneous procedures (Jewett et al., 1990). Both advocates and critics of clinical ecology reached agreement on the methodology and validity of the study. The procedures used in clinical practice were carefully duplicated and implemented by seven experienced clinical ecologists. The subjects were 15 female and 3 male patients who had previously responded to unblinded clinical testing in a consistent manner. To allow for the possibility that the patient might not be in the optimum adapted or de-adapted state to ensure reliable responding, the physicians had the opportunity to withdraw patients from the study, both after the initial selection and on the day of the testing session. Patients were retested in the same office, by the same technician with the same extracts at the same dilutions with the same diluent but under double-blind conditions with unmarked syringes. Enough tests were conducted on each individual so that statistical analysis could be applied to the results. As expected, a host of multi-system, general malaise symptoms were reported by patients in response to the challenges. Across all subjects, the number of challenges appraised positive was 27% for active injections and 24% for placebo injection, not a statistically significant difference. For all 18 patients, individual responses to active and placebo challenges were no different from chance performance as determined by the chi-square statistical test. Of seven patients who were given neutralization doses by their physicians after a positive response, symptoms were relieved equally whether the patient had been exposed to active agent or placebo. The results unequivocally demonstrated that the symptoms were placebo responses that were generated and remitted spontaneously. The investigators concluded that the technique works only if practiced unblinded under the influence of direct or indirect suggestion.

Despite the agreed-upon protocol, when a true positive hypersensitivity patient could not be produced, advocates offered a variety of opinions about the implications and delayed the publication of the study. Jewett concluded that the absence of evidence for that single true positive patient is essentially a passive admission that the provocation method claimed to demonstrate hypersensitivity to environmental agents is fatally flawed (Jewett, 1992b). Nonetheless, the use of these procedures did not cease. Markedly different "expert" opinions were presented to the public, and Jewett was deemed *persona non grata* among clinical ecologists.

In a public forum sponsored by ATSDR and hosted by the California Department of Health Services, held in Berkeley in May 1994, I witnessed the confrontation between Jewett and Rea, who was an invited speaker but had agreed to speak only under the condition that no questions could be asked of him, a condition announced to those in attendance by the moderator as he introduced him. During his presentation, Rea touted the virtues of provocation-neutralization procedures, claiming clinical evidence to support his assertions. He ignored the negative findings from scientific studies. Jewett had no recourse but to disrupt the proceedings and confront Rea on the veracity of his assertions. While the disruption created a strained atmosphere of embarrassment, in hindsight one can question the wisdom of the decision to allow the condition of no questions in a federally funded scientific forum.

Sauna depuration

Depuration or purification treatments include dry saunas, exercise, and massage (Rea, 1996). Historically, these methods were developed and originally described by the acclaimed science-fiction writer and founder of Scientology, L. Ron Hubbard (Hubbard, 1990). The treatment is coded and billed as physical therapy and costs from $3500 to $4300 for a 4-week course. Some patients undergo multiple courses of treatment. The procedure is conducted up to three times a day, totaling 4 to 5 hours and including 2 hours of heat (140 to 160°F), 2 hours of exercise, and a massage after. A strict protocol is said to facilitate the clearing of poisons from the body. In an insightful review of this procedure, Kurt (1995a) questioned the appropriateness of the use of niacin in the depuration methods. Niacin is administered six times, before and after each of three daily saunas, to "tolerance", sometimes in total doses of 3000 mg. Niacin promotes histamine release. The side-effects from niacin listed in the *U.S. Pharmacopoeia Drug Index* include many of the complaints of EI patients: feeling of warmth; flushing or redness of skin, especially on face and neck; headache; diarrhea; dizziness or faintness; dryness of skin; nausea or vomiting; and stomach pain. Because patients are warned that they will temporarily experience toxic symptoms as the substances stored in body fat are mobilized during the sauna-depuration, the side-effects from the niacin reinforce the perception that the poisons are being cleared, and therefore prior poisoning must have occurred.

What is the attraction of sauna and massage therapy? Europeans and Asians have flocked to spas through the ages and enjoyed these relaxation exercises. Clinical ecologists take advantage of this and add an iatrogenic twist — unsubstantiated therapies involving vitamins, minerals, and supplements with the ability to purify the body of accumulated toxins — despite the lack of an identified physiological mechanism that could transport the toxins from the tissue in which they are stored to the sweat glands.

Imprinted water and EMF

Of all the unsubstantiated methods I have seen passed off as having scientific merit for diagnosis and treatment of EI patients, imprinted water stands out as being far beyond the pale of plausibility. The issue is not with what exactly imprinted water is, but with how the concept is misused. The description offered by a clinical ecologist in a deposition (District Court of Alaska, Civil Action No. A93-107) may serve as a singular example of circle speak employed by a practitioner seemingly lost in a ritual unfettered by reason. This physician had diagnosed the patient with multiple chemical sensitivities and electromagnetic sensitivities. The court transcription reads as follows:

> Q. [attorney]   And then you have in your note, "Put her imprinted water in a cup and drank it and felt better." What does that mean?

A. [ecologist] Well, we were on an experiment with [a colleague] from England on electromagnetics and he was trying some stuff about imprinting water with electromagnetic frequencies.

Q. What does "imprinting water with electromagnetic frequencies" mean?

A. Well, he beams different frequencies into the water and it apparently holds it.

Q. The water holds electromagnetic frequencies?

A. Yeah.

Q. And what about her imprinted water, is there something about ...

A. Well, it would be for her pattern that he worked out on her.

Q. Explain to me a little bit more what this means, what patterns is he working out?

A. Well, he has them imprint water and then he has a measuring device that he uses to read the different frequencies on it, and then he uses other balancing frequencies for these. Like I say, it's just an experiment on this yet, it's nothing proven.

Q. And how much water did she drink?

A. Well, it says a cup.

Q. Okay. And then you also say in your note, "Also slept on top of imprinted water vials last night and felt better." What does that mean?

A. Just what it says.

Q. Well, what's the procedure when you say you sleep on top of imprinted water vials?

A. Oh, I don't know why she was sleeping on top of them. Probably she was given some vials with her imprint on it and had them in her bed clothes or whatever.

Q. And is that something that this doctor who is working with you recommends that people do?

A. Well, it's some of the experiments he's carrying on, yeah.

Q. That people sleep on top of vials of water and they feel better?

A. No, I don't think it was sleeping on top of any vials; they just put them in their pocket.

Q. Okay. And describe for me the process ...

A. I just did.

Q. Could you give me a little more detail?

A. Well, he takes a frequency generator — you know what a frequency generator is?

Q. No, I don't.

A. Well, that probably is the problem then. In physics there's a generator that generates frequencies through the spectrum, for example, like one tenth hertz, 1 hertz, 2 hertz, 20 hertz, 50 hertz, 100 hertz, 5 megahertz, you've heard of the different frequencies?

Q. Right.

A. Okay. Well, it will generate those, okay? And his experiment is that he feels that he can take different frequencies and that water will hold them, and he's not alone in this; there are a lot of people in the world that have done that, and it's been observed that some people get a mild clinical benefit out of that if they hold it or if they keep it next to their skin.

Q. Okay. And is this procedure approved by the Food and Drug Administration as an experimental protocol?

A. I don't know whether it is or not.

Q. Okay. Is it done only in your clinic?

A. No.

Q. Where else is it done?

A. Well, he's done it in several clinics across the country.

Q. Where else?

A. I can't really tell you.

Q. All right. Have you obtained any approval form the FDA to test people with this imprinted water?

A. No.

Q. Did she tell you —

A. Why would we?

Q. I don't know, sir.

A. Well, you asked the question.

Q. Did she tell you in what ways she felt better after sleeping on top of imprinted water vials?

A. It just says she felt better, I don't know.

Q. Have you recommended that she continue sleeping on top of vials of imprinted water?

A. I don't really recall. As I say, we were running an experiment. She complained to be a little sensitive to electricity, so we thought we'd try it and see.

Q. Was it your notion or belief or idea that drinking imprinted water would help her sensitivity to electromagnetic field?

A. I didn't know. Like I say, it was an experiment. We try it on some of these people so we can get a feeling to whether it's worthwhile pursuing or not.

Q. And given her response, do you think it's worthwhile pursuing?

A. I don't really know.

Q. You have her diagnosed as "electromagnetic field sensitivity, chemical sensitivity." How did you decide that she had electro-magnetic sensitivity?

A. Well, apparently just from the testing. It wasn't a thing; if you notice, we didn't put that down on any of our diagnosis or anything except for in this progress note here. We didn't put it on any of the front sheets or anything.

Q. Okay. But my question is how did you decide she had electro-magnetic field sensitivity.

A. Well, I said, from the testing.

Q. Of her?

A. Yeah.

Q. And what testing did you do on her?

A. I just told you.

Q. Did he hook up electrodes to her or did he beam frequencies at her?

A. He beamed frequencies at her.
Q. He beamed frequencies at her?
A. Yeah.
Q. And how did she respond?
A. I don't recall.
Q. Is there any record made of the test?
A. No. He's got it in his research book I'm sure.

This account is self explanatory. It highlights the irrationality of a diagnostic and treatment method which presupposes that EMF can cause EI and histrionic symptomatology. While this hypothesis has no supporting evidence, there is evidence to refute it. Gyuk et al. (1989) presented a case of alleged EMF sensitivity tested by double-blind provocation challenges. A young male complained of multiple sensitivities including tin, stainless steel, a number of foods, the color orange, and electrical equipment which elicited severe reactions such as uncontrolled spasms and fainting. Driving under a transmission line purportedly caused neck spasms and other responses. His beliefs were iatrogenic, and he carried a vial of imprinted water as protection against severe attacks. The first trial was open, utilizing a 9-volt battery on which the man placed his thumb across the terminals. Uncontrolled, violent convulsions started immediately, lasting approximately 20 seconds. As an objective test, a series of double-blind provocations were devised employing mild DC currents, magnetic fields at various frequencies, and application of permanent magnets. Seated in a comfortable chair, the stimuli were applied with both subject and technician blinded. Responses were remarkable, with convulsions, uncontrolled hand waving, and somnolence states being typical. Analysis of the time series showed no correlation between stimuli and responses. Appraised of the results, the subject discontinued use of magnets and imprinted water, and the symptoms became minimal.

Inappropriate use of diagnostic methods

Table 3.2 lists several recognized diagnostic methods that are inappropriately used to confirm diagnoses attributed to EI. The immunological tests have been thoroughly reviewed

Table 3.2 Diagnostic Tests Inappropriately
Applied in the Diagnosis of EI

Immunological tests
Pulse test
Serum IgG antibodies
Total serum immunoglobulin
Lymphocyte subset counts
Lymphocyte function assays
Cytokine and cytokine receptor assays
Body chemical analysis
Food immune complex assay

Brain imaging tests
Positron emission tomography (PET)
Single photon emission tomography (SPECT)
Neuropsychological testing

(Terr, 1993c; Salvaggio, 1996) and will not be addressed here. I will limit my discussion of diagnostic tests to those employed in support of CNS-related diagnoses, e.g., toxic encephalopathy. I will limit my review of the inappropriate use of treatment methods to acupuncture.

Radiology brain-imaging tests

Nuclear radiologist Mayberg (1994) reviewed the reliable and valid application of positron emission tomography (PET) and single photon emission tomography (SPECT) imaging studies in patients with well-defined clinical deficits following focal neurological insults. Group comparison studies on patients with well-characterized and often pathologically confirmed diseases aided the diagnosis and treatment of individual patients with epilepsy, brain tumors, dementia, stroke, movement disorders, and head trauma. Brain scan abnormalities have also been identified in groups of patients with psychiatric diagnoses, including depression, panic disorder, schizophrenia, attention deficit, anxiety and obsessive-compulsive disorder, alcoholism, and substance abuse, among others. Specifically addressing EI, Mayberg emphasized that the sensitivity and specificity of these patterns have not been determined. Mayberg's observations and cautions apply equally to newer brain-imaging methodologies including magnetic resonance imaging (MRI) and functional magnetic resonance imaging (fMRI) that are likely to be presented as confirmatory evidence of EI.

Uncontrolled brain imaging studies such as PET and SPECT have been presented as confirmatory evidence to validate the diagnosis of EI in individual cases (Callender et al., 1993; Heuser et al., 1993; Mena et al., 1993; Simon, 1992c). These studies have been criticized for methodological and interpretive errors (Mayberg, 1994; Moser et al., 1995; Society of Nuclear Medicine Brain Imaging Council, 1996). The specificity of functional abnormalities for particular symptoms or specific causal agents have not been evaluated. Although preliminary studies report multifocal defects in patients purported to be affected by EI, no clear lesion-deficit correlations have been demonstrated. Similarly, studies have yet to correlate specific clinical complaints or objective neurological or neuropsychological deficits with specific functional imaging patterns.

Neuropsychological testing

Specific cognitive impairments associated with specific neurobehavioral test performance decrements are reported in the scientific literature with high-level exposure to numerous neurotoxins such as lead, manganese, trichloroethylene, mercury, styrene, toluene and paint solvents, organophosphate pesticides, carbon monoxide, and carbon disulfide (Bolla et al., 1995; Cherry et al., 1985; Gamberale, 1985; National Research Council, 1992b; Rosenstock, et al., 1991; Spencer and Schaumburg, 1985; White and Proctor, 1992, 1997). These patients have specific true positive effects in cognitive impairment with true negative, unpaired performance on tests that measure other cognitive functions not related to specific toxic effects. In short, toxicologic effects detected by neurobehavioral testing show both sensitivity and specificity. Without a hypothesis about specific effects, testing results that show an impairment on one among many specific tests can be a statistical anomaly, a chance occurrence under the probability of a Type I error. The problem with testing which is not hypothesis driven is summarized as (Gamberale, 1985):

> "Thus, too many investigations have been performed in which the
> researchers had no manifest hypotheses concerning the type of

behavioral performance impairment to be expected as a result of the condition under study. Researchers in this field have frequently shown a profound lack of knowledge of the fundamentals of performance measurement and of the interpretation of psychometric results."

Neuropsychological tests are effort dependent and susceptible to conscious malingering or unwitting influences. Performance effects may also be affected by mood states such as depression and anxiety. The contaminating effects of these factors were summarized as (Benton, 1994):

"It is a truism that all specific behaviors, including the responses elicited by a neuropsychological test, have multiple determinants. In addition to the factor of cerebral abnormality, which is the neuropsychologist's focus of interest, defective test performances may be produced by inadequate cooperation or effort, medication, deliberate simulation or exaggeration of disability, disorders of mood, hostility or mistrust, and poor task adjustment on the part of culturally deviant subjects. Neuropsychologists generally consider the possible influence of these factors on level and quality of performance before they judge what the performance indicates about the state of the brain."

The effects of psychiatric disorders on neuropsychological performance were summarized in the position paper of the American Academy of Neurology (1996) as:

"Neuropsychiatric disorders may profoundly affect neuropsychological performance. Anxiety, depression, psychosis, apathy, and irritability all have an impact on the patient's ability to cooperate with testing and may directly affect cognition. Anxiety and depression impair performance on effort-demanding tests and have less effect on tests of overlearned skills. Memory complaints are common manifestations of depressive disorders, and severe depression is commonly accompanied by psychomotor slowing, impaired attention, decreased cognitive flexibility, and poor retrieval memory."

Caution must be used in attributing an etiology for any observed decrement in neurobehavioral test performance, as these techniques are extremely sensitive but not specific. No neuropsychological tests have been shown to have consistent diagnostic validity (American Academy of Neurology, 1996).

Clinical ecologists often rely on the results and interpretation of neuropsychological testing to support diagnoses attributed to EI. The results and interpretations of such neuropsychological assessments are invalid because they do not conform to practices generally accepted among the neuropsychological community and are not specific for a diagnosis of toxic brain syndrome. I base this opinion on my review of the literature on neuropsychological testing and my knowledge of the misuse of these techniques by EI advocates acquired in my involvement in civil court cases.

EI patients report difficulty with many different areas of neurobehavioral functioning including attention/concentration, memory and accuracy and speed of problem solving. In three independent neurobehavioral studies in which EI patients were compared to normal controls, few, if any, decrements in performance were observed (Bolla, 1996; Fiedler et al., 1992; Simon et al., 1993).

Neuropsychologist EI advocates also employ tests that have not been validated according to accepted methods. These methods have been described as (The American Academy of Neurology, 1996):

> "Reliability of tests refers to the consistency with which the same information is obtained if the test is given by different examiners, by the same examiner on different occasions, or to the same patient on different days. There are several types of validity regarding neuropsychological tests, including construct validity (do memory tests assess memory?), concurrent validity (do new tests come to the same conclusion as established tests?), localization validity (do tests actually diagnose disease?), and ecologic validity (do test results predict real-life performance?)."

The Harrell-Butler Comprehensive Neuropsychological Screen is an example of one such test used by clinical ecologists to confirm impairment on brain-related abilities attributed to EI. To date, this questionnaire has not been published in the scientific literature, precluding reliability/validity studies using methods accepted among the neuropsychological community. In conclusion, the testing techniques employed by neuropsychologists to validate the diagnosis of EI have no recognized scientific or neuropsychological validity and are outside the acceptable practices of the neuropsychological community.

It has become commonplace for some examiners to predict the pre-morbid status of an EI patient without the benefit of pre-morbid testing data. Such data are retroactively created by asking the patient to remember what he or she felt like before the precipitating event. The questionnaires or interview checklists typically have items that are transparent in content and intent, and investigators tend to accept scale scores at face value. In a typical EI evaluation scenario involving clinical ecologists, a patient tested subsequent to an alleged chemical exposure is determined to have decrements in one or more cognitive or intellectual parameters. Then based on assumptions usually garnered from the history or supplied by the patient, conclusions are made with respect to the pre-morbid condition of the patient. This conclusion is then used in an attempt to establish injury on the basis of a change in performance status after the alleged exposure. Matarazzo (1990), a dean of psychological and neuropsychological assessment, has pointed out the flaw in this kind of speculative comparison and cautions examiners on the limitations involved. Objective comparisons often can be made when preemployment, military, or educational measurements of intelligence and cognitive status are available. However, estimates of pre-morbid abilities based on education, occupation, or socioeconomic level have been found to be highly inaccurate in clinical assessment (White and Proctor, 1992). In the absence of reliable and valid estimates of pre-morbid function, conclusions drawn from postmorbid tests are, at best, speculative with respect to cause and effect attribution to a specific environmental factor.

Self-report of psychological symptoms

Brewin (1987) has cautioned that the results of self-report symptom checklists as well as the report of symptoms during a structured interview are difficult to interpret because the experimenter/therapist does not know whether it represents:

1. How the individual experiences himself or herself
2. How the individual would like to see himself or herself
3. How the individual thinks the examiner would like to see him or her

Failure to recognize the limits of self-report, be it by paper-and-pencil instrument or structured interview, leads to the erroneous conclusion that the absence of reported symptoms means the absence of a psychological disorder. All structured diagnostic procedures have limitations, and most trade off the presumed validity of the clinical interview for an increase in reliability. Convenience and apparent truth should never be mistaken for facts (Andrews et al., 1990). In the assessment of EI patients, the evaluator needs to be familiar with the kind of information that is frequently omitted, suppressed, or falsified to substantiate toxicogenic beliefs. Thorough medical record review invariably finds evidence of psychologic symptoms, even in those EI patients who do not show any on structured interview or psychologic testing (Terr, 1993a).

My critique of these issues (Staudenmayer and Selner, 1995) was prompted in response to a particular study by Fiedler et al. (1992) in which standardized psychiatric and psychological assessment instruments were modified so that EI patients were specifically asked about their symptoms before onset of EI. Based on the assumption of the validity of this retrospective self-report, these investigators concluded:

> "Also, part of the purpose of the present study was to determine whether a more purely defined subset of MCS patients exists whose symptoms did not arise from preexisting psychiatric conditions. Therefore, based on the present patient group, the symptoms of a subset of chemically sensitive patients cannot be explained by premorbid psychiatric conditions. ...With regard to current psychiatric conditions ... patients uniformly report that these symptoms are sequelae to their chemical sensitivity ... Whether chemicals act directly to cause psychologic distress or psychologic distress is secondary to other aspects of the exposure remains to be determined."

Without going into the details of our critique, we expressed the following objection (Staudenmayer and Selner, 1995):

> "This conclusion is made despite the lack of substantiating evidence for the occurrence of exposure, reliability of the chemical sensitivity, or any objective signs of biologic harm. The only discernible basis is the subjective opinion of the subjects."

Our critique focused on general problems (Rosen, 1956):

> "It is widely if not universally accepted that the statements a person makes about his own behavior and personality traits are related to his [her] perceptions of the desirability and acceptability of these behaviors and traits."

There are at least six issues to consider in evaluating patient self-reports (Table 3.3). First, self reports are subject to the patients' ability and willingness to self-disclose in an unbiased manner. The stronger the belief about a particular attribution, the greater the likelihood of selective inattention to symptoms that imply alternative explanations. EI

Table 3.3   Factors To Consider
in Evaluating Self-Report

1.   Willingness to self-disclose
2.   Primary and secondary gain
3.   Contextual bias
4.   Limits of memory
5.   Psychophysiologic arousal and amplification
6.   Illusion of mental health

patients may choose complaints from a large pool of symptoms, but the symptoms they explicitly choose to report may point away from the true causes and toward an ill-defined and inconsistent set of symptoms and causes. A simple example is the denial of psychological symptoms clearly associated with psychiatric disorders, while the associated somatic symptoms are endorsed.

Second, self-report symptom checklists cannot distinguish between the patient's actual experience and his or her conscious or explicit attempts to exaggerate distress or to dissimulate (Mendelson, 1987). When compensation or disability are motives, the patient's presentation often highlights contrasts to pre-morbid functioning. Another form of motivation is called primary gain in psychiatry, defined as the motivation to avoid underlying personal anxieties. The preservation of toxicogenic beliefs is essential to the EI patient essential when primary or secondary gain are involved.

Third, context and demand characteristics inherent in structured instruments may result in inaccurate estimations of symptoms. For example, self-report questionnaires restrict patients to answering specific questions that may have to do with such particulars as time periods (e.g., symptoms of depression in the past week or two), or situations or events (e.g., before the time of exposure). In many clinical contexts, such specificity is useful and informative, as in the assessment of reactive depression. However, beliefs hold over multiple contexts and context specificity may only serve to obscure judgments about causes of symptoms. The contextual factors affecting memory of symptoms are different before and after the onset of EI. After onset, symptoms become the focus of attention and are elaborated and processed extensively. Before onset, symptoms are more likely to be ignored and minimized rather than amplified.

Fourth, self-report may suffer from limits on memory processes. Symptom checklists require recall, one of the least sensitive measures of memory. Patients simply may not be able to remember and report on all the relationships among symptoms and their perceived causes. The difficulty is compounded when memory is compartmentalized by psychological defenses such as repression, suppression, denial, or dissociation. To preserve the inhibition of these memories from emerging into awareness, symptom links to them are inhibited.

Fifth, perceived environmental exposure may affect perception and psychophysiologic mediation of symptoms. Cognitive amplification of bodily sensations is characteristic of somatization (Barsky, 1979; Barsky and Klerman, 1983).

Sixth, self-report measures cannot distinguish between genuine mental health and the facade of mental health created by psychological defenses. Shedler et al. (1993) described people who deny their psychological vulnerability as having illusory mental health. They studied patients with infrequent self-report of psychological symptoms. Experienced clinicians also rated each of these patients on level of distress and divided them into high (illusory mental health) and low (genuine mental health) groups. The high-distress group

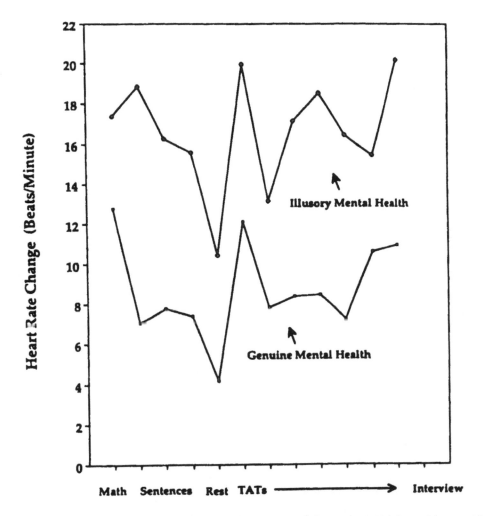

Figure 3.1 Mean heart rate reactivity during various laboratory tasks for subjects with genuine mental health and subjects with illusory mental health. (From Shedler, J. et al., *Am. Psychol.*, 48(11), 1117–1131, 1993. With permission.)

showed significantly more autonomic nervous system stress physiology such as heart rate while performing various cognitive tasks, as illustrated in Figure 3.1. The findings demonstrate that low self-report of psychological symptoms measures mental health in some people, but for others reflects a degree of psychological defense.

In conclusion, mediating processes can be inferred from self-report data only if it can be assumed that the patient has a valid capacity for introspection. That assumption is invalid when beliefs can be shown to influence sensory perception and symptom reports. Another serious limitation of self-reporting is that the individual has only a limited awareness of his or her psychological state. Also, we cannot ignore the observations that come from the psychodynamic tradition, that unconscious ideas also can affect and distort self-reporting. The lack of validity of self-reports is summarized by Brewin (1987): "The value of self-reports would appear to be more in their relation to intentional future actions than in any insight they might provide into complex feeling states or into the contingencies governing past behavior."

Inappropriate use of treatment methods

Acupuncture

EI has associations to alternative medicine practices, and acupuncture has been suggested as a treatment for EI. In Sweden, Arnetz et al. (1995) conducted a case-controlled study of deep vs. superficial acupuncture to evaluate the effects of a practiced version of the procedure (deep) and a placebo (superficial). Twenty subjects with typical EI symptoms believed to be caused by electromagnetic force fields from visual display units were randomly divided into the two groups, one receiving deep acupuncture and the other superficial acupuncture. Dependent measures included self-report symptoms and certain biological parameters including serum cortisol. The results showed no significant differences between the groups, with the exception of increased levels of cortisol in the deep acupuncture group. However, the self-report data indicated that both interventions were associated with improved well-being. Subjects also showed more resilience to EMF exposures in that they reported fewer reactions when working in front of visual display units. These findings indicate that any alleviations of symptoms associated with EI and attributed to the healing properties of acupuncture are no more than a placebo effect, summarized by the investigators (Arnetz et al., 1995):

> "By offering a neutral treatment method, acupuncture, hope is rein-
> stated in the patients. ... The end effect may be a weakening of the
> strength of the conditioned reflex, that is, a deconditioning from
> allocating cause of bodily symptoms to the external environment."

Diagnoses incorrectly attributed to EI

Numerous medical and psychiatric/psychological conditions have been incorrectly attributed to foods or environmental chemicals (Table 3.4). I shall focus on two of these: attention deficit/hyperactivity disorder (ADHD) in children, and *Candida albicans* hypersensitivity syndrome (the "yeast connection").

Attention deficit/hyperactivity disorder

Scientific opinion favors a neurological explanation of genetic origin for most cases of attention deficit/hyperactivity disorder. Differential diagnosis are complex, including behavioral disorders, learning disabilities, emotional disorders, and family dysfunction. A neurological explanation of ADHD does not necessarily exclude hypotheses about alternative causes or precipitating factors, such as common dietary components to which some persons are thought to be abnormally sensitive (Kinsbourne, 1994). Beginning in the 1960s, foods and food additives were hypothesized to cause ADHD. The relationship to foods was entrenched in the uninformed public with Feingold's (1975) best-seller, *Why Your Child Is Hyperactive*, largely a book of opinions and testimonials. Feingold claimed that 50% of children with hyperactivity or learning disabilities improved when placed on a salicylate-free diet. Salicylate is a salt of salicytic acid used in making aspirin, as a preservative, and in the external treatment of dermatologic conditions such as eczema. He suggested that hyperactivity was a relatively recent phenomenon associated with the increased use of additives in the American food supply. Actually, Still (1902) described hyperactivity in children in the medical literature at the turn of the century, and its incidence in the

Table 3.4 Some Medical and Psychiatric
Diagnoses Incorrectly Attributed to EI

Medical
Allergy
Asthma
Autoimmune deficiency syndrome (AIDS)
Candida albicans hypersensitivity syndrome
    ("yeast connection")
Chronic fatigue syndrome (CFS)
Endocrine disorders (premenstrual syndrome)
Fibromyalgia
Gait disorder
Hypoglycemia
Paradoxical vocal cord dysfunction
Peripheral neuropathy
Porphyria
Seizures
Toxic encephalopathy

Psychiatric
Anxiety
Attention deficit/hyperactivity disorder (ADHD)
Bipolar affective disorder
Delusions
Depression, dysthymia
Hallucinations
Insomnia
Impulse-control disorders
Mania, hypomania
Panic
Personality disorders
Psychoses
Sexual dysfunction

classroom based on the judgment of some school teachers has remained constant at about 3% (Forbes, 1982).

Feingold recommended eliminating all food dyes and restricting preservatives and all other artificial food additives regardless of chemical structure. This recommendation alluded to studies showing that food coloring caused some aspirin-sensitive asthma patients to develop bronchoconstriction (but not hyperactivity). Actually, double-blind, placebo-controlled studies showed that food dyes and preservatives are an uncommon cause of clinically significant bronchoconstriction, even in moderately severe perennial asthmatics (Weber et al., 1979). Aspirin sensitivity is unrelated to hyperactivity. Terr (1993c) noted two ironic observations about the Feingold diet. First, the salicylates that were the original culprits in Feingold's speculations are found in virtually all plants. Second, with the shift in emphasis to food additives, adherents to the Feingold diet concentrated their efforts on the additives and ignored the presence of salicylates in their diet, the issue that prompted the diet in the first place.

Articles in the *Wall Street Journal* and *Reader's Digest* advised parents of ADHD children to request dietary advice from their physicians. Scientific skepticism and caution and even an outright challenge of these unsubstantiated practices, found in a poignant critique by Werry (1976), could not stem the tide of a populist movement. In 1982, in response to this confusion, the National Institutes of Health's Office of Medical Applications for Research published a consensus report based on the scientific findings and concluded that there was no evidence that toxicity, idiosyncrasy, or allergic hypersensitivity from food additives causes direct effects on the central nervous system (Consensus Conference, 1982). Parents' reports that their children become restless, irritable, hyperactive, and intractable in reaction to certain foods and additives may have been biased. When the hypothesis was tested, unblinded provocation challenges were supportive, but blinded studies were not. While Feingold's claims of epidemic proportions were deemed unreasonable, the possibility that a very few children may be shown to have attentional disorders resulting from food additives was left to further study.

The improvement observed in children on the Feingold diet is explained by placebo and behavioral factors. It offered hope to distraught parents, and the rigors of adhering to the strict diet cannot help but change the family dynamics. Most importantly, the child must surely perceive a greater commitment to his or her well-being (Forbes, 1982).

Sugar sensitivity was also hypothesized to cause ADHD. Crook (1975) reported an open-challenge study supporting parental observations that sugar "sets their children off." Hypothesized causal agents of behavioral disorders were sugar in its natural form — sucrose — as well as the artificial sweetener aspartame, commonly found in diet carbonated beverages in the early 1980s (Wurtman, 1983). By the mid 1980s, the Food and Drug Administration (FDA) received hundreds of anecdotal accounts of seizures associated with aspartame which prompted at least two controlled studies by Tollefson (1993). A double-blind, placebo-controlled, cross-over study of seizure activity in children who had well-documented seizures demonstrated that aspartame does not provoke seizures or increase epileptiform discharges; the companion study with adults showed the same negative results.

Reports of the effects of sugar on hyperactivity and other behavioral problems showed a consistent pattern: open studies conducted by advocates of the "sugar connection" affirmed a relationship (Crook, 1975; Rapp, 1978b), but controlled studies by others did not (Behar et al., 1984; Milich and Pelham, 1986; Milich et al., 1986; Wolraich et al., 1985).

Wolraich et al. (1994) conducted a definitive double-blind, placebo-controlled study of the effects of diets high in sucrose or aspartame on the behavior and cognitive performance of children described to be sensitive to sugar. The experimental group included 23 school-age children judged sugar sensitive by their parents, meaning they exhibited hyperactive behavior proximal to ingestion of sugar. The controls were 25 preschool children without complaints of sugar sensitivity. Using a double-blind, cross-over design, each child ate three diets. One was high in sucrose, another contained aspartame as a sweetener, and the placebo diet contained saccharin as a sweetener. All diets were essentially free of additives, artificial food coloring, and preservatives; the amount of sweetener was regulated; and plasma glucose levels were monitored. Psychological assessment was conducted at baseline. Thirty-nine behavioral and cognitive measures were evaluated that had been proven sensitive to hyperactivity, attention deficits, and the effects of medications and foods in earlier research (Barkley, 1990). The measures included attention, impulsivity, hyperactivity, aggression, oppositional behavior, mood, cognition, motor performance, academic performance, activity level, and somatic symptoms. Results showed that for the children described as sugar sensitive, there were no significant differences among the three diets. There were some differences among the diets for the control subjects, suggesting the opposite of what was expected: instead of being associated with hyperactivity, ingestion of sucrose had a slight calming effect. There were no adverse responses to sucrose or aspartame in either group. The investigators concluded that even when intake exceeds typical dietary levels, neither sucrose nor aspartame affects children's behavior or cognitive function.

It had been recognized that there is no evidence that sugar or aspartame can turn a child with normal attention into a hyperactive child (Kinsbourne, 1994). The Wolraich study extended this conclusion, showing that the symptoms of children diagnosed with ADHD were not exacerbated by sugar or aspartame.

Parents' despair of coping with their hyperactive children is compounded by physician attitude (positive or negative); absence of a data-driven, recognized cause for ADHD; and the absence of effective treatments (Forbes, 1982). Biological psychiatry hypothesizes neurotransmitter imbalance. Many ADHD children have benefited from treatment with stimulant drugs including methylphenidate (Ritalin), dextroamphetamine (dexedrine), or magnesium pemoline (Cylert). But, the drugs are not a long-term panacea and there are risks of adverse side-effects. Selective use of drugs where demonstrated to be effective remains problematic. The parents' and adolescents' negative attitudes toward drugs in general often is not discriminating, creating problems of compliance.

Biofeedback practitioners have observed slow frequency brain wave abnormalities in quantitative EEG spectral patterns from ADHD children (Mann et al., 1991). A spectral pattern is an x,y plot representing an average over time (e.g., 30 seconds) of the magnitude or voltage (y axis) of EEG frequencies (x axis). Little is known about mechanisms underlying quantitative EEG (QEEG) parameters. Diagnostic criteria are based on QEEG statistically derived parameters which are evaluated in comparison to normative QEEG databases compiled from large population samples. QEEG assessments are suggestive but not conclusive. Comparisons to normative data bases are highly sensitive (likely to detect deviations on multiple parameters) but not specific (deviations are not unique to any disorder including ADHD). Clinical reports suggest improvement of ADHD symptoms with EEG biofeedback, although controlled outcome studies have yet to be completed which correlate improvement to expected changes of QEEG parameters toward homeostasis.

Candida albicans *hypersensitivity, the "yeast connection"*

Candida albicans is a microorganism that normally exists on the skin and in the gastrointestinal tract and vagina. Systemic candidiasis or chronic mucocutaneous candidiasis is a well-recognized disease associated with immune irregularities that manifest as opportunistic infections (Kirkpatrick, 1989). "*Candida albicans* hypersensitivity" is also called "*Candida* hypersensitivity syndrome" and postulates that colonization of *Candida albicans* results in the release of a toxin that leads to immune reactions including immune suppression, autoimmunity, and food allergy (Truss, 1986). These events were hypothesized to cause systemic illness and multisystem symptoms including fatigue, depression, loss of libido, premenstrual syndrome, muscle aches, constipation, and diarrhea. Sensitization could spread to non-food chemicals, making it another manifestation of EI. It was popularized and given a new name with the publication of the book, *The Yeast Connection* (Crook, 1986). A recent printing of *The Yeast Connection* offers the enticement, "If you feel sick all over, this book could change your life." The book contains many anecdotal reports but lacks any scientific data or clinical trials to differentiate the claimed effects of treatment from placebo effects or spontaneous recovery.

Proponents of the "yeast connection" say that there are no diagnostic physical abnormalities, no laboratory test abnormalities, and no evidence of *Candida albicans* overgrowth locally or systemically. Nevertheless, the "yeast connection" is postulated to be a predisposing factor for multiple sclerosis, arthritis, psoriasis, schizophrenia, cancer, AIDS, depression, and various emotional and behavioral problems. Populations susceptible to systemic candidiasis disease include individuals undergoing cancer chemotherapy or prolonged antibiotic therapy, drug addicts, those suffering from conditions such as AIDS, chronic alcoholism, or diabetes mellitus, or individuals who are otherwise immuno-compromised (Blonz, 1986). Advocates of the "yeast connection" suggest that yeast problems are a cause rather than a complication of these conditions.

Recommended treatment for the "yeast connection" includes diet restrictions on in-take of sugar, yeast, and foods containing molds and encourages nutritional supplements of vitamins and minerals and certain medications including the oral antifungal agents nystatin and ketaconazole. No clinical trials have evaluated and demonstrated the effec-tiveness of diet or nutritional supplements. Terr (1993c) has presented a compelling argument refuting the hypothesis that a diet high in sugar is the mechanism that induces "*Candida* hypersensitivity syndrome." A diet free of sugar has not been shown to be effective in treating patients with well characterized candidiasis.

One carefully controlled study failed to show a therapeutic effect for nystatin treat-ment. Dismukes et al. (1990) conducted a randomized, double-blind, cross-over study with 42 premenopausal women diagnosed as suffering from *Candida* hypersensitivity syn-drome. The study compared oral nystatin, vaginal nystatin, or a combination of the two to placebo. As expected, the active nystatin therapies were significantly more effective than placebo in relieving vaginal symptoms. But both nystatin and placebo therapies reduced psychological symptoms and global indexes of distress equally, arguing against the advo-cates' hypothesis that depression and fatigue are alleviated by nystatin therapy. The investigators concluded that long-term nystatin therapy for women with complaints of the "yeast connection" appears to be unwarranted.

In an editorial accompanying the Dismukes et al. publication, Bennett (1990) warned that this study would not end the controversy. Negative outcomes invariably prompt criticism about how the study might have been better designed or that the "key" catalyst variables necessary to potentiate the effect of nystatin were not controlled. For one, advocates contend that diet was not controlled, despite any evidence that diet has an effect. Also, since there is no clear definition of the "yeast connection," any study will be criticized that the wrong subjects were included, even if they were selected by advocates of the phenomenon. In short, advocates reject any non-confirming results.

Lastly, from the perspective of a psychologist, the most absurd claim about the "yeast connection" is that it is an underlying cause of schizophrenia. Schizophrenia is a devastat-ing disease of unknown cause, and has a prevalence of 1% in the U.S. with an annual cost estimated from $10 to $20 billion. Its protean symptoms range from pervasive blunting of affect, thought, and socialization to florid hallucinations and delusions. Despite much speculation and much research effort, no neurotoxin or neurotransmission abnormality has been proven to be a primary feature of schizophrenia (Mesulam, 1990).

The American Academy of Allergy and Immunology Position Statement (1986) on the *Candida* hypersensitivity syndrome states that the concept is "speculative and unproven" and elements of the proposed treatment program are potentially dangerous. This reserved statement is consistent with conduct expected of the scientific community. Others who feel less restrained have referred to these practices in more pointed terms, with Pearson (1985) calling it quackery and Barrett (1993) calling it an epidemic of nonsense.

Barrett, a board member of the National Council Against Health Fraud, notes that under federal law, any product intended for the prevention or treatment of disease is a drug, and it is illegal to market new drugs that do not have FDA approval. The health-food industry manufactures claimed that their products were not medical treatments, but rather dietary supplements, including products labeled CandiCare, Candida-Guard, Candida

Cleanse, Candistat, Cantrol, Yeast Fighters, Yeast Guard, Yeaststop, Yeasterol, and Yeast-Trol. The FDA did not agree and in 1989 issued a letter indicating that it was illegal to market vitamin concoctions intended for the treatment of yeast infections. The courts upheld the FDA's position, and in 1990 a federal judge ruled that Yeaststop was a drug and ordered its manufacturer to pay for the seizure and destruction of Yeaststop initiated by the FDA, as well as legal costs. Additional cases of illegally marketing can be found in the original article (Barrett, 1993).

Conclusion

> "Propositions cannot be proven by revelation, intellectual intuition, or experience." (Popper, 1968)

The elaborative nature of hypothesis, illusiveness of testable methods, and the adherence to a theory in the face of contradictory evidence characterize the toxicogenic theory. The theory is unsubstantiated, keeps changing, and clashes with what is known. The diagnoses attributed to EI are often nebulous, and the tests used in support of it are unsubstantiated or applied inappropriately. These problems are not new to science because they characterize its antagonist, pseudoscience.

chapter four

Studies supporting
the psychogenic theory

Beginning in the late 1970s, EI patients came to the attention of traditional physicians for a variety of reasons. These included: (1) cost containment from insurance committees that perceived over utilization, (2) workers' compensation claims, (3) civil litigation against chemical manufactures, and (4) questions about iatrogenic illness. By 1980, psychiatrists and psychologists skeptical of clinical ecology became more involved with EI patients, with a few clinical practitioners working independently and apparently in isolation from each other, notably in Britain, Canada, and the U.S. The initial scientific publications were typically clinical case series or individual case reports, usually selected because of circumstances that brought these patients to the attention of conventional physicians. EI patients did not seek out mental health care practitioners on their own. The common element in these cases was that the patients believed that they suffered from EI, a belief most often resulting from contact with a clinical ecologist. In the U.S., EI patients were more likely to be involved in litigation or with workers' compensation disability claims, while in Britain they tended to reflect more the British fascination with alternative health-care practices (although that distinction is certainly not definitive).

In any new scientific research program, the initial phase is descriptive for the purpose of identifying the problem. Characteristic of a progressive research program, descriptive studies were followed by case-control studies in which EI patients where compared to non-EI patients, either normals called "negative controls" or patients with psychiatric disorders called "positive controls". Studies of this type are correlational and by design cannot demonstrate causality. Also, because of the small number of subjects and the lack of a well-defined population, they lack epidemiological validity. Nevertheless, the studies showed a general progression in the sophistication of the research designs, and some turned out more than satisfactory, given that EI patients had no identifiable objective findings, no common set of symptoms, no common set of putative environmental toxins, and no accepted criteria for diagnosis. As more studies were reported, the toxicogenic theory remained unsubstantiated while evidence accumulated in support of the psychogenic theory.

Commonalities were consistently reported among the otherwise heterogeneous EI patients. Typically, there was a prevalence of females, and the reported age of symptom onset was in middle age. They tended to be well-educated, above average in intelligence, economically self-sufficient, and predominantly Caucasian. Upon evaluation, physicians attributed symptoms to medical diagnoses unrelated to EI. The co-morbidity of psychiatric disorders was found to be almost universal, leading to hypotheses about cause or effect. Psychiatric disorders or psychophysiologic stress disorders pre-morbid to the alleged onset of EI were commonly found, limited only by the depth and thoroughness of assessment.

Theoretical postulation and research direction tended to bifurcate into toxicogenic and psychogenic considerations. Toxicogenic theory suggested that such pre-morbid conditions lowered the threshold of susceptibility to acquire toxically induced EI. Psychogenic theory proposed that psychological factors were invariably necessary and sufficient to account for the phenomenon.

The first two clinical psychiatric studies on the psychology of EI were published in 1983, one in the U.S. by Brodsky (1983) and the other in Britain in two parts by Pearson et al. (1983, 1984). In 1985, the first of Stewart's articles appeared in Canada (Stewart and Raskin, 1985), in which she identified somatization common among EI patients who had other quite heterogeneous psychiatric disorders.

That same year, Selner and I published the first in a series of articles in the allergy literature. We focused on methods to differentially diagnose EI patients evaluated in an inpatient Environmental Care Unit during the years 1980 to 1985 (Selner and Staudenmayer, 1985). The five cases reported were not EI patients, and we chose these in contrast to many EI patients we had evaluated because we wanted to demonstrate the need for thorough medical evaluations of patients complaining of chemical intolerances. Each of the five patients complained of intolerance to a specific chemical agent. In three, objective measures were observed to correlate with symptoms of respiratory distress or irritation under provocation challenges. In the other two cases, medical evaluation and provocation challenges revealed no pathophysiological basis for the symptoms and reflected psychogenic etiologies. One involved an aircraft painter who presented with respiratory symptoms and symptoms of withdrawal associated with alcohol abuse. The case was resolved with alcohol rehabilitation treatment. The other case involved a woman with a life-long psychosis documented by a history of psychiatric hospitalizations who was alleging sensitivities from an underground gasoline spill. After hearing of the spill, she and her husband had moved into the neighborhood from another state, purchased a home with a down payment of $1, and joined as plaintiffs in a lawsuit against the gasoline company. She filed a complaint against the psychologist in an attempt to discredit him and to disqualify his testimony. Her complaint was dismissed.

In 1986, Selner and I published an overview of psychological factors involved in several aspects of EI, including suggestions for evaluation and treatment of EI patients in inpatient and outpatient settings (Selner and Staudenmayer, 1986). Schottenfeld (1987) contributed an article on the psychiatric approach to diagnosis and treatment of EI patients seen in occupational medicine clinics and presented many of the complex psychological and social factors influencing EI.

By 1990, several of the above-mentioned investigators published additional articles elaborating on certain psychological factors such as contagion in mass psychogenic illness (Brodsky, 1988; Stewart 1990; Staudenmayer and Selner, 1987) or unique symptom presentations such as vocal-cord dysfunction (Selner et al., 1987). In addition, Simon et al. (1990) reported on a sample of plastic assembly workers in the aerospace manufacturing industry who were alleging intolerances to low-level chemical exposure. The importance of looking beyond Axis I clinical syndrome diagnoses and into Axis II personality disorders was emphasized in a case report by Rosenberg et al. (1990) of a man with obsessional/paranoid traits. Black et al. (1990) reported the first of two studies on a group of EI patients who were assessed by standardized psychiatric interview and psychological testing and compared to

Table 4.1　Summary of Demographics and Percentage of Cases
With Psychiatric Symptoms Across Studies of EI Patients

Study author and year	N	% Female	Mean age (range, +s.d.)	% with psychiatric/ psychologic disorder
Brodsky, 1983	8	88	"Middle age"	100
Pearson et al., 1984	19	68	39.1 (21–67)	94
Stewart and Raskin, 1985	18	83	38 (23–61)	100
Selner et al., 1986–92[a]	13	69	45.9 (28–68)	100
Terr, 1989	90	70	39.5 (20–63)	42[b]
Staudenmayer et al., 1990, 1993	63	71	M 43.6 (+13.3)	100
			F 39.0 (+10.1)	100
Black et al., 1990	26	88	49.1 (27–78)	87
Simon et al., 1990	13	—	—	92
Fiedler et al., 1992	11	73	M 42	72
			F 43 (28–57)	
Simon et al., 1993	41	85	46.4 (+9.5)	44
Witorsch et al., 1995	61	82	47.3 (27–66)	70
Black, 1996	4	50	43.7 (39–52)	100
Altenkirch et al., 1996	8	—	47 (31–62)	100
Totals and averages	375	77	42.8 (20–78)	

[a] Composite of Selner and Staudenmayer (1986, 1991, 1992a,b) and Selner et al. (1987).

[b] 38 of 90 cases had pre-exposure psychiatric diagnoses in their medical records; no psychiatric assessments were conducted.

carefully matched non-psychiatric control subjects. In 1993, they reported the results of personality testing on this sample of EI patients and noted a high incidence of Axis II disorders of all kinds (Black et al., 1993). Later, Black (1996) reported additional cases of iatrogenic EI, and Witorsch et al. (1995) reported on a case-series of EI patients involved in litigation.

Reports of EI in Europe were prevalent in Scandinavian countries, mostly in the form of "sick building syndrome" (Gothe et al., 1995). Recently, complaints of problems from exposures to low-level environmental toxicants appeared in Germany (Altenkirch et al., 1996; Koppel and Fahron, 1995). These patients had identified co-morbid psychological disorders, although only a small number had associated complaints and attributions characteristic of EI.

Demographics and psychiatric/psychological conditions

The demographic data and general findings for some of these studies were summarized previously by Sparks et al. (1994). I have added studies not included in that review, as well as those published more recently (Table 4.1). The percentage of cases with psychiatric symptoms is not a statistic that is comparable across the different studies. In some studies, cases with alternative medical diagnoses that could explain the symptoms were excluded before analysis of the series; in others, they were not. Some studies relied solely on patient self-report which could explain the lower percentages of psychiatric diagnosis. A strict distinction of psychological or psychophysiological learned responses from psychiatric symptoms associated with psychopathology is not always made, and in some studies it is not clear whether they were inclusive or exclusive symptoms. And finally, the strength of the patient's conviction in his/her belief about the etiology of EI is not specified. EI patients run the continuum from true believers ("don't bother me with the facts") to those unsure of their supposed diagnosis, accepting the label simply because "that's what the doctor says I have."

The demographic data of age and gender were averaged across all of the studies, with mean age weighted by the number of cases in each study. A total of 375 cases were included in this summary, with all but 21 having a gender designation. In the remainder, the prevalence of females was 77%. The overall mean age, averaged across gender and based on 340 cases with available data, was 42.8 years, with a range from 20 to 78 years. Because these EI patients were not randomly selected from the population, these parameters cannot be construed as epidemiological estimates. Of these 375 carefully evaluated, EI-labeled patients, not a single patient had symptoms that could be causally related to an environmental exposure.

Next, I will briefly highlight some important findings from these studies which reflect the heterogeneity observed in EI patients. The presentation of these studies is grouped into two general experimental designs: (1) clinical case reports and case series limited to descriptions of patients who were usually self-selected or referred for workers' compensation or disability evaluation, and (2) case-control studies in which groups of EI patients were compared to control groups on specific measures. These cases were usually consecutive admissions to a clinic or seen for workers' compensation evaluation.

Clinical case reports

Brodsky

Brodsky (1983) reviewed eight workers' compensation claims for injury ostensibly caused by allergic response to substances in the workplace unsubstantiated by physical evidence. There were no clear specific exposure events and the alleged injuries were supported only by the opinions of clinical ecologists. Brodsky called it "allergic to everything." The eight individuals showed avoidance behaviors such as withdrawal from work, a life-style engineered to avoid exposure to putative noxious substances, and an identity as a disabled person. Brodsky noted that an affiliation with clinical ecology, what he called a medical cult, appealed to patients with a history of chronic psychiatric symptoms who could not find a place for themselves in traditional social channels. He placed part of the blame for this phenomenon on the shoulders of psychiatrists for failing to provide explanations and regimens that could benefit these patients. In effect, they were abandoned to any alternative that might be available to the patient. Brodsky's conclusions, though based on a sample of only eight cases, have withstood the test of time with independent replication.

Stewart and Raskin

Stewart and Raskin (1985) reported on 18 EI patients referred to a university psychiatric consultation liaison service in a teaching hospital in Canada. While all agreed to a psychiatric interview, 10 did so reluctantly, and 16 refused psychological testing. All were diagnosed with a psychiatric condition. Three case presentations identified the heterogeneity and severity seen in these patients, including schizophrenia, anxiety panic disorder, and somatization disorder. Ten of the patients presented with well-recognized psychiatric conditions (psychoses or affective or anxiety disorders) that usually can be treated effectively. Seven of the remaining eight patients presented with somatoform disorders which

are notoriously difficult to treat. Stewart concluded with a historical review of similar conditions dating to the ancient Egyptians and later wrote a scholarly treatise on the changing faces of somatization (Stewart, 1990).

Selner et al.

The series of 13 clinical cases by Selner et al. were compiled from five different publications (Selner and Staudenmayer, 1986, 1991, 1992a,b; Selner et al., 1987). Selner and Staudenmayer (1986) evaluated five EI patients in an inpatient facility at a hospital in Denver, an environmental care unit (ECU). One of these patients had cancer which was not recognized in prior evaluations. She was severely depressed, adjusting to dying from what she was led to believe was EI. The other four cases clearly involved psychiatric conditions which were pre-existing to the alleged onset of EI, including dissociative disorders and severe personality disorders. The sixth case, which was not an EI patient, involved a woman who had reliable dermatologic symptoms in response to exposure to insecticides but who had been misdiagnosed as a somatoform disorder.

The second study (Selner and Staudenmayer, 1991) reported two new female EI patients who focused on food intolerance after evaluation by clinical ecologists. Upon medical evaluation, one was found to have severe gastrointestinal disease unrelated to foods which accounted for her symptoms. The other was a mother of two who was profoundly malnourished and at risk of starvation because of her belief in unfounded food intolerances. Both suffered from severe dissociative disorders and associated childhood incest which they had not disclosed to anyone previously. The second one was undergoing continued solicitation from her father which she had kept from her husband. Addressing these issues in psychotherapy resolved all food intolerances.

Two EI patients with unusual psychogenic etiology were reported in the third study by Selner and Staudenmayer (1992a). The first involved litigation by a woman who was alleging EI resulting in multi-system complaints including anxiety and depression from exposure to a domestic application of organophosphate pesticides. Upon review of records, an entry in the medical chart by her treating clinical ecologist suggested that the patient's husband was threatening the patient by brandishing a shotgun and behaving like a "paranoid schizophrenic." The patient's husband insisted through their attorney that he be present at all interviews we conducted with his wife. There was an alternative explanation to organophosphate poisoning causing her symptoms, namely, terror resulting from anticipated harm from her husband accompanied by obvious signs of acute anxiety when interviewed. The validity of this concern was confirmed by the discovery of existing counseling records that detailed the husband's threats to the patient.

The second case involved a 56-year-old male silkscreen painter complaining of laryngeal obstruction symptoms at the worksite. He was employed by a major electronics manufacturer where he worked with chemicals with known irritant properties. Symptoms had spread to other workers, and the section was shut down. Provocation challenges with the silkscreen chemicals that had been identified as existing in the workplace failed to produce his symptoms. However, a pharyngeal muscle spasm lasting minutes was directly observed, triggered by the slightest contact of the examining flexible fiberoptic rhinoscope with the epiglottis. This spasm could be relieved by verbal reassurance. Because of a history of gagging that was induced by tight shirt collars, we were able to reproduce the gag phenomenon simply by cinching the shirt collar with his necktie. The pharyngeal spasm was observed by rhinoscopy. With the cooperation of the employer, who granted a dispensation from wearing ties and shirts with collars (apparel required of him since a recent promotion), the symptoms resolved and he returned to work. This man had no identifiable psychiatric conditions at the time of evaluation. Upon taking a complete history, the origin of these suffocating sensations were traced to a drowning experience in Lake Michigan from which he was resuscitated without immediate identifiable adverse effects. He did not manifest flashbacks or numbing symptoms in response to the drowning and therefore did not meet criteria for post-traumatic stress disorder. His reflexes represented a learned response which was a sequelae of a near-death experience. This case highlights the difference between a psychiatric and a psychological explanation. Upon his return to work, his co-workers' symptoms promptly resolved, thus avoiding a potentially costly closing of a manufacturing division due to mass psychogenic illness.

The fourth study (Selner and Staudenmayer, 1992b) presented the case of a retired insurance agent who had been living in isolation in a downtown hotel room. Treatment by clinical ecologists included vitamins and supplements and removal of all his dental amalgams. The man was financially well off in retirement but was intolerant of his upscale automobiles and lived sequestered from his home in the foothills and his charming wife. A composite of psychotherapeutic treatments including relaxation with biofeedback, antidepressant medication, behavioral desensitization to the odor of alcohol, education, and cognitive therapy resulted in resolution of symptoms. He was able to return to his wife, house, and autos and to resume his favorite activity, international travel, all without further incidence of EI.

The fifth study in this series involved three women documented to have vocal cord dysfunction by fiberoptic rhinoscopy (Selner et al., 1987). The pre-morbid psychodynamics of all three were consistent with the somatoform disorder of conversion reaction. One patient presented as particularly sensitive to the odor of corn. By the end of psychological treatment, she was eating corn. The dynamics of her conversion disorder were not as easily resolved. She returned to a rural town where continued psychiatric treatment was not readily available. She returned to live with her parents, who were assessed to be contributory to her somatization. Her primary care physician was still convinced she had asthma. Upon follow-up several years later, she had regressed to the point EI was the focus of her life. She had been surgically treated with a permanent tracheotomy.

The second patient, a 32-year-old woman who had onset of stridorous inspiratory and expiratory obstruction after a penicillin reaction, exemplified what psychiatrists call primary and secondary gain motivation. During a brief psychotherapy, the patient related an isolated event at age 5 when she was inexplicably nearly strangled by her father during an impulsive rage resulting from her inability to open a locked door quickly. A terrified child found herself apologizing for her inadequacy to an impassive father. Subsequent interactions with her father were unremarkable and amicable until his death when she was 17 years of age. Therapy revealed her vulnerability to the need for achievement and perfection. When her strategy for securing her mother's attention through great accomplishment (over-achievement) failed, she discovered laryngeal stridor, as her mother had always responded to signs of threatening physical illness. A classic scenario of secondary gain was revealed in analysis of stridorous events which resulted in supportive maternal response: the penicillin episode, a potentially fatal river-rafting accident, and a stridorous chemical reaction experienced during a methacholine challenge to assess asthma. For instance, between challenge doses used in methacholine bronco-provocation, she was writing her mother a letter detailing her "illness in response to chemicals" represented by the challenge drug. In therapy, she became aware that all of her life she had viewed her siblings as favored, and she saw herself as the Cinderella of the family. She recognized that attempts to buy maternal affection with educational or social perfection were not successful. After witnessing her mother's tendency to canonize her father after his death, she even had the ideation that by becoming not only a hero but a martyr (suicide) she might achieve maternal attention. Laryngeal stridor ceased completely after she gained insight into the obvious role of secondary gain with mother. The role of primary gain associated with anxieties from the interaction with her father was more difficult for her to process. Childhood traumatic events defined the target organ for symptoms (larynx) and also initiated the suppression of emotions reinforced by a perceived absence of maternal love which she was unable to protest.

Terr

Terr (1989) reported 90 EI cases for which he had conducted complete medical histories and physical examinations; most were cases referred by the employer or workers' compensation insurance carrier. In all 90 cases, EI had been diagnosed by one or more of 26 different clinical ecologists based on clinical history or a procedure of provocation-neutralization. This case series included 40 reported previously (Terr, 1986a). Physical examinations gave completely normal results in 77 of the 90 cases, and non-environmental, alternative diagnoses could account for the abnormalities in the rest. Laboratory tests did not distinguish these patients as a group from normal individuals except for several patients who had immunoglobulin (IgA) and lymphocyte levels above the normal range, reflecting a history of infections. There was no instance of susceptibility to opportunistic infections or neoplasms that are characteristic of immune deficiency. Review of medical records indicated a history of other illness prior to the reported onset of EI in 75 patients.

Terr classified these patients into three groups: (1) patients who had physical illness or physiological responses causally unrelated to environmental agents, including asthma, rhinitis, and eczematous dermatitis — conditions that are typically aggravated by environmental allergens and irritants; (2) patients who had no evidence of any physical or emotional disease or even any symptoms; and (3) patients who had long, extensive, involved histories of multi-system complaints who had previously been either undiagnosed or diagnosed with a host of different disorders. These patients had clinical features of hypochondriasis, somatization, conversion hysteria, anxiety, depression, and obsessive behavior. Medical records indicated pre-morbid psychiatric diagnoses by non-clinical ecology practitioners in 38 cases. No additional psychiatric assessments were conducted at the time of his evaluation. Terr concluded that the patients in this series constituted a heterogeneous group with no identifying pattern of symptoms and no significant consistent physical or laboratory abnormalities. The subgroup of patients with multiple symptoms most likely of psychological origin correspond to the type of patient described repeatedly in the clinical ecology literature.

Fiedler, Maccia, and Kipen

Fiedler et al. (1992) published the findings of 11 cases selected based on Cullen's definition of MCS. Some reported not to have identifiable pre-morbid psychiatric conditions based on their self-report on a modified structured psychiatric interview. That study led to a commentary about the pitfalls of self-report and illustrated how the reported patient histories were not consistent with denial of psychological symptoms by self-appraisal (Staudenmayer and Selner, 1995).

Altenkirch, Hopmann, Brockmeier, and Walter

In 1993, 64 cases of chronic pyrethroid intoxication, usually from pesticides or wood preservatives, were reported to the Federal Health Office in Germany which shortly thereafter resulted in an "epidemic of thousands" as reported by the media (Altenkirch et al., 1996). Pyrethroids exhibit a high insecticide effect and low human toxicity. Data on the acute human toxicity of pyrethroids come from occupational exposure and include symptoms of facial paresthesia; systemic effects such as dizziness, nausea, and headache; disturbances of consciousness; muscle fasciculation; seizures; and coma. The occupational medicine literature does not describe any irreversible central or peripheral nervous system impairment. Neurological, neuroradiological, and laboratory investigations including pyrethroid values in blood and urine were conducted in 23 of the cases in whom an actual pyrethroid exposure as well as clinical symptoms seemed most plausible. The demographics included 13 females and 10 males with a mean age of 47 years and a range of 31 to 62. There was not a single case in which evidence for irreversible peripheral nervous system or central nervous system lesions could be found. Eight cases were classified as EI (demographics not provided), none of whom showed objective signs of alternative diseases to explain symptoms, and none had abnormal assays for pyrethroids. Treatment recommendations were referral to a psychologic clinic.

Witorsch, Ayesu, Balter, and Schwartz

Witorsch et al. (1995) compiled 61 EI cases they had evaluated in the context of plaintiff litigation. Retained by the defendants, they conducted independent medical exams (IMEs) in 40 cases and/or medical record review in 20. More than half (33) involved cases of "sick building syndrome" and in no case was there evidence of exposure to any chemical at toxicologically significant levels. In no case were there any objective physical or laboratory findings that correlated with subjective multi-system complaints, nor could any pathophysiological link be demonstrated between chemical exposures and the patients' conditions. Of the 61 cases, 43 were evaluated for a psychiatric diagnosis (DSM-III-R) and at least one such diagnosis was made in all 43 cases. The diagnoses included somatoform disorder (36), mood disorder (27), anxiety disorder (10), alcohol abuse (1), and personality disorder (16). Psychiatric disorders pre-morbid to the alleged exposure and onset of EI were identified in 28 cases. Witorsch's group concluded that EI is best characterized as a manifestation of one of several primary psychiatric disorders.

Black

The next year, Black (1996) presented four cases which illustrated how hypochondriasis was encouraged and promoted by clinical ecologists. The first was a 52-year-old divorced woman with a 20-year history of anxiety and depression, including one psychiatric hospitalization. After undergoing "challenge" testing in a clinical ecology facility in Chicago, she was absolutely convinced her problems were the result of chemicals. The second case was a 39-year-old divorced man with a history of psychological problems that began in his teens with a suicide attempt. He attributed his problems to worksite exposure to chemicals and was in litigation against his employer. The third case was a 42-year-old man accidentally sprayed with malathion who was involved in litigation against the city. Under the influence of clinical ecologists, he became convinced he was now chemically poisoned and the effects had spread to numerous chemicals. He became isolated and depressed. The fourth case was a 42-year-old woman who reported chemical sensitivities developing from exposure to gasoline fumes at a convenience store where she worked who also was involved in litigation. Told to avoid exposure to chemicals, she became socially isolated at home and was being treated pharmacologically for depression. All four patients were preoccupied with EI and believed the diagnosis and treatment offered by their environmental physicians were correct and appropriate.

Case-control studies

These case control studies were designed to test hypotheses about specific mechanisms that could explain the symptoms in EI patients. Because of the influence of the clinical ecologists, allergic and immune mechanisms were among those receiving the most attention, and two such studies both showed that allergic mechanisms were not plausible and inconsistent with the toxicogenic theory. The most plausible alternative explanation was psychogenic. The other two studies provided additional evidence supporting the psychogenic theory.

Pearson, Rix, and Bentley

Pearson et al. (1983) focused on "pseudo food allergy" in 23 patients evaluated by double-blind, placebo-controlled challenges, and they presented their results in two companion studies. In the first study, of the 23 patients who presented to a allergy clinic, only four patients demonstrated proven food hypersensitivity with classic atopic conditions such as asthma, eczema, and urticaria. They were remarkable in that none had psychological symptoms in the course of reactions to food, and they were also assessed as not having concurrent psychiatric conditions. The other 19 patients who complained of a multitude of general malaise and cognitive symptoms which they attributed to foods had no objective findings to substantiate their belief. The second study added a positive control group of 20 outpatient referrals to a psychiatric clinic (Rix et al., 1984). Of the 19 patients with non-allergic food intolerances, 18 were administered a structured psychiatric interview, as were the psychiatric control subjects. The two groups were indistinguishable in regard to their symptoms, with the exception that the patients reporting food intolerances showed significantly less anxiety than the psychiatric controls. There were no significant differences between the psychiatric outpatients and the patients with unconfirmed allergy in terms of psychiatric diagnosis. Only one of the 19 patients complaining of food intolerances was not given a psychiatric diagnosis.

Pearson's group made several insightful psychodynamic observations. They suggested that the EI patients showed less anxiety because they used psychological defenses of repression, dissociation, and the conversion of psychological conflicts into physical symptoms. EI patients were hostile to the suggestion that they had psychological problems. They often had a strong belief in an organic etiology, an overvalued idea which was sometimes maintained in the face of evidence to the contrary. They were highly suggestible, as reflected by improvement on elimination diets or false positives to placebos.

In 1990, Ferguson (1990) wrote a frank and timely editorial on self-deception in food sensitivity which accompanied Jewett's study (reviewed in Chapter 3) in the *New England Journal of Medicine*. She quoted May (1980), a pioneer in the use of blinded food challenges: "It is essential to realize that the reactions occur in human beings and are therefore enmeshed in the interplay between body, mind, and soul." Ferguson concluded her editorial by saying:

> "Self-deception affects doctors as well as patients, and through
> kindness and enthusiasm many of us may be doing a great disser-
> vice to ill persons anxiously seeking a non-psychiatric diagnosis. If

we apply the wrong label with conviction, and then treat the symptoms with suggestion and placebo, relief is likely to be only transient,
and psychopathology will probably reemerge."

Simon, Katon, and Sparks

Simon et al. (1990) evaluated a group of 36 workers exposed to chemicals in an aerospace
equipment manufacturing plant. The 13 who developed EI and submitted workers' compensation claims scored significantly higher on self-report measures of psychological
symptoms, and had more psychiatric diagnoses by structured interviews compared to the
23 who did not develop EI. Even though pre-morbid psychiatric conditions were estimated
retrospectively by interview, allowing for under-reporting, a history of psychiatric morbidity prior to the worksite exposure to chemicals was a much stronger predictor for the
onset of EI. Simon et al. concluded that the findings of these EI patients are consistent with
those found amont other populations — fibromyalgia, chronic fatigue syndrome, and
chronic pelvic pain, all of which suggest that the development of idiopathic symptoms are
related to a pre-existing history of psychological distress.

Simon, Daniell, Stockbridge, Claypoole, and Rosenstock

Simon et al. (1993) evaluated 41 EI patients recruited from a clinical ecology practice in the
Seattle community, with a matched control group of 34 chronic musculoskeletal pain
patients obtained from university clinics. An extensive panel of immunological measures
was employed and conducted by the same reference laboratory, with samples blinded. The
only difference noted between the groups was attributable to laboratory method. These
findings unequivocally supported the clinical interpretations of Terr which were criticized
by EI advocates because they had been done as part of evaluations commissioned by the
worker's compensation authority or insurance companies.

Simon et al. also conducted neuropsychological testing, as the EI patients also complained of cognitive dysfunction including difficulty concentrating and poor memory. The
results showed no significant differences between the EI patients and the chronic pain
patients. Because chronic pain patients do not typically report cognitive dysfunction, the
absence of significant findings in this study suggests that the EI patients amplified their
symptoms. Other controlled neurophysiological studies have shown an absence of evidence for performance impairment compared to normal controls (Bolla, 1996; Fiedler et al.,
1994). Furthermore, cognitive test results from EI patients have been found to be significantly different from positive control groups, such as those of solvent-exposed patients
(White and Proctor, 1997).

The structured psychiatric interview and self-report symptom checklists were specifically selected to reduce the potential for interviewer bias in assessment of psychiatric
disorders, seen as a shortcoming in previous studies. However, this methodology does not
control for the bias of self-report expected in a population engaged in compensation
proceedings and known to be defensive about psychological diagnoses. I would note that
the clinical syndromes assessed by the structured interview were limited to panic disorder,
generalized anxiety, and major depression. Notably absent from this list is one of the most
commonly hypothesized class of disorders, the somatoform disorders. The EI patients had
significantly elevated scores on self-report measures of somatization on the SCL90-R;
furthermore, the structured interview somatization symptom count was elevated for both
current symptoms and pre-existing symptoms. Axis II personality disorders were not

assessed. Judgments of pre-existing symptoms were made retrospectively by patients who attributed their symptoms to EI at time of assessment. In light of these assessment limitations, the estimate that 44% of the EI patients had a current psychiatric condition and that 47% had a pre-existing one is remarkable.

Black, Rathe, and Goldstein

Black et al. (1990) published the first of two companion studies on 26 EI patients with demographically matched subjects from the community (non-psychiatric patients) employing standardized psychiatric and psychological assessment methods. The EI patients reported significantly greater depression, anxiety, and somatization. In 1993, Black et al. (1993) published additional personality assessment data from these 26 EI patients, having found them to have significantly more personality disorders than the controls, with about three fourths of the EI patients showing at least one DSM-III-R diagnosis, with a total of 40 diagnoses for all 26 patients. The most frequently diagnosed was histrionic personality disorder in 11 (47.8%) of the patients. Compulsive personality disorder was diagnosed in 7 (30.4%), dependent personality disorder in 5 (21.7%), avoidance personality in 5 (21.7%), and the remainder distributed among all the other personality disorders. Black et al. concluded that EI patients may suffer from unrecognized psychological distress, which may account for some or all of the symptoms attributed to EI. However, the study was not designed to explain the etiology of these distressing symptoms.

Sparks et al. (1994) criticized this study for using a community control group which may have artificially deflated the rate of the control group and contributed to the significance of the statistical effects. I believe that criticism is arguable, as EI patients profess to have no long-standing trait disorders existing prior to the onset of EI, suggesting they were no different from a community control. More important, the absolute number incidence of personality disorders in the EI sample was remarkable, irrespective of the control group.

Staudenmayer, Selner, Buhr, and Selner

Staudenmayer and Selner (1990) reported the first of three companion studies on a sample of 63 consecutive EI patients seen in private practice for evaluation or treatment from 1985 to 1988. Neuromuscular measures of scalp tension (EMG) and quantitative EEG brainwave data (QEEG) were obtained in a case-controlled study of 58 EI patients (5 did not have data) who were evaluated for basal characteristics under conditions of relaxation. The study had a negative control group of 55 medical patients psychiatrically asymptomatic and a positive control group of 89 patients with heterogeneous psychological disorders seen in my practice who did not attribute their symptoms to EI. EEG frequencies below 15 Hz were subjected to discriminant function analysis which yielded eight spectral pattern categories (Figures 4.1, 4.2, and 4.3) into which each subject's basal QEEG was classified. The distribution of subjects in the control group was significantly different from the distributions of both EI patients and positive psychological controls, which were not different from each other (Figure 4.4). Most subjects in the asymptomatic control group were classified into the unimodal alpha categories illustrated in Figure 4.1, while most EI patients and psychologic controls were classified into the other spectral categories in Figures 4.2 and 4.3, which reflect QEEG patterns associated with attention abnormalities. This finding demonstrates that EI patients have the same underlying attentional disorders seen in patients with overt psychologic symptoms without beliefs in EI.

EEG frequencies between 15- and 25-Hz labeled beta are independent of the lower frequency categories and are associated with anxiety and other non-specific factors. Each

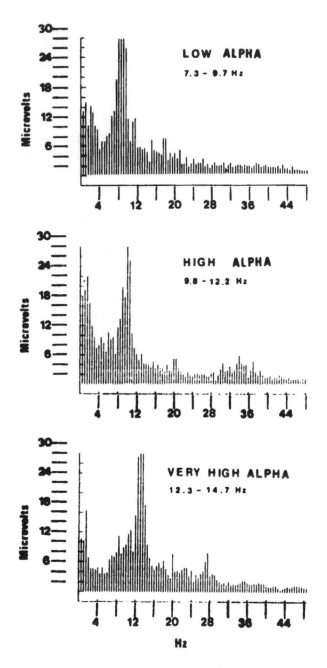

Figure 4.1 Three EEG spectral patterns with monomodal frequency peaks in the alpha range: low alpha (7.3 to 9.7 Hz), high alpha (9.8 to 12.2 Hz), and very high alpha (12.3 to 14.7 Hz). (From Staudenmayer, H. and Selner, J.C., *J. Psychosomatic Res.*, 34(6), 259–270, 1990. With permission.)

subject was categorized as high or low on the power of beta. The percentage of subjects in the high category was significantly smaller in the control group compared to the percentage in the EI and psychological groups (Figure 4.5). This finding suggests that EI patients show neurophysiological correlates of anxiety which were expected in the psychological positive control patients.

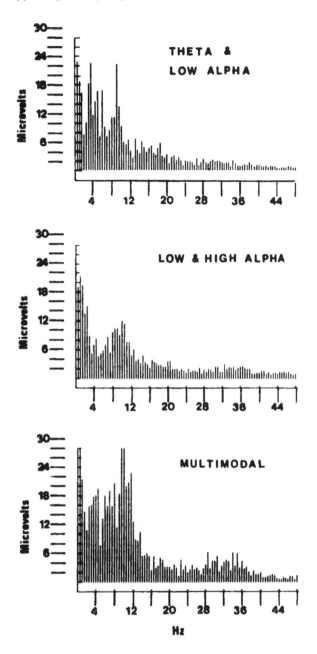

Figure 4.2 Three EEG spectral patterns with multimodal frequency peaks below 15 Hz, theta (3.8 to 7.2 Hz) and low alpha, low and high alpha, and multimodal. (From Staudenmayer, H. and Selner, J.C., *J. Psychosomatic Res.*, 34(6), 259–270, 1990. With permission.)

Each subject was also categorized as high or low based on the power of EMG recordings in the lower frequency range for muscles (26 to 48 Hz). Significantly more EI patients were classified into the high-power scalp EMG category compared to both asymptomatic controls and positive psychological controls (Figure 4.6). Elevated levels of EMG are consistent with bracing, a defensive response seen in a variety of neuromuscular stress disorders.

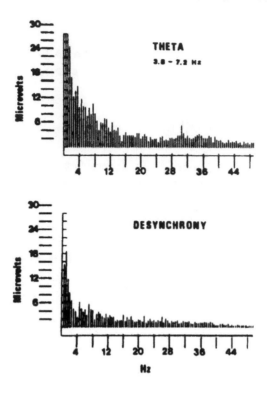

Figure 4.3 Two EEG spectral patterns without significant power in the alpha range, theta, and desynchrony. (From Staudenmayer, H. and Selner, J.C., *J. Psychosomatic Res.*, 34(6), 259–270, 1990. With permission.)

In 20 of the total sample of 63 EI patients, comprehensive medical evaluations could not identify alternative medical conditions to explain symptoms attributed to EI. In addition, these patients did not entertain a psychological explanation. These 20 cases were studied individually with double-blind, placebo-controlled chamber challenges based on their unique hypothesized provoking agents (Staudenmayer et al., 1993a). Not one of these studies showed a reliable pattern of responses to putative active agent or placebo across a series of challenges, as shown in Chapter 5. In all 63 cases, non-environmental medical disorders, psychiatric diagnoses, or psychological factors were identified which could explain the symptoms incorrectly attributed to EI.

Thirty of the 63 EI patients and 36 of the 89 psychological control patients underwent psychotherapy with me, allowing for more comprehensive assessment of their social and family history. The self-reported incidence of childhood physical and sexual abuse in 20 female and 10 male EI patients and 25 female and 11 male psychological control patients (Figure 4.7) was reported in the third companion study (Staudenmayer et al., 1993b). There were no statistical differences between the groups of males, with a low incidence of abuse reported in both groups. Significant differences in the incidence were found between the EI females and the controls on physical abuse perceived to be life threatening (50 vs. 12%) and sexual abuse (60 vs. 25%).

While childhood abuse and trauma are not necessary conditions for adult psychopathology, literature on childhood abuse clearly shows that abuse is a sufficient condition for psychopathologic and neurophysiologic sequelae which can explain the multi-system complaints of EI patients, topics addressed in Chapter 11. We hypothesized that the

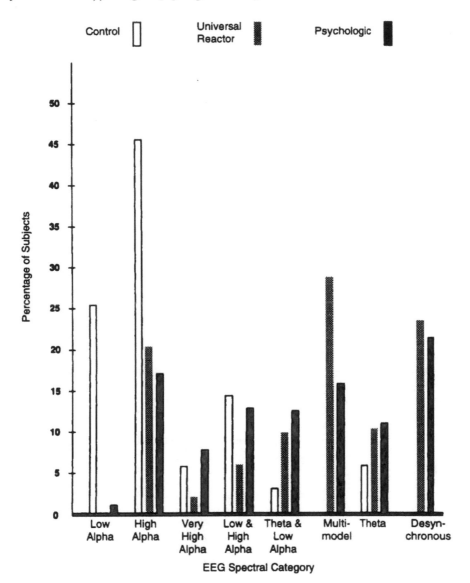

Figure 4.4 Percent of subjects in each group classified into EEG spectral pattern categories recorded from the right parietal (P4A1) during eyes-closed relaxation. Frequency bands are theta = 3.8 to 7.2, low alpha = 7.3 to 9.7 Hz, high alpha = 9.8 to 12.2 Hz, very high alpha = 12.3 to 14.7 Hz. (From Staudenmayer, H. and Selner, J.C., *J. Psychosomatic Res.*, 34(6), 259–270, 1990. With permission.)

etiology of the complaints of some EI patients may be found in childhood trauma and its associated biological abnormalities, independent of environmental intolerances for which there was no scientific evidence in these patients.

A cohort study of chronic fatigue

Some EI patients have co-morbid chronic fatigue syndrome (CFS). CFS was first operation-ally defined as an illness consisting of persistent or relapsing fatigue which is debilitating and

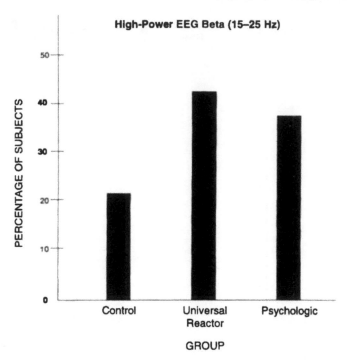

Figure 4.5 Percent of subjects in each group classified as having high EEG beta power (>8.5 microvolts/sec/Hz) recorded from the scalp (P4A1) during eyes-closed relaxation. (From Staudenmayer, H. and Selner, J.C., *J. Psychosomatic Res.*, 34(6), 259–270, 1990. With permission.)

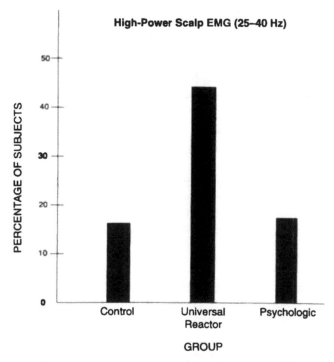

Figure 4.6 Percent of subjects in each group classified as having high power (>8.5 microvolts/sec/ Hz) in a low-frequency (25 to 48 Hz) EMG recorded from the scalp (P4A1). (From Staudenmayer, H. and Selner, J.C., *J. Psychosomatic Res.*, 34(6), 259–270, 1990. With permission.)

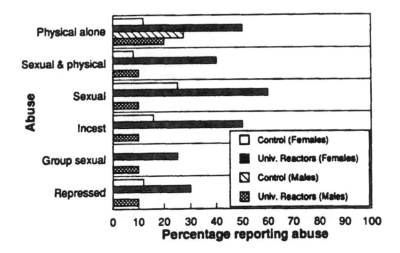

Figure 4.7 Percentage of reported childhood abuse for 20 female and 10 male EI patients (Universal Reactors) and 25 female and 11 male control patients. (From Staudenmayer, H., Selner, M.E., and Selner, J.C., *Ann. Allergy*, 71, 538–546, 1993. With permission.)

of at least 6 months' duration in the absence of any identifiable medical diagnosis (Holmes et al., 1988). This definition was exapnded in 1994 by the Centers for Disease Control and Prevention (CDC) to include more specific symptoms unrelated to exercise, such as cognitive dysfunction, and having new or definite onset (Fukuda et al., 1994). The initial speculations that CFS reflected a form of Epstein-Barr viral infection have been refuted by a series of controlled studies of seroepidemiology (Horwitz et al., 1985; Holmes et al., 1987; Buchwald et al., 1987) and antiviral therapy (Straus et al., 1988), although the role of persistent viral infection has not been completely ruled out.

There is evidence of impaired activation of the HPA axis in CFS patients, and specifically, decreased CRH secretion during times of fatigue (Demitrack et al., 1991). Abnormalities in the HPA axis are postulated to underlie chronic fatigue syndrome, similar to effects seen in depression, possibly mediated by cytokines (Ur et al., 1992). There are many similarities between CFS and EI patients with respect to demographics, symptoms, locus of control (Buchwald and Garrity, 1994), co-morbid psychiatric disorders (Abbey, 1992, 1993; Abbey and Garfinkel, 1991), and elevated incidence of a self-reported history of childhood sexual abuse (Boyce, 1992; Jones, 1995). Numerous studies have shown that psychiatric illness preceded the onset of CFS (Gold et al., 1990; Hickie et al., 1990; Kruesi et al., 1989; Manu et al., 1989; Taerk et al. 1987).

Chronic fatigue syndrome is more amenable to scientific study than other forms of EI because: (1) it is a medical diagnosis with inclusion criteria, and (2) the hypothesized precipitating environmental event, namely, clinical viral infection, can be objectively identified. These two factors allow for a subject selection in a cohort epidemiological study. The advantage of a cohort study is that psychological symptoms and psychiatric disorders can be more readily assessed both pre- and post-onset incident (i.e., infection) through medical records, interview, and self-report questionnaires.

Wessely et al. (1996) conducted such a prospective cohort study in Britain, carried out in three stages (Figure 4.8). In the first stage, a postal survey was conducted of all adults ages 18 to 34 years who were registered in six general medical practices (n = 15,283). Stage 2 consisted of the 12 months following the survey, during which all patients who were

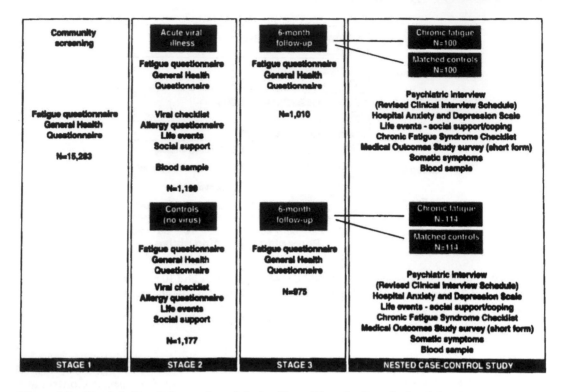

Figure 4.8   Postinfection fatigue: plan of study. (From Wessely, S. et al., *Am. J. Psychiatry*, 153, 1050–1059, 1996. With permission.)

suspected of having a viral infection by the medical staff were invited to participate in the study, and matched controls without viral infections were selected. Stage 3 consisted of following up the cohort with n = 1010 in the group with known infection and n = 975 in the non-infected control group.

Stage 3 also included a nested case-control study. All subjects recruited at stage 2, whether the initial illness was viral or nonviral, were sent a further questionnaire 6 months later to assess chronic fatigue. Those with chronic fatigue and without alternative diagnoses, regardless of having acute viral illness in stage 2, were classified as chronic fatigue cases and were matched with non-fatigue controls. A total of 428 subjects participated in the case control study and underwent psychological assessment beyond that obtained on the general health questionnaire completed by the entire cohort.

The first analysis of the study compared subjects with viral infections to those without and found no reliable differences, leading Wessely et al. to conclude that viral infection is not an etiological factor in CFS. According to the data on the general health questionnaire, subjects with chronic fatigue were 5.5 times more likely to have a current psychiatric disorder. Further comparisons followed the nested case-control design in stage 3 which employed psychiatric interviews on 378 of the 428 subjects. Those with chronic fatigue were at greater risk than those without chronic fatigue for current psychiatric disorder assessed by standard interview (60 vs. 19%). Chronic fatigue subjects were more likely to have experienced psychiatric disorder in the past and to have been treated with psychotropic medications. There was a trend for previous psychiatric disorder to be associated with co-morbid rather than non-co-morbid chronic fatigue. Wessely et al. (1996) concluded:

"Both chronic fatigue and chronic fatigue syndrome are associated with previous psychiatric disorder, partly explained by high rates of current psychiatric disorder. The symptoms thought to represent a specific process in chronic fatigue syndrome may be related to the joint experience of somatic and psychological distress."

Perplexing questions

Why middle-age?

The existence of commonalities in demographics among these studies raises some perplexing questions. First, "Why onset in middle age?" One intriguing hypothesis is that the events in middle age are the culmination of the life experiences that preceded. This hypothesis has been advanced by literary figures, but, more importantly, also by psychiatrists and neuroscientists, who study the interactive effects of biology and stress.

In the last century, the French psychiatrist Janet was among the first to note that a long time might pass between the occurrence of a traumatic event and its full-blown psychopathological expression. Janet postulated that the latency of expression is determined by the time required to perceive the inescapable reality of the event, the time necessary for the individual to expend useless efforts to fight the inescapable, and the occurrence of other stressful events that deplete the individual's last resources (van der Kolk and van der Hart, 1989). Janet's postulate is consistent with both models of the stress response and "learned helplessness" or depletion of coping strategies explanations of depression.

Many traumatized patients show a slow, progressive decline in their ability to deal with stressors, old or new. For example, in the context of a demanding job, impaired cognitive processing ability can lead to additional stress because of failure to meet expectations externally or internally imposed. Eventually, coping capacity breaks down, and a feeling of helplessness sets in accompanied by a pervasive desire to "get away from it all." When the coping capacity breaks down, the patient may withdraw into social isolation and experience a state of helplessness which is expressed through depression or somatization, as in EI patients, and distorted beliefs about attribution of harm to environmental agents.

An alternative hypothesis based on animal conditioning models of post-traumatic stress disorder proposed by Yehuda and Antelman (1993) is that time-dependent sensitization better describes the bipolar symptoms that are sequelae of acutely traumatic events, and that the delay in onset of symptoms can only be accounted for by time-dependent sensitization. Irrespective of which model, or some combination of the two, will evolve to explain the multi-system symptoms of EI patients, both models have in common the hypothesis that there is only so much time that an organism can resist the disruptive effects of chronic stress responses once initiated.

The years between 28 to 38 seem to be characterized by an awareness of the limitations and imperfections of one's chosen paths and a sense of the narrowing of one's horizons (Back and Gergen, 1968; Levinson, 1978). The average age — 32 to 38 years in many retrospective studies of adult patients who are incest victims — suggests they too enter a period of adulthood reflection. Interestingly, many of these patients enter psychotherapy with presenting problems of psychophysiologic stress disorders, depression, or anxiety without insight about the implications of their childhood sexual abuse. A sizable number of patients may mention it in the intake assessment, but not bring it up again until later in therapy when a therapeutic relationship has been established. The therapy process, including the availability of a secure relationship with an adult, may stimulate the recollection process and promote development (Herman and Schztzow, 1987).

Why female prevalence?

The second perplexing question is, "Why are EI patients predominantly female?" The first order of explanation is usually biological. Female hormones are postulated to contribute to the greater prevalence of certain psychiatric disorders among women, particularly anxiety, depression, and somatoform disorders (Leibenluft, 1996). EI advocates hypothesize a predisposing neuroendocrine factor which heightens susceptibility to acquire EI, although no such biological marker has been identified in EI patients (Sparks et al., 1994). There is a finding that appears inconsistent with the neurologic complaints and diagnosis of toxic encephalopathy in female EI patients. Estrogens are neuroprotective with respect to neuronal degeneration, growth, and susceptibility to toxins (Seeman, 1997).

Several hypothesis have been proposed to explain gender differences which are consistent with the psychogenic theory. First, a PET brain-imaging study comparing men and women's cortical activation responses to self-induced sadness showed women activated a greater portion of their limbic system compared to men (George et al., 1995). This may help explain women's greater risk for depression (Leibenluft, 1996). A second hypothesis is that men and women react differently to stress, with women's responses putting them at greater risk for depression (Nolen-Hoeksema, 1990). And a third hypothesis is that women are more likely to be sexually abused as children (Cutler and Nolen-Hoeksema, 1991).

Symptoms without disease?

The most perplexing question of all may well be, "How does one explain symptoms without disease?" The answer may lie in the high prevalence of psychiatric or psychological disorders identified in these studies. The manifestations of multi-system general malaise complaints can be explained by psychological and psychophysiological mechanisms which have been studied in the context of the placebo response and learned sensitivity, systems of the stress response, and psychiatric disorders. All of these have heterogeneous etiologies, but the biological sequelae of childhood abuse and post-traumatic stress are strikingly similar to the symptoms of the most severely dysfunctional EI patients. These are topics covered in detail in the second half of this book.

chapter five

Assessment
of the toxicogenic
research program

> "The natural unit of methodological appraisals is not a single theory, but a succession of theories or a research program, and we do not judge the state in which a research program finds itself at a particular moment; we judge its history, preferably in comparison with the history of rival programs." (Feyerabend, 1975)

The purpose of this chapter is to assess the research programs associated with the toxicogenic theory. I begin with the position statements from several medical associations which have reviewed the procedures and practices of clinical ecology. Shifting to the rulings on civil litigation by state and federal courts, I review some opinions excluding EI advocates as expert witnesses on the grounds of practicing or advocating "junk science". Next, I present the double-blind, placebo-control provocation challenge protocol used with an environmental chamber to assess the hypothesis of an individual EI patient and provide results from our studies. Lastly, I address questions to be asked in the assessment of any scientific research program and evaluate the toxicogenic theory.

Critical reports from medical associations

Many prominent medical associations, scientific groups, and government panels, listed in Table 5.1, have concluded that the diagnostic and treatment procedures practiced by clinical ecologists are unproven and that the theory is unfounded. For example, the California Medical Association's (1986) position statement concluded:

> "No convincing evidence was found that patients treated by clinical ecologists have unique, recognizable syndromes, that the diagnostic tests employed are efficacious and reliable or that the treatments used are effective. Even though clinical ecology has existed for approximately 50 years, only a few studies have been conducted that are scientifically sound. Most have such serious methodological flaws as to make their conclusions unacceptable. Those few studies that

Table 5.1 Organizations and Panels with Position Statements Critical of EI

American Academy of Allergy (1981, 1986)
American College of Occupational [and Environmental] Medicine (1990)
American College of Physicians (1989)
American Medical Association (1992)
California Medical Association (1986)
Committee on Environmental Hypersensitivities [Canada] (Thompson et al., 1985)
International Society of Regulatory Toxicology and Pharmacology (1993)
Report of the Advisory Panel on Environmental Hypersensitivity [Canada] (1986)
Royal College of Physicians [Britain] (1992).

used scientifically sound methods have provided evidence that the effectiveness of certain methods used by clinical ecologists is based principally on the placebo response."

The American College of Physicians' position paper (1989) concluded:

"The clinician who sees a patient who has been given a diagnosis of 'environmental illness' by a clinical ecologist should first establish the presence or absence of any physical disease or psychological factors in the patient's illness."

The American College of Occupational Medicine (1990) endorsed the position papers previously written and specifically the following summary:

"Clinical ecology lacks scientific validation, and the practice of 'environmental medicine' cannot be considered harmless. Severe constraints are placed on patients' lives, and, in many cases, invalidism is reinforced as patients develop increasing iatrogenic disability. Treatment by clinical ecologists frequently creates a severe financial burden for patients and imposes significant costs on health insurers and workers' compensation systems."

And (Kahn and Letz, 1989):

"At present, serious ethical problems are raised by the use of treatments before they have been proved safe and effective."

Junk science excluded from the courtroom

"I have had medical testimony before me that was shockingly suspect. Had a lawyer given equivalent misleading information I would have brought the matter to the attention of the disciplinary authority. What realistic threat exists against doctors? As to unlicensed economists and statisticians, the matter is now hopeless."

— Judge Jack Weinstein

This quote reflects the frustrations of the Chief Judge, U.S. Court Eastern District, NY, presiding over the highly emotional 1985 trial for injury claims allegedly caused by exposure to Agent Orange, a defoliant sprayed in Vietnam. The alleged harm came from dioxin, a contaminant in the defoliant, for which there is no scientific evidence of ill-health effects (Franklin, 1994). Federal rules of civil procedure say that an opinion given by an expert "will be disregarded where it amounts to no more than mere speculation or a guess from subordinate facts that do not give adequate support to the conclusion reached." The principle seems intuitively clear, but not easy to apply in cases such as Agent Orange where the basis of expertise is changing almost daily and the "subordinate facts" are not much clearer than the opinions (Marshall, 1986).

The frustrations of Judge Weinstein reflect the subjectivity surrounding the admissability of evidence, usually left for the jury to decide. For years, federal and state courts applied the Frye test (Frye v. U.S., 293 F. 1013, 1923) in determining whether expert testimony and other scientific evidence should be admitted at trial (Locke, 1995). Under the Frye test, a method or theory must be shown by the proponent to be "generally accepted" within the relevant scientific field before it is deemed sufficiently reliable to be presented at trial. The Frye case had involved a criminal defendant who was convicted, but the conviction was appealed on the grounds that an expert's testimony on a systolic blood pressure deception test [a "lie detector" test] had been erroneously excluded. The appellate court ruled in favor of the lower court's decision that the test was unreliable. While the Frye test seemed reasonable in requiring peer acceptance of methodology and theory, it was criticized as unduly restricting admission of new and emerging theories and techniques. As a result, the decision of what defined peer acceptance was often left to a jury who were misled by the "aura of infallibility" sometimes perceived as accompanying expert testimony.

In 1994, the U.S. Supreme Court, in Daubert v. Merrell Dow Pharmaceutical (113 S. Ct. 2786, 1993), concluded that judges, not the jury, should evaluate the relevancy and reliability of all scientific testimony and evidence. Judges were to be the gatekeepers. The issue adjudicated in Daubert was whether birth defects were caused by Bendectin, an anti-nausea drug taken by pregnant mothers. The court noted that none of the work of the plaintiffs's experts, particularly the methods for selection and analysis of data, had been subjected to peer review and publication or other scientific scrutiny within the field (Locke, 1995). The Supreme Court noted four factors the trial court should take into account in the evaluation:

1. Whether a theory or technique can be and has been tested
2. Whether the theory or technique has been subjected to peer review and publication
3. The known or potential rate of error
4. The general acceptance of the theory within the relevant scientific community

Farley (1993) suggested more detailed legal standards for the admissability of evidence obtained with the use of a scientific technique, listed in Table 5.2. These guidelines suggest that a scientist who cannot practice in the courts of science cannot practice in the courts of law. There is also a trend in the opposite direction — scientists who cannot testify in the courts do not practice science.

But even before Daubert, courts rejected the methodology of clinical ecologists using the Frye test, a test still used in some state courts. In line with excluding "junk science" from the courtroom (Hubert, 1991), magistrates and judges have disqualified EI advocates as expert witnesses in several cases, preventing them from testifying about their

Table 5.2   Standards for the Use of a Scientific Technique

1.   The technique's acceptance in the field
2.   The expert's qualifications and stature
3.   The use that has been made of the technique
4.   The potential rate of error in using the technique
5.   The existence of specialized literature
6.   The novelty of the invention
7.   The extent to which the technique relies on the subjective interpretation of the expert
8.   The existence and maintenance of professional standards
9.   The presence of safeguards
10. Analogy to results of other scientifically admissible techniques
11. The nature and breadth of the inference adduced
12. The clarity and simplicity with which the technique can be described and its results explained
13. The extent to which the basic data are verifiable to the trier of fact
14. The availability of other experts to test and evaluate it
15. The probative significance of the evidence in the case
16. The care with which the technique is employed in the case

Source: Adapted from Farley, M.A., *Forensic Science Handbook, III,* Prentice-Hall, Englewood Cliffs, NJ, 1993.

unsubstantiated methodology or declaring their treatment practices unnecessary. The list of these cases is becoming longer, and they are summarized in Appendix B (excluded testimony of experts). Some of these cases seem more important because of the specific assessments made by the court.

In Bradley v. Brown, Rea and Johnson, two senior physicians at the Environmental Health Center in Dallas, were disqualified from testifying as experts in Indiana. The court held that the methodology employed by Rea and Johnson was anecdotal and speculative and that scientific knowledge about the etiology of EI had not progressed beyond hypothesis. This decision was affirmed by the Court of Appeals for the 7th Circuit in Chicago.

Exclusion of the expert testimony by psychologists, in addition to physicians, occurred in two notable cases. In Kathleen Frank et al. v. State of New York et. al., the court ruled that the psychological experts' testimony "fails to meet the standard of evidentiary reliability established in Daubert." Similarly, in Summers and Potts v. Missouri Pacific Railroad System, the court excluded the psychologist's testimony because it relied on a psychological test that had not been validated.

Bell's testimony on olfactory limbic kindling was excluded in Bahura et al. v. S.E.W. Investors et al. Bell conceded that the olfactory limbic kindling model is not generally accepted in the fields of psychiatry or neurology and that low-level exposure to everyday chemicals does not cause permanent injury. Her own testimony showed that no published data support the plaintiffs's theory that low-level exposures cause permanent toxic encephalopathy in human beings.

Testable hypotheses: provocation challenges

To be scientific, a theory must generate hypotheses testable by experimentation, the results of which must be reliable, reproducible, and specific. Furthermore, a theory must identify outcomes predicted not to occur (Popper, 1968). EI lacks a definition that meets these criteria. Indeed, as noted by Cullen et al. (1993), "until there is an agreed-upon case definition, the possibility of performing reproducible and interpretable clinical studies is nil."

In science, it is true that one cannot prove that a phenomenon does not exist, i.e., one cannot prove the negative. Statistically this is called "not being able to prove the null hypothesis." Nonsignificant statistical tests "fail to reject the null hypothesis" meaning that the observed effects are not different from chance occurrence. Research does not test a theory or postulate directly. Rather, it tests for specific effects one expects to observe if a hypothesis implicated by the theory is true. Statistically, the hypothesis predicting an effect in an experiment is the alternative to the null hypothesis of no effect. When a statistical test is significant, the null hypothesis that the observed results are a chance occurrence is rejected. This adds to the corroborative evidence that the alternative hypothesis is true, even though it does not prove the theory. But no such corroborative evidence has been demonstrated for the toxicogenic theory, and quite the opposite was found when certain procedures were tested. This appears to create a dilemma. Since a theory cannot be proven false, regardless of the amount of negative evidence, should the search for confirming evidence continue indefinitely? And during this time are we to accept, as EI advocates would have us believe, that EI exists because it has not been been proven *not* to exist? Are we to be left imprisoned in this labyrinth of circular logic? Not necessarily!

Even if we define EI in the context of the subjective impressions and symptoms suggested by EI advocates, each EI patient has a testable hypothesis about the causal relations between chemical exposures and ill-health effects. Despite all of the theoretically illusive formulations, despite the "yin-yang" postulates, despite the alluded-to mechanisms without objective signs, despite the auxiliary postulates that may be invoked *post hoc* to rationalize the toxicogenic theory, each EI patient presents with a specific account of symptom appraisal and exposure attribution. When established medical diagnoses have been ruled out through comprehensive assessment (Selner, 1989), the validity of the patient's complaint can still be tested with provocation challenges. Idiopathic environmental intolerances (IEI) are demonstrated only if the results are reliable (IPCS, 1996):

> "Human research is urgently needed to determine the nature (e.g., psychogenic, toxicogenic) of EI since the outcome will influence public policy and clinical practice for EI prevention and treatment, respectively. The key question is whether subjects with EI are able to discriminate in double-blind, placebo-controlled challenge studies between reported environmental (e.g., chemical) triggers and placebos. Ability to discriminate suggests a toxicologic (i.e., chemical-receptor) mechanism. Inability to discriminate would suggest a psychogenic (e.g., conditioned or learned) mechanism."

Patients deserve adequate assessment for alternative medical diagnosis and toxicological explanations, as exemplified in the field of allergy-immunology which has employed human challenge testing procedures to test hypotheses about causal effects of airborne allergens and chemicals with allergic sensitization potential. For example, foods treated with the antioxidant sodium metabisulfite will release the gas SO_2 when chewed. This gas accumulates in the oropharynx where respiratory mechanics require its mixture with the gases of expired air. This results in the production of several gases, including ammonium sulfate, which is an irritant to the airway and potentially a fatal situation for asthmatics (Selner, 1996). According to Selner (1996), there is need for scientific studies on EI:

> "The causal relationship of chemical exposure and myriad clinical syndromes is a very contentious issue. The completion of the challenge chamber facility at EPA's Human Exposure Research Facility presents a grand opportunity for government investigators to work harmoniously with other nongovernment investigators in an effort to bring the redeeming spotlight of scientific discipline to the testy considerations of multiple chemical sensitivity, chronic fatigue, and Gulf War illness phenomena."

Chamber studies make sense when alternative assessments of toxic disease are inconclusive, the patient remains unconvinced by alternative medical or psychiatric/psychological explanations, or in some cases of civil litigation. In other cases, chamber challenge studies would not be cost-effective (Montanaro and Bardana, 1992).

A protocol for provocation chamber challenges

Double-blind, placebo-controlled provocation challenges conducted in an environmental chamber are employed in the assessment of some EI patients. The studies conducted at the Allergy Respiratory Institute of Colorado followed the decision-making protocol presented in Figure 5.1. Several decisions are made before initiating these costly and time-consuming studies. Most important, is there a rationale for such studies in the history. During step #2, the pre-challenge evaluation, a specific agent to test must be identifiable whether by patient attribution, clinical ecology evaluation, or manufacturer's safety data sheets. If a testable hypothesis is identified, an experimental design is formulated with a competent chemist consultant to allow for presentation of the agent without it being detectable. In step #3, if there is any uncertainty about the reliability of the historically identified agent, the patient must demonstrate symptoms upon exposure to it under an unblinded or open challenge. If there is no response, there is usually no reason to proceed. There are circumstances when a single chemical may not be clearly identifiable, and challenge studies may be initiated to assess combinations of chemicals, as I will illustrate shortly with a case presentation. The second consideration is identification of a placebo condition, either clean air when the agent is presented at concentrations below detection thresholds or an olfactory masker when presented above detection thresholds. In step #4, the patient must remain symptom free (or not show exacerbation of baseline symptoms) when placed in the environment chamber free of the chemical which allegedly causes symptoms.

In step #6, the patient is tested with the placebo under single-blind conditions (patient is not told which agent is used). If the patient responds, step #7 allows for psychological intervention to overcome testing anxiety or anticipatory anxiety. This process may be repeated several times until the patient can demonstrate non-response to placebo. In step #9, the patient begins a series of double-blind, placebo-controlled challenges carried out over several days. The sequence of the agent and the placebo over repeated challenges is known only to the chemical technician. The staff members who interact with the patient and the supervising physician are blinded. Upon completion of the studies, the data is analyzed and recommendations are made to the referring physician or agency. If the results demonstrated reliable responses to the agent, environmental recommendations are made. If the results were unreliable, recommendations for psychological intervention are made.

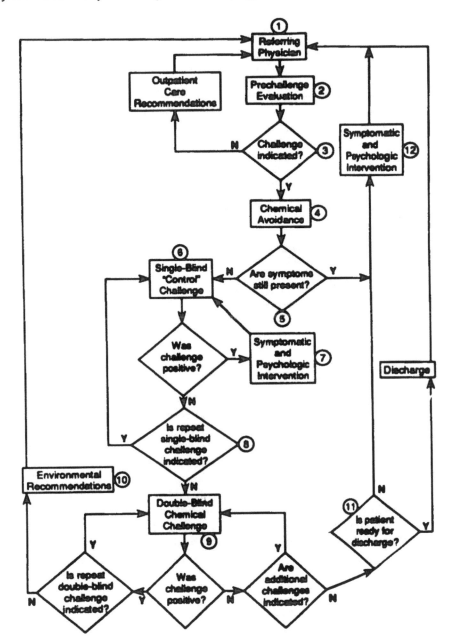

Figure 5.1 Decision-making algorithm used in the evaluation of suspected environmental illness. (From Staudenmayer, H. et al., *Regul. Toxicol. Pharmacol.*, 18, 44–53, 1993. With permission.)

The challenge chamber

A challenge chamber is a sealed area that enables air to enter through one portal and exit through another. Figure 5.2 is a schematic illustration of the chamber at the Allergy

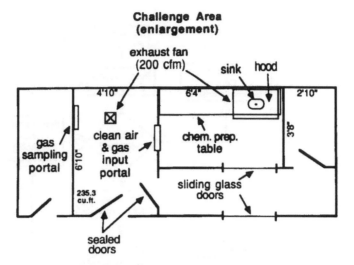

Figure 5.2   Schematic illustration of the chamber challenge facility. (From Selner, J.C., *Regul. Toxicol. Pharmacol.*, 24, S87–S95, 1996. With permission.)

Respiratory Institute of Colorado (Figure 5.3), which is constructed of porcelainized steel, glass, and aluminum; sealed with a silicone cement; and supplied by air filtered through a HEPA device and activated charcoal. The chemicals used for testing are piped through a Dectronics® multiflow metered device and delivered to a rheostat-controlled squirrel cage, where they are mixed with filtered clean air and delivered to the chamber. Air sampling is taken from the patient's immediate breathing space. This chamber proved adequate for many challenges in our clinical practice (Selner, 1996).

The EPA Human Exposure Research Laboratory located at the University of North Carolina has more sophisticated chambers with added climate controls for humidity, temperature, pressure, and light, as well as more advanced air filtration and purification systems. This technology best represents today's state-of-the-art for engineering and design (Selner, 1996). These chambers are used for physiology, indoor air, particle clearance, cardiovascular, and neurophysiology studies. Unfortunately, they have not been used to test EI patients.

Methods of blinding the exposure

Consider an event in which the putative chemical agent cannot be discriminated from background air levels. This implies that appraisal of a reaction must rely on the interpretation of symptom responses, however produced. Odoriferous agents can be masked with other odorants commonly used as flavorings in foods (e.g., peppermint spirits, almond spirits, lemon spirits, etc.). These maskers have similar chemical structures related to substituted isoprene and terpene units that have a low threshold of odor (easily smelled) but a relatively high threshold for irritation (do not irritate the respiratory tract) (Lehninger, 1970). The effects of ambient air exposure to these odors is insignificant by toxicological standards. Nevertheless, tolerance is assessed in an open challenge before any blinded testing is initiated. If a patient can demonstrate tolerance of the odor, there is no basis to argue that the masker itself produced a toxic effect later in the double-blind studies. Even

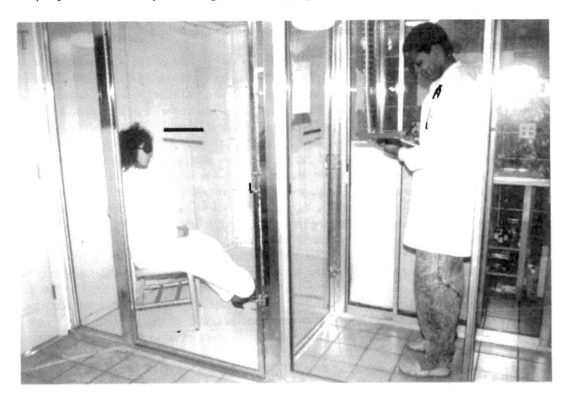

Figure 5.3 Photograph of chamber challenge facility. (Courtesy of the Allergy Respiratory Institute of Colorado.)

allowing for sensitization to the odor with repeated exposure, the effect would be consistent across different exposure contingencies — at least according to accepted scientific models of toxicology (Waddell, 1993). The placebo condition is defined as the masking odor presented without the agent. The active condition is the masking odor presented with the agent.

Placebos are defined as agents or procedures that cannot act through a pharmacologic, toxicologic, or physiologic route or mechanism directly. If an agent thought to be a placebo is subsequently demonstrated to have such a biochemical effect, the conclusions about the placebo effect attributed to that agent would have to be modified. Lactose, for example, was used as the inert agent in food capsule challenges until it was noted that certain individuals with lactate intolerance reacted to it. EI advocates argue that olfactory maskers, used to camouflage the odor of an agent to be tested, should not be employed in provocation challenges because they are also chemicals. They also argue that the masker may interact with the test agent, altering it or creating a new compound to which the patient may be sensitive. Logically, there is truth to this argument in that the irritant properties of different chemical agents are additive, as are the odor properties. If this argument were universally valid, olfactory maskers could not be used to test any putatively harmful agent that has a characteristic odor. However, this contention can be empirically evaluated.

To control for the possibility that there may be an interaction between the masker and the agent that produces a new toxic vehicle, the agent may be combined with different maskers for which the patient has demonstrated tolerance. If the combination does not produce consistent symptoms and true positive (TP) appraisals during challenges, it demonstrates that the masker is not a confounder.

To rule out the possibility of a learned response associated with the masking odor, it has to be established that the patient has not previously associated it with the onset of symptoms. Tolerance under an open, unblinded exposure, indicates that the masker is free of such associations. If the subject subsequently responds to the masking odor during a blinded provocation challenge, there is no basis on which to argue that this is the result of a conditioning effect.

There is the possibility that an EI patient will not tolerate any masking agent. This suggests that the patient chronically responded in every environment unless it was totally odor free. In our experience, such a history is typically not seen in EI patients. Nevertheless, there is an alternative procedure to using masking odors which precludes learned responses, i.e., present the agent at doses below detectable odor and irritant thresholds.

The elimination of sensory cues also has implications for the conditioning theory which suggests that after an initial exposure, the individual will subsequently react to the conditioned stimuli associated with the exposure. Under blinded conditions, is there a stimulus if the subject cannot perceive it? Toxicologically, yes, because the basis of response is an unconditioned toxic stimulus which invariably elicits a toxicodynamic or toxicokinetic response. Behaviorally, no, because the basis of response is the sensory conditioned stimulus, the odor, that is not perceptible because it cannot be discriminated from the masking odors. The masking procedure essentially precludes Pavlovian classical conditioning.

Statistical analysis

The reliability of responses to the alleged toxic agent and a benign placebo can be tested by statistical criteria. The data from all the challenges is tabulated using a 2×2 contingency table (Figure 5.4). There are two contingencies that reflect correct responses. In the first, the patient correctly appraises symptoms in response to the agent, coded as a true positive (TP) or a "hit". In the second, the patient correctly appraises the absence of symptoms in response to the placebo, coded as a true negative (TN) or "correct rejection". The other two contingencies reflect errors. In the first error contingency, the patient appraises symptoms in response to placebo, coded as a false positive (FP) or "false alarm". In the second error contingency, the patient does not appraise symptoms in response to the agent, coded as a false negative (FN) or "miss".

To be defined as a "true positive" responder, the patient should consistently develop symptoms when challenged with the active agent (TP) and not show symptoms in response to a placebo (TN). Data in the form of the 2×2 contingency table shown in Figure 5.4 can be analyzed statistically using Fisher's exact test for small numbers, or the chi-square test when there is a larger number of challenges (Fisher, 1956).

Additional statistics may be computed to identify a patient's response bias or strategy across several challenges, defined in Table 5.3 (Galen and Gambino, 1975). For example, a patient with a bias to appraise mostly positive reactions across a series of challenges will have high sensitivity because the "hit" rate for the chemical agent will be high, but low specificity because there will be more "false positives" to placebo. Similarly, a patient with a bias to appraise no reaction will have high specificity because of few "false alarms" to placebo, but low sensitivity because there will be more "misses" of active agent. Efficiency is the percentage of correctly appraised active and placebo challenges and is an estimate of chance performance.

Provocation Challenge

Subject's Appraisal	Putative Chemical Agent	Placebo
Reaction	TP Hit	FP False Alarm
No Reaction	FN Miss	TN Correct Rejection

Abbreviations: TP — True Positive
FP — False Positive
FN — False Negative
TN — True Negative

Figure 5.4 Classification of responses from double-blind, placebo-controlled provocation challenges. (From Staudenmayer, H., *Regul. Toxicol. Pharmacol.*, 24, S96–S110, 1996. With permission.)

The case of a true positive responder

This case report illustrates how double-blind, placebo-controlled chamber challenges were employed at the Allergy Respiratory Institute of Colorado to establish the reliability of subjective symptoms in response to provocation with a combination of chemicals in a patient who showed no objective signs of reactivity (Selner, 1996).

A 42-year-old woman was exposed to multiple volatile organic chemicals in the course of her employment as an assembler of components of renal dialysis machines. Symptoms, which began 2 years prior to our evaluation, included voice loss, drowsiness, cognitive dysfunction, severe headaches, dizziness, chest tightness and shortness of breath, and numbness and tingling of the extremities. Most symptoms cleared within hours of leaving the workplace, but some took a week or more to resolve. Worsening was noted in the spring and the summer. She described her work area as usually being stuffy with a very unpleasant odor, and she reported that ventilation, humidity, and temperature were either too much or too little. Methylene chloride and various solvent combinations were incriminated in episodes of apparent acute airway obstruction that were diagnosed as asthma and led to an emergency room visit. These symptoms resolved without a specific diagnosis

Table 5.3   Definition of Sensitivity, Specificity, and Efficiency
With Incidence Expressed as a Percentage

Sensitivity is the percent of true positive results
Sensitivity = TP/(TP + FN) × 100

Specificity is the percent of true negative results
Specificity = TN/(TN + FP) × 100

Efficiency is the percentage of active and placebo challenges correctly identified
Efficiency = (TP + TN)/Total × 100

Note: TP = number of true positive challenges; FN = number of false negative
challenges; TN = number of true negative challenges; FP = number of false
positive challenges.

being clearly established, but medical records suggest that gastroesophageal reflux was considered as a causative factor. Although workplace nursing staff and emergency room physicians observed no objective signs of illness, her subjective symptoms resulted in a disruption of her work schedule and job reassignment, potentially threatening her continued employment.

A 5-mg/mL methacholine challenge produced a change in pulmonary function measurements of forced expiratory volume in the first second (FEV_1) from 3.16 L at baseline to 2.15 L, demonstrating lower airway hyperreactivity consistent with asthma. Because the patient was exposed to a large variety of chemicals in the workplace, 16 separate double-blind, placebo-controlled olfactory-masked chamber challenges were performed. When a mixture of methylene chloride and other volatile organic compounds was employed, the patient's symptom scores, although increased from baseline, were no different from those seen with peppermint-masked clean air challenges. However, when a combination of propanol, trichlorofluoromethane, methylene chloride, 1,1,1-trichloroethane, and ethylene glycol monoethylether acetate was employed, there was a substantial increase in symptom scores, as well as an obvious change in voice quality. Fiberoptic laryngoscopy showed no change in the larynx or hypopharynx following this challenge. Symptoms lasted for approximately 40 minutes and cleared spontaneously. A repeat challenge reproduced these symptoms with similar intensity and duration. Exposure to polyvinyl chloride plastic cement (a mixture of tetrahydrofuran and cyclohexenone) resulted in voice-quality changes and increased nasal resistance measured by rhinomanometry in three separate challenges. Psychological assessment did not suggest that psychological factors or psychophysiologic stress response were involved.

The consistent voice-quality changes were not observed when blinded challenges with only the peppermint masker were employed. Based on consistent true positive responses to the chemical compound and true negative responses to the placebo, it was concluded that this patient was experiencing chemically induced adverse reactions at work. A recommendation for avoidance strategies within the workplace was made, and relocation resulted in relief of the patient's symptoms, consistent work availability, and a harmonious solution to a potential employment dispute.

Because of the subjective nature of the complaints, this form of idiopathic environmental intolerances could only be reliably demonstrated by double-blind, placebo-controlled provocation challenge testing. This patient reacted to a combination of chemicals that did not provoke symptoms when presented individually. This case sharply contrasts

Table 5.4 Results of Chamber Challenges
Across 20 EI Patients

Patient response	Chemical agent		Total
	Active	Sham	
Positive	TP	FP	
	21	31	52
Negative	FN	TN	
	41	52	93
Total	62	83	145

Note: Sensitivity = TP/(TP + FN) × 100 = 33.9%. Specificity = TN/(TN + FP) × 100 = 62.7%. Efficiency = (TP + TN)/Total × 100 = 50.3%. This table has been corrected for a counting error reflected in the original publication. Five challenges had been incorrectly classified as TN; of these, 3 were FN, and 2 were TP. The net effect is that there were 3 more erroneous subject appraisals than originally reported. The effect of this correction on the summary statistics was inconsequential: sensitivity from 33.3 to 33.9%; specificity from 64.7 to 62.7%; efficiency from 52.4 to 50.3%.

Source: Staudenmayer, H. et al., *Regul. Pharmacol. Toxicol.,* 18, 44–53, 1993. With permission.

with EI patients who allegedly respond to low-level multiple chemicals alone or in combination.

Controlled challenges with EI patients

Using the protocol for double-blind, placebo-controlled studies, 20 patients were evaluated individually. These patients were selected from a group of consecutive EI patients complaining of universal reactivity to multiple chemicals, evaluated between 1985 and 1988 at the Allergy Respiratory Institute of Colorado (Staudenmayer et al., 1993a). A total of 145 challenges (62 active, 83 placebo) were conducted for the 20 subjects. Each challenge was classified as a response or non-response based on the subject's appraisal of a reaction and symptoms and/or objective response, e.g., a drop in pulmonary functions.

As a group, the patients' appraisals were no different from chance performance (Table 5.4). None of these EI patients demonstrated consistent responses as reflected in measures of sensitivity, specificity, and efficiency (Table 5.5). This study clearly demonstrated that the toxicogenic hypotheses of these EI patients were incorrect.

EI advocates criticized this study on the grounds that the use of the olfactory masker created an interaction effect with the putative chemical agent which created synergistic and/or antagonistic effects which masked or exacerbated symptom response (Grace Ziem, personal communication). The biologic and toxicologic issues associated with risk assessment of chemical mixtures are summarized in a recent review (Mumtaz et al., 1993). According to accepted toxicologic principles (Waddell, 1993), if additive or interactive toxicologic or biologic effects occur among chemicals, the effects should remain consistent. How masking odors with limited variance in the chemical structure can have such interactive effects with a host of putative chemical agents with diverse chemical structure

Table 5.5   Sensitivity, Specificity, and Efficiency of Response Patterns for Each of 20 EI Patients

Case	Challenge number 01	02	03	04	05	06	07	08	09	10	Sensitivity	Specificity	Efficiency
01	FP	TN	TP	TP	TP	TN	FP	FN	TP	FN	66.6	50.0	60.0
02	FN	TN	TN	TN	FN	TN	FN	TN	FN	TN	0	100	60.0
03	TN	TN	TN	FN	FP	FN	TP	FN	TP	TN	40.0	80.0	60.0
04	FP	FP	TN	TP	TN	TP	TP	TN	TP	TN	100	66.6	80.0
05	TN	TP	TN	FP	FP	FP	FP	TN	TN	FN	50.0	50.0	50.0
06	FP	TN	FN	TN	FN	TN	FP	FP			0	50.0	37.5
07	TN	FN	TN	TN	TN	FN	TN	FN			0	100	62.5
08	TN	TN	TN	TN	FN	FN	FN	TN			0	100	62.5
09	TN	FN	FN	TN	FN	FN	FP				0	66.6	28.5
10	FN	TP	FP	TN	FN	TN	FN				25.0	66.6	42.8
11	TP	TN	TN	TP	FP	FN	FN				50.0	66.6	57.1
12	TP	TN	TN	TP	FN	FN	TP				60.0	100	71.4
13	TN	FP	FP	TP	FP	FP					100	25.0	33.3
14	TP	FP	TN	FN	TN	FN					33.3	66.6	50.0
15	FN	FN	FP	FP	TN	FN					0	33.3	16.6
16	TN	FP	FP	FP	FN						0	25.0	20.0
17	FP	FN	FP	FN	TP						33.3	0	20.0
18	TN	TN	FP	TN	FN						0	75.0	60.0
19	TN	FN	FN	TN	FN						0	100	40.0
20	TN	FP	TP	FP	FP						100	25.0	40.0

Source: Staudenmayer, H. et al., *Regul. Pharmacol. Toxicol.*, 18, 44–53, 1993. With permission.

employed in this study and how these effects can come and go at random during a series of provocation challenges, I leave to the speculations offered by EI advocates.

Challenges unanswered

Some critical position statements mentioned that the hypotheses and treatments of clinical ecology could be valid, even though no evidence has been found to support them — for example, the California Medical Association (1986):

> "Undoubtedly, some patients suffer from illnesses that cannot be readily diagnosed and for which only supportive treatments exist. It may even be true that some or all of the hypotheses and treatments proposed by clinical ecologists are valid, but we found no evidence to support them. These hypotheses and treatments should be subjected to modern, scientific methods of evaluation."

And, the American Medical Association (1992):

> "The fact that the diagnostic tests and therapy recommended by clinical ecologists are largely unproven by controlled clinical studies does not necessarily establish the lack of scientific validity. Well-controlled studies could validate and provide a scientific basis for many of the tests and therapies associated with multiple chemical sensitivity."

Table 5.6 Judging the Scientific Merit of a Research Program

1. How many new facts has the research program produced?
2. How great was the research program's capacity to explain refutations in the progressive growth of the research program?
3. In the progressive generation of hypotheses, did newer ones have more empirical content than their predecessors?
4. When a theory finally gets eliminated, the "precise" measurements carried out within the discarded framework may look rather foolish.

These challenges posed by the continuing presence of EI patients remain scientifically unanswered. The lack of response is taken into account when appraising a research program using the questions presented in Table 5.6. For the toxicogenic theory, these questions are addressed in three general areas:

1. The misattribution of environmental factors to diagnoses with well-established alternative explanations
2. Unsubstantiated diagnostic practices
3. Unsubstantiated treatment methods

How many new facts? Despite a half century of theorizing about it, EI advocates have identified no pathognomonic physical signs, laboratory abnormalities, tissue pathology, or natural course (Terr, 1992, 1993a). To date, not one case of EI has been substantiated in the scientific literature.

How have refutations been explained? The list of controlled studies showing negative findings which fail to support the toxicogenic theory grows longer. The toxicogenic research program has not explained negative evidence. The strength of refuting evidence is ignored or dismissed with the rationalization that the research was not conducted properly. If this defense fails, EI advocates deflect focus from the evidence with attempts to discredit the investigators. Scientists whose research does not support EI postulates have been accused of bias and, and more contemptible, falsely accused of ethical misconduct or threatened with lawsuits or personal harm in attempts to drive them from the field.

Did newer hypotheses have more empirical content? No, newer hypotheses have no more empirical content than their predecessors. Speculations about CNS mechanisms, unsupported by data, include olfactory limbic kindling, and neural sensitization. The foundation of this hypothesis is an allusion to mechanisms underlying epileptogenisis and clinical seizures. Brain-imaging studies with non-specific findings do not warrant a diagnosis of toxic encephalopathy. Hematic disorders of porphyrin metabolism (Ellefson and Ford, 1996) are alluded to in a document circulated by the EI advocacy group MCS Referral & Resources (Donnay and Ziem, 1995). In a comprehensive review of environmental exposures and disturbances in heme synthesis, Daniell et al. (1997) concluded that there is currently no convincing evidence that EI is mediated by a disturbance of heme synthesis.

Have methods shown to be ineffective for diagnosis and treatment been discarded? Unsubstantiated and refuted diagnostic and treatment methods such as provocation testing and neutralization therapy continue to be practiced and defended tenaciously. No clinical ecology procedure has been acknowledged to be useless by those who practice it. Only the threat of regulatory action has been a deterrent. A good example is the cytotoxic test, an *in vitro* immunologic test first employed in the 1950s for the diagnosis of food allergy. Terr noted that proponents of the cytotoxic test claim that morphologic changes in peripheral blood leukocytes in contact with allergen *in vitro* indicate that the patient is allergic to that particular antigen. The spectrum of allergy claimed to be diagnosed by this method encompasses rhinitis, asthma, headache, gastrointestinal symptoms, skin disease, hearing disorders, genitourinary diseases, and obesity. The procedure entails placing processed blood from the patient on a microscope slide that had previously been coated with dried extract of the food to be tested. Any changes visually observed in the appearance of the blood cells are used as evidence for "cytotoxicity" (Terr 1993c). Controlled evaluations (Benson and Arkins 1976; Lehman, 1980; Lieberman et al., 1974) failed to substantiate the value of cytotoxic testing for the diagnosis of adverse reactions to airborne allergen, foods, or chemicals as claimed by its proponents (Boyles, 1977; Updegraff, 1977). Only after federal agencies threatened possible legal action against interstate trafficking of materials used in cytotoxic kits did clinical ecologists attempt to distance themselves from this test (Selner and Condemi, 1988).

Conclusion

The International Society of Regulatory Toxicology and Pharmacology concluded that the clinical ecology theories claiming to unify EI as a toxicological-mediated disorder transgress basic principles of toxicology and clinical science (Gots et al., 1993). I agree. There is no scientific evidence in support of the toxicogenic theory. I would now make one additional argument: the logic of EI advocates is inconsistent with respect to foundation of their theory.

There is also increasing evidence from controlled challenges which refutes the hypotheses of individual EI patients. This evidence has been dismissed by EI advocates. The logic of one of these criticisms is particularly illuminating. Miller (1997) has suggested that double-blind, placebo-controlled studies are not valid because they did not take into account the baseline state determined by several properties of adaptation/deadaptation, something that could only be validly assessed in a hospital ECU. She argued that EI patients' errors of appraisal-exposure (false positives and false negatives) are in fact not errors but reflect pre-challenge potential which is realized during the challenge. For example, a FP to placebo really means the patient is reacting to the delayed effects of an earlier exposure such as a cup of coffee possibly days before, the perfume of someone encountered in the street on the way to the testing facility, etc. The FN would mean that the patient does not appraise a reaction to the test agent because the effects have been "masked" by other ambient exposure. If this were true, any patient's appraisal of a particular chemical inducing a reaction in the real world could never be valid.

For the moment, I wish not to focus on the implausibility of these hypothesized cyclical events affecting "masking/unmasking" but on something more fundamental. Reliability of EI patients' self-report and appraisal of events is the essential presupposition of the toxicogenic theory. All of the intricate postulates and speculative mechanistic hypotheses are proposed to explain specific phenomena described by EI patients. The argument against provocation challenge data is based on the premise that patients' appraisals were not accurate. Ironically, that argument by EI advocates refutes the foundation of their own toxicogenic theory.

chapter six

Psychogenic theory

> "Physical concepts are free creations of the human mind, and are not however it may seem, uniquely determined by the external world."

Albert Einstein, *Evolution of Physics*

Psychogenic theory postulates that environmental exposure is not required for the "acquired loss of tolerance" or eliciting an "environmental reaction". Psychogenic theory presupposes that symptoms, and psychophysiologic responses are associated only with perception or belief of exposure and appraised harmful effects. The physiologic responses are mediated through central activation of the pathways and systems of the stress response (detailed in Chapter 9). Brodsky and others described how beliefs are often iatrogenic, acquired by suggestible patients from clinical ecologists or EI advocates (Black, 1993; Brodsky, 1983; Staudenmayer and Selner, 1987). Patient motives also contribute to acquiring EI beliefs (Selner and Staudenmayer, 1986; Staudenmayer et al., 1993b). Motivational factors include factitious disorders, malingering, secondary gain defined as unconscious means of obtaining attention or benefit, or primary gain representing unconscious projection of deep-seated anxieties. The psychogenic theory postulates that the iatrogenic effect and the patient's motivations interact dynamically to produce florid symptoms that have nothing to do with the physical environment. Once acquired, EI beliefs, like any belief, bias perception and interpretation of information to be congruent with EI knowledge. Attitudes and behaviors adopted to cope with the stress response or underlying psychopathology may also interact with culturally shaped illness behavior to formulate EI beliefs about attribution of causality. Beliefs are dynamic in the sense that they generate or amplify physiological responses which are then appraised as symptoms attributed to environmental exposure leading to actions unwarranted by their physical condition.

A disorder of belief

In the past 200 years, the types of cognitive distortions seen in EI patients have been at the root of dynamic psychiatry, studied by such preeminent figures as Rush, Charcot, Bernheim, and Briquet (Carlson, 1986; Ellenberger, 1970). Rush wrote about dissociative experiences in 1812. Charcot was known for his studies of hysteria and also for having taught two students who elaborated on his ideas and became giants in the field in their own right, Sigmund Freud and Pierre Janet. Bernheim took the animalism and pseudoscience of Mesmer and made the study of hypnosis and suggestion credible. Briquet described a syndrome which bears his name and evolved into the somatoform disorders in the *Diagnostic and Statistical Manual of Mental Disorders* (DSM-IV), published by the American Psychiatric Association in 1994. These prominent clinicians all postulated that disorders of belief and the associated somatization originate in traumatic or stressful experiences which overwhelm the coping skills of the victim. To survive, experiences are inhibited from consciousness by complex psychological defenses that displace the cause for the somatic and emotional sequelae of such experiences outside the self.

Overvalued ideas are defined as closed belief systems that defy rational argument. McKenna reviewed the history of overvalued ideas in psychiatry, and some of his key points are presented here (McKenna, 1984). At the turn of the century, Wernicke defined the overvalued idea as a solitary belief that came to determine an individual's actions to a morbid degree, while at the same time being considered justified and a normal expression of his/her nature. Overvalued ideas were accompanied by strong affective responses that took precedence over all other mental activity. The belief was generally associated with an abnormal personality and grew out of adverse experience in a way that made it comprehensible. Overvalued ideas are different from delusions in that delusions usually have an alien quality about them, whereas overvalued ideas are more like normal beliefs. Delusions also do not have the conviction and the control over behavior that overvalued ideas possess. The phenomenology of the overvalued idea seems to combine the unlikely elements of non-delusional conviction, non-obsessional preoccupation, and non-phobic fear (McKenna, 1984).

The onset of an overvalued idea follows an incident that is often trivial but is viewed as a slight or injustice; the patient becomes convinced that he/she has been treated unfairly and begins making complaints. Appeasement seems impossible. Any accommodation, no matter how generous and genuine, is felt to be unsatisfactory; litigation usually follows. The process is often acrimonious, with the individual becoming more embittered as his/her beliefs are rejected by authorities. In 1905, Kraeplin, another pioneer in psychiatry, commented on overvalued ideas (McKenna, 1984):

> "The innumerable petitions which the patient has drawn up in the course of the last few years, chiefly at night, are exceedingly long-winded, and always allege the same thing in a rather disconnected manner. In their form and mode of expression they incline to the legal document, beginning with 'concerning,' going on throughout with 'evidence' and concluding with 'grounds.' They abound in half or wholly misunderstood professional expressions and paragraphs of totally different laws. Often they are careless and appear to have been written under excitement, contain numerous notes of exclamation and interrogation, even in the middle of a sentence; one or more underlinings, some in red or blue pencil, marginal notes and addenda, so that every available space is made use of. Many of the petitions are written on the backs of judgements and refusals of other courts."

The condition generally arises in individuals with preexisting personality disorders including obsessional, paranoid, sociopathic, and histrionic (McKenna, 1984). Not surprisingly, disorders with overvalued ideas have a chronic course with poor prognosis, unresponsive to any form of treatment. McKenna also notes that a number of well-documented disorders conform to the definition of overvalued idea and continue to defy classification as anything else, an aspect characteristic of EI. Despite its importance, "overvalued idea" is not a diagnostic category in DSM-IV, nor is it even listed in its index.

The term "true believer" comes from the sociologist Hoffer (1989) who commented on the passionate hatred and fanaticism often expressed by individuals holding overvalued ideas. He described how "true believers" structure and regulate their beliefs: "True believers shut themselves off from facts, ignoring a doctrine's validity while valuing its ability to insulate them from reality." Distorted beliefs can also affect interpretation of the stress response (Edwards and Cooper, 1988):

> "To clarify our understanding of the impact of stressors and life events on health, we must include the individual's appraisal of these events as a critical intervening variable. This may be accomplished by asking the individuals to provide their own evaluation of each life event relative to some internal standard, such as values, desires, expectations, abilities, or whatever standard is important to the individual and consistent with the researcher's theoretical focus."

And (Lipsitt, 1970):

> "'True believers' are difficult to treat and can elicit countertransference reactions from caregivers who view them as 'crocks'."

Gomez et al. (1996) utilized psychometric scaling techniques to create memory network representations of EI beliefs. They compared the memory networks abstracted from EI patients to several control groups including allergists, asthma and allergy patients, and asymptomatic college students. The memory representation of EI patients was significantly different from that of the other groups, characterized by the concept "multiple chemicals" in the center of the attribution pattern linking it to all symptoms (Figure 6.1), as would be expected of a "true believer". The network abstracted from asthma patient controls did not have a core belief concept, and the concept "multiple chemicals" was only peripheral in the memory network (Figure 6.2). Based on psychometric measures derived from ratings of the individual items on the questionnaire, EI patients' belief system could be reliably discriminated from the other groups.

Psychogenic and psychosomatic illnesses

The terms "psychogenic" and "psychosomatic" are often used interchangeably outside the specialties of psychiatry and psychology, but the distinction is important to theories of EI. "Psychogenic" means the symptoms are caused solely by psychological and psychophysiological mechanisms and there is no underlying medical disease that disrupts these same mechanisms or exacerbates psychological symptoms.

"Psychosomatic" means that biological and psychological mechanisms and processes interact such that psychological factors exacerbate physical illness and vice versa. In a psychosomatic disorder a person's predisposition to disease may be multiple, not just psychological, ultimately traceable to a biochemical, physiological, immunological, or genetic source. Weiner's (1977) psychobiologic theory is an example. Weiner postulated that stressful life events, maladaptive cognitive processing style, misperception of physiologic symptoms, affective reaction triggered by intrapsychic fear, abnormal behavioral reactivity, and defective physiological processes interact to activate the stress response and

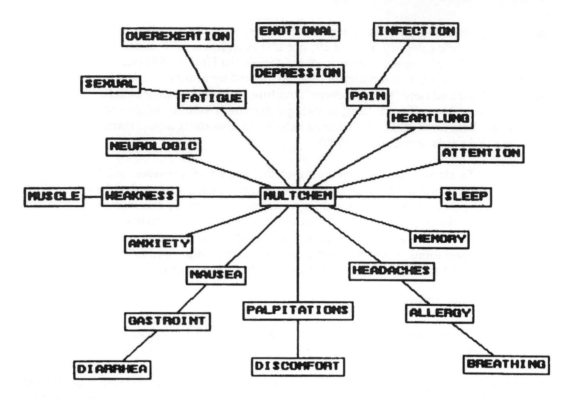

Figure 6.1 Average EI patient network. (From Gomez, R. et al., *J. Health Psychol.*, 1(1), 107–123, 1996. With permission.)

produce symptoms. It allows for the complexity of different patterns of abnormalities to take into account individual differences in temperament and genetic susceptibility as well as the heterogeneity among somatizing patients. For example, one individual might be more affected by physiologic disturbance, while another individual might be more affected by psychological disturbance, each of which contributed to symptom exacerbation or symptom maintenance. Weiner's theory was timely, coming when different theoretical factions and disciplines in medicine, psychiatry, and psychology were not communicating effectively. The scope of the theory was grand enough to be a guiding model for rapprochement in the study of diseases in which pathophysiologic mechanisms were identified, including: peptic ulcer, essential hypertension, bronchial asthma, Graves's disease, rheumatoid arthritis, ulcerative colitis, Crohn's disease, cardiac disease, and now, acquired immune deficiency syndrome (AIDS). Despite its complexity, the theory generated testable hypotheses and stimulated progressive research programs for the mind-body interaction (Weiner, 1985). For example, in the 1970s, Kinsman and colleagues applied the psychobiological model to bronchial asthma (Dirks et al., 1977, 1979; Kinsman et al., 1977; Staudenmayer et al., 1978).

The distinction between psychosomatic and psychogenic is important to differentiate toxicogenic and psychogenic theories of EI. One form of the toxicogenic theory, the interaction explanation, proposes that EI is like a psychosomatic disorder in that there is a disease of toxic etiology, genetically predisposed and exacerbated by psychological factors. There is a subtle reversal of the role of predisposing factors, a reversed causality. In psychosomatic disease, it is the biological and genetic factors that predispose an indi-

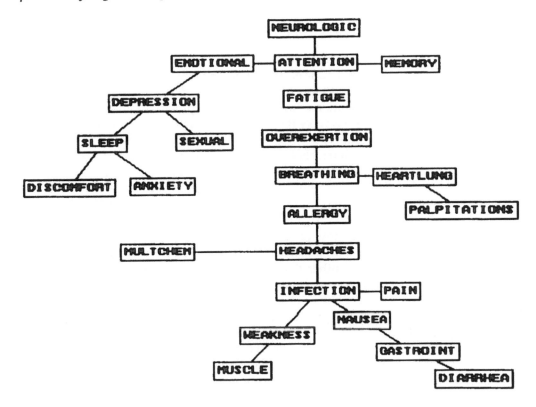

Figure 6.2 Average asthma patient network. (From Gomez, R. et al., *J. Health Psychol.*, 1(1), 107–123, 1996. With permission.)

vidual to acquire a medical disease. Psychological and psychophysiological factors do not predispose or cause a disease; they act upon the medical disease to exacerbate symptoms. The toxicogenic interaction explanation postulates that the psychological and psychophysiological factors predispose susceptible individuals to acquire EI, which is assumed to be a toxic disease. Thus, certain predispositional biological and genetic factors associated with personality traits (e.g., shyness) or neurotransmitter abnormalities (e.g., endogenous depression) are postulated to increase the risk of acquiring EI. Life experiences (e.g., trauma or childhood abuse) are postulated to exacerbate genetic predisposition. Once EI is acquired, all of the predisposing psychological factors may exacerbate it. And, any additional psychological symptoms which manifest are attributable to EI.

The essential difference between psychosomatic disorders and EI is the absence of an identified defective toxicogenic mechanism, be it immunologic, neurologic, hematologic, or electrical polarity. Without identifiable pathophysiology, EI does not qualify as a psychosomatic disorder. Calling EI a psychosomatic disorder alludes to a biologic and toxic origin that has not been demonstrated. The psychogenic theory postulates that psychological and psychophysiological components of the theory are sufficient to explain EI. Psychogenic explanations need not be "by exclusion" as EI advocates would have it. They contend that psychiatric or psychophysiologic disorders can only be diagnosed if and when no other medical reason can be identified or *speculated*. The scientific research I shall present in the following chapters demonstrated grounds for inclusive diagnoses of psychiatric disorders.

Psychogenic illness mistaken as asthma

Historically, psychogenic theories of medical illness originated in psychoanalysis and culminated in the work of Alexander et al. (1950). The empirical work of Alexander and his colleagues was an attempt to overcome valid criticism of psychoanalytic theories as suffering from gross generalizations and *post hoc* interpretations that did not generate testable hypotheses. In a bold but risky theory, Alexander postulated that specific psychological mechanisms were uniquely associated with specific diseases — for example, that asthma patients have a different psychology from cardiac patients. His work lost favor for at least two reasons. First, empirical studies showed that specific psychological conflicts and personality defenses where not uniquely mapped onto specific medical diseases. The postulate of specific conflicts causing or exacerbating specific diseases was empirically disproved. And second, with the advance of immunologic science starting in the 1950s, biological mechanisms were discovered for some of the diseases for which he postulated a purely psychogenic mechanism, particularly asthma. To date, the purely psychological etiology of identifiable disease, other than psychiatric illness, has not been convincingly demonstrated. But what of non-disease that is mistaken for real disease? Some of Alexander's clinical observations may have been put to rest prematurely (Horowitz, 1986).

As an example, consider the phenomenon labeled "paradoxical vocal cord dysfunction" (VCD) presenting as asthma. Christopher et al. (1983) demonstrated that VCD was a respiratory disorder of the upper rather than the lower airway, and therefore it had been mistakenly diagnosed and treated as asthma. The psychogenic nature of VCD and its diagnosis as a somatoform conversion disorder have been replicated in several studies (Brown et al., 1988; Freedman et al., 1991; Selner et al., 1987; Snyder and Weiss, 1989). VCD masquerading as exercise-induced asthma has also been identified in elite athletes who "choke" during athletic activities, providing an excuse for not meeting self-imposed or parental expectations of perfectionism (McFadden and Zawadski, 1996).

Let us briefly return to the time of Alexander. He lucidly described the psychodynamics of the conversion symptoms he observed in patients diagnosed with "asthma". Whether these were truly asthma cases or misdiagnosed cases of VCD seems impossible to assess retrospectively from the symptom data available in Alexander's reports. While his psychologic generalizations were too broad given what is known about the heterogeneous biologic nature of asthma today, his psychological observations accurately describe the patients who have been diagnosed with VCD in the past 20 years, after his death. The following case-study of VCD illustrates many of Alexander's insights.

> In a workers' compensation case, a middle-aged woman (medical records showed three different birth dates, years apart) complained of delayed laryngeal symptoms and was diagnosed to have paradoxical vocal cord dysfunction (VCD) brought on by a single high-level exposure to formaldehyde resulting from an alleged worksite spill several years prior to assessment. She reported that the spill occurred on a weekend when she was alone in her office, but company records showed that the only documented spill had occurred several months prior. Her account described how she was overwhelmed by the fumes to the point she passed out before being able to put on a respirator. She described being found by two co-workers but refused to identify them as witnesses. She alleged onset of chronic debilitating EI.
>
> The pulmonologist involved in the discovery of VCD conducted

an independent medical examination and determined that the diagnosis of VCD could not be established based on the available rhinoscopy evidence, a video tape taken at the initial assessment at a national pulmonary research center. The plaintiff's attorney refused to have his client undergo any further rhinoscopic examination. In a blinded study, two other physicians experienced in rhinoscopy and VCD independently judged a series of laryngeal videos which included a normal case (a negative control), another case of indisputable VCD (a positive control), and the plaintiff's. Both independent judges accurately identified the negative and positive control case, and both judged the plaintiff's video as not characteristic of VCD.

This had two implications, one forensic the other psychological. Failure to demonstrate the characteristic vocal "chink" abnormality precluded the diagnosis of "toxic-induced VCD" and obviated the medical argument over whether such a phenomenon existed. Psychologically, it raised the question of malingering.

Medical records prior to the incident and standardized psychological personality testing (Minnesota Multiphasic Personality Inventory-2 [MMPI-2] and Millon Clinical Multiaxial Inventory-II [MCMI-II]) after the alleged incident both indicated premorbid personality disorder including traits associated with borderline/narcissistic personality disorder and somatoform disorders, particularly primary gain and secondary gain. Accounts of childhood incest were documented in the records and also repeated during psychological interview.

The important psychological issue in this case is that she responded well to speech therapy intervention which employed behavioral breathing exercises. From a psychodynamic perspective, how can it be explained that an individual with underling personality disorder and motivation for primary and secondary gain improved with speech therapy? From a disability litigation point of view, the secondary gain for compensation was not at risk because resolution of laryngeal symptoms did not compromise the contention of initial harm. Furthermore, the claim for disability alleged permanent injury in the form of EI.

The resolution of laryngeal symptoms in the context of behavioral speech therapy did not undermine the primary gain from conversion for two reasons. First, a psychogenic conversion mechanism is not considered in behavioral theory because it is founded on the principle of being atheoretical, leaving the question of the underlying anxiety and its etiology unasked. Second, hysterical conversion symptoms in general tend to disappear when they are exposed as such (lack of corroboration by experts), or other less dramatic symptoms serve equally well to preserve the inner secrets, the primary gain. In this case, the underlying anxieties that were initially converted to the laryngeal symptoms reappeared as other general malaise symptoms (symptom substitution) to maintain her sick role, the cause of which was projected onto the physical environment.

This case illustrates how intrinsic motivation and extrinsic demand characteristics can influence both the induction of a symptom and its "cure". The positive effects of speech therapy appear to have been a non-specific effect in that the breathing exercises were directed at an idiopathic hypothesized mechanism underlying VCD, which in fact appeared to be an incorrect diagnosis. It is more likely that a placebo effect was operative. The positive results of speech therapy may have been due to the expectation of being better from undergoing therapy, mediated by the patient's implicit inference or the therapist's explicit suggestion that the behavioral techniques would control the symptoms. Or, it may have been the self-persuasion that these techniques justify a cure, a cure which could be allowed consciously and unconsciously without compromising either secondary gain or primary gain issues.

The sick role

In many cultures including our own, functional illness is often misconstrued as not having biological correlates ("its all in your head"), and the stigma of malingering is attached to it. The discovery of the biological and physiological effects of the stress-response have helped to curtail these misconceptions and dampened some of the associated stigma, but unfortunately some long-standing attitudes about psychological dysfunction are hard to overcome. Depending on the prevailing cultural ethics and morality of any historical period, psychological illness has been portrayed in stereotypes reflecting demonic influence, wrong political thinking, malingering, a bad attitude, weak character, or simply being a bad or undesirable person. The individual afflicted with mental illness "lost face".

Throughout history, few societies have been willing to face the embarrassment and unpleasantness associated with the etiology and personal suffering of the mentally ill, often incarcerating them as unfortunate chattel to be mistreated and subjected to psychiatric experimentation in institutions (Gilman, 1982). In recent times, they were released onto the streets to swell the ranks of the homeless under misguided regulations purporting to protect their civil rights. These factors may have contributed to the sensitization of having a mental disorder in those who have lost their ability to cope.

In contrast, most societies offer sanctions to individuals who are physically ill or disabled. There is sanctioned relief from the responsibility of daily life and permission to be passive (e.g., bed rest) or indecisive (e.g., catered care). The ill person is not expected to engage in difficult interpersonal or social interactions. Physical illness is accepted as involuntary and the individual is absolved of personal responsibility for its onset. But, there are also responsibilities attached to persistence of physical illness. Society supposes the patient is motivated to get well, expects the patient to cooperate with medical treatment, and to function at his/her level of maximal medical improvement. Rejection of efficacious treatment and nonadherence to a medication regimen are not sanctioned. The disabled are expected to maximize their functional abilities through rehabilitation. In summary, society expects the ill and the disabled to relinquish the sick role, not to embellish it (Alexander, 1982).

Under many liability, compensation, and disability laws in the U.S., it is necessary to demonstrate that a physical injury has been inflicted in order to claim secondary psychological harm and suffering. Ironically, most EI patients claiming disability on the basis of chemical sensitivities resist any suggestion of disability on the basis of psychiatric disorders even in those circumstances in which it would be compensable. Calling EI a toxicogenic illness fosters the belief that it is an objective, validated physical disease. The healthcare behavior of many EI patients is inconsistent with what is expected from those with a physical illness. Their beliefs and actions maintain their illness.

Overgeneralization of fearful conditions is a common observation within cognitive-behavioral psychology, described as follows (Sarason, 1985):

> "According to cognitive-behavioral theorists, thinking disturbances that occur only in certain settings or in relation to specific problems are often the primary sources of anxiety. This would be true of obsessive and phobic individuals. These types of thoughts include unrealistic appraisals of situations and consistent overestimation of their dangerous aspects; the degree of harm and the likelihood of harm may both be exaggerated. These thoughts can be viewed as vulnerability factors that interact with the characteristics of situations. From this point of view, precipitating events elicit or magnify an underlying attitude of fear and give rise to hypervigilance for signs of danger and/or other components of the anxiety syndrome. As this attitude strengthens, danger-related thoughts become more easily activated by less specific, less avoidable situations. As a result, the anxious individual may continually scan internal and external stimuli for danger.

But, unlike so many physical illnesses, problems of living cannot be treated or cured impersonally through medically unsubstantiated rituals. The "frightening psychological therapies" that EI patients dread and avoid are directed at painful self-reflection. Psychotherapy with EI patients should be initiated to identify and resolve conflicts that have generated demoralization and undermined self-esteem and to replace distorted beliefs and false defenses with effective coping processes to facilitate functioning in the real world.

EI patients, like other patients who express psychological conflict and distress through somatic symptoms (Lipowski, 1988), are psychologically heterogeneous. Most have overt signs and symptoms of depression and anxiety. For many, these are only the tip of the iceberg, beneath which lies the psychopathology that results from disrupted personality development in childhood.

History of psychogenic illness: neurasthenia

> "Hysterical symptoms do not in any way present a copy of the anatomical conditions of the nervous system. It may be said that hysteria is as ignorant of the science of the structure of the nervous system as we ourselves before we have learnt it." (Freud, as quoted in Macmillan, 1990)

EI represents an old phenomenon that has had various psychogenic manifestations throughout history, the only apparent difference being the attribution of causality (Abbey and Garfinkel, 1991; Wessely, 1990). Some of the diversity among somatization illnesses was presented in Shorter's (1992) absorbing historical book, *From Paralysis to Fatigue*. Despite the diversity of such theories, they had at least two common elements. First, there was no empirical basis for the hypotheses, but this did not curtail the practice of the physicians of the period. Second, the patients who were attracted to these fad "diagnoses" and treatments had common, multi-system complaints characterized as general malaise.

An interesting theme was the association to toxic effects. At the turn of the century, autointoxication was hypothesized to occur when intestinal toxins were produced at a rate that exceeded the capacity of the liver or kidneys to detoxify their products (Hudson, 1989).

Treatment included high-risk surgery, including removal of the entire large intestine. In a concise review of autointoxication, Gots (1993b) noted that one advocate of autointoxication was the chief of surgery at the prestigious Guys Hospital in London around the turn of the century, and "contributing to his zeal and influence was his personal anecdote of his own suffering from autointoxication and its reversal by laxatives and diet."

Neurasthenia was attributed to effects from the industrial revolution, analogous to EI being attributed to environmental exposures. I limit this review to neurasthenia for the following three reasons:

1. The multi-system symptoms are identical to those presented by EI patients.
2. General anxiety neurosis was proposed as an alternative diagnosis by the psychiatry community of the 19th century.
3. Randolph proposed EI to be separable and distinct from neurasthenia, an assertion not demonstrated to date.

The term "neurasthenia" originates from a Greek word meaning "lack of nerve strength". In the 19th century, Beard (1880, 1881), an American neurologist, popularized the condition and it quickly became the fad diagnosis in North America and Europe (Rosenberg, 1962). Beard (1880) described neurasthenia as a functional disease of the nervous system without structural organic changes, but distinct from mental illness or insanity:

> "Neurasthenia is a chronic, functional disease of the nervous system, the basis of which is impoverishment of nervous force, waste of nerve-tissue in excess of repair; hence the lack of inhibitory or controlling power — physical and mental — the feebleness and instability of nerve action, and the excessive sensitiveness and irritability, local and general, direct and reflex. The fatigue and pain that temporarily follow excessive toil, or worry, or deprivation of food or rest, are symptoms of acute neurasthenia, from which the chronic form differs only in permanence and degree. ... The vague and multitudinous symptoms that accompany neurasthenia are largely the result of reflex irritation that takes place, not only through the ordinary motor and sensory nerves, but through the sympathetic system and vaso-motor nerves."

Beard postulated that neurasthenia resulted largely from environmental factors, defined as non-biologic or non-genetic, as it is commonly used in psychiatry and psychology, which allows for psychosocial stress factors.

The demographics of neurasthenics presented by Beard include the fact that it is more common in women, young adults and early middle life, and among the educated and wealthy. Treatment modalities included diet, hygiene, massage, medication, electric stimulation, and appropriate use of rest, work, and change of work. Neurasthenic patients were often difficult to cure, symptoms persisted, and both doctor and patient were aware of the illness' social function of sanctioning withdrawal from and manipulation of social relationships without conferring blame on the patient, although neither could address the issue openly.

Doctors were reluctant to broach psychological issues because patients had such a strong aversion to the implications of psychiatry. The psychological etiology of neurasthenia was stated as (Ballet, 1908):

> "The depressing emotions, that is to say, vexation, anxiety, disap-
> pointments, remorse, thwarted affection, in a word all states of
> sorrow and disquiet — these are the usual causes of nervous ex-
> haustion."

Amplification of somatic complaints and suppression of psychological troubles were problems for the clinician treating neurasthenia patients, described as follows (Ballet, 1908):

> "Some [neurasthenic patients] describe minutely with unwearying
> insistence symptoms of a secondary order, and hardly mention those
> of real importance. Others speak abundantly of their headache and
> their muscular weakness, but deliberately conceal their emotional-
> ism, their childish fears, their state of anxiety, and the powerlessness
> of their intellectual faculties, all symptoms which it would offend
> their self esteem to confess."

Neurasthenia is retained as a diagnosis in the World Health Organizations's (1992) International Classification of Diseases (ICD-10) and is defined as being:

> "A neurotic disorder characterized by fatigue, irritability, headache,
> depression, insomnia, difficulty in concentration, and lack of capac-
> ity for enjoyment [anhedonia]. It may follow or accompany an infec-
> tion or exhaustion, or arise from continued emotional stress."

DSM-IV does not have a diagnostic code for neurasthenia, but lists it as an undifferentiated somatoform disorder.

Neurasthenia flourishes in some non-Western cultures, most notably China, described in a medical anthropological account by Kleinman (1982) and summarized as follows (Kleinman, 1986):

> "Neurasthenia patients are neither prisoners of society nor of the
> biology of medical disorder; instead, they are trapped in the iron
> cage of inner personal pathology. They are prisoners of a personality
> that is simultaneously illness generating, self defeating, and socially
> alienating and even stigmatizing."

How did Beard's contemporary psychiatric community react to his contention that neurasthenia was an all-encompassing diagnosis with a biological mechanism of etiology? One of Freud's first papers published in 1894, was entitled "The justification for detaching from neurasthenia a particular syndrome: the anxiety-neurosis" (Freud, 1959). Freud not only pointed out a central role for anxiety in understanding neurasthenia, but he also addressed questions of differential diagnosis of verifiable diseases that are to be separated from neurasthenia (Freud, 1959):

> "It is difficult to say anything of general validity concerning
> neurasthenia so long as we allow this name to cover all that Beard
> included under the term. In my opinion, nothing but gain to neuro-
> pathology can result if we make an attempt to distinguish from

neurasthenia proper all those neurotic disturbances of which the symptoms, on the one hand, are more closely related to one another than to the typical symptoms of neurasthenia (headache, spinal irritation, and dyspepsia with flatulence and constipation) and on the other hand, show in their etiology and their mechanism essential differences from typical neurasthenia. If we accept this plan we shall soon obtain a more or less uniform picture of neurasthenia; and shall be in a position to differentiate more sharply than had hitherto been possible between neurasthenia proper and various kinds of pseudo-neurasthenia, such as the clinical picture of the organically determined nasal reflex neurosis, the nervous disorders of the cachexias [meaning a depraved condition of the body in which nutrition is defective, or depraved habit of mind or feeling] and arterio-sclerosis, the early stages of general paralysis of the insane and of the psychoses. Further, it will be possible to eliminate many of the nervous conditions of the hereditarily degenerate; and shall also find good reason to include under melancholia many neuroses (especially intermittent and periodic types) which are to-day called neurasthenia. But the most decisive change of all will be introduced if we decide to distinguish from neurasthenia the syndrome I here propose to describe, which fulfills with unusual completeness the conditions set forth above. The symptoms of this syndrome are clinically much more closely related to one another than to those of neurasthenia proper (that is, they frequently appear together and replace each other during the course of the illness), while the etiology and mechanism of this neurosis are essentially different from what remains of true neurasthenia after this subtraction has been made from it. I call this syndrome 'Anxiety-Neurosis', because all its component elements can be grouped round the central symptom of 'morbid anxiety' and because individually they each have a definite connection with this."

Freud's comments are as germane to EI today as they were to neurasthenia a century ago. Freud allowed for the possibility that there is some objective disease underlying neurasthenia but argues that identifiable medical and psychiatric/psychological disorders be removed from under the classification. Once these disorders were removed, however, history shows us that there is nothing left of neurasthenia. This leads us to ask the compelling question, "Will EI meet the same fate, once objectively identifiable, chemically related illness (e.g., industrial asthma) and organic diseases, as well as psychiatric/psychological disorders, are separated from it?"

Primary and secondary gain

"A person in distress wants action — rational action, if possible, of course, but irrational action, if necessary, rather than none at all."
(Findley, 1953)

Why would someone believe in EI? This is a rhetorical question which I shall address from various perspectives throughout the remainder of this book. First, let me exclude from

Table 6.1 Secondary Gain Motivation
and Reinforcement

Interpersonal
Sympathy, attention
Affection from family and friends
Assurance of importance
Rationalization of failure (avoid humiliation)
Protection from criticism
Explanation for not trying
Avoidance of difficult relations

Monetary
Disability insurance
Workmans' compensation
Medical insurance
Litigation

Exemption
Responsibility
Work and employment
Achievement
Interaction with the outside world

Power
Manipulation and control in relationships
Employee empowerment
Community problems
Political cause

discussion a subset of EI patients who are not motivated by monetary or psychological gain, patients who are looking to get well. I have seen many EI patients with stress-response symptoms who had fallen into the hands of clinical ecologists, and their involvement with EI was purely iatrogenic. Often, an explanation of the unsubstantiated theory and practice of clinical ecology is sufficient to dispel their misconceptions about the appraisal of their symptoms. When the EI patient has something to gain from being ill and remains ill despite appropriate medical intervention, thereby adopting the sick role, other motivations must be considered.

If labeled as a disease, EI legitimizes withdrawal from responsibility of work, obligation, and interpersonal relations. Like one of its historical predecessors, neurasthenia, EI sanctions isolation, depression, anxiety, and demoralization. While secondary gain can be conscious and function like malingering, clinical experience suggests that, more often than not, it is unconscious in EI patients. Self-deception is often mediated by psychological defenses intended to reduce anxiety, especially in patients with personality disorders which preclude insight or lack sensitivity toward others. Because the beliefs are irrational from a medical standpoint, many doctors underestimate the resistance this unconscious motivation can muster. Several manifestations of secondary gain can occur alone or in combination (Table 6.1).

The social aspects of secondary gain derived from neurasthenia in the last century were portrayed in the following commentary on the American psychologist William James, who is said to have suffered from neurasthenia (Feinstein, 1984):

> "In mid-nineteenth-century New England, [neurasthenic invalid-
> ism] coalesced from a romantic and puritanical matrix into a durable
> social role. Salvation through work, condemnation of idleness, sus-
> picion of pleasure, and a belief that suffering leads to grace flowed
> from the puritan source. Insistence on self-expression, a high valua-
> tion of leisure, and the admiration of delicacy and acute sensibility
> issued from the romantic. In such vigorous crosscurrents, illness had
> considerable utility. It provided social definition, sanctioned plea-
> sure, prescribed leisure for health, protected from premature respon-
> sibility, forced others to care, and expressed inadmissible feelings
> while protecting vital personal ties."

Secondary gain is a powerful influence on some EI patients (Brodsky, 1983, 1988). Others show little evidence of secondary gain, and their motivation originates in the psychological conflicts about self — called primary gain (Staudenmayer et al., 1993b; Selner and Staudenmayer, 1986).

The psychodynamic tradition defines primary gain as intrapsychic conflict; the cognitive-behavioral psychology approach defines it in terms of overload of coping resources. Frank and Frank (1991) identified five factors associated with primary gain:

1. Trauma and resulting posttraumatic stress disorder
2. Cultural or economic deprivation
3. Social alienation
4. Demoralization
5. Unbearable loss

Discussion

> "As Sherlock Holmes advised Dr. Watson, 'When you eliminate the
> impossible, whatever remains, however improbable, must be true.'"
> (Hill, 1965)

This chapter has introduced the psychogenic theory and its presuppositions and presented its distinction from psychosomatic illness. I introduced a question about what motivates EI patients to acquire the sick role. In the following eight chapters I shall attempt to answer this question by reviewing topics with mechanisms in place, including: placebo, learning, stress and the mechanisms of the stress-response, anxiety disorders and panic attacks, trauma and post-traumatic stress disorder, emotions and the limbic system, affective disorders, personality disorders, and iatrogenic exploitation. A recurrent theme shall be the biological and psychological sequelae of trauma.

For EI patients with a history of premorbid trauma, EI beliefs and associated ritualistic behaviors serve as defenses to inhibit trauma memories which, if brought into conscious awareness, would create extreme anxiety or overload coping resources. EI patients have been found to report grotesque memories of early childhood physical and sexual abuse experiences (Staudenmayer et al., 1993b). Intrafamilial trauma invariably has associated psychological abuse that contributes to the disruption of personality development (Bowlby, 1980) and undermines the formulation of a healthy, cohesive, integrated concept of self (Kohut, 1971). Sexual and physical trauma in childhood can also create permanent biological and physiological effects which alter and compromise coping resources (Schwarz and

Perry, 1994; van der Kolk, 1994). The basis for the overload of coping resources in other EI patients may be found in stressful life events such as losses due to divorce or alienation, loss of job, job stress, spouse abuse, or self-imposed unrealistic expectations. These reactions can often be traced to beliefs originating in childhood that have become self-defeating in adulthood.

Beliefs about EI serve the psychological defense mechanism of projection by which factors external to the self are held responsible for the illness; psychological conflicts and anxieties are inhibited (Staudenmayer and Camazine, 1989). The character trait of projecting blame has been identified as a common characteristic in individuals with any kind of personality disorder (Cloninger et al., 1993). Projection often interacts with other defense mechanisms, particularly conversion reaction, which diverts underlying anxiety into more dramatic somatic symptoms associated with somatization. These two psychodynamic defense mechanisms are often seen in EI patients who meet diagnostic criteria for somatoform disorder and co-morbid personality disorders.

Similar ideas are discussed in terms of attribution of causality (Kelley, 1967) associated with a personality trait called "locus of control" (Rotter, 1966). Individuals with an internal locus of control tend to attribute favorable experiences and life events to their own efforts and positive personality traits, and they attribute unfavorable experiences and life events to lack of effort or negative personal traits. Individuals with an external locus of control tend to project responsibility outward and attribute experiences and life events to external factors such as fate, chance, the actions of other people, and, in the case of EI patients, environmental factors.

Psychiatric symptoms, specifically those of the panic and other anxiety disorders, have been identified in manifestations of both mass psychogenic illness and EI (Brodsky, 1988; Pearson and Rix, 1987). These symptoms are associated with hyperventilation and include: muscle aching, nondescript pains, syncope, light-headedness, shortness of breath, dizziness, heaviness in the chest, weakness, headaches, blurred vision, and paresthesia. With the outbreak of mass psychogenic illness, anxiety can be anticipated as a manifestation of the fear of the unknown. When patients and their doctors do not know the cause of symptoms, it is very alarming. Anxiety is often interpreted to be an understandable consequence of this kind of uncertainty about the future (May, 1979). There are documented cases in which panic attacks are acquired after a toxic exposure (Shusterman and Dager, 1991). Nevertheless, in some cases this kind of reactive anxiety, or "state anxiety" as it is usually called, camouflages pre-morbid, long-term anxiety, or "trait anxiety" as it is called. With every effort made to account for an external explanation, very often the first suggestion is that the patient has been injured by a physical agent, and nonspecific treatment follows.

chapter seven

Placebo and somatization

A few years ago, a truck driver for the Texas Department of Transportation related a case he wanted me to explain. A local resident professing EI complained that roadside spraying of herbicides for weed eradication was making her ill. She could no longer go to town or visit friends because she could not travel along roads that had been sprayed at any time. To accommodate her wishes, certain roads to and about town were not sprayed. This led to a potential for vegetation overgrowth and impaired visibility along the roadway, a problem managed locally by mowing until she complained of gasoline fumes from the equipment. Mowing by hand seemed implausible, but she insisted it was necessary and that she also required such a chemically free corridor to Dallas, approximately 500 miles away, to see her physician. Whenever she saw the large, yellow truck drive by her house she would phone in a complaint that she had been made ill. The complaints were the same whether spraying occurred or not. Most of the time, the truck was simply passing by on the way to a distant site. The truck driver was perplexed by the pattern of her complaints, which made no sense to him.

One of most common effects in research and clinical practice is the placebo effect characterized by experienced changes in symptoms in response to a perceived trigger. Depending on the suggestion accompanying a placebo, the effects can result in either reported improvements of symptoms or adverse side effects. I begin with common understandings of the term "placebo", how it is used in research as a control, and how it is employed clinically when no other treatment is available. Physician and patient factors are briefly identified, and the power of suggestion and its physiological effects are illustrated. Lastly, I shall describe the psychological characteristic of EI patients who are most vulnerable to the suggestion of illness in terms of somatization and somatoform disorders.

Definitions

"Sugar pill" is a familiar understanding of the term "placebo", formally defined as (Shapiro, 1964):

> "Any therapeutic procedure (or that component of any therapeutic procedure) which is given deliberately to have an effect, or unknowingly has an effect on a patient, symptom, or disease, but which is objectively without specific activity for the condition being treated."

In medicine, placebos are utilized in several ways. First, in clinical pharmacology research, the placebo is used to control for nonspecific effects on bodily states as well as the effects of expectation or suggestion on symptom report during the course of a clinical trial of a new drug. In these experiments, the subjects are informed that placebos are being used and why, but not when, in the course of the study. In a case-control study (different subjects get the active agent or placebo), the effects of expectations are assessed by the magnitude of effects in the placebo group. In cross-over studies, the subjects get both the placebo and the active drug during different phases. The differences in symptom measures between the placebo and the drug periods are used to assess the reliability and the magnitude of the effects of the drug. The cross-over design is essential if the number of subjects is small or it is a single-case design (Kazdin, 1976).

The second use of placebo is in clinical practice, in which the placebo is administered to promote healing, especially in psychogenic illness and stress-related disorders (Roberts et al., 1993). In this case, the patient expects that the maneuver will be beneficial but is unaware that it is a placebo. It is this use of the term which is closest to the translation of the Latin for "I shall be pleasing or acceptable". According to the Oxford English Dictionary, this positive use of the term refers to the doctor administering a placebo to comfort and reassure the patient when no other remedy is available or to overcome an acute anxiety episode.

The third use is better described by the Latin term *nocebo*, meaning "to harm". In the Oxford English Dictionary this use of the term "placebo" means to play the sycophant, to flatter, or to be servile or time serving. In its most pejorative meaning, placebo was used by Chaucer to refer to a sycophant, a parasite.

A distinction is made between a placebo and the placebo effect, defined as any psychologic or psychophysiologic effect attributable to a pill, potion, procedure, ceremony, or religious ritual, but not to its pharmacologic or specific properties (Wolf, 1959; Frank and Frank, 1991). A placebo can be an inert agent or a procedure or ritual portrayed to have a specific pharmacologic effect but in fact does not. Any treatment modality can exert a placebo effect, and patient reactivity is often related to the potency attributed to the treatment (Brody, 1980). The doctor/therapist/healer can administer the placebo unknowingly (believing it to be effective on some basis other than heightened expectation) or knowingly (aware of the deception). The motive for using the placebo can be positive, as in therapy to promote healing, or negative to exploit the patient, in which case the term *nocebo* may be more appropriate, although I am reluctant to use it because it is unfamiliar in the English language.

The placebo effect as nuisance

"There appears to be nothing predictable about placebo effects."
(Wolf, 1959)

In the industrialized world, the lay public is largely biased to a biomedical model to explain symptoms, with psychologic etiology and exacerbation relegated to secondary consideration. The idea of a biological mechanism as an explanation (valid or not) for which one can take a pill is certainly more appealing than the alternative of a psychologic disorder that requires delving into one's soul. Medical institutions have contributed to that attitude (Brody, 1980):

"In medicine, underlying paradigms include theories and assumptions about the nature of disease and therapy and about laws and regularities in human pathophysiology. The present-day, Western medical paradigm emphasizes causal mechanisms affecting organs, tissues, cells, chemical factors, and physical phenomena. Theories relating psychological and sociological factors to disease and therapy are generally less well developed and held in lower esteem — as though one might feel that they will have to do until "real" explanations in physical-chemical terms become available through further research. Within such a paradigm, the fact that a chemically inert pill can change symptoms and organic bodily states constitutes a significant anomaly. ... A discovery such as the placebo effect is likely to arouse consternation among medical scientists."

Many medical scientists pushing back the frontiers of immunology, for example, do not wish to be bothered with pseudoscientific models. I have witnessed some of these eminent scholars become livid with frustration when faced with spending time on EI issues. Somewhat ironically, the EI phenomenon may force medical scientists to reconsider and appreciate the placebo effect rather than dismissing it as a nuisance.

Medical science, following the model of clinical pharmacology in which placebos are used as controls, defined the placebo effect by exclusion, something to be controlled for and preferably eliminated in assessing "real" biologic effects. Brody (1980) raised an interesting question, "Why not an inclusive definition, such as one attributing the placebo effect directly to the psychological component of the healing [or illness reinforcing] intervention?" This would allow for the separation of EI beliefs about the harmful effects of chemical exposure from nonspecific stress-responses associated with tension, vigilance, anxiety, poor diet, lack of sleep, and a host of other identifiable factors. Inclusion of psychologic factors known to affect physiologic systems that may result in sensations that are appraised as symptoms by EI patients would seem preferable to the exclusion of such factors from consideration until, as EI advocates would have it, all toxicogenic hypotheses are exhaustively evaluated.

Beecher (1955) estimated the prevalence of the placebo response to be 35.2%, based on studies, chosen at random, that included 1082 subjects. He concluded that the constancy of the placebo effect, as indicated by the small standard error of the mean (2.2%) in a fairly wide variety of conditions, suggests a common fundamental mechanism is operating (Beecher, 1955). Of clinical note, the placebo effects were greatest in individuals under the greatest stress (anxiety or pain). Another estimate of the prevalence of the placebo effect was based on the number of primary-care physician contacts with underlying emotional rather than physical causes, characterized as manifestations of somatization. The number of such visits were surprisingly high, estimated to be 50 to 80% (Shapiro, 1964; Katon et al., 1991).

The placebo response is generally inconsistent from one situation to the next. Most individuals who respond to a placebo in one situation will fail to respond reliably in another, even in the course of the same study (Lasagna et al., 1954). The symptoms of the placebo effect are not specific in that they may reflect any of a number of multi-system complaints, often in line with the expected effects of the placebo. Furthermore, the symptoms reported by the patient may be different from trial to trial, even with the same placebo. The variability of placebo effects is strikingly similar to the symptom effects postulated by the toxicogenic theory. In contrast, toxicologic disease demonstrates specificity of symptoms and reliability of responses to exposure (Hill, 1965; Schaumburg and Spencer, 1987; Waddell, 1993).

Placebo, ritual, and provocation challenges

Provocation challenges have been used throughout history, long before toxicology became a science, and long before double-blind, placebo-controlled studies became the "gold standard" for identifying environmental intolerances. Provocation challenges were conducted openly or with different degrees of blinding of both the subject and the investigator. There are historical accounts of single-blind studies in which placebos were used to dispel unfounded beliefs. One early account of blinded provocation challenges involved magistrates unhappy with the events of the Inquisition (Carus, 1974):

> "In 1588, Martha Brossier, a French peasant girl, claimed to be possessed of a devil. The excitement was great and the pulpits resounded with alarming denunciations apt to renew all the terrors of former witch prosecutions. But Bishop Miron of Angers, and Cardinal De Gondi, Archbishop of Paris, retained their tranquillity, and had the case investigated not only according to a truly rational method, but even in a spirit of humor. When the never-failing tests with exorcisms through sacred books and holy water were administered, Bishop Miron so arranged matters that the possessed girl was induced to draw wrong conclusions, and lo! simple spring-water and the reading of a line from Virgil regularly brought on epileptic fits, while neither the old reliable exorcisms nor the holy water produced any effect when the girl did not apprehend the sacred texts. Believers in Satanic possession were not satisfied with Bishop Miron's experiments, for they regarded them as proof of the cunning of the Devil who thus slyly deceived his enemies. The case was brought before Archbishop De Gondi, but he, too, proved skeptical and declared after some judicious experiments that the demeanor of the possessed girl was a mixed result of insanity and simulation. In spite of the sound judgment shown by these and other prelates, the prosecution of witches continued."

The question not addressed in this account is, "Why did this woman profess to be possessed by a devil?" I will address this question extensively later.

There are many historical examples of the inappropriate use of invasive medical procedures. In a review of surgical treatments which were historically considered to be efficacious by their proponents but no longer considered effective based upon later controlled trials, the majority of reported successful treatment outcomes were due to both the healer and the patient believing in the efficacy of a treatment (Roberts et al., 1993).

Beecher (1955) noted that a placebo can also produce "toxic" reactions with symptoms similar to the multi-system complaints reported by EI patients (Table 7.1). Another study on 25 patients showed that placebos produced toxic side effects similar to those of drugs (Honzak et al., 1972). The findings included (Brody, 1980):

> "Somnolence (10 cases), palpitations (9), irritability and insomnia (8),
> weakness with drop in blood pressure of more than 20 mm mercury

Table 7.1 Percentage of Subjects in Double-Blind Studies
Reporting Toxic Effects to Placebos

Symptom	Percentage (%)	Number of subjects
Drowsiness	50	72
Headaches	25	92
Fatigue	18	57
Sensations of heaviness	18	77
Difficulty concentrating	15	92
Sleep	10	72
Nausea	10	92
Relaxation	9	57
Dry mouth	9	77
Warm glow	8	77

Source: Adapted from Beecher, H.K., *JAMA*, 159(17), 1602–1606, 1955.

(5), temporal headache (4), diarrhea (3), collapse (2), and itching (2). In addition, three of these patients developed dependence on the placebo and demonstrated withdrawal symptoms when the pill was stopped."

Some of the observed effects deserve comment. First, two of the patients collapsed, indicative of hysterical conversion symptoms. Second, five patients had a drop in blood pressure. Unlike the symptoms reported by the patient, this is an objective sign. Physicians tend to put greater weight on objective signs in their decision-making, especially one such as a drop in blood pressure as it is potentially fatal if taken to extreme. However, Engel (1978) reported that death resulting from vasovagal syncope can be psychologically induced through fear — the so-called "voodoo death". Symptoms and signs that are mediated by the physiological systems of the stress-response (see Chapter 9) may be induced by emotion and/or cognition. Organ systems most often affected include the nose, bladder, esophagus, colon, heart, blood flow through the kidneys, and behavior of the stomach and duodenum during nausea (Wolf, 1959). With the development of brain-imaging techniques such as PET, brain functioning may be added to this list (Posner and Raichle, 1994; Roland, 1994; Roland and Friberg, 1985). It is important to realize that *placebo effects are not imaginary*, but rather the effects are psychophysiologically mediated with an imaginary origin (Kojo, 1988).

This point was clearly demonstrated by Luparello and colleagues in a series of double-blind, placebo-controlled studies on the effects of suggestion on airway resistance in asthmatics (Luparello et al., 1968; McFadden et al., 1969). Bronchoconstriction in response to saline occurred in about 50% of subjects who were told they were inhaling the nebulized allergen to which they attributed their asthma attacks. Bronchoconstriction was measured with changes in airway resistance, a measure obtained in a whole-body plethysmograph. Airway resistance is less effort dependent and therefore less likely to be affected by the subject than some other pulmonary function measures. These investigators also identified a mechanism — activation of efferent cholinergic pathways — through which the phenomenon was mediated by demonstrating that they could prevent the bronchoconstriction response to suggestion by intravenous atropine sulfate.

Table 7.2 Interactive Effects of Subject Expectation and
Medication on Airway Resistance Changes in Asthmatics

Suggestion/ expectation	Bronchodilator (isoproterenol)	Bronchoconstrictor (carbachol)
Dilation	+ 39.6%	−12.8%
Constriction	+ 20.1%	−22.3%

Source: Adapted from Luparello, T.J. et al., *Psychosomatic Med.*, 32(5), 509–513, 1970.

Luparello et al. (1970) also demonstrated the powerful interactive effects suggestion can have on two pharmacologically active substances that affect the airways in different ways, namely isoproterenol (produces bronchodilation and ease of breathing) and carbachol (produces bronchoconstriction and shortness of breath). In a double-blind, four-way, cross-over design with 20 asthmatics, each of the two medications was paired with the suggestion that it was a bronchodilator or a bronchoconstrictor (Table 7.2). The dependent variable was the ratio of airway resistance to thoracic gas volume obtained by means of a whole-body plethysmograph. The expectation of the subject significantly influenced airway responses to the bronchodilator isoproterenol. Combined with the suggestion of bronchodilation, the improvement was +39.6%, compared to +20.1% with the suggestion of bronchoconstriction. Subject expectation also influenced the airway responses to the bronchoconstrictor carbachol. Combined with the suggestion of bronchoconstriction, the breathing impairment was −22.3%, whereas it was only −12.8% with the suggestion of bronchodilation. The interactive effects of expectation and pharmacological agent showed that the greatest effects on airway reactivity occurred when the suggestions and the pharmacological action were consonant.

Factors contributing to the placebo effect

From a theoretical viewpoint, many valid psychological mechanisms have been hypothesized to explain the placebo effect, including belief, expectation, faith, hope, suggestion, transference, and conditioning. How much each of these factors contributes to a specific manifestation of the placebo effect depends on the context and the personality of the individual in question. Behavioral conditioning theory makes no reference to the mental state of the individual, which has led to the following criticism (Brody, 1980):

> "Such an account would have to construe all the antecedent and consequent events in terms of publicly observable behavior. Problems may arise in two areas, however, giving a behaviorist account of subjective symptoms such as pain and believing that one is in a healing [or unsafe] context."

Brody hypothesized that the placebo effect is most likely to occur when the following three conditions are optimally met:

1. The patient is provided with an explanation for his illness that is consistent with his pre-existing view of the world.
2. A group of individuals assuming socially sanctioned caring roles is available to provide emotional support for the patient.

3. The healing intervention leads to the patient's acquiring a sense of mastery and control over the illness.

The third point has different implications for a positive and negative effect of placebo. Loss of control is the feeling experienced by many patients with serious medical diseases and psychiatric disorders. The goal of psychotherapy, even when it involves only a nonspecific effect, is to enhance coping skills to overcome the sense of lost control. By contrast, in psychogenic illness loss of control is the cause rather than the effect of the symptoms. The patient is often more concerned with finding an explanation for the loss of control, preferably one that does not implicate self. Overcoming the feeling is usually a secondary motivation.

The doctor-patient relationship plays a critical role in the diagnosis and treatment of non-disease related illnesses, or what I refer to as psychogenic illness. When the doctor-patient relationship is solidified, the placebo can be used to promote wellness or illness. Placebos can help alleviate acute anxiety or provide a temporary strategy to facilitate entry into more appropriate psychotherapy (Oh, 1994). On the other hand, they can be used to exploit the patient's need for emotional catharsis by reinforcing illness beliefs, or worse, instilling them.

The psychological issues the patient may be trying to address through interaction with the doctor (Adler and Hammett, 1973) are the essence of what psychiatrists call "transference". Transference is the process whereby the patient's feelings that are unconsciously attached to significant persons in the past (usually parents) are displaced onto the doctor. In a positive transference relation, the doctor is seen as the good (real or idealized) parent to which the patient entrusts his or her care. In the context of EI, that transference appears positive for the patient when the doctor fulfills the patient's expectation of a physical ailment by presenting an unsubstantiated "diagnosis" of EI.

When the patient and doctor share a belief about the attribution of the illness, the patient is more likely to have faith in the doctor and the proposed remedy (Ross and Olson, 1981). If the doctor challenges that belief, as is often the case when traditional physicians treat an EI patient, he in effect plays the correcting parental role of presenting reality. The patient may feel at risk if the belief in the unsubstantiated malady represents a psychological defense against unresolved anxieties or conflicts. If the loss of the defense threatens access to anxiety-provoking memories, the patient may reject the incongruent information presented by the doctor through psychological defense mechanisms of denial, reinterpretation, or discrediting the doctor and his brand of medicine. Such alienation may also elicit negative feelings toward the doctor, emanating from long-standing feelings about a non-supportive or abusive parent or significant other. The patient may displace that anger and show overt resistance and hostility, or withdraw and internalize the anger.

The most frequently reported characteristic of placebo reactors is free-floating anxiety (Beecher, 1959; Lasagna et al., 1954; Frank and Frank, 1991). This effect refers mostly to the positive, anxiety-reducing qualities of administering a placebo to a patient in acute distress. There is a second group of patients characterized as poor placebo responders, those who do not manifest anxiety but present a more stoic resolve to accepting illness (Shapiro, 1964). These patients tend to have a more flat affect and are more obsessional in their thinking with little, if any, insight about their psychiatric disorders. They have conviction in their belief. They are quite prone to hysterical conversion when they think that they have been exposed to a harmful environmental agent. In that sense, they are quite suggestible and agreeably respond to a placebo which reinforces their underlying defenses and motivation; they are reliable *nocebo* responders. However, they tend not to be good responders to any psychotherapeutic interventions. Patients with psychiatric disorders who are free of manifest anxiety generally have a poor prognosis. In EI patients, this may be associated with their vested interest in maintaining the sick role.

Self-persuasion is one of the most effective means of changing attitudes and behaviors and may be a major source of placebo effects. Self-persuasion through role playing is also effective in altering attitudes (Janis and King, 1954). Suggestion and self-persuasion may create unusual beliefs and changes in patient expectations and behaviors that may have no objective basis. The unsubstantiated methods practiced by clinical ecologists are often adhered to with ritualistic fervor by EI patients. For example, some may hang their mail on a clothesline for hours, waiting for it to "offgas toxins" before reading it. Letters and newspapers are read under glass in so-called reading boxes. Oxygen and vitamins are used as rescue agents when going forth into an environment believed to be generally toxic. Victimization can take on a life of its own, and the more it is accommodated or reinforced by family, friends, and societal institutions, even though well intended, the more entrenched the beliefs become and the worse off the patient.

Food intolerance: physiology and immunology

Food intolerance as a basis for unfounded theories of ill health is not unique to EI. At the turn of the century there was the autointoxication phenomenon. Foods were a focal point in the clinical ecology speculations of Randolph in the 1940s. Nonetheless, there were also systematic research programs which discriminated immune mediated food intolerances from psychogenic aversion. Two research programs that deserve recognition as exemplary are those of Wolff and Pearson and their respective colleagues.

Shortly after World War II, Wolff (1950) conducted double-blind, placebo-controlled, provocation challenge studies to test beliefs about the adverse effects of foods in both normal controls and patients with psychological disorders. For example, a double-blind study involved four physicians who believed they could provoke migraine headaches in themselves by eating chocolate in any form and in minimal amounts (Graham et al., 1950). They were each given envelopes containing 8 g of powdered chocolate or lactose (placebo) in eight black capsules that were indistinguishable. A within-subject, cross-over design was employed over a 4-month period, with the subjects carefully logging their symptoms daily and also noting when they ingested the capsules. Most migraine attacks occurred without reference to the ingestion of capsules. Migraines followed the ingestion of lactose just as frequently as they followed the ingestion of chocolate.

In another demonstration, Wolff and associates showed how symptoms commonly seen in allergy could be manipulated with emotional material and suggestion. In one patient it was possible to alter the response of the skin to histamine and pilocarpine from negative to positive and then to reverse it again by changing the topic of discussion. For example, mention of the patient's sister triggered a physiological response (Figure 7.1).

In another single-blind demonstration, symptoms of gastric and duodenal activity responded to suggestion. A 50-year-old woman presented with multi-system complaints including pressure in her head, poor thinking and memory, vision not clear, dizziness, abdominal cramps nausea, hives, and abdominal bloating. She stated that 4 drops of milk in a glass of water induced symptoms 10 to 20 minutes after ingestion. The onset of attacks dated from 11 years ago after she had a Cesarean section. The patient had never liked milk yet had consumed a great deal during the pregnancy, gaining 60 pounds. In the experiment, balloons were introduced into the stomach and duodenum to monitor gastric motility. After establishing a baseline, 50 cc of whole milk were introduced into the stomach via a feeding tube to which the patient was blinded. She was told she was being

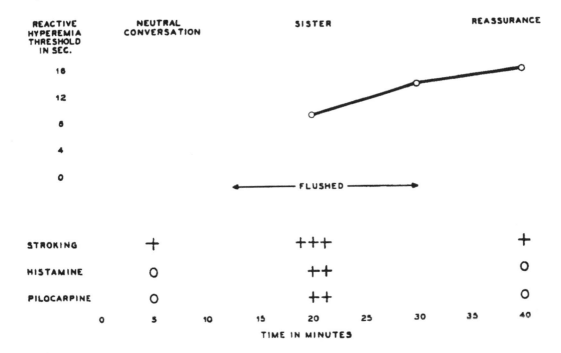

Figure 7.1 Simultaneous changes in the response of the skin to histamine, to pilocarpine, and to stroking during a stressful interview with a woman patient with hives (histamine acid phosphate 0.001% and pilocarpine hydrochloride 1% at 10 µA for 2 minutes over 1 cm²). There was no response to isotonic sodium chloride solution (placebo) applied in the same way before, during, or after the period of stress. (From Graham, D.T. et al., *J. Allergy*, 21, 478–486, 1950. With permission.)

given water as a preliminary testing procedure. There was no significant change in the duodenal motility pattern and the patient exhibited no symptoms. Two weeks later the experiment was repeated, but this time she was given 50 cc of water and was told that milk was being introduced. She developed nausea and abdominal discomfort following this suggestion, associated with a change in the duodenal motility pattern (Figure 7.2). Wolff and associates concluded, "... Unless the circumstances of administration of the agent preclude all opportunity for conditioning factors to operate, it may not be inferred that such reactions are allergic in nature."

In Britain, Pearson and Rix (1987) studied the effects of foods on psychologic or cognitive symptoms. In their study of 23 patients presenting to an allergy clinic, only four patients had objective food sensitivity based on double-blind, placebo-controlled challenges (Pearson et al., 1983). Pearson's group also demonstrated how some of the remaining patients with non-allergic food intolerances report false positive responses during open food challenges. For example, with open challenges, a middle-aged woman reported significant psychological symptoms to a multitude of foods that she believed she was sensitive to. Double-blind studies showed that her coffee and sugar intolerances were not organic. Pearson et al. (1983) concluded, "... It is no longer adequate to accept the presence of an organic basis for such symptoms without the demonstration of repeatability under double-blind conditions." This position had long been held by the scientific allergy community for the assessment of food sensitivities in the general population (May, 1976; Lessof et al., 1980).

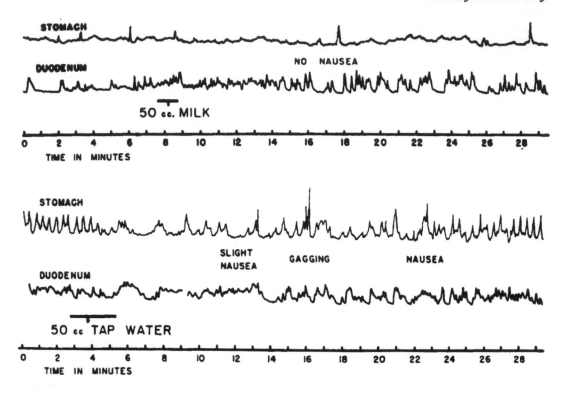

Figure 7.2 Changes in symptoms and gastric and duodenal activity in the patient after the administration of milk and of tap water with suggestion that the opposite food was being administered. The results show objective responses to suggestion of milk but no response to actual milk. (From Graham, D.T. et al., *J. Allergy*, 21, 478–486, 1950. With permission.)

Somatoform disorders

Many of the unsubstantiated reactions observed in EI patients are consistent with the response observed with placebo challenges. As recognition of the wide range of physical symptoms resulting from anxiety and depression grows, somatic symptoms (somatization) are recognized with increased frequency (Katon, 1986; Katon et al., 1982a,b; Rix et al., 1984; Stewart, 1990). Somatization has been correlated with increasing distress, disability, and maladaptive illness behavior (Katon et al., 1991). In this section, I shall present the psychiatric conditions in which the placebo response seems to play the most significant role, the somatoform disorders.

The somatoform disorders have at least five common, essential dimensions (Table 7.3). The different diagnosable disorders emphasize the relative contribution of illness worry, number of symptoms, the types and dramatic nature of symptoms, and age of onset. Multisystem symptoms unsubstantiated by objective tests and laboratory data qualify as criteria for the diagnosis of somatoform disorders. If symptoms are not attributable to alternative medical or psychiatric diagnoses, diagnosis of a somatoform disorder in EI patients is appropriate (Stewart, 1990).

Table 7.3 Dimensions Common to the Somatoform Disorders

1. Physical symptoms suggest a physical disorder, but no organic findings or known physiologic mechanisms can explain them.
2. Evidence, or a strong presumption, links symptoms to psychological factors or conflicts.
3. The prevalence is significantly higher among women than men, varying in ratios of 2:1 to 10:1.
4. Symptom production is not intentional (exclude malingering, factitious disorders).
5. Symptoms are clinically significant in that they require medical treatment or cause impairment in normal functioning.

Source: Diagnostic and Statistical Manual of Mental Disorders, 4th ed., American Psychiatric Association, Washington, D.C., 1994. With permission.

Hypochondriasis

The essential feature of hypochondriasis is preoccupation with fears of having a serious disease based on the misinterpretation of one or more bodily signs or sensations. The belief persists despite appropriate medical evaluation and reassurance, and while there may be poor insight, it has not reached delusional levels. A co-existing physical disorder may be present, but the fears of serious illness transcend it.

The cognitive distortions of hypochondriacal patients include preoccupation with bodily functions, amplification of sensations, and distortions of perception. There is often co-morbid obsessive-compulsive personality disorder as indicated in these patients' meticulous and lengthy presentations of their medical history and environmental attribution, most of which is irrelevant. There is usually co-morbid depression and anxiety. These are difficult patients to manage medically. They often have a history of doctor shopping, show poor adherence to prescribed treatments, and express dissatisfaction with prior care, and their attempts to manipulate and control the diagnosis and treatment may strain the doctor/patient relationship.

Somatization disorder

The essential feature of somatization disorder is a pattern of recurring, multiple, clinically significant somatic symptoms which begin before age 30. Patients describe their complaints in colorful, exaggerated terms, but specific factual information is often lacking. They are inconsistent historians so that a checklist approach to interviewing or self-report symptom checklists are less reliable than a thorough review of medical history, preferably with medical records, including treatment and hospitalizations to document a pattern of frequent somatic complaints. Laboratory test results are remarkable for the absence of findings to support the subjective complaints. Physical examination is remarkable for the absence of objective findings associated with the complaints. The patients' lives are as chaotic and complicated as their medical histories, and personality traits common to these patients include histrionic, borderline, and antisocial features. As might be expected, it is a chronic disorder that fluctuates with conflicts associated with the personality disorders and rarely remits completely (Ford, 1983, 1986; Lipowski, 1987, 1988; Oken, 1989; Quill, 1985). Somatization disorder is difficult to differentiate from general anxiety disorder and

Table 7.4   Symptom Criteria
for Somatization Disorder

Four pain symptoms from different sites
1.   Head
2.   Abdomen
3.   Back
4.   Extremities
5.   Chest
6.   Rectum
7.   During menstruation
8.   During sexual intercourse (dyspareunia)
9.   During urination

Two gastrointestinal symptoms
1.   Vomiting (exclude pregnancy)
2.   Abdominal pain
3.   Nausea
4.   Bloating (gassy)
5.   Diarrhea
6.   Food intolerance

One sexual symptom
1.   Sexual indifference
2.   Erectile or ejaculatory dysfunction
3.   Irregular menses
4.   Excessive menstrual bleeding
5.   Vomiting throughout pregnancy

One pseudoneurological symptom
1.   Impaired coordination or balance
2.   Paralysis or localized weakness
3.   Difficulty swallowing or lump in throat
4.   Aphonia
5.   Urinary retention
6.   Hallucinations
7.   Loss of touch or pain sensation
8.   Double vision
9.   Deafness
10.   Seizures
11.   Dissociative symptoms, e.g., amnesia
12.   Loss of consciousness other than fainting

Source: Diagnostic and Statistical Manual of Mental Disorders, 4th ed., American Psychiatric Association, Washington, D.C., 1994. With permission.

may coexist with panic disorder. Unlike persons with depressive mood disorders, whose somatization symptoms are exacerbated with depressive mood, individuals with somatization disorder have recurrent symptoms throughout their lives regardless of mood state.

The symptom criteria in DSM-IV have changed from DSM-IIIR in that the total number has been reduced from 13 to 8, but specific numbers of symptoms are required in each of four systems (Table 7.4). Individuals who have many of the other characteristics of somatization disorder but fewer than the number of required symptoms may be diagnosed with undifferentiated somatoform disorder.

Conversion disorder

> "The hysterical conversion reaction may provide the patient with a
> defense against anxiety, enable him to maintain his self-respect, and
> at the same time accomplish some purpose, the achievement of
> which respect for self would otherwise have been forbidden. It may
> make possible an escape from an intolerable situation; it may afford
> an exoneration for oneself, or enable one to evade some duty, shun
> a responsibility, express some spite, or realize some purpose that
> would bear the scrutiny of consciousness. ... The hysteric is always
> desirous never to disclose to himself the real nature of his illness, an
> effort that is successful, since the conscious personality has no access
> to the cause of the illness. Universal, too, is the tendency to project
> one's difficulties upon something tangible. Similarly, any physical
> basis for an explanation of the symptom is welcomed and assigned
> an important place." (Kolb, 1973)

Hysterical conversion reaction encompassed two diagnostic classifications, separated in
DSM-IV into conversion disorder and the dissociative disorders. Kolb (1973) noted that the
psychology of hysterical conversion reaction has been reviewed in many textbooks on
modern clinical psychiatry, and the formulations have not changed substantially since
Freud and Janet. Psychoanalytic explanations suggest that conversion symptoms serve to
prevent or lessen any consciously felt anxiety and usually symbolize the underlying
mental conflict that is productive of anxiety. Primary and secondary gain are the mechanisms for conversion reaction. Primary gain is the defense against conscious awareness of
anxiety-producing conflicts which are converted to symptoms. Secondary gain is a material, social, emotional, or other advantage that is contributed by the symptom. DSM-IV,
unlike its predecessors, does not imply that the symptoms necessarily involve such constructs, but it does require that psychological factors be associated with symptom onset.

One characteristic of the conversion reactor is that placebos or suggestion may relieve
or induce symptoms. This is helpful diagnostically with EI patients when symptoms may
be brought on and/or relieved by suggestion. Furthermore, it offers some additional
understanding as to why conversion reactors are attracted to certain controversial and
unsubstantiated medical practices, and why they are attracted to "messianic-healer" figures who give them the assurance or the illusion of an explanation and cure for their
ailments.

The overlap between the symptoms of conversion reactors and many EI patients is
striking. Historically, the sensory symptoms of the conversion reactor included anesthesia,
paresthesias, and disturbance of special sense organs, such as blindness and deafness.
DSM-IV has created specific subtypes based on symptoms associated with motor deficit,
sensory deficit, and seizures or convulsions (Table 7.5). Of all the multi-system complaints
of EI patients, the conversion symptoms are the most dramatic.

Each of the DSM-IV diagnostic criteria (Table 7.6) deserves comment in reference to EI
patients. First, EI patients claim impairment attributed to chemical sensitivities in two of
the most rudimentary cognitive processes, attention and memory. However, controlled
studies on neurobehavioral performance tasks have failed to show the reported impairment, suggesting that the symptoms were not associated with organic brain abnormalities
(Bolla, 1996; Fiedler et al., 1992; Simon et al., 1993). Traditionally, pseudoseizures were
considered to be another clear manifestation of a conversion disorder (Liske and Forster,

Table 7.5   Conversion Disorder Subtypes

1. *With motor symptom or deficit*
   Impaired coordination or balance
   Paralysis or localized weakness
   Difficulty swallowing or "lump in the throat"
   Aphonia
   Urinary retention

2. *With sensory symptom or deficit*
   Loss of touch sensation
   Loss of pain sensation
   Double vision
   Blindness or deafness
   Hallucinations

3. *With seizures or convulsions*
   Voluntary motor or sensory components

4. *With mixed presentation*
   Symptoms of more than one category

Source: Diagnostic and Statistical Manual of Mental Disorders,
4th ed., American Psychiatric Association, Washington, D.C.,
1994. With permission.

1964) resulting from abnormal emotional states (Volow, 1986). A conversion reaction is likely to involve a single symptom during a given episode but may vary in site and nature if there are subsequent episodes. By comparison, objective physical illness is usually characterized by replicability of specific signs and symptoms.

Second, psychological factors associated with an episode of conversion symptoms are evident when they enable the EI patient to obtain social support that otherwise might not be forthcoming. This could be from a spouse or children. It can also come in a physician's office during an evaluation, especially a provocation challenge. Most difficult to judge is the third criteria, whether symptoms are feigned. EI patients meet the fourth criteria by definition in that objective medical and laboratory tests are normal.

The fifth criteria is particularly relevant to the Americans with Disabilities Act (ADA) which prohibits discrimination against the disabled and entitles reasonable accommodations. EI advocates have lobbied legislatures, particularly in New Mexico, to have EI recognized under ADA. And well it should be. But conversion disorder, not chemical sensitivities, is the appropriate basis for the disability. Desire for medical confirmation of their beliefs in EI is often the primary motivation for EI patients to seek consultation. The patient often presents as debilitated mentally, physically, and socially. In the most extreme cases, the EI patient is sequestered in the home or withdraws to a hermitage existence in the mountains or desert where little feedback from reality is available.

EI patients meet the sixth criteria by virtue of having multi-system complaints. While there is often sexual disturbance in the more severe cases (especially those who report sexual abuse as children), the symptom pattern is not limited to sexual dysfunction. Whether symptoms are better explained by another psychiatric disorder is open to clinical judgment.

Reported rates of conversion disorder vary widely, with about 1 to 3% being diagnosed in outpatient referrals to mental health clinics. There is a puzzling statement about

Table 7.6 Criteria for Conversion Disorder

1. One or more symptoms or deficits affecting voluntary motor or sensory function suggest a neurological or other general medical condition.
2. Psychological factors are judged to be associated with the symptom or deficit because the initiation or exacerbation of the symptom or deficit is preceded by conflicts or other stressors.
3. The symptom or deficit is not intentionally produced or feigned (as in factitious disorder or malingering).
4. The symptom or deficit cannot, after appropriate investigation, be fully explained by a general medical condition, or by the direct effects of a substance, or as a culturally sanctioned behavior or experience.
5. The symptom or deficit causes clinically significant distress or impairment in social, occupational, or other important areas of functioning or warrants medical evaluation.
6. The symptom or deficit is not limited to pain or sexual dysfunction, does not occur exclusively during the course of somatization disorder, and is not better accounted for by another mental disorder.

Source: Diagnostic and Statistical Manual of Mental Disorders, 4th ed., American Psychiatric Association, Washington, D.C., 1994. With permission.

prevalence in the American Psychiatric Association DSM-III-R which was not carried over into DSM-IV:

> "Apparently common several decades ago, it is now rarely encountered. Most cases are seen on neurology or orthopedic wards and in military settings, especially in time of warfare."

What is most striking about this description is the disappearance of the disorder several decades ago. Where did it go? Most of the earlier cases were seen on neurology or orthopedic wards, and this is consistent with some of the symptoms of EI patients. For many years, EI patients have been primarily seen, not on neurology or orthopedic wards (although some who present with chronic pain symptoms are still treated by orthopedic physicians), but rather in allergy clinics and practices. Until recently, the toxicogenic postulated allergic and immunological mechanisms and allergists have come to face an ever-increasing number of cases of what was believed to have disappeared several decades ago. With the expansion of the toxicogenic theory to include hypotheses about brain mechanisms, neurologists may find more of these patients at their door again.

An EI case of pseudoseizure

A 44-year-old mother presented to us with a history of 60 multi-system complaints including generalized tonic/clonic and myoclonic jerks, feelings of numbness and tingling in the extremities, disorientation and loss of concentration, slurred speech, and feelings of "semi-unconscious" and "semi-paralysis", all of which were self-reported to be triggered immediately after ambient exposure to specific fragrances (Staudenmayer and Kramer, in preparation). She had been off work for eight years, classified as disabled by the workers' compensation authority based on a clinical ecologist's diagnoses of hypoglycemia, premenstrual syndrome, chronic fatigue syndrome, fibromyalgia, intolerance of all foods that were not "organic", and toxic encephalopathy, all secondary to chemical sensitivities. She self-dated the onset of chemical sensitivities to age 34 after she had read about chemical sensitivities and was influenced by a similarly afflicted friend. The association of her own symptoms to chemicals came while she was putting on lipstick and concurrently started a migraine. Upon further reflection, she dated the onset to age 12 when she had difficulty with comprehension and reading in school and an infection of mononucleosis which she attributed to her first dental mercury-amalgam fillings. She also attributed onset of pre-menstrual syndrome at age 32 to pesticide exposure. Onset of "seizure" symptoms induced by chemical sensitivities was dated to age 37, three years after the initial onset involving migraine.

Video-EEG monitoring with a standard, clinical 17-channel EEG was conducted in a hospital neurodiagnostic laboratory with an epileptologist in attendance. Results were normal. An open challenge followed with an air deodorant presented as reliably inducing a reaction. Within 20 seconds of spraying the air deodorant in the room, she demonstrated a dramatic clinical response with generalized tonic/clonic and myoclonic jerks, body rigidity, and body tremor followed by head jerking and diffuse tremulousness with decreased communicative responses.

After recovery, another open challenge was conducted with a self-selected perfume. Within 20 seconds of taking several whiffs she responded with diffuse tonic/clonic jerking as well as diffuse tremulousness of the arms, legs, and head. She then developed diffuse twitching and clenching of the mouth followed by focal motor twitching involving the right upper extremity, generalized tonic/clonic jerking with decreased responses, slurred speech, and inability to repeat numbers given to her. She later reported feeling weak on the right side and "semi-paralyzed" and had difficulty moving her right arm.

She appraised her responses to the air deodorant and the perfume as "typical sei-zures". Throughout the entire time of these provocative challenges the EEG remained entirely normal. Standardized psychological testing using the Minnesota Multiphasic Personality Inventory-2 (MMPI-2), and the Millon Clinical Multiaxial Inventory-II (MCMI-II) showed long-standing psychological maladjustment and personality disorders includ-ing paranoia, dependency, and obsessive-compulsiveness. Her MMPI-2 results suggested a somatoform disorder consistent with her firm conviction in the overvalued ideas about EI.

Social history revealed a psychologically abusive mother and a stepfather who made sexual advances toward her. She had her first child at 16 by a boyfriend whom her mother alienated. At the time of the alleged onset of chemical sensitivities at age 34, she was pregnant with her youngest daughter and had a stressful job running a boys' home with her husband. She resigned from the job because of her first daughter's health, who she alleged also suffered from EI and was home-schooled by her mother. Psychological vulner-ability and primary and secondary gain explained how she came to accept beliefs about EI and why she was susceptible to iatrogenic influence of the clinical ecologists. The pseudoseizures and their attribution to fragrances facilitated projection of her anxieties away from herself onto a physical external environment which she perceived as danger-ous, allowing her to withdraw into a delusional world into which she pulled her daughter.

Conclusion

The placebo response is woven into the history of modern medicine. For every biological or physiological mechanism hypothesized, psychological mechanisms must be ruled out. For every physical cause speculated, psychological mechanisms must be controlled. For every explanation of EI, human aspects of the patient and doctor must be assessed before valid conclusions about the toxicogenic effects of an environmental agent can be incriminated. When all controls are in place, toxic mechanisms considered, and alternative medical diagnoses ruled out, there is a wealth of psychiatric and psychologic insight that can explain what seemed so intuitively obvious to the truck driver who did road-side spraying for the Texas Department of Transportation.

chapter eight

Learned sensitivity

This chapter presents two types of learning that have been postulated by the psychogenic theory: Pavlovian conditioning (also called "classical conditioning") and higher order abstract thought processes, generally called "thinking". Conditioning is a much simpler form of learning applicable to animals, but limited in explaining human behavior. Also presented are human studies in which conditioning has been suggested as an explanation for overgeneralized responses in cases with well-documented environmental exposures. I will demonstrate why conditioning cannot explain most cases of EI and why cognitive learning is a better explanation. Last, I will present learning and other processes associated with phobias to illustrate the complexity of the avoidance and environmental reaction responses demonstrated by EI patients.

Pavlovian classical conditioning

Pavlovian classical conditioning (Figure 8.1), as applied to the responses of EI patients, defines a symptom as a conditioned response (CR) to a conditioned stimulus (CS), usually an odor, which was associated with a documented toxic exposure (UCS) that resulted in an unconditioned symptom response (UR), e.g., sinus irritation. Classical conditioning is often explained with a dog because Pavlov, the Russian physician and physiologist, was studying gastric activity in dogs and applied conditioning to activate the response without feeding. For example, a dog given meat (UCS) orients, shows anticipatory motor responses, and salivates (UR). During the acquisition phase of conditioning, a bell (CS) is rung immediately before the dog is fed and repeatedly over several trials to establish the link between the bell and the meat, the CS-UCS association. During the conditioned response phase, the bell is rung without the food and the dog still salivates (CR), but not as much as to the food. With repeated trials of the bell without food, salivation decreases until it is extinguished. During the conditioned response phase, the dog responds to the bell but not to every other sound in the laboratory; stimulus generalization is quite limited. The dog limits his response to salivation and the orienting and motor responses characteristic of feeding; the conditioned response is quite specific.

Animal conditioning of the immune system

Conditioned immunomodulation is recognized as being an important line of evidence for the involvement of psychologic and psychophysiologic factors in what was traditionally

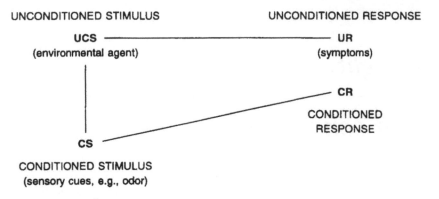

Figure 8.1 Pavlovian classical conditioning model for an EI reaction.

believed to be a domain of immunology proper (Ader, 1981; Ader et al., 1991). Animal-conditioning studies show that the central nervous system (CNS) affects immune system responses at virtually every level, including antibody production, lymphocyte proliferation, delayed hypersensitivity, natural killer cell activity, graft-vs.-host response, total white blood cells count, T-helper cell:T-suppressor cell ratio, arthritic inflammation, and systemic lupus erythematosus (Djuric and Bienenstock, 1993). The brain and the immune system form a bidirectional communication network mediated by intercellular messengers called cytokines, with interleukins which communicate between white blood cells being an example (Maier and Watkins, 1998). Histamine release, a mechanism of hypersensitivity, can be centrally mediated. In one series of animal studies, Pavlovian conditioning experiments showed increased plasma histamine levels in response to a conditioned stimulus (Russell et al., 1984). Since histamine is produced by many tissues and cells other than mast cells, additional studies were necessary to link histamine specifically to mast cells to show conditioning of a hypersensitivity response. In one particularly elegant study, Bienenstock and colleagues showed conditioned increase of rat mast cell protease II (RMCP II), the enzyme that is restricted to mucosal mast cells unique to the immune system (MacQueen et al., 1989). A summary of the study follows (Djuric and Bienenstock, 1993):

> "In this study, rats were sensitized to egg albumin. Exposure to audiovisual cues (CS) was followed by a subcutaneous antigen challenge (UCS). Eleven days after 3 weekly CS-UCS pairings a test trial was performed. Exposure to audiovisual cues followed by an injection of saline [placebo] (presentation of CS alone) resulted in an increase of serum RMCP II that was comparable to the increase observed in the group of animals challenged with antigen but significantly higher than the levels of RMCP II found in the appropriate control groups [Figure 8.2]. The finding of conditioned mast cell degranulation implicates a functional relationship between the elements of the central nervous system and the elements of the immune system that are involved in immediate hypersensitivity. The information that was processed and stored by the central nervous system was used by the organism in order to modulate mast cell function."

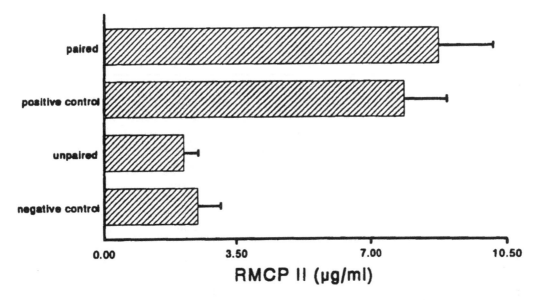

Figure 8.2   Results of the conditioning experiment by MacQueen et al. (1989). The positive control group was challenged with antigen and the conditioning stimulus. The unpaired group consisted of animals in which the conditioning stimulus was separated from unconditioned stimulus by a period of 24 hours in order to prevent conditioning. The negative control group was presented with the conditioning stimulus together with saline injection to control for the injections themselves. The paired group received only the conditioning stimulus without antigen (but with saline injection). RMCP II was measured in serum radioimmunoassay. Mast cells from paired group of animals released mediators in response to the CS. (From Djuric, V.J. and Bienenstock, J., *Ann. Allergy.*, 71, 5–14, 1993. With permission.)

Conditioning after documented exposure

Pavlovian conditioning is sufficient to describe the learned responses to perceptual stimuli associated with harmful toxic exposure. Studies have demonstrated conditioning of odors associated with acute overexposure to irritant or acutely toxic chemicals (Table 8.1). Some of these studies proposed new labels for conditioning specific to the observed learned effects (Table 8.2).

Tabershaw and Cooper (1966) reported on patients with documented overexposure to pesticides who experienced recurrent headaches and nausea in response to odors of solvent-containing products. They called this phenomenon "acquired intolerance to pesticides". Schottenfeld and Cullen (1986) describe an industrial worker who was repeatedly overexposed to irritant levels of phosphoric acid vapors that left him responding with complaints of respiratory irritation and chest pain when exposed to a multitude of odorant chemicals both in the work place and in the home. They labeled this learned effect "atypical post-traumatic stress disorder", a term that may sound intuitively appealing but has little explanatory value and is not a diagnosis because of the modifier "atypical". Shusterman et al. (1988) described workers who developed intolerance to odors of phosphine gas and formaldehyde after acute overexposure Subsequently, these workers showed conditioned responses (chest pain, paresthesias, lightheadedness, air hunger) to an odorant concentration of the particular gas or vapor, but without generalization to multiple other odorants.

Table 8.1 Studies of Conditioning Effects After Documented Exposure

Study	Chemicals UCS	Symptoms CR	Triggers CS
Acute exposure			
Tabershaw and Cooper (1966)	Pesticides	Headache, nausea	Odors of solvents
Schottenfeld and Cullen (1986)	Vapors of phosphoric acid	Chest pain, respiratory irritation	Multiple odorants
Shusterman et al. (1988)	Phosphine gas Formaldehyde	Chest pain, paresthesias, lightheaded, air hunger	Specific to gas and formaldehyde
Chronic exposure			
Bolla-Wilson et al. (1988)	Pesticides Solvents	Headache, fatigue, memory problems, irritability, nausea	Similar odors
Dager et al. (1987)	Organic solvents	Panic	Organic solvents

They called this response "behavioral sensitization to irritant/odorants", an appropriate description of a learned effect.

Pavlovian conditioning has also been proposed to explain chronic exposure to low levels of chemicals, absent of an acute overexposure. Bolla-Wilson et al. (1988) described conditioning of odor-triggered symptoms, including headaches, chest pain, nausea, memory problems, fatigue, and irritability triggered by odors from common environmental substances such as pesticides and solvents. Dager et al. (1987b) reported panic-like symptoms among workers exposed to organic solvents. While onset occurred in the workplace, symptoms were subsequently triggered in other contexts as well. Shusterman and Dager (1991) have proposed the descriptive term "odor-triggered panic attacks" to be applied to symptoms of autonomic nervous system arousal elicited by odors, regardless of exposure history, documented acute overexposure, or chronic low-level exposure. In summary, all of these terms refer to acquired intolerances to a variety of chemicals, with symptoms triggered by detectable odors that were associated with identifiable exposure, allowing for generalization to similar irritant/odorants. In these studies there were identifiable UCS-UR toxicologic links which created conditioned associations between perceptual cues such as odor, sight, and sound (CS) and the pain and fear responses suffered (CR) that meet the criteria of a conditioned panic disorder (Wolpe and Rowan, 1988).

Thinking and learning

"Although inborn reflexes and fixed action patterns play some role in human behavior, they are over-shadowed by learning as a basis for guiding action. Through symbolic, associative processes, stimuli that are inherently meaningless come to exert profound influences over the emotional systems of the brain." (LeDoux, 1987)

Table 8.2 Labels for Learned Effects Observed After Toxic Exposure

Study	Label
Tabershaw and Cooper (1966)	Acquired intolerance to pesticides
Schottenfeld and Cullen (1986)	Atypical post-traumatic stress disorder
Shusterman et al. (1988)	Behavioral sensitization to irritant/odorants
Bolla-Wilson et al., (1988)	Classical, Pavlovian conditioning
Shusterman and Dager (1991)	Odor-triggered panic attacks

In addition to experiential conditioning, human learning is mediated by cognitive processes which create ideas and regulate the memory links among them to allow for creative thinking. The neuronal encoding of Pavlovian classical conditioning is different from that of associative learning (Bailey and Kandel, 1993; Kandel and Hawkins, 1992; McNaughton, 1993). Learned sensitivity is a term used to describe expressions of environmental intolerances through physiologic mechanisms that are centrally mediated but not restricted to the Pavlovian conditioning model.

Conditioning onset of asthma

Conditioning was hypothesized to explain some cases of asthma exacerbated by emotions because clinical histories of asthmatics often contain examples of asthmatic attacks following emotional events. In the 1950s, Dekker et al. (1957) conducted several studies of classical Pavlovian conditioning with asthmatic patients using emotional stimuli from individual patient histories as the conditioned stimulus. In one study of 12 asthmatics, six failed to react at all and three reacted only with transient decrease in breathing capacity, but three did respond with frank asthma attacks. The great individual differences did not fit well with their expected conditioning effects. In a related study of two patients who reacted to an emotional stimulus and who were studied in detail, the authors concluded that while there were some learning effects that seemed in line with a conditioning mechanism, they did not represent classical examples of such a mechanism. In the discussion of the results of the patients who reacted with asthma attacks in the earlier study, there was a rather revealing observation (Dekker and Groen, 1956):

> "Consequent to the provocation tests these patients related traumatic life experiences, emotional phantasies, and disturbing dreams. No attempt will be made to give a psychiatric interpretation of these observations, but in a number of instances it was quite obvious that the environmental asthmatogenic stimulus was associatively related to former traumatic life experiences. The clinical picture of the patients during these psychogenic attacks was indistinguishable from 'spontaneous' attacks or from attacks provoked by the inhalation of allergens. They could be aborted by the administration of thiacinamine or isoprenaline."

Conditioning had been proposed as an alternative to the prevailing psychoanalytic explanations during the 1930s and 1940s (French and Alexander, 1941) for emotional effects on asthma. The results of these conditioning studies did not disprove the psychoanalytic theories. Rather, they demonstrated how both theories were struggling to model cognitively mediated learned effects which could not be explained by the postulates and mechanistic hypotheses of either theory.

Cognitive learning

Animal models of classical Pavlovian conditioning rely on sensation, a mechanism which has its evolutionary origins in the reptilian brain. Humans have some of the same capacities as animals, and to that degree the mechanisms and laws applicable to animal conditioning also apply to human learning. Through the evolution of a neocortex, thinking and complex reasoning are the prerogatives of humans. Humans have the unique capacity for abstraction, which allows them to contemplate their own thoughts, to plan intentional behavior, to bemoan their own condition and destiny, and to have an existential crisis. This uniqueness requires different conceptual mechanisms and laws of learning to explain behavior, both overt and covert, which have come under the purview of cognitive science (Sperry, 1988, 1993).

Pavlovian classical conditioning cannot explain the unsubstantiated, inconsistent, and overgeneralized reactions observed in EI patients. For many of the purported reactions, the UCS-UR pairing believed to have initiated "chemical sensitivity" never occurred, or if there was an exposure the dose was too low to elicit an UR, at least according to toxicologic criteria. If there is no UCS, there cannot be a CS-UCS pairing. When the CS, usually an odor, elicits a symptom, it is unrelated to the attributed UCS, at least by any pairing as defined in classical conditioning. The UCS-CS pairing is nonspecific, and the CS involves stimuli that do not even remotely resemble anything in the initial exposure environment. The CS-CR pairing is without limits; any environmental stimulus can elicit any number of multi-system complaints, in any combination over repeated trials. Classical conditioning cannot explain the complexity of human learning and memory involved in the appraisal and attribution of a "reaction". Pavlov never intended it to.

Less known about Pavlov is his interest in a higher order, symbolic form of learning which he called the "second-signalling-system", or what today is referred to as language and cognition. Pavlov (1951) succinctly described the central role beliefs have in learned sensitivity:

> "Before the appearance of the family of *Homo sapiens* the contact of the animals with the surrounding world was effected solely by means of direct impressions produced by its various agents which acted on the different receptor mechanisms of the animals and were conducted to the corresponding cells of the nervous system. They were the sole signals of the external objects. In the future human beings, there emerged, developed and perfected, signals of the second order, signals of these initial signals, in the shape of speech — spoken, auditory, and visible. Ultimately these new signals began to denote everything taken in by human beings directly from the outer, as well as from the inner world; they were used not only in mutual intercourse, but also in self-communion. This predominance of the new signals was conditioned [learned], of course, by the tremendous significance of speech, although words were and remain but second signals of reality. We know, however, that there are large numbers of people who, operating exclusively with words and failing to base

> themselves on reality, are ready to draw from these words every possible conclusion and all knowledge, and on this basis to direct their own life as well as the lives of others."

Modern theories of Pavlovian classical conditioning postulate that cognitive mediation is essential to learning. Contingency between stimulus and response is no longer necessary nor sufficient to establish a learned response. Rather, perception of such a contingency is sufficient to establish a link. Conscious and automatic cognitively mediated processes — such as expectations, beliefs, contextual contingencies, and especially inferences about causal relationships — are the processes that define conditioning (Rescorla, 1988). The cognitive reframing of classical conditioning reflects the evolution of learning theory in psychology. It does not appear well known outside psychology in that the original behavioral version of Pavlovian classical conditioning was proposed to explain EI (Kurt, 1995b; Spyker, 1995).

The logical error of applying behavioral explanations of Pavlovian conditioning to explain EI may stem from overgeneralizing the analogy between documented conditioning effects in human cases of identifiable toxic exposure, presented above, to the usually undocumented exposures and unreliable symptom responses of EI patients.

1886 historical example

In humans, the association of the CNS with mast cells suggests that psychoneurological mechanisms could produce non-allergic (non-IgE-mediated) mast cell mediator release (Bienenstock, 1991). There is an early historical account in the medical literature which corroborates this hypothesis. MacKenzie (1886) summarized cases (mostly in Latin and as early as the 16th Century) in which odors triggered hysterical reactions that even included pseudoseizures. For example, as early as 1557, Amatus Lusitanus related the case of a Dominican monk, who, whenever he perceived the odor of roses or saw them at a distance, immediately fell to the ground unconscious.

MacKenzie also presented the case of a 32-year-old woman who believed she was sensitive to numerous environmental factors (Table 8.3), particularly the odor of roses and hay. While on a visit to the country, she caught sight of a distant hayfield and was immediately overcome with a reaction. The patient dated the onset of her trouble to age 6 to 8. She described the general temperament of her family members as "nervous". For several days prior to the onset of a reaction, she suffered from depression and general malaise. The focal complaints of sinus symptoms were invariably followed with multi-system complaints, including general depression of the spirits and system, constant malaise, incapacity for mental work and household duties, and general irritability. During one of her reactions, she became unconscious, with generalized muscle rigidity, followed by tremors [an account consistent with a pseudoseizure]. She had tried almost every treatment known to the medical profession and the laity, including a host of quack specifics, all to no avail. She had even spent several summers at a well-known "hay fever resort" without the slightest improvement in her condition.

Doubting that pollen could trigger this woman's symptoms, and being scientifically minded, the good doctor proceeded to conduct a single-blind provocation study under deception. For the purpose of the experiment, he obtained an artificial rose so exquisite that it presented a perfect counterfeit of the genuine flower. To exclude the possibility of environmental contamination, each leaf was carefully wiped, so that not a single particle of foreign matter was hidden within the convolutions of the artificial flower. Upon arrival

Table 8.3   Triggering Agents Reported by a Patient
With "Rose Cold"

1.   Heated, sultry condition of the atmosphere
2.   Sudden changes in the temperature
3.   Dampness
4.   Exposure to "night air"
5.   Electrical disturbances of the atmosphere
6.   Exposure to a high wind
7.   Sudden excitement
8.   Fright
9.   Physical overexertion
10. Worry
11. Overloading the stomach, but has a voracious appetite
12. Inhalation of common dust, cinders, and gases,
     especially those given off in the combustion of coal
13. Presence of odor of strong perfumes, as, for example,
     articles of the toilet; the smell of tobacco smoke —
     anything, in fine, that has a pronounced or penetrating
     and heavy odor, as the tuberose, lily, etc.
14. Almost any stimulant, taken internally, such as
     brandy, whisky, beer, etc.
15. Handling peaches
16. Ingestion of fruit; quinia and morphia
17. Presence of odor of hay and roses — especially active in
     exciting the attack

Source: Adapted from MacKenzie, J.N., *Am. J. Med. Sci.*, 91, 45–57, 1886.

for her next visit to MacKenzie's office, the patient had no symptoms, and a medical exam revealed no abnormal signs. Sitting in front of her, the doctor took the artificial rose from behind a screen, held it in his hand, and continued the conversation. Immediately, she experienced symptoms and signs including itchy throat, congestion, hoarseness, itchy eyes, redness of the conjunctivae, tearing, redness of the auditory meati, runny nose, and shortness of breath. [Today we know these are consistent with mast cell mediator release.] When told that the rose was an artificial one, her amazement was great, and her incredulity was only removed upon personal examination of the counterfeit flower. She returned to the doctor's office a few days later with a genuine rose in which she buried her nose and inhaled the pollen without incident. MacKenzie (1886) concluded:

> "The chief lesson to be derived from the study of this particular case
> (i.e., so far as the psychical element is concerned) is that it opens our
> eyes to the fact that the association of ideas sometimes plays a more
> important role in awakening the paroxysms of vasomotor coryza
> [rhinitis] than the alleged vital property of the pollen granule."

Learning to control allergies

MacKenzie's example of "rose cold" illustrated not only the power of suggestion on induction of symptoms, but also how the symptoms can be overcome with a change in cognition. This observation was corroborated by Cohen et al. (1993–94) in a controlled study of patients with allergy skin-tests positive to ragweed pollen, indicating an IgE-mediated response. Patients were matched and randomly assigned to a control group or an experimental group receiving intervention with a 3-week program comprised of various relaxation and guided imagery procedures to promote relaxation and psychological integration. To modulate cellular activity of the immune system, subjects were shown 35-mm slides depicting cellular activity during a non-allergic immunological response to ragweed pollen. Nasal challenges to ragweed-pollen abstract were conducted before and after intervention. Dependent measures were subjective symptoms (runny nose, nasal congestion, and itchy nose or throat) and objective immunologically generated biochemical mediators in nasal secretions, histamine, and TAME (tosylarginine methyl ester)-esterase. Histamine release is indicative of mast cell degranulation and influences sneezing. TAME-esterase release is indicative of vascular permeability, a manifestation of an allergic response and a more consistent indicator of allergic response. Results showed significant lower levels of TAME-esterase pre to post for the group receiving behavioral intervention. No significant changes were observed in histamine levels or on symptom report. These results show concurrent validity with two established treatments of hay fever, immunotherapy and corticosteroids, which affect TAME-esterase in nasal secretions more than they do histamine in secretions and symptom report. The altering effects of stress on immunomodulation are well known, but this pilot study suggests that allergic responses can be controlled by cognitive intervention and relaxation.

Fear and anxiety

The different implications of conditioning and cognitive/belief explanations of learning are apparent in the study of fear and anxiety. Conditioning paradigms of aversion with animals study fear responses to well-defined and perceivable unconditioned stimuli (e.g., shock or high-dose pharmaceuticals) that invariably elicit discomfort or pain, the unconditioned response. In the experimental laboratory procedure, the conditioned stimulus is readily perceivable to the animal, usually a tone, light, taste, or odor that is paired with the UCS during acquisition trials. The study of acute fear responses is the focus of these experimental procedures, and no distinction is made between fear and anxiety (Rachman, 1990), as it is in studies of humans.

The conditioning paradigm is much better suited to the study of fear than anxiety. Symbolism and abstraction are beyond the scope of classical Pavlovian conditioning theory. Until relatively recently, cognition, conceptual schema, and causal attribution were not constructs in behavioral learning theory. Whether a cognitive-behavioral spin can revive conditioning theory or whether cognitive processing theories of learning will replace behavioral, stimulus-response formulations is a theoretical issue which need not concern us here. The point I wish to emphasize is that classical Pavlovian conditioning alone cannot explain the fears, anxieties, and phobias of most EI patients. Higher order cognitive processes, both conscious and unconscious, are involved. These include perception and appraisal, reasoning and judgement, conceptual schema, psychologic defenses, and false belief systems about attribution. The distinction between learned fear, induced in experimental conditioning paradigms with animals, and general anxiety neurosis in humans is essential to understanding not only the psychology of EI patients, but also the psychology of other aspects of EI.

The actions taken to avoid frightening situations are not considered irrational or unusual. In Cannon's fight-or-flight model, running from the bear would seem a rational, although not necessarily the most effective, solution according to current forestry wisdom. Regardless, the volitional decision to stand still and remain calm is different from the freezing terror response associated with physiologic collapse mediated by parasympathetic activation. When fight, flight, or freeze physiologic responses occur spontaneously in the absence of an identifiable threatening object in humans, acute anxiety or panic may be a more appropriate theoretical construct than fear, even though the behavioral and physiologic responses may be indistinguishable.

Anxiety is defined by worry and feelings of apprehension without well-defined stimuli. Anxiety patients usually report feelings of uneasiness, hypervigilance, and general arousal without knowledge of what it is that they are reacting to. The basis for fear often remains amnesic or inhibited. Anxiety is the emotional and somatic reaction to the expectation of danger, which is mediated cognitively. The apprehension of threat and the expectation of harm, characteristics of anxiety, are not the result of an immediate danger, nor are they always well defined as it for fear (May, 1979).

Learning and phobia

> "The predominant feature of phobias is a persistent avoidance behavior secondary to irrational fears of a specific object, activity, or situation. Although fears are ubiquitous to the human condition, phobias are unreasonable and unwarranted fears given the actual dangerousness of the object, activity, or situation avoided." (Nemiah and Uhde, 1989)

The main determinants of phobias are learned, and the proposed systems and mechanisms are the same as those suggested for learned sensitivity. The three ways to acquire phobias are conditioning, observational learning or modeling (e.g., mass psychogenic illness), and through verbal or written information which encompasses suggestion and iatrogenic influence. Like learned sensitivity, phobias cannot be explained by conditioning alone. A common feature of phobias is their tendency to spread to objects, situations, and functions which have no common features on which to base stimulus generalization (Nemiah and Uhde, 1989), similar to what EI advocates call "spreading". Psychodynamically, phobias do not eliminate emotional symptoms from consciousness, but they project their etiology onto agents external to the self, creating a self-illusion that the danger lurks elsewhere.

EI patients more often voice and displace their anger and hostility onto fantasized "polluting perpetrators". The role of the victim is often played out dramatically. The self-awareness or insight that the phobias are irrational is lost to "true believers". The selectivity of belief reflected in trendy, fad diagnoses and unsubstantiated treatments can be explained only by sociocultural influences and by higher order abstraction and thinking. The politics of victimization intersect with the distorted beliefs in EI and can be seen played out in several arenas including courtrooms of civil litigation, subcommittee hearing rooms of legislators, compensation and insurance hearings, before media cameras that seem riveted on sensationalism, and within the annals of pseudoscience (including the Internet) promoting junk science and misinformation.

Janet described three major categories of circumstances associated with phobic anxiety: phobias of objects, situations, and function. Janet's classification is reflected in DSM-IV (American Psychiatric Association, 1994) as specific phobia, agoraphobia, and social phobia. Some EI patients may have manifestations of any one or all three in their history prior to their alleged EI. Once they believe they suffer from EI, they tend to manifest chemophobic ideation and avoidance behaviors that do not meet the diagnostic criteria of any one of the phobias, but resemble all three. The gender feature of phobias, as with all of the anxiety disorders, underlines the prevalence in females.

DSM-IV diagnostic criteria for specific phobia include an excessive, unreasonable fear to a specific object or situation that invariably provokes an immediate, intense anxiety response. The phobic situation is avoided or endured with extreme anxiety, disrupting normal life, especially freedom of action. The resulting disability may be as crippling as a major physical illness. The intensity of the fear may not always relate predictably to the phobic stimulus, suggesting cognitive mediation. This is analogous to EI reactions which also show variability in the intensity of response to a putative chemical agent over repeated exposures, as well as variability in the type and intensity of symptom expression. Phobias can explain the same variability as the toxicogenic postulates of adaptation/deadaptation.

Specific phobias frequently co-occur with other anxiety disorders. The fear associated with specific phobias is limited to specific objects and situations which makes them different from the reactions to multiple chemicals, the effects of which are ever changing and spreading to new chemicals. One could argue that chemicals as a class of objects could be a specific phobia analogous to rattlesnake phobia, which generalizes to all species of snakes. But "chemophobia" was not included as an example in DSM-IV and such an inclusion seems unwarranted given the dynamics of EI patients. It has been my clinical experience that behavioral desensitization techniques, which are effective in overcoming simple phobia, are not usually effective with EI patients. However, successful behavioral treatment has been reported in EI patients with odor aversion (Amundsen et al., 1996).

Agoraphobia is popularly interpreted as being a fear of open spaces, but it has wider implications. Agoraphobic patients are generally thrown into a state of trepidation and severe anxiety over symptoms that may be incapacitating or embarrassing, leaving them feeling helpless or out of control with help unavailable. So infrequently is agoraphobia observed without panic attacks in clinical practice (less than 5%) that some investigators have questioned its validity as a distinct clinical syndrome (Nemiah and Uhde, 1989). Some of the aversions and avoidance behaviors displayed by EI patients have similarities to agoraphobia, but the associated distorted beliefs about environmental dangers go beyond it. It seems that EI patients have filled the empty spaces characteristic of agoraphobia with a host of perceived environmental poisons. Both patients with agoraphobia and EI report histrionic symptoms which leave them feeling as if they were "falling apart".

Central to social phobia is a concern of appearing shameful, stupid, or inept in the presence of others, resulting in embarrassment. Patients who suffer social phobia fear that their behavior will be the focus of scornful scrutiny by those around them, an expectation that often further impairs their performance. To compensate, patients may develop perfectionistic habits, even in minor aspects such as appearance. Shyness and blushing have been suggested as early variants of the illness. For EI patients, social phobias may be further rationalized as manifestations of their chemical sensitivities, often to justify avoidance of family, friends and loved ones.

The psychodynamics of phobia

Clinical manifestations of phobia are often difficult to discriminate from other conditions such as hypochondria or obsessive-compulsive anxiety disorder. When a learning explanation is not sufficient to explain symptoms and behaviors, psychodynamic explanations are often more helpful. Responses to a large quantity of unassimilated information are primarily of two kinds: withdrawal, which is characteristic of phobias, and accommodation, which is characteristic of delusions and overvalued ideas. Selective inattention, denial, and selective avoidance of stressful issues and situations are defenses used to some extent by all people. However, these processes are excessive and unreasonable in EI patients, given the unsubstantiated nature of their appraisals of harm and attribution to the physical-environment. Withdrawal, selective inattention, selective avoidance, apathy, and resignation are behaviors characteristic of phobias that may be employed to prevent an increase in the quantity of unassimilated ideas. Rather than facing an overwhelming reality, the EI patient seems prone to avoidance.

In general, persons with phobias tend to avoid situations likely to elicit memories and experiences in their own unique areas of limited coping resources. For example, EI patients may highlight selective avoidances, such as social relations by living a hermetic existence, sexual relations by alleging semen sensitivity, competitive or demanding work situations with professing "sick building syndrome" scenarios, and so on. In simple phobias, avoiding the specific situations unique to the phobia serves to mitigate anxiety.

In EI patients, the phobias are often more generalized, representing unassimilated concepts in several areas of psychosocial functioning. The cumulative effect of unassimilated ideas in several conceptual categories creates high levels of anxiety. Cognitively, uncertainty about which phobic object or situation is threatening can create generalized fear to all objects and situations. Such a generalized dysfunction creates a precarious position for an EI patient, as it is impossible to avoid all of the feared objects and situations. To accommodate, distorted and overvalued beliefs are created, and elaborate cognitive and behavioral rituals are practiced such as reading a newspaper under glass to ward off the perceived ill effects the "offgasing" print chemicals. As for most phobic disorders, the avoidances and reactions of EI tend to be chronic, with frequent recurrences of symptoms. Agoraphobia is notoriously difficult to manage and may cause the patient to lead a severely constricted life of psychologically induced incapacitation.

Freud viewed neurotic phenomena such as phobias as being the result of heightened arousal caused by the forces of repression, which blocked normal avenues of emotional discharge and substituted physical and psychological symptoms. Freud made a clinical separation of hysteria from phobias and obsessions. In hysteria, the affect associated with an idea unacceptable to the ego was converted into a somatic symptom, and both the idea and its related affect disappeared from consciousness. In phobias and obsessions, the offending affect (e.g., anger, remorse, shame, anxiety) was not converted into a somatic symptom, but continued to be consciously experienced. It did not, however, remain attached to the unacceptable idea, but was transferred to another seemingly harmless conscious idea, object, or situation. That idea, object, or situation was related in special ways to the original affect-producing idea, which now disappeared from consciousness. What remained in the patient's awareness was an apparently insignificant idea, object, or situation that, through the mechanism of projection, became associated with a painful affect (Nemiah and Uhde, 1989).

In his theoretical conception of phobia formation, Freud went even further away from instinct in the direction of psychology. The repressed affects, he postulated, were at least partially projected onto the external world. The person could now find relief from the painful anxiety by avoiding the object that produced it. The anxiety attached to the object was subjected to a further psychological process, which he called "displacement" or "projection". The anxiety associated with the original object was displaced or projected onto another object that was connected to the first by associative links that allowed the second object to symbolize the first, but without the threat of repercussion. This theoretical construct may be applicable to the clinical ecology postulate of "spreading".

Recurring obsessions or compulsions constitute a distinct obsessive-compulsive anxiety disorder. Obsessions are persistent ideas, thoughts, impulses, or images that are experienced as intrusive and inappropriate and that cause marked anxiety or distress. One of the most common obsessions is repeated thoughts about contamination from germs or dirt by contact (e.g., becoming contaminated by shaking hands). This type of obsession is similar to that expressed by some EI patients who complain of being constantly poisoned by chemicals.

Compulsions are repetitive behaviors or mental acts, the goal of which is to prevent or reduce anxiety or distress, not to provide pleasure or gratification as is the case in hedonistic acts. The patient's insight that the obsessions or compulsions are excessive or unreasonable occurs on a continuum. When insight and reality testing are lost, the obsession may reach delusional proportions and the presence of psychotic features may indicate an additional diagnosis of delusional disorder.

Discussion

Both EI advocates (Ziem and Davidoff, 1992) and critics (Terr, 1986a, 1989; Staudenmayer et al., 1993a) have observed that evidence for significant toxic exposure that can be objectively corroborated to account for the symptoms (the unconditioned stimulus) is invariably lacking in EI patients. In those EI patients who did suffer such an exposure, many of the symptom reactions to alleged exposure are dissimilar to the unconditioned responses (usually irritant effects on the eyes, sinuses and airway) elicited by the toxic agent, the UCS. Also, the unconditioned stimuli that allegedly initiated the hypersensitivity to further reactions have no identifiable gradient of stimulus generalization, chemical or perceptual, with the multiple agents which subsequently trigger reactions.

Classical Pavlovian conditioning cannot explain *ex post facto* (after-the-fact) attribution of a triggering chemical, especially if it was not perceived. This is often characterized by the rationalization, "I reacted so I must have been exposed." Contrary to the expected effect predicted by the principle of extinction in Pavlovian conditioning theory, the EI patient's belief is strengthened when the putative offending stimulus was not perceived. This logic, in effect, is, "Because I didn't smell the chemical, it proves I was not conditioned to respond and therefore the attack was due to a chemical reaction." This deduction might be correct if in fact there had been a toxic chemical exposure at a dose capable of triggering a reaction. However, when no such exposure exists, the conclusion of this deductive reasoning is fallacious.

Ex post facto attribution can affect physiological conditioned responses, as demonstrated by Loftis and Ross (1973) in an experimental conditioning study. A light (CS) was conditioned to elicit a galvanic skin response, a sign of autonomic nervous system arousal. After the completion of the acquisition trials, subjects where told to misattribute the source of the arousal, the light experienced during acquisition, to background white noise constantly heard throughout acquisition. Subsequent to this cognitive restructuring, extinction of the galvanic skin response to the light was facilitated, i.e., subjects no longer responded to the light as a CS because of the suggestion that white noise was the real CS. Pavlovian conditioning cannot explain *ex post facto* attribution. A learning model perfectly valid within its own context of application is not valid in contexts in which other types of learning unique to higher order species also apply and offer a better explanation for EI.

EI patients usually identify odor as a significant trigger for an environmentally induced reaction. What the odor is associated with is not clear. Is it the actual toxic chemical, as in the case of a hypothesized Pavlovian conditioned response to a documented exposure of low-level neurotoxins? Or is it a fear response associated with the belief that this chemical will induce symptoms? What is the meaning of this fear, what is its origin, and how does it link to the myriad of concepts in the labyrinth of memory networks? Could it be associated with a fragmented memory of a sensory or emotional sensation triggered by the stress response after the stimulus is perceived to be harmful (Miller, 1951)? Is the reaction a reenactment of emotional and sensorimotor memory which lacks links to contextual knowledge of who, what, when, and where, making the experience incomprehensible in the present context? In evaluating the effects of learned sensitivity on the reactions of EI patients, not only must we consider the stimuli associated with the environmental agent, but also the mediating psychological and psychophysiological processes.

chapter nine

The stress-response

The stress-response represents normal physiologic adaptation which enables an individual to cope with challenging or threatening situations with heightened attention and arousal. The neuroendocrinology and the psychoneuroimmunology of the stress-response have been extensively reviewed (Chrousos, 1995; Chrousos and Gold, 1992; Chrousos et al., 1995; Stanford and Salmon, 1993). The stress-response initiates central and peripheral nervous system processes that function to preserve internal equilibrium or homeostasis when a stressor of any kind exceeds a threshold (Chrousos et al., 1988). The stress-response is an adaptive process necessary for survival. As such, the physiological systems involved are viewed to be dedicated processes that have evolved for the coordination of responses to maintain homeostasis. When the internal equilibrium is disrupted, symptoms usually result, and if the dysregulation is chronic stress disorders result.

In discussion of the stress-response, two types need to be differentiated: (1) a disturbance in the tonic level of homeostatic regulation resulting in a persistent, chronic basal state of either hyperarousal (vigilance, sympathetic arousal, anxiety) or hypoarousal (diffuse concentration, fatigue, parasympathetic collapse); and (2) an acute and transitory state disturbance or phasic reactivity from the basal level in response to external stimuli or internal thoughts or emotions. The latency of recovery is usually a matter of minutes but may be prolonged in certain individuals with personality disorders or post-traumatic stress disorder. Classical theories of non-specific stress assume the mechanisms are similar in acute and chronic stress. However, there is evidence suggesting that, in humans, psychosocial stress has different mechanisms for acute and chronic responses. For example, catecholamine changes are more apparent in acute stress-responses, but less so in chronic conditions where lack of cognitive-behavioral coping and chronic anxiety are the primary mechanisms (Christensen and Jensen, 1995).

The stress-response can account for multi-system complaints (Rose, 1980) and has been proposed as one psychogenic theory (Jewett, 1992a). This chapter focuses on central nervous system (CNS) complaints: difficulty concentrating, attentional difficulties, memory loss or poor memory, learning difficulties (inappropriately portrayed as dyslexia), difficulty expressing oneself or understanding others (portrayed as aphasia), feelings of being out of touch, depersonalization or derealization, fatigue, and irritability. These symptoms are what EI patients refer to collectively as "brain fog" or "brain fag". They are mediated by disturbances of central neuronal pathways that regulate arousal, alertness, and vigilance, which in turn modulate the general capacity and effectiveness of cognitive processing involved in learning and memory (Craik and Lockhart, 1972; Ellis et al., 1984). These symptoms are also on the menu of criteria that define anxiety disorders, particularly panic attacks in DSM-IV.

Chrousos and Gold (1992) described the three main physiologic systems involved in the stress-response:

1. Corticotropin-releasing hormone (CRH), which affects the brain directly and also initiates a cascade of responses along the hypothalamic-pituitary-adrenal (HPA) axis
2 Norepinephrine (NE) systems associated with the locus coeruleus (LC) referred to as LC-NE
3. Autonomic nervous system (ANS), primarily the sympathetic branch.

These three systems interact with inhibitory neurotransmitter systems to dampen the stress-response (Figure 9.1). They also interact with the endocrine and immune systems to create organ-specific and systemic effects. Each of these systems is defined with discussion focused on the stress-responses consistent with multi-system complaints seen in EI patients.

Hypothalamic-pituitary-adrenal axis

The HPA axis is the major neuroendocrine stress regulatory system in the body (Krishnan et al., 1991). Its clinical manifestations vary, depending on the extent and severity of the pituitary hormone deficiency and the specific organs targeted by the endocrine mediators. In the extreme, hypopituitarism may result from either pituitary or hypothalamic disease (Vance, 1994). Of note, dysregulation of the HPA-axis may result in nonspecific symptoms of fatigue and malaise.

Corticotropin-releasing hormone (CRH) is a hypothalamic peptide widespread throughout the brain but best characterized in the hypothalamus, specifically in the paraventricular nucleus (PVN). In animal studies, exogenous stimulation with CRH has been associated with hypercortisolism, sympathetic activation, and behavioral activation and intense arousal (Sutton et al., 1982). Larger doses of CRH administered directly to the CNS produce effects that can be construed as frankly anxiogenic. This includes hyperresponsiveness to sensory stimuli, assumption of the freeze posture, decreased exploration in unfamiliar environments (avoidance), and enhancement of conditioned fear responses during aversive stimuli (Dunn and Berridge, 1990). In humans, parallel symptoms are defining of post-traumatic stress disorder (PTSD). For example, Vietnam combat veterans with PTSD have been found to have higher concentrations of cerebral spinal fluid CRH (Bremner et al., 1997).

In lower doses, exogenous administration of CRH sets into motion a coordinated series of physiological and behavioral responses that are adaptive during stressful situations. The brain is directly affected as reflected in activation of the EEG brain waves (Friedman and Irwin, 1995). There is selective stimulation and regulation of peptides along the HPA-axis, such as pituitary adrenocorticotropic hormone (ACTH) secretion, which in turn stimulates the adrenal cortex to release glucocorticoids, principally cortisol (Aronin et al., 1986). The circulating levels of cortisol complete inhibitory negative feedback loops to modulate the activity of CRH in the hypothalamus and ACTH in the pituitary. Attenuated plasma ACTH responses to exogenous infusion of CRH reflect dysregulation of the HPA axis and have been observed in a large proportion of patients diagnosed with major depression. Elevated concentrations of CRH in the spinal fluid were found in clinically depressed patients. This suggests that endogenous CRH hypersecretion in these patients resulted

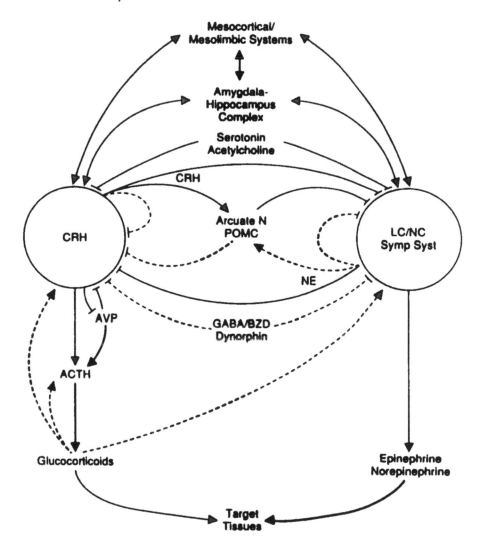

Figure 9.1 Physiological systems of the stress-response. (From Chrousos, G.P. and Gold, P.W., *JAMA*, 267, 1244–1252, 1992. Courtesy of U.S. Government.)

from an abnormality of the CNS (Gold et al., 1988). The hyperactivity of the HPA axis in depressed patients is expressed by increased basal cortisol concentrations and urinary-free cortisol excretion of a magnitude resembling the hypercortisolism of Cushing's disease (Gold et al., 1988a,b; Lesch et al., 1988). In contrast, PTSD patients show evidence of a highly sensitized HPA axis characterized by decreased basal cortisol levels and increased negative feedback regulation (Yehuda et al., 1993, 1995a). These differences have implications for EI patients who show signs of depression and PTSD but often complain of side effects from antidepressant medications.

A wide variety of nonspecific physical and mental stressors, actual or perceived, can activate a state stress-response via the HPA axis. Pathways lead from the neocortex to areas of the hypothalamus involved with CRH secretion (Gann et al., 1981). CRH elicits an immediate and marked increase in ACTH secretion, followed within minutes by significantly increased adrenocortical secretion of cortisol. It has been suggested that this is a survival response because cortisol causes rapid mobilization of amino acids and fats from their cellular stores, making these available both for energy and for synthesis of other compounds needed by the different tissues of the body (Munck, 1971; Sapolsky, 1992). This is one way the body can maintain heightened physiological arousal necessary to cope with a stressor that elicited the stress-response in the first place. However, with excessive stimulation from chronic stress, there is a depletion effect and a loss of equilibrium. Evidently CRH plays a major role in coordinating the endocrine, autonomic, behavioral, and immune-mediated inflammation responses to stress, through actions in the brain and in the periphery (Rosler, 1994). For example, the principal physiologic effect of hypocortisolism in the immune system is the loss of the anti-inflammatory response capacity necessary to fight off infection, heal from trauma, control autoimmune disease, and other immune functions (Chrousos, 1995; Claman, 1972).

The importance of the HPA axis in explaining EI patients' symptoms of odor intolerance and aversion, especially hyperosmia, is suggested by the ability of glucocorticoids to penetrate the blood-brain barrier where they have complex actions on the brain, including effects on brain enzymes. Patients with Addison's disease, which is characterized by inadequate production of adrenal hormones, have reported increased smell and taste sensation (Baxter, 1988).

Psychologic effects of ACTH and corticosteroids

Excessive administration of pharmacologic doses of corticosteroids can cause euphoria or dysphoria in the short-term, which can lead to "steroid psychosis" in individuals predisposed by certain personality traits, e.g., schizoid, obsessional, or cyclothymic (Rome and Braceland, 1952). The mechanism for this effect may involve stimulating the production of CRH by the central nucleus of the amygdala and the production of dopamine by neurons of the mesocorticolimbic system (Chrousos, 1995). Inadequate corticosteroid levels may also have CNS effects, sometimes experienced by patients after sudden cessation of medications after high-dose bursts of prednisone; there is a delay before the endogenous adrenal cortisol production starts up again. Over a prolonged period of steroid use, mood tends to be more depressed, disposition is more irritable with associated lowering of frustration tolerance, physical fatigue and exhaustion set in with less exertion of mental and physical effort, and concentration is more difficult to maintain. Chrousos (1995) suggested that these symptoms, which are characteristic of atypical depression, are due to the suppression of hypothalamic CRH, based on studies which showed low CRH concentrations in the cerebrospinal fluid.

The history of the use of corticosteroids medications serves as a good model for psychological effects resulting from neuroendocrine dysregulation. Prednisone was made available in 1950 and was initially regarded as a miracle drug for certain autoimmune disorders such as arthritis and some respiratory disorders such as asthma. Physicians quickly noticed not only the physical side effects (Cushingoid signs) but also the psychological effects on mood and cognition. Rome and Braceland (1952) summarized the early observations with ACTH and prednisone treatment and graded the responses into four categories of cumulative psychological side effects (Table 9.1). In general, at lower doses thinking is accelerated, creative thinking is enhanced with better qualitative results, and mood is elevated. These effects did not appear to be mediated psychologically from feeling better, because the feelings of euphoria preceded the remission of breathing difficulty in asthmatics. This led Rome and Braceland to conclude that the effects were mediated

Table 9.1 Psychological Responses to Corticosteroids

Grade 1 (less than 100 mg Prednisone, daily)
  Stimulation, well-being, feeling more vigorous, less fatigue, able to concentrate better

Grade 2 (more than 100 mg Prednisone daily)
Grade 1 alterations, plus ...
  Mood effusive and expansive in affirmation of well-being
  Increased physical activity, increased muscle strength
  Restlessness, insomnia, increased appetite
  Cognitive processing reflects hypomania, flight of ideas, imprecise expression
  Mood elevation impairs judgment and perspective

Grade 3
Grade 2 alterations, plus ...
  Psychologic expression of non-psychotic disorders
  Anxiety ascribed to events that are facades shielding underlying psychologic conflicts
  Anxiety frequently expressed as phobia
  Rumination and obsessional preoccupation
  Mood swings of hypomania and depression

Grade 4 (reserved arbitrarily for the grossly psychotic reactions)
  Personality structure seems most important

Source: Adapted from Rome, H.P. and Braceland, F.J., *Am. J. Psychiatry*, 108, 641–650, 1952.

physiologically in the central, autonomic, and endocrine systems. With larger daily doses, there appears to be a potentiating or catalyst effect which either enhances or disinhibits the underlying personality structure. In the extreme, certain individuals with pre-morbid pathological personality structure may have psychotic reactions. These individuals were often found to have had a prior history of psychiatric treatment.

Adrenocorticotropic hormone and corticosteroids were also reported to have disinhibiting effects on memory such that memories of conflicts that had only been partially resolved became conscious. These early observations have been corroborated by research on the involvement of hormonal and neuromodulary systems in regulation of memory storage (McGaugh, 1989). This unexpected flooding of anxiety-provoking material may explain some of the more terrified reactions characterized as "steroid psychosis".

The psychological significance of the patient's symptoms were also observed with these suddenly effective treatments. The abrupt removal of the sick role that accompanied symptom resolution can tax psychological defenses and coping skills. When the illness has been the object of displaced personal anxieties, its resolution can precipitate decompensation. The anxiety of facing the real world and the personal relations in it may be too much for someone who fundamentally believes that "It's not a safe world."

Stress and aging: the glucocorticoid cascade hypothesis

Based on animal studies, Sapolsky suggested that chronic stress can accelerate the aging process through adjustments of the HPA axis brought about by cascading effects of hypersecretion of glucocorticoids (McEwen and Sapolsky, 1995; Sapolsky et al., 1984; Sapolsky, 1992). This is consistent with historical intuition that physiological age in humans may be better measured by the rate of living rather than chronological age (Pearl, 1929). As the aging process accelerates, a second factor, decreased adaptiveness to stress, contributes to deteriorating health.

In humans, decreased adaptiveness is exemplified in psychophysiologic stress profiling as a delay in recovery from a short-term or state stress-response, e.g., a noise startle (Lacey and Lacey, 1958). For example, the electrodermal response (EDR) in humans, characterized by sweaty hands, may remain activated longer than expected from the initial sympathetic system activation, implicating a second mechanism to sustain activation, prolonged adrenal glucocorticoid secretion.

Elevated basal cortisol levels have been associated with cognitive deficits in humans and correlated with aging. In one study, older subjects showing both a significant increase in cortisol levels with years of life and a high current basal cortisol level were impaired on tasks measuring explicit memory and selective attention (Lupien et al., 1994). Furthermore, subjects presenting a decrease in cortisol levels with aging performed as well as young healthy subjects with regard to cognitive performance. Impaired cognitive performance was associated with recent evidence of HPA axis perturbation and elevated basal cortisol levels.

Long-term basal psychophysiologic abnormalities may also reflect brain abnormalities caused by cumulative exposure to glucocorticoids, particularly in the hippocampus (Sapolsky, 1985; Sapolsky et al., 1984). Anatomically, the hippocampus is a large, heterogeneous structure containing many different neuron types and comprised of distinct cell fields, with different portions of the structure associated with different functions (Isaacson and Pribram, 1975). The lost sensitivity for inhibition of the HPA axis from stress is mediated by decreased hippocampal binding of glucocorticoids, damage to the structure of the hippocampus, or both. The damage is specific to the hippocampus, with only minimal effects detectable in the amygdala (Sapolsky et al., 1986).

Animal studies have demonstrated that the chronic stress-response causes release of excessive amounts of glucocorticoids which destroy receptors and mediate neural death in the hippocampus. Hippocampal damage disrupts normal inhibition of the HPA axis to maintain homeostatic functioning. There is hypersecretion of the chemical transmitters along the HPA axis, including CRH from the hypothalamus and ACTH from the anterior pituitary (Landfield et al. 1978; Sapolsky et al., 1985). However, there is some dampening of the effects in that the hypersecretions are associated with less sensitivity of response along the HPA axis. For example, excessive secretion of ACTH is not associated with proportional amounts of glucocorticoid secretions from the adrenal cortex, although the basal levels of glucocorticoids may be elevated. Similarly, the pituitary shows a dampened response of ACTH secretion to hypersecretion of CRH, although basal levels of ACTH may be elevated. The glucocorticoid cascade hypothesis facilitates an understanding of some of the stress-response effects observed in aging individuals with a history of chronic stress or trauma (McEwen and Sapolsky, 1995).

In a rare human study on four patients in whom depth electrodes were placed in the temporal lobes to control intractable seizures, direct electrical stimulation of the hippocampus and amygdala produced different effects on the HPA axis (Mandell et al., 1963). Stimulation of the hippocampus resulted in initial inhibition of adrenocorticoid secretion, followed 30 minutes later by an increase above control levels, suggesting that a compensatory mechanism had been activated. In contrast, stimulation of the amygdala was followed in every instance by an increase in blood adrenocorticoids.

The observations of Sapolsky et al. (1986) showed that the neuron loss in the hippocampus was not due to toxins, exogenous insults, slow viruses, or autoimmune attack, but rather to cumulative exposure to normal concentrations of cortisol, a hormone that is essential to life. These findings suggest that the causal mechanisms for many of the symptoms attributed to EI may instead be psychogenic, mediated through well-defined cognitive processes and physiological and biochemical systems which activate the stress-response.

Damage to the hippocampus from the combined effects of aging, chronic stress, and social abuse may explain two aspects of EI. First, the puzzling but reliable finding in several studies shows that the mean age of reported onset of chemical sensitivities, particularly in women, is in middle age (about 40). Second, the hypersensitivity to environmental agents reported by EI patients has been observed in animal studies in which hyperresponsiveness to environmental stimuli was associated with damage to the hippocampus (Altman et al., 1973). This evidence serves to support the validity of the symptom complaints of EI patients. It also demonstrates the nonspecificity of the cause of these effects.

Role of psychological factors

A deterministic view hypothesizing a linear causal relation between an environmental stimulus and a physiological response does not fully capture the complexity of the relationship between human beings and their responses to the physical environment. Humans are not passive organisms, but rather respond to the environment in a manner that may attenuate or increase the influence of the stressor. The personality factors, anxiety, palliative coping strategies (e.g., cigarette smoking), and the psychosocial context within which an environmental stressor functions can interact with that stressor and modulate its impact on health and wellness (Christensen and Jensen, 1995; Evans et al., 1989).

Most of the early environmental stress research examined the direct effects of ambient, physical conditions on animal and human behavior and health. In animal research, a variety of stressful stimuli including extreme temperatures, exercise, and reduced caloric intake did not always show a reliable dose-response activation of the HPA axis, especially when the circumstances of the experimental situation were such that the animal did not experience distress or novelty (Mason, 1975). For example, chronically stressed rats had more rapid endocrine response to novel stimuli, even though they failed to show any adrenocortical responses to the original stressor after extinction from repeated re-exposure (Sakellaris and Vernikos-Danellis, 1975).One of the primary qualities of a stressful experience for an animal is exposure to a novel, strange, or unfamiliar environment. As the organism becomes accustomed to a new stressor, learns to control it, or learns to cope with it, the stress-response habituates. What explained the animals' responses in these studies was the psychological relevance of the stimulus rather than the particular physical quality of the stressors to which they were exposed.

From a cognitive perspective, the discussion of the stress effects of novelty would be incomplete without noting that it is not novelty *per se* that elicits the stress-response. Only novel stimuli that are perceived as threatening or harmful elicit the response. This context-dependent discrimination has been observed in both animals and humans. For example, in one study with freely moving cats in which activation of LC-NE neurons was monitored, subjects showed significant activation when confronted with a dog or an aggressive cat, but other novel stimuli such as a passive cat did not increase neuronal activation (Levine et al. 1990).

According to the interactive toxicogenic theory of EI, shyness is postulated to be a genetic predisposition for susceptibility to chemical sensitivities (Bell, 1994b). However, if shyness is defined as an abnormal response to novelty, clinical observations of shy children leads us to conclusions similar to those of animal studies. Shy children do not react to all novel stimuli with anxiety, but only to novel situations and people that they perceive to be beyond their ability to cope with. Furthermore, once shy children become familiar with someone, they are usually less withdrawn. It would seem that these types of social interactions are more readily explained by familiarity rather than toxicogenic theories of adaptation/deadaptation to chemicals in the physical-environment.

Human psychophysiologic research has demonstrated habituation to a novel stressor reflected in diminished biological and behavioral responses with repeated exposure. For example, parachute jumpers showed a very large endocrine response to initial jumps, which habituated in subsequent jumps for most individuals (Ursin et al., 1978). Experienced parachutists also showed a brief anticipatory response in cortisol levels but it returned to basal levels before the jump actually occurred. This psychophysiologic research corroborates the psychogenic theory of EI in that expectation of danger or harm initiates the systems of the stress-response, but also that repetitive exposures should produce "tolerance". The exception involves exposure to repeated, novel, and intense manipulations more characteristic of trauma than stress (Rose, 1980, 1984). Habituation also fails to occur in many EI patients, and some show an opposite effect — sensitization. In animal studies, lower concentrations of a pharmaceutical stressor will elicit avoidance responses after an acute dose. In humans, sensitization has been identified in amphetamine abusers (Antelman et al., 1980) and in patients with affective disorders (Post et al., 1995).

Catecholamine system

The second major biochemical system of the stress-response involves the catecholamines. The catecholamines are progressive metabolites of tyrosine and include dopamine (DA), norepinephrine (NE; also called noradrenalin in the British literature), and epinephrine (EPI; also called adrenalin). Catecholamine effects have been localized in specific brain regions in the limbic system, the cerebral cortex, and the midbrain (mesencephalon). Regions in the midbrain are associated with regulation of arousal, pain threshold, emotion, and memory. Catecholamines may also modulate the effect of other neurotransmitters (Kupfermann, 1979).

Without oversimplifying, for the most part the different catecholamines have similar stimulating or inhibiting effects on the central nervous systems. Relative to basal levels, higher concentrations of the catecholamines have a stimulatory effect, and lower concentrations have a sedative or inhibitory effect. The effects on learning and memory performance follow a non-linear dose-response curve (Figure 9.2) for the relationship between arousal and performance. Cognitive performance is usually poorer below homeostatic levels of a catecholamine because of hypoarousal, facilitated at homeostatic levels due to alertness and disrupted at higher levels because of hyperarousal.

Catecholamines are particularly germane to discussion of the evaluation of EI patients because they are objective biological markers. Concentrations of the catecholamines NE, EPI, DA, and their metabolites such as normetanephrine, vanillylmandelic acid (VMA), and 3-methyoxy-4-hydroxyphenyl-glycol (MHPG) levels can be measured in 24-hour urinary concentrations determined with gas chromatography and mass spectrometry (Kopin et al., 1983; Goldstein, 1995). Increased levels reflect increased activity within the peripheral and central aminergic systems (Maas et al., 1987). The sensitivity of the tests for the combined catecholamines alone is usually sufficient to detect an alteration, without the additional component contributed by their metabolites. Also, if the effect of a measurable change from baseline catecholamine level is not observed, it is unlikely to be observed in their metabolites (Chester Ridgeway M.D., Director of Laboratory Endocrinology, University of Colorado Health Sciences Center, personal communication).

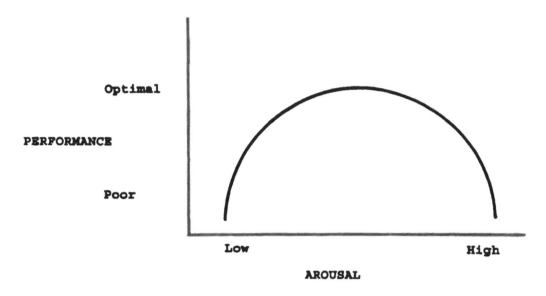

Figure 9.2 Performance vs. arousal.

Changes in catecholamines may reflect subtle effects not observable with less sensitive behavioral or psychometric measures. For example, a task requiring focused attention may elicit catecholamine changes if it is carried out in a stressful environment, even though performance is not impaired. In a study which manipulated a noisy environment, performance was not affected, but catecholamines and blood pressure increased relative to a quiet environment (Lundberg and Frankenhaeuser, 1978). Fatigue may also be a long-term after-effect of coping with stressors. After a work day, latency for catecholamine levels to return to baseline in the evening was longer in workers with more demanding jobs than in workers with less demanding work conditions (Frankenhaeuser and Johansson, 1986).

Norepinephrine

The major norepinephrine nucleus is the locus coeruleus, which is part of the medulla located in the midbrain. The LC has been shown to innervate both cortical and limbic areas directly involved in the perception and discrimination of meaningful and specifically fear-related stimuli (Wolfe and Charney, 1991). Discharge of NE from LC-NE neurons plays a role in attentional vigilance, selective attention, and orienting responses (Aston-Jones et al., 1991a; Foote et al., 1991). Peripheral effects may result directly, or through the activation of the HPA-axis by stimulating the release of CRH in the PVN. LC-NE release may be related to the appraisal of an event and may have specific or generalized effects on the memory networks in certain contexts. The presence of NE can facilitate memory storage; its absence can lead to amnesia (Zornetzer and Gold, 1976). NE may also modulate the excitation or the inhibition produced by another neurotransmitter (Gold and Zornetzer, 1983; Zornetzer, 1978) and increase anxiety in clinical patients (Teicher, 1988).

Central dyscontrol in the LC-NE system is believed to contribute to panic attacks, causing intermittent release of catecholamines which can significantly alter ANS function. During an acute attack, patients with panic disorder show evidence of increases in NE levels (Ko et al., 1983; Charney et al., 1984). Between attacks, patients with panic disorder have significantly elevated heart rates as well as increased NE levels (Nesse et al., 1984). However, there is little evidence to suggest that panic attacks occur as a consequence of widespread overactivity of the LC-NE system *per se*. Rather, the LC-NE system plays a modulatory role for other etiologic factors such that decreased levels of NE protect against panic attacks, and increased levels are indirectly inductive (Teicher, 1988). Medications that decrease noradrenergic turnover and inhibit LC-NE firing (e.g., clonidine, used to treat hypertension) are effective in suppressing panic attacks (Charney and Heninger, 1986).

When the stress-response is chronic, NE levels in the brain may decrease below homeostatic levels because of release and increased metabolism of NE prior to resynthesis of the amine. Under prolonged stress, characterized by depression, the depletion of NE may be offset by antidepressant medications which block the reuptake of NE in the neuronal synapses (e.g., imipramine), thus maintaining homeostatic levels of NE and alleviating symptoms of depression associated with NE depletion such as attentional deficit (Charney and Heninger, 1985).

The second brain region is the amygdala/hippocampus complex in the limbic system which is activated during stress by NE neurons or by an emotional stressor, such as learned fear, which could be generated from memory as a thought or a perception (Gray, 1991). The hippocampus plays a major inhibitory influence on the amygdala and the CRH system (Squire, 1987). The hippocampus is also involved in memory processing, especially for the knowledge and contextual components of memory (O'Keefe and Nadel, 1978). Normal recall of life experiences, referred to as episodic memory, depends on the hippocampus. The hippocampal system receives projections from many neocortical areas. This region may be involved in relaying information between various neocortical areas so that memory storage and retrieval can be efficient. It has been suggested that amnesia is caused by disruption of the interaction of the hippocampus and the amygdala with structures in the neocortex and forebrain (Mishkin, 1982).

Activation of the amygdala is important for retrieval and emotional analysis of information pertinent to the stressor and, if the stressor is perceived to be threatening or arousing, for stimulation of the activity of the CRH and LC-NE/sympathetic systems (LeDoux, 1987, 1994). The amygdala is involved in the response to sensory input from peripheral receptors for visual, auditory, taste, and olfactory stimuli. Olfactory inputs reach the amygdala through direct projections from the olfactory bulb. The amygdala also has effector pathways to the sensory areas such that it can affect or modulate sensory experience. It also has reciprocal pathways to the prefrontal cortex and serves an inhibitory function for arousal and anxiety (Davis, 1992).

Epinephrine

The sensations associated with the effects of EPI include sensations of precardial or epigastric palpitation, diffuse arterial throbbing, oppression in the chest and tightness in the throat, trembling, chilliness, dryness of the mouth, nervousness, malaise, and weakness (Cannon, 1927). All of these sensations are represented among the multi-system complaints of EI patients. Peripheral adrenergic activity correlates with cognitive processing efficacy. The pathway to the brain is indirect as EPI and other catecholamines are largely blocked by the blood-brain barrier (Weil-Maleherbe et al., 1959). However, the reticular activating system in the upper midbrain is sensitive to EPI and mediates its CNS arousal effects (Jasper, 1958). When EPI is injected into experimental animals after exposure to a stressor, the effect on cognition is dose related and also interacts with the intensity of the stressor. The net effect follows a non-linear dose response curve (Figure 9.2) with enhanced memory performance at moderate levels followed by amnesia at higher doses. Stress affects the endogenous EPI plasma levels which correlate with perceived stressor intensity. The plasma concentrations below or above the homeostatic levels are correlated with poorer memory performance and amnesia (Gold and Zornetzer, 1983).

The catecholamine system has peripheral effects through release of NE and EPI from the adrenal medulla of the adrenal glands, located at the superior poles of the kidneys. Direct NE and EPI stimulation of the sympathetic nerves in many physiologic systems elicits different peripheral effects. For example, both NE and EPI increase systolic blood pressure. There are also differential effects due to these neurotransmitters. For example, NE increases diastolic pressure, increases peripheral vascular resistance, and cools skin temperature while EPI lowers diastolic blood pressure, enhances peripheral vascular dilatation, and warms the skin temperature. EPI increases heart rate while NE produces a reflex slowing of the heart secondary to vasoconstriction and elevated blood pressure (Lykken, 1968). Skeletal muscle tension is affected through blood vessel constriction by NE and blood vessel dilation by EPI (Lader, 1975).

The different peripheral physiologic effects of NE and EPI are correlated with different emotional reactions. Responses of psychiatric patients after NE injection were characterized by aggressiveness and outward expression of anger (Funkenstein, 1955; Funkenstein et al., 1952, 1957). Other studies with NE have shown that it is not particularly anxiogenic (Frankenhaeuser and Jarpe, 1962; Vlachakis et al., 1974). In contrast, the response to EPI was associated with inward-directed anger or anxiety and fearfulness. There are numerous studies which have shown that infusion of EPI induces greater anxiety in patients with a history of anxiety symptoms compared to nonanxious controls (Basowitz et al., 1956; Lindemann and Finesinger, 1938; Pollin and Goldin, 1961). The physiological differences and associated emotional responses with NE and EPI may be remembered by the following anecdote:

> The next time you are in a bar and you inadvertently spill beer on the biker fellow next to you, watch the pallor of his face and his body language. If the fellow blushes (vasodilatation) and seems flustered, you are seeing an EPI response usually associated with embarrassment or avoidance. You are probably not in danger from retaliation. Apologize, offer to buy him a drink, and finish your beer. On the other hand, if he turns pale (vasoconstriction), muscles brace, and he looks you square in the eye, you are seeing a NE response and signs of the fight response. Time to leave.

Dopamine

Acute stress increases dopamine (DA) secretion and metabolism in a number of specific brain areas, including neurons that originate in the substantia nigra which is located in the midbrain, extending from the pons to the subthalamus region. The terminations of these neurons are mainly in the basal ganglia which is located in the cerebrum (Moore and Bloom, 1978). The DA innervation of the medial prefrontal cortex, which is considered part of the limbic system, appears to be particularly vulnerable to stress and has a lower threshold for DA release than other brain regions (Charney et al., 1993; Deutch and Roth, 1990).

DA and its metabolites such as 3-4-dihydroxyphenylacetic acid and homovanillic acid substantially reflect events in the brain, whereas EPI and its metabolites tend to reflect the activity of the adrenal medulla. Stimulation with drugs which enhance DA (e.g., amphetamines and cocaine) may result in attentional hypervigilance and feelings of paranoia

(Satel et al., 1991). Elevated DA metabolism may be an adaptive response to environmental stress (DeBellis et al., 1994b), but may also reflect hypervigilance, panic attacks, and other anxiety symptoms associated with stress-induced hyperactivity of central DA systems. DA levels below homeostasis are associated with reduction in activity and absence of pleasure. Extreme abnormalities in DA concentrations are associated with schizophrenia (high levels) and Parkinson's disease (low levels) (Walker and Diforio, 1997).

The prefrontal cortex has the most delayed ontogeny of any brain region, with myelination of major projections to it delayed until adolescence (Fuster, 1989). Psychologically, this maturation correlates with development of abstract thinking and formal reasoning (Piaget, 1972). Data suggest that dopamine secretion induced by stress and trauma may activate the prefrontal cortex and alter its development, either by producing precocious development and maturity (e.g., the "parentified" child) or by arresting development of this region and preventing it from reaching full adult capacity (Teicher et al., 1994).

Pupillary reflex

> "'What big eyes you have, Grandmother!' said Red Riding Hood. 'A
> mere epiphenomenon of my general state of arousal,' replied the
> Wolf." (Lykken, 1968)

Pupillary response has been used as a measure of chemical sensitivities under provocation challenge by the clinical ecologists (Rea, 1992, 1995). Pupillary reflexes are of two kinds — those that respond to light and those associated with accommodation or visual focusing of the eye. Pupillary diameter is affected by the LC-NE systems and the ANS and mediated through nuclei in the reticular formation (Gray, 1973), making pupillary response subject to cognitive and emotional states. Animal research has shown that pupil size varies directly with cortical activation (Naquet et al., 1960). In humans, pupil size has been found to vary with the interest value of a visual stimuli (Hess and Polt, 1960). Pupils also dilate in proportion to the difficulty of a mental arithmetic problem being solved by the subject (Hess and Polt, 1964). Increases in pupil size during arousal by startle, orienting to a pleasant or unpleasant stimulus, focused attention during detection of an anticipated stimulus (particularly if threatening), or when actually faced with a threatening stimulus are consistent with Cannon's (1929) model for the fight-or-flight response.

The use of such a sensitive measure which reacts to placebo factors of suggestion, anticipation, and expectation cannot be validly interpreted unless double-blind testing has been achieved. Provocation challenges conducted by the clinical ecologists were found to lack true double-blind conditions (McCourtie, 1990). Independent replication of the claimed effects must be established scientifically before the validity of pupillary measurements is accepted.

Autonomic nervous system

The third physiological system involved in the stress-response is the autonomic nervous system. There are numerous biological markers of the stress-response that can be measured in several physiologic pathways mediated by the autonomic nervous system, including neuromuscular, cardiovascular, neurohumoral, and interneuronal (Straight and Vogt, 1997). The hypothalamic CRH systems form a feedback loop with the LC-NE systems, and this loop affects the ANS (Aston-Jones et al., 1991). NE can profoundly affect numerous types of endocrine and autonomic regulation, including cardiovascular function and respiration. This has been well established in animal studies in which NE is applied to brain sites that are the targets of these NE cells (Guyenet, 1991). Similar effects can occur in humans through endogenous pathways of the stress-response. Mayberg (1992) cautioned that these effects must be taken into account when evaluating symptom report or brain-imaging studies in forensic situations involving EI patients. With respect to clinical psychiatry, many isolated but seemingly attractive brain-imaging findings, employing a wide variety of methodologies, have resulted in speculative musings, rather than the construction of a significant, testable hypothesis and validation with an independent sample (Brodie, 1996).

The descending pathways of the ANS have two branches, the sympathetic and parasympathetic. The sympathetic nerves originate from neurons located in the intermediolateral column of the spinal cord; the parasympathetic nerves are axons of cells located in the dorsal motor nucleus of the vagus in the medulla (LeDoux, 1987). The sympathetic postganglionic endings primarily secrete NE; the parasympathetic postganglionic endings secrete acetylcholine (Ach). NE and Ach mediate the effects on the end organs innervated by the ANS and are referred to as sympathetic (adrenergic) or parasympathetic (cholinergic) mediators. In general, adrenergic mediators facilitate responses associated with arousal, and cholinergic mediators facilitate responses associated with inhibition and relaxation. The effects of these transmitters are not always direct because they may modulate other transmitters and have interactive effects. For example, an inhibitory effect of Ach on another system may cause an arousal response in an end organ.

The functioning of the parasympathetic branch of the ANS is closely linked to that of the sympathetic branch. Its function is primarily to counterbalance arousal in the sympathetic branch and maintain homeostasis. Inhibition of the parasympathetic system can produce effects analogous to sympathetic activation. For simplicity, discussion is restricted to the net effect of sympathetic activation or inhibition.

Until fairly recently in the history of psychophysiology, it was believed that visceral and glandular responses controlled by the ANS were not subject to learning. Miller (1969) was instrumental in demonstrating that this theory was incorrect. The results of his experiments had implications not only to learning theory, but also for individual differences in ANS responses, for the cause and cure of abnormal psychosomatic symptoms, and possibly for the understanding of normal homeostasis. The ANS innervates psychophysiologic stress-response systems that have measurable clinical signs including heart rate and blood pressure, electromyography (EMG) for muscle tension, electroencephalography (EEG) for attentional states, increased respiration for anxiety, pupil size for vigilance, electrodermal response (EDR) for autonomic activation, peripheral temperature (TEMP) for skin sympathetic activity associated with cold and emotional provocation (Mark, 1990), and peripheral blood flow (photoplethysmography) for vascular constriction. These measures are employed to assess tonic baseline and state reactivity during psychophysiologic stress profiling (Cannon, 1927; Lacey and Lacey, 1958). There are also influences on the brain arising from increased or decreased activity of sympathetic end organs. For example, there are reflex influences of systemic blood pressure upon reticular formation activity (Baust et al., 1963) and regional changes in cerebral blood flow mediated by the sympathetic ganglia (Bentman et al., 1978). These same measures have been inappropriately applied in toxicogenic attributions of EI (Rea, 1995).

Inhibitory systems

The effects on the ANS are so closely linked with the LC-NE systems that the two are often referred to in conjunction as the LC-NE/sympathetic systems (Chrousos and Gold, 1992).

The two principle components of the general adaptation response system — the CRH systems and the LC-NE/sympathetic systems — interact in reverberating feedback loops, making the picture more complex (Figure 9.1). In addition to the arousal response, both components respond to inhibitory systems. Two such neurotransmitter systems involve gamma-amino-butyric-acid (GABA) and the opioid peptides which respond to signals from the pituitary. Another inhibiting system involves the glucocorticoids released from the adrenal cortex, which have a profound immunosuppressive effect (Claman, 1972). Acetylcholine from the cholinergic system also affects cognitive processing synergistically. For example, nicotine is a cholinergic agent which increase NE and serotonin (5-HT) such that there is a calming effect on the CNS but attention remains intact, facilitating cognition (Christensen and Jensen, 1995).

Serotonin

Serotonin (5-HT) is secreted from raphe neurons which have their origin in the median raphe nuclei located in the pons and medulla of the brain stem. Nerve fibers from these nuclei spread widely in the reticular formation and upward into the limbic system. They extend downward into the spinal cord where they serve to inhibit pain. Decreased 5-HT activity in central neurons reduces the inhibitory influence on the LC-NE system. This is one mechanism hypothesized for increased anxiety and panic (Lingjaerde, 1985).

Serotonin has an inhibitory effect on arousal and pain and also helps control the mood of the individual. van der Kolk (1994) hypothesized that low levels of 5-HT are associated with PTSD based on the findings of decreased CNS 5-HT levels that developed in animals subjected to inescapable shock (Valzelli, 1982). The effects observed in animals typify an exaggerated startle response and include hyperirritability, hyperexcitability, hypersensitivity, and an exaggerated emotional arousal and/or aggressive display to relatively mild stimuli. Irritability can result from similar disinhibitory influences in the prefrontal cortex (Teicher et al., 1994). These effects are also characteristic of some EI patients who are suffering from PTSD unrelated to their beliefs about chemical sensitivities.

The role of 5-HT in memory and cognition is not well understood. Nevertheless, CNS effects reflected in the EEG have been observed with 5-HT depletion. For example, decreased 5-HT levels in the forebrain resulted in hypersensitivity to environmental stimuli and patterns of persistent EEG activation (Kostowski et al., 1968). 5-HT may also play a role in the degree of inhibition and disinhibition of memory associated with the hippocampus, including the anxiogenic, dissociative, amnestic, and disinhibitory aspects and symptoms of PTSD (Teicher et al. 1994). van der Kolk (1994) hypothesized that 5-HT plays a role in the capacity to monitor the environment flexibly and to respond with behaviors that are situation-appropriate, rather than reacting to internal stimuli that are irrelevant to current demands.

Medications that are hypothesized to inhibit the reuptake of 5-HT in the neural synapses — selective serotonin reuptake inhibitors (SSRIs), e.g., Prozac (fluoxetine hydrochloride), Zoloft (sertraline hydrochloride), and Paxil (paroxetine hydrochloride) — have proven effective in amelioration of symptoms of depression and anxiety. Obsessive thoughts and compulsive behaviors in particular are reduced by SSRIs (Jenike et al., 1990), an effect consistent with reduced hyperarousal and decreased hypervigilance.

GABA

Gamma-amino-butyric-acid is a major central inhibitory neurotransmitter which is pervasive throughout the CNS and has been estimated to reside in about one third of all neural synapses (Enna and Gallagher, 1983). It appears that the effectiveness of benzodiazepine medications (e.g., Valium, Xanax) in mitigating symptoms of anxiety works through activating the inhibitory effects of GABA (File et al., 1979; Insel et al., 1984; Mason and Fibiger, 1979). The stress activation of DA neurotransmission is inhibited by stimulation of GABA and benzodiazepine receptors (Costa, 1985). One implication of this regulatory effect is that DA is also affected if there is a dysregulation of the GABA system. For example, decreased functioning of the GABA system and associated loss of the inhibiting effect over the benzodiazepine system will result in increased symptoms of anxiety.

The inhibitory activity of GABA-ergic interneurons is highly influential in modulating the overall level of neural activity of the cerebral cortex. GABA-containing interneurons are modulated by other neurotransmitters, including the catecholamines and serotonin. For example, with the development of micro-imaging techniques, the anatomical basis has been shown for neuronal interactions between a dopamine-containing axon making contact with a GABA-containing neuronal cell body in the prefrontal cortex (Benes and Tamminga, 1994).

Opioids and endorphines

Projections of CRH neurons from the hypothalamus to proopiomelanocortin-containing neurons in the arcuate nucleus in the brain stem promote the release of corticotropin and β-endorphin from the latter, which serve to inhibit CRH secretion (Howlett and Rees, 1986). The neurons are connected to the LC-NE system and to the raphe nuclei that control the release of 5-HT. The release of the opioid proopiomelanocortin (POMC) from the arcuate nucleus affects learning and memory and the limbic system by its modulating effects on the LC-NE system and the raphe-5-HT system (Figure 9.3). The paraventricular nucleus (PVN) releases arginine vasopressin (AVP), which effects the release of ACTH, and β-endorphin, which modulates the LC-NE system and the raphe-5-HT system.

Release of endogenous opiates raises the pain threshold and creates analgesia (Hemingway and Reigle, 1987; Pitman et al., 1990). Panic attacks have been observed in drug addicts undergoing opiate withdrawal which suggests that opiates and endorphins inhibit LC-NE activation (Liebowitz et al., 1984). Endogenous opiate systems also have an inhibitory effect on memory processing. It has been suggested that this may account for rapid forgetting or retrograde amnesia after a learning experience, physical trauma, or a stressful event (Gold and Zornetzer, 1983). Disturbance of endogenous opioids may contribute to avoidance and numbing symptoms reported by patients with PTSD (Charney et al., 1993; Pitman et al., 1990), as well as EI patients. Analgesia accompanies neutral stimuli that previously have been paired with aversive stimuli (Fanselow, 1986) consistent with learned sensitivity.

These inhibitory systems serve to induce stress-related analgesia (and perhaps downregulate the emotional tone) and cognitive processing (Gold and Zornetzer, 1983). That characterization describes what psychiatrists call *la belle indifference*, defined as showing a lack of concern over having an illness believed to be debilitating and incurable.

Psychoneuroimmunology

The immune system is directly linked to the CNS both biochemically and structurally and is profoundly influenced by the effectors of the stress-response (Bienenstock, 1991; Chrousos, 1995; McCann et al., 1998). In a review, Djuric and Bienenstock (1993) present anatomical, histologic, electrophysiologic, and psychologic evidence that relates immediate hypersensitivity to the functioning of the nervous system.

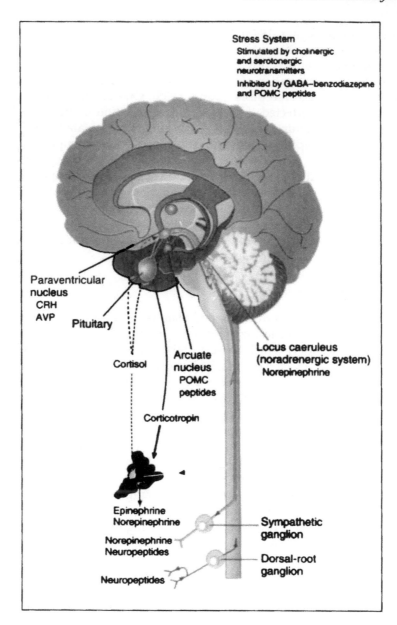

Figure 9.3 Major components of the central and peripheral stress systems. The paraventricular nucleus and the locus caeruleus (noradrenergic system) are shown, along with their peripheral components, the pituitary-adrenal axis, and the adreno-medullary and systemic sympathetic systems. Hypothalamic corticotropin-releasing hormone (CRH) and central nervous system noradrenergic neurons innervate and activate each other, whereas they exert presynaptic autoinhibition through collateral fibers. Arginine vasopressin (AVP) from the paraventricular nucleus acts synergistically with CRH in stimulating corticotropin secretion. Both components of the central stress system are stimulated by cholinergic and inhibited by gamma-aminobutyric acid (GABA) — benzodiazepine and arcuate nucleus proopiomelanocortin (POMC) peptides. These peptides are directly activated by the stress system and are important in the enhancement of analgesia that takes place during stress. Corticotropin (solid arrow) stimulates the adrenal cortex to produce cortisol. Cortisol (broken arrow) inhibits the production of CRH, AVP, and corticotropin. (From Chrousos, G.P., *New Engl. J. Med.*, 332(20), 1351–1362, 1995. With permission.)

The modulating effect of the endocrine system on the immune system (Maclean and Reichlin, 1981) and particularly the immunosuppressive and anti-inflammatory effects of exogenous glucocorticoids on the immune system are well recognized (Barnes and Adcock, 1993; Bateman et al., 1989; Claman, 1972). What is often overlooked is that these effects can be elicited by cortisol levels induced by associated stress-responses (Chrousos, 1995; Fauci, 1978).

The influence of the stress-response on immunity is mediated not only by glucocorticoids but also by other neurotransmitters including catecholamines, endogenous opioids, pituitary hormones, and cytokines (Frankenhaeuser, 1975; Maier and Watkins, 1998). The sensitivity of the immune system to stress is an indirect consequence of the regulatory reciprocal influences that exist between the immune system and the central nervous system (Dantzer and Kelley, 1989). There is abundant neural hard-wiring of both primary (bone marrow, thymus) and secondary lymphoid organs (spleen, lymph nodes, gut-associated lymphoid tissue) (Felten, et al., 1987; Felten and Felten, 1991). NE, the classical neurotransmitter of the LC-NE and ANS systems, has been established as a neurally derived molecule influencing the immune response. Other neuropeptides, particularly substance P, have also been detected in the nerves and organs of the immune system and have been shown to modulate immune function (Bienenstock, 1991; Chancellor-Freeland et al., 1995). The neocortex also appears to be involved in modulation of the immune system, and cerebral hemispheric lateralization effects on immune functioning have been suggested (Neveu, 1988).

It is clear that a variety of environmental and psychosocial factors can influence immune functions through the CNS. For example, blood counts, T and B cell populations, and lymphocyte-mitogen response can be altered by suggestion (Smith and MacDaniel, 1983). The literature has become voluminous since 1981 when Ader published his classic book, *Psychoneuroimmunology*, a title that helped launch a new field of interdisciplinary investigation simply referred to with the acronym, PNI (Ader, 1981; Ader et al., 1991; Ader and Cohen, 1993; Solomon, 1995).

There are many stressors affecting immune responses that have been studied in animals and humans in the past 30 years (Table 9.2). The effects on the immune system are nonspecific. The infancy status of this emerging interdisciplinary field called PNI has been summarized as follows (Maier et al., 1994):

> "Indeed, it is difficult to think of an aspect of immunity that has not been found to be altered by some stressor. ... In addition, the details of the mechanisms involved are largely unknown."

Effects of the stress-response on cognition

Physiological systems have both specific and general effects on cognition. For example, an endogenous chemical modulator may have direct dose-related effects on memory storage and retrieval (McGaugh, 1989, 1992). Or, it can have indirect effects by altering other cognitive processes such as sensation, perception, attention, encoding, or motor integration that are necessary for concept formation, memory consolidation, and executive function (Zelazo et al., 1997).

During homeostasis, there is a background appropriate level of neural activation for a specific time in the brain following a stimulus or an experience that may be necessary for assimilation of information into a conceptual schema and memory consolidation (Yerkes and Dodson, 1908). When arousal levels exceed homeostasis (hyperarousal) or fall below it (hypoarousal), performance may be affected (Figure 9.2).

Table 9.2 Some Stressors
that Affect Immunity

Animals (acute)
Exposure to electric shock
Social defeat
Maternal separation
Rotation
Odor of a stressed cohort
Immersion in cold
Restraint
Handling
Intraperitoneal injection of saline
Loud noise

Animals (chronic)
Crowding

Humans (acute)
Final examination
Battle task vigilance
Sleep deprivation

Humans (chronic)
Divorce
Bereavement
Alzheimer caregiving

Source: Adapted from Maier, S.F. et
al., *Am. Psychol.*, 49(12), 1004–1017,
1994.

There may be associated attentional disorders that mediate the cognitive difficulties presented by EI patients. For example, states of attentional hypervigilance have been associated with neuronal activation (Lindsley, 1951), whereas states of attention deficit have been associated with hypoarousal (Mann et al., 1991). Throughout the 60-year history of brain wave recording with the electroencephalogram (EEG), numerous theories and findings have suggested neurophysiologic parameters that reflect these experiential attentional states and associated clinical conditions.

Shortly after a learning experience, a stressful event, or a trauma there is a certain amount of time during which the integration of memory can be modulated (Zornetzer and Gold, 1976). Memory modulation refers to conditions under which memory storage can be altered after information has been acquired, with implications to later difficulty in recall (Squire, 1986). Some set of biologically adaptive responses to the experience, such as alterations in arousal level (e.g., norepinephrine), autonomic function (e.g., epinephrine), or neuroendocrine activity (e.g., cortisol) may facilitate the encoding of traumatic memories, impair memory storage and other cognitive processes, or affect the response to the experience (d'Ydewalle et al., 1985; van der Kolk et al., 1985). Experiences that alter biological mechanisms through the modulation of endogenous neurochemicals may also affect the response to the experience, including the effectiveness of cognitive processing. For example, arousal level may be affected by changes in neurotransmitter concentrations (e.g., increased norepinephrine or decrease in serotonin); autonomic function may be elevated by increases of epinephrine; immune-inflammatory reaction may be initiated by histamine, leucocytes, or substance P; and neuroendocrine activity may be stimulated by cortisol.

The effect may be observable in several ways, including the magnitude of response, the time-delay of the response, or recovery from the response. To illustrate this line of reasoning, consider the differential effects of a mild stressor and severe trauma. After mild stress, the endogenous neurochemical responses may return to basal, tonic levels within the expected time course of that physiologic system. For example, the electrodermal response will show state activation mediated by perspiratory eccrine glands that are exclusively innervated by sympathetic fibers in response to an unexpected startle (e.g., a loud noise) that normally returns to baseline or basal levels within 30 to 120 seconds (Venables and Christie, 1980). With a severe stressor or a trauma, the EDR activation may show delayed latency in return to tonic levels because secondary hormonal mediators released through the endocrine system maintain prolonged ANS activation. The individual's perception, amplification, and appraisal of such peripheral nervous system effects may stimulate arousal systems that affect higher order cognitive processes such as learning, memory, and thinking or emotional responses such as fear and panic (Pribram and McGuinness, 1975).

Alternatively, heightened emotional arousal may facilitate memory directly, without affecting attention. A recent study in humans on the long-term memory for emotionally arousing material (such as a child involved in a gruesome car accident) demonstrated that memory storage is directly modulated by β-adrenergic mediators (Cahill et al., 1994). In this study the β-adrenergic receptors were blocked with propranolol, administered one hour before the memory acquisition task. One week later, subjects receiving propranolol showed significantly impaired recall of the emotionally arousing story. Control subjects who received propranolol but recalled an emotionally non-arousing story showed no such memory impairment and performed at the same level as placebo control subjects. These findings were interpreted to support a substantial body of evidence from animal studies which suggests that enhanced memory associated with emotional arousal results from an activation of β-adrenergic stress hormone systems during and after an emotional experience (McGaugh, 1992).

Conclusion

There are numerous neurotransmitters, neuropeptides, and hormones that can modulate and dysregulate physiologic homeostasis and the effectiveness of cognition. There is ample evidence that adrenergic, cholinergic, gabaergic, opioid peptidergic, and serotonergic systems can affect cognitive processes in complex ways by modulating non-specific physiologic responses. These effects can be brought on by the stress-response triggered by heterogeneous factors including environmental agents, the appraisal of threat, or internal fears inadvertently triggered by perception of environmental cues associated with memories of trauma. There exist biochemical and physiological systems and mechanisms that are affected by cognition, and numerous studies have demonstrated these interactive effects. This ever-growing body of research offers an empirical foundation for the psychogenic theory of multi-system complaints and demonstrates a progressive research program.

chapter ten

Panic attacks and anxiety disorders

Emotions are physiologically mediated by many neurological systems, and in this chapter I shall focus primarily on structures in the limbic system and the reticular activating system. The most extremely dramatic anxiety reaction is panic attack, which is also mediated by the limbic system. I begin the chapter summarizing two historical physiological theories of emotion to illustrate corresponding differences between the toxicogenic and psychogenic theories. Next, panic and panic attacks are described and illustrated with studies of EI patients in which panic attacks were induced by perception and lactate infusion. Panic attacks are symptoms of several anxiety disorders, including agoraphobia, panic disorder, and post-traumatic stress disorder (PTSD). PTSD involves many EI patient characteristics and symptoms and therefore it is presented in greater detail.

Psychophysiological theories of emotion

The first theory of emotions that relied on physiologic measurement was the James-Lange theory (James, 1884, 1890; Lange, 1922), which proposed that organic processes in the periphery affect the experience of emotion (James, 1884, quoted in Goldstein, 1968):

> "An object falls on a sense-organ and is apperceived by the appropri-
> ate cortical center. ... Quick as a flash, the reflex currents pass down
> through their pre-ordained channels, alter the condition of muscle,
> skin, and viscus [viscera], and these alterations, apperceived like the
> original object, in as many specific portions of the cortex, combine
> with it in consciousness and transform it from an object-simply-
> apprehended into an object-emotionally-felt. ... My thesis ... is that
> the bodily changes follow directly the Perception of the exciting fact,
> and that our feeling of same changes as they occur Is the emotion."

According to the James-Lange theory, the direction of effect of fear is from the peripheral visceral or vasomotor sensations to the central subjective experience of fear; hence the often-quoted example, "I see the bear, I run; therefore, I am afraid."

Table 10.1 Cannon's Evidence for the Primacy of Cognition

1. Total separation of the viscera from the central nervous system does not alter emotional behavior. This was based on Sherrington's experiment on the transection of the spinal cord which denervated the viscera (Cannon, 1931).
2. The same visceral changes occur in very different emotional states and in non-emotional states.
3. The viscera are relatively insensitive structures.
4. Visceral changes are too slow to be a source of emotional feeling.
5. Artificial induction of the visceral changes typical of strong emotions does not produce them.

Source: Adapted from Cannon, W.B., *Am. J. Psychol.*, 39, 106–124, 1927.

Cannon (1927) criticized this peripheral theory of emotion as "the hypothesis of reaction or response as the basis of all mental life." A contemporary version of the peripheral theory focuses on the feedback from the facial musculature during emotional expression (Gellhorn, 1964). Although facial expression is surely an important aspect of emotional behavior, LeDoux (1987) noted that the facial feedback hypothesis is seriously weakened by the fact that patients with facial paralysis appear to suffer no dulling of emotional experiences. In the model developed by Cannon (1927, 1929), perception and appraisal of the situation are primary, for which he presented five categories of evidence (Table 10.1).

Cannon's psychophysiological model has been incorporated into cognitive theories of emotion (e.g., Lazarus, 1966). Experiments on exogenous stimulation with epinephrine which elicited diffuse physiologic arousal showed a broad range of reported emotional experiences (Cantril and Hunt, 1932). Two social psychologists, Schacter and Singer (1962), also injected normal subjects with epinephrine but were able to influence the emotional experiences of anger and euphoria by suggesting misleading expected effects of the injection. Another way they manipulated the response was to sequester the subject after injection with an experimenter collaborator who behaved in a jovial or angry manner, modeling the suggested response. The amplification of physiological signals and misplaced appraisal of causality are learned (Schachter, 1964):

> "Given a state of physiological arousal for which an individual has no immediate explanation, he will label this state and describe his feelings in terms of the cognitions available to him."

This research demonstrated the processes of amplification and attribution characteristic of somatoform disorders. How cognition applies to the appraisal of symptoms was summarized by Bem, another social psychologist, (Bem, 1967, quoted by LeDoux 1987):

> "Individuals come to 'know' their own attitudes, emotions, and other internal states partially by inferring them from observations of their own overt behavior and/or circumstances in which this behavior occurs. Thus, to the extent that internal cues are weak, ambiguous, or uninterpretable, the individual is functionally in the same position as an outside observer."

The importance of attention is its unique role in connecting consciousness and the mental and social processes "from above" with the anatomical and biochemical events within the organism "from below". The primacy of mental over physical systems is the essence of the break in cognitive psychology with the long-established, behaviorist, stimulus-response tradition which unfortunately still has theoretical influence in neuroscience and medicine (Sperry, 1993). The central role that mental concepts and beliefs play in understanding brain function is described by Sperry (1988):

> "Control from below upward is retained but is claimed not to furnish the whole story. The full explanation requires that one take into account new, previously nonexistent, emergent properties, including the mental, that interact causally at their own higher level and also exert causal control from above downward."

Psychophysiological theories of emotion addressed some of the same issues being discussed to explain EI. There is an analogy between the James-Lange peripheral theory of emotion and the toxicogenic theory — a physiologic sensation elicited by an odoriferous chemical activates an emotional experience without cognitive mediation. Likewise, there is an analogy between Cannon's central theory and the psychogenic theory — a thought or belief about exposure to a putatively harmful agent modulates the emotional experience and its physiological effect.

Anxiety

Breuer and Freud (1973) and Freud (1936) conceptualized anxiety as a reaction of the ego to perceived threatening situations. Freud viewed anxiety as an etiological manifestation of any affect, conscious or unconscious, that has been unable to find an avenue to expression. Factors leading to a blocking of discharge of the affects were viewed as the result of the psychological mechanism of repression. There is debate among psychiatrists as to whether or not repression is the best or only mechanism of inhibition, but this does not alter the basic principle that psychogenic anxiety results from unresolved conflict or incongruent representations in memory.

I begin with reference to Freud to distinguish psychogenic anxiety from anxiety induced by a medical condition or toxic substance. The first DSM-IV diagnosis which does not apply to EI patients is anxiety disorder due to a general medical condition. The essential feature of this diagnosis is clinically significant anxiety, panic attacks, or obsessions or compulsions that are judged to be due to the direct physiological effects of a general medical condition for which there is evidence from the history, physical examination, or laboratory findings. Since there is no evidence to date that EI is a medical condition, the diagnosis is not valid based on EI. However, it is possible that some EI patients have an alternative medical diagnosis having nothing to do with EI which can account for their anxiety. Such alternative diagnosis could account for the anxiety experienced by patients for whom no pre-morbid anxiety can be documented, and onset of anxiety dates to the onset of an alternative medical diagnosis. Our typical clinical experience has been that when such an alternative medical condition can be treated, the anxiety symptoms remit.

The second of the DSM-IV anxiety diagnoses which does not apply to EI patients is called substance-induced anxiety disorder. The essential feature of this disorder is prominent anxiety symptoms that are judged to be due to the direct physiological effects of a substance. In addition to drugs and alcohol, that substance can be an environmental agent such as a heavy metal or toxins, including volatile compounds or gases such as petrochemicals, organophosphate insecticides, carbon monoxide, and carbon dioxide. These effects pertain to well-documented toxicologic illness which follows principles of dose-response and time course of onset and recovery to maximal medical improvement. These principles have not been demonstrated for EI.

EI patients invariably have symptoms of anxiety. Some could be diagnosed with more than one anxiety disorder, e.g., phobia and panic, while the most severe cases could be diagnosed with all of them were it not for the hierarchical decision-making algorithm used in DSM-IV. The anxiety disorders have some features in common. Demographically, prevalence is higher in females than in males. There is marked distress which interferes with the conduct of daily activities. By definition, the person recognizes, at least with some insight, that the fears, phobias, obsessions, compulsions, or panic attacks are excessive and unreasonable. If that awareness and insight is lacking, delusional and other psychotic disorders must be considered in assessment.

One of the more common anxiety disorders seen in a general psychiatric or psychologic practice is general anxiety disorder. The essential features of generalized anxiety disorder are excessive anxiety and worry which are chronic and pervasive, are about a number of events or activities, and frequently occur without precipitant. The person finds it difficult to control the worry. The symptoms of anxiety include restlessness or feeling keyed up, fatigue, difficulty concentrating or mind going blank, irritability, muscle tension, and sleep disturbance. Three or more of these symptoms are required to make a diagnosis of generalized anxiety disorder. Two of the most common complaints presented by EI patients are difficulty concentrating and fatigue. Muscle tension, even during attempted relaxation, is also common (Staudenmayer and Selner, 1990), as are the associated aches and pains including headaches. On the basis of these symptoms alone, general anxiety could be diagnosed in all EI patients. However, the DSM-IV hierarchical decision tree precludes the diagnosis of generalized anxiety when criteria for other anxiety disorders are met, e.g., panic disorder, which is usually the case with EI patients.

Panic attack

The most dramatic manifestations of anxiety — panic attacks — occur in agoraphobia, panic disorder, and post-traumatic stress disorder. Panic attacks best describe the acute state reactions EI patients attribute to chemical sensitivities, the primary topic of this chapter. The Oxford English Dictionary defines panic as:

> "Originated from the name of the Ancient Greek god Pan, a creature half man and half goat, symbolizing the anthropomorphism of the devil or evil, the shadow side of human nature. Sounds heard by night on mountains and in valleys were attributed to Pan, and hence he was reputed to be the cause of any sudden and groundless fear. Panic is defined as groundless fear, terror; sudden foolish frights without any certain cause; exhibiting unreasoning, groundless, or excessive fear."

The myth alludes to forces of the unconscious which may reflect repressed trauma masked in fantasy. Panic also may be used to describe interpersonal influences, or contagious emotion associated with mass hysteria. Panic, along with other processes associated with mass psychogenic illness, must be assessed in cases of "sick building syndrome" or toxic spills. The uncertainty inherent in EI about what it is that is feared can make the fear greater.

Table 10.2   Criteria for Panic Attack (Four Required)

1.   Palpitations, pounding heart, or accelerated heart rate [tachycardia]
2.   Sweating
3.   Trembling or shaking
4.   Sensations of shortness of breath [dyspnea] or smothering
5.   Feeling of choking
6.   Chest pain or discomfort
7.   Nausea or abdominal distress
8.   Feeling dizzy, unsteady, lightheaded, or faint
9.   Derealization (feelings of unreality) or depersonalization (being detached from oneself)
10. Fear of losing control or going crazy
11. Fear of dying
12. Paresthesias (numbness or tingling sensations)
13. Chills or hot flashes

Source: *Diagnostic and Statistical Manual of Mental Disorders*, 4th ed., American Psychiatric Association, Washington, D.C., 1994. With permission.

EI also warrants consideration of the definition of the agents who amplify panic and mass hysteria. The Oxford English Dictionary defines panic-monger as one who endeavors to bring about or foster a panic, an alarmist. Agents of iatrogenic influence and political advocates consciously or unwittingly assume this role. The harm to the patient of falsely suggesting injury from low-level chemical sensitivities may have long-lasting impact, including panic attacks.

Psychodynamic theories emphasize unconscious associations to experiences in memory and propose that the sensations underlying somatization are triggered by frightening thoughts, images, sensations, and affects associated with frightening fantasies of being trapped by and/or separated from powerful others. Minor physical sensations are said to provoke panic because they originate from an overly threatening negative affect that immature defenses are inadequate to control (Shear et al., 1993).

Panic attacks resemble acute discharges of the sympathetic system or withdrawals of the parasympathetic system (Roy-Byrne et al., 1986), the net effect of which leads to acute episodes of dysphoric hyperarousal and anxiety. To rationalize these acute reactions, EI patients attribute them to chemical sensitivities (Leznoff, 1997), and often a triggering agent is identified *post hoc* to suit the circumstances. Neurotransmitter mechanisms are involved in panic reactions (Dager et al., 1995), which may be culmination of a more complex anxiety disorder involving anticipatory anxiety.

The phenomenological characteristics of a panic attack are sudden onset of intense fear or terror, coming unexpectedly "out the blue", a sense of being overpowered or out of control, mental confusion or inability to think, and a strong urge to flee (Shear, 1988). During the discrete periods of up to 10 minutes in which panic attacks occur, feelings of impending doom are associated with symptoms such as shortness of breath, palpitations, chest pain or discomfort, choking or smothering sensations, and fear of "going crazy" or losing control. There are DSM-IV criteria for panic attack (Table 10.2), and an attack involving fewer than four of the symptoms is called a limited symptom attack. A panic attack is considered a sign or symptom that can occur in the context of several different anxiety disorders (panic disorder or agoraphobia), but panic attack is not a coded disorder *per se.* The prevalence of panic attacks in the U.S. is estimated to be 15% (Eaton et al., 1994), a figure similar to the alleged prevalence of EI as estimated by EI advocates.

A full-blown panic attack can be conceptualized as an emergency, fight-or-flight, acute stress-response to situations or experiences perceived to be of immediate threat with catastrophic consequences, beyond the coping capacity of available resources. Panic is a normal response to conditions where personal coping capacities are overwhelmed in the face of immediate danger. The panicking person utilizes last-ditch, emergency coping behaviors that are primitive, automatic, simple, and rigid. These behaviors consist of efforts to flee to safety and/or recruit immediate help. If escape is blocked, freezing or aggressive behavior may occur. The experience of panic is highly aversive and thus stimulates conditioned avoidance of associated situations. In short, panic can be conceptualized as an organism's total, single-minded commitment to escape a situation appraised as being highly threatening and potentially lethal and to avoid it in the future. Panic may be considered adaptive when the appraisal of catastrophic danger is accurate and when primitive behaviors such as flight or freezing are the best coping alternatives, arguments for cool, calm action under fire not withstanding. Psychologically, panic becomes maladaptive when it occurs in response to the misperception or the fantasy of being in harm's way in the absence of real threat.

Shear (1996) observed that the object of fear need not necessarily be in the immediate environment, but can be an abstraction such as the feeling of being unsafe, often a long-term sequelae of an attachment disorder originating in childhood. The centrally mediated stress response may include cognitive, affective, sensory, and action concepts in memory. Cognitive processes such as attention and short-term memory are activated, and their limited capacity may become overloaded and disrupt executive functions and other decision processes required to formulate and execute alternative actions (Warburton, 1979). Autonomic nervous system activation associated with a hypervigilant state response leads to peripheral physiological sensations (e.g., accelerated heart rate and shortness of breath) which may be cognitively amplified and perceived to be lethal. Patients with co-morbid PTSD are particularly vulnerable to affective feelings associated with anxiety (e.g., uncertainty, depression, loss of control, and hopelessness) which may serve as access links to memories of prior traumatic events and their associated anxieties. Motor programs and cognitive "scripts" to action (fight, flight, or freeze) may also be activated and contribute somatic sensations to a cascading feeling of anxiety.

Unexpected, sudden onset with recovery latency of less than 10 minutes differentiates panic from general anxiety, which is chronic and pervasive. Lack of threatening environmental triggers differentiate panic from simple phobia. While these differences may serve classification criteria, it is not obvious that they represent unique clinical syndromes that have distinctive etiologies. Rather, they may be symptoms of a pervasive anxiety or other neuroses that manifest through different systems of the stress response.

Scientific theories of panic attack have biological and physiological evidence to support the postulated mechanisms. In a series of studies employing the method of magnetic resonance spectroscopy imaging (MRI), Dager and colleagues have demonstrated that panic patients are more prone to react with panic to lactate infusion than non-panic comparison subjects (Dager et al., 1987a, 1994). There is evidence that the effects of lactate are occurring directly in the brain in that hyperventilation induced significantly greater brain lactate elevations in panic patients than in non-panic controls (Dager et al., 1995). Shortness of breath (hypocapnia) induced by hyperventilation is a potent stimulus for decreasing cerebral blood flow and may be the mechanism responsible for brain lactate elevations in response to hyperventilation. Decreased cerebral blood flow is also associated with under-activation of cognitive processes which could explain some of the symptoms of central nervous system dysfunction such as difficulty concentrating and poor memory.

Another mechanism for the relationship between cerebral blood flow and anxiety involves the basilar artery, which is the primary supply to the CNS regions implicated in panic, such as the locus ceruleus and the autonomic centers in the brain stem. The symptoms associated with decreased basilar blood flow are visual disturbances, lightheadedness, and unsteadiness. Panic attacks often involve dizziness, numbness, and other symptoms that suggest transient cerebrovascular disturbances. Ball and Shekhar (1997) tested this hypothesis by measuring basilar artery flow using transcranial Doppler ultrasonography during rest and after hyperventilation in 16 patients with panic disorder and eight normal controls. The patients with panic disorder demonstrated greater reduction in flow rates and greater increases in dizziness which suggests that the basilar artery is a mechanism involved in neurological symptoms during panic attacks.

Panic attacks are spontaneous in the sense that there are no consistent identifiable environmental stimuli that set them off. This apparent spontaneity supports the suggestion that panic is caused by processes within the individual, with either psychological or physiological etiology. Klein (1981) hypothesized a unique pathophysiologic mechanism without psychologic significance. He specifically suggested carbon dioxide hypersensitivity, which is misinterpretation by a biological suffocation monitor that misfires a suffocation alarm system (Klein, 1993; Papp et al., 1993). It has also been hypothesized that panic attacks result from a cognitively mediated anxiety disorder, and that work on biological causes and pharmacological treatments for panic attacks should take cognitive factors into account (Gelder, 1986). There is room for both psychological and physiological mechanisms to account for psychophysiological effects, although the debate as to which might contribute more is likely to continue. For example, a study of 176 panic disorder patients with primarily respiratory symptoms found an increased incidence of a history of traumatic suffocation (19.3%) compared to controls (6.7%) (Bouwer and Stein, 1997). While the association suggests a causal relationship, there is another possibility. An episode of actual suffocation acts physiologically to sensitize a presumed suffocation alarm, which is then more easily triggered at a later point (Bouwer and Stein, 1997).

Recently, an entire issue of the *Journal of Psychosomatic Research* (Vol. 44, No. 1, 1998) was devoted to panic disorder. Reviews of the physiologic systems and their symptoms identified the multi-system complaints presented by EI patients. State-of-the-science papers on theoretical perspectives of etiologic factors, mechanisms, and treatment recommendations were also presented.

Pearson (1985, 1986) observed hyperventilation and many of the hyperventilatory physical symptoms in EI patients during the course of an alleged chemical or food sensitivity reaction. Two recent studies demonstrated how physiological mechanisms involved in the onset and response of panic attacks can explain reactions in EI patients. Leznoff (1997) demonstrated the induction of hyperventilatory physical signs in open provocation challenges using self-identified chemical agents in 15 EI patients while monitoring blood gases. Carbon dioxide concentration and partial pressure (PCO_2) in the lungs are determined by the rate of excretion of CO_2 from the blood into the alveoli and the rate at which CO_2 is removed from the alveoli by exhalation. Two thirds of the subjects reproduced their symptoms upon challenge, and all showed clinical evidence of acute hyperventilation with a rapid fall in PCO_2 (indicative of exhaling excessive carbon dioxide) associated with no change or rise in oxygen saturation (SaO_2). The symptoms and signs were consistent with panic attack with hyperventilation.

In the second study, Binkley and Kutcher (1997) induced panic attacks in five EI patients in a single-blind, placebo-controlled provocation challenge study with sodium

Table 10.3 Questions About a Panic Attack (or EI Reaction)

1.   What is the mechanism of production of symptoms?
2.   What accounts for the onset of a panic/reactivity episode and the particular quality of the symptoms?
3.   What terminates a panic episode?
4.   What are the mechanisms involved in vulnerability to panic/reactivity, both before and after the initial panic episode?
5.   What processes are involved in the development of anxiety and avoidance responses?
6.   What mechanisms block panic/reactivity?
7.   What processes enhance or mitigate panic/reactivity frequency, anxiety, and phobic avoidance?

Source: Adapted from Shear, M.K., in *Panic: Psychological Perspectives*, Rachman, S. and Maser, J.D., Eds., Earlbaum Associates, Hillsdale, NJ, 1988.

lactate infusion. All five patients exhibited a positive symptomatic response to sodium lactate compared to placebo (saline) infusion. Standardized psychiatric assessment identified four of the five patients as meeting DSM-III-R criteria for panic disorder along with other depressive and/or anxiety related disorders. These scientific studies contributed significantly in demonstrating that the psychogenic theory is supported by a progressive research program to identify psychophysiological mechanisms underlying perceived chemical reactions by EI patients (Staudenmayer, 1997).

Panic disorder

The pathological syndrome of panic disorder includes vulnerability to recurrent panic episodes and/or maladaptive consequences of panic. The typical course includes a prodromal period which is often associated with major life stresses, an initial panic episode, recurrent panic, and associated anticipatory anxiety and phobic avoidance. This time course is consistent with the histories of many of the clinical histories of EI patients. The DSM-IV notes that panic disorder and agoraphobia with panic attacks together have a worldwide prevalence rate between 1.5 and 3.5%, and are two to three times more frequently diagnosed in women as in men. The clinical syndrome of panic disorder begins with an initial panic attack, typically in late adolescence or the mid-30s. The initial attack is unexpected and sudden, and the long-term effects are rather dramatic. For example, one study showed that 81% of patients with panic discontinued their activities and 25% contacted a medical facility at the time of the first attack (Barlow and Craske, 1988).

The panic disorder syndrome is a psychopathological state in which panic attacks are the central pathognomonic feature. In order to understand the pathogenesis of panic disorder, it is useful to ask a number of questions (Table 10.3). The same questions should be asked of alleged reactions attributed to low-level exposure to chemicals.

Shear and her associates have developed a model of panic disorder that integrates psychodynamic, neurophysiological, and cognitive-behavioral theoretical formulations and clinical approaches, and they emphasize the importance of early childhood experiences of trauma and emotional deprivation (Shear, 1996; Shear et al., 1993; Busch et al., 1991). In a controlled study, patients with anxiety disorder had a significantly higher incidence of history of childhood physical and sexual abuse than a community sample, with a particular association between sexual abuse and panic disorder in women (Stein et al., 1996).

Table 10.4 Illustrative Life-Events Preceding Onset of Panic Disorder

1. Feeling trapped in a relationship
2. Resentment at perceived unfair treatment by employers or loved ones
3. Frustration over dissatisfaction with work
4. Loss of a relationship

Source: Adapted from Roy-Byrne, P.P. et al., *Am. J. Psychiatry*, 143(11), 1424–1427, 1986.

Shear et al. (1993) illustrated their model with observations on nine (six female, three male) patients who had panic attacks. In the initial interview, all nine patients denied having anxiety prior to the onset of the panic attacks. They also believed that stress was unrelated to the onset of panic. Later in the interview, stressors and significant life events were identified which preceded the onset of panic attacks, a finding consistent with other studies (Roy-Byrne et al., 1986a; Pollard et al., 1989). Heightened frustration and resentment characterized these types of stressful events (Table 10.4). Furthermore, the occurrence of a prodrome of generalized anxiety in panic disorder has also been reported (Garvey et al., 1988).

Shear et al. (1993) present a case that would read like a vignette of an EI patient were "EI reaction" substituted for panic episode:

> "One patient initially disavowed being nervous or anxious before the onset of panic [EI reactions]. However, later in the interview, she explained that although she never thought of herself as an anxious person, other people said she was. She noted that she had always been 'claustrophobic' in situations such as traffic jams or department stores, where she would become anxious and feel that she was trapped. She further described herself as a 'worrier, always thinking the worst.' The patient initially stressed that things were fine before her first panic episode [EI reaction]. However, in the course of the interview she revealed that she was chronically unhappy at her job, where she felt unfairly treated — overworked, underpaid, and emotionally drained. She described her boyfriend of 3 years as 'very cold'. She felt shortchanged by him and often felt hurt and useless. She finally said, 'I think that was one of the reasons I got so nervous.'"

Like this panic patient, many EI patients reveal stressful life events and experiences of frustration and anxiety upon establishing rapport with a doctor or therapist. These patients are often more amenable to referral for stress management treatment and have a good prognosis. In other cases (often involved in litigation) where the belief system is closed and the patients are well defended with denial of psychological factors in their lives, evidence of emotional disorders, loss and other stressful life events, and poor relationships with others is often revealed upon examination of their prior medical/ psychological records. Most important for the comparison of panic and EI patients, the origin of the anxiety which manifests as panic attacks precedes the alleged event of EI onset, often dating to childhood.

Of the nine panic patients interviewed by Shear et al., all described at least one parent as angry and frightening, critical, or controlling. Adulthood discomfort with aggression emerged as a theme in seven patients. Each patient complained about aspects of a relationship or job situation and indicated having chronic feelings of resentment and a sense of injustice.

However, in each case the patient felt too guilty or anxious to confront the situation, fears that imply an earlier etiology. Seven patients described feelings of low self-esteem or prominent negative self attributes. Complaints and aggressive feelings were regularly accompanied by contradictory comments. For example, one patient described her family as very tense but quickly added that they were close and loving. Such neurotic symptoms are best explained by predisposing personality factors (Shear et al., 1993):

> "The interviews revealed themes of early life anxiety and shyness, unsupportive parental relationships, and a chronic sense of being trapped and troubled by frustration and resentment. Each patient described frightening, controlling, or critical parents and most described feelings of inadequacy and/or self reproach. Aggression was prominent, uncomfortable, and often managed with efforts to contradict complaints and/or turn negative into positive attributes. This consistency was impressive and led the interviewers to revise their view that the pathogenic mechanism of panic was primarily physiological."

Time course

The course of panic and a perceived "chemical sensitivity reaction" are similar in that the initial reaction is followed by development of a panic-prone state. There are recurrent panic attacks and mild or aborted forms of panic limited-symptom attacks, near-panic, or partial panic. Patients begin to fear having another attack. Much of the morbidity of the panic disorder syndrome is determined by panic consequences. Patients develop fear and begin avoiding situations where panic has occurred or where they would not be able to leave or get help if panic did occur. The experience of recurrent panic may lead to anxiety and self-criticism in panic patients or projection of blame to objects external to the self in the case of EI patients. Co-morbidity for panic includes demoralization and depression (Breier et al., 1985), which are frequently observed in EI patients.

Non-panic anxiety may be related to anticipation of a panic episode, fear of entering a phobic situation, or general worries and insecurities, which are typical of generalized anxiety states. Non-panic anxiety accounts for much of panic disorder morbidity. In a 24-hour monitoring study of 23 panic disorder patients, all patients had symptoms of anxiety and/or panic on the day of the recording (Shear et al., 1987). Seventy percent had non-panic anxiety, 65% had partial panic, and 21% had panic attacks. These findings also describe the clinical phenomenology of many EI patients who obsess about their symptoms and general problems of living but rarely talk about the good periods of a day. Recurrent chemical reactions, like recurrent panic, may occur either unexpectedly or be expected, in a cued or noncued setting. Phobic avoidance is typical in panic disorder patients, as it is in EI patients. It is not clear what factors contribute to the development of phobias in some panic patients but not in others. Analogously, it is also not clear why some EI patients have more generalized phobias than others. In the panic literature, it has been suggested that individuals with dependent personality disorder have greater avoidance behavior; dependent personality traits are also common among EI patients. These observations suggest that personality disorders may help to explain some of the anxiety characteristics of EI patients who panic in the face of an environmental exposure appraised to be harmful.

Panic patients sometimes report feeling prodromal apprehension before a full-blown attack. These apprehensions seem to have no apparent stimuli in the environment to

Table 10.5 Criteria for Post-Traumatic Stress Disorder

A.
  1. Experiencing, witnessing, or being confronted by event(s) that threaten or involve death or serious injury to self or others
  2. Intense response of fear, helplessness, or horror

B. Re-experiencing criteria (one):
  1. Recurrent, intrusive, distressing recollections
  2. Recurrent distressing dreams
  3. Flashbacks
  4. Intense psychological distress upon exposure to symbolic cues
  5. Physiological reactivity upon exposure to symbolic cues

C. Avoidance and numbing criteria (three):
  1. Avoidance of associated thoughts, feelings, or conversations
  2. Avoidance of associated activities, places, or people
  3. Inability to recall an important aspect of the trauma
  4. Markedly diminished interest or participation in significant activities
  5. Feeling of detachment or estrangement from others.
  6. Restricted range of affect
  7. Sense of foreshortened future

D. Arousal criteria (two):
  1. Difficulty falling or staying asleep
  2. Irritability or outbursts of anger
  3. Difficulty concentrating
  4. Hypervigilance
  5. Exaggerated startle response

E. Duration at least one month

F. Clinically significant distress or impairment of functioning

Source: Diagnostic and Statistical Manual of Mental Disorders, 4th ed., American Psychiatric Association, Washington, D.C., 1994. With permission.

trigger them and may be described as "coming out of the blue". Agoraphobics characteristically experience panic attacks when they are in open spaces or when they leave what they perceive to be a safe environment, e.g., their home or office. But agoraphobics also have panic attacks in the safety of their homes, even after *in vivo* exposure directed to extinction of phobic responses and restructuring of cognitive distortions (Michelson et al., 1985).

EI patients show similar patterns of spontaneous reactivity which may be associated with a perceived exposure. At other times, they report no perceivable exposure associated with a reaction. When not identifiable, the triggering agent may be appraised as having been below sensory threshold. Also, a *post hoc* search for a triggering candidate will invariably identify one.

DSM-IV criteria for PTSD

Panic attacks are also a symptom of post-traumatic stress responses, commonly seen in EI patients. Models of PTSD have been hypothesized as part of the toxicogenic theory to explain the bipolarity of symptomatology expressed by EI patients. Specific criteria have evolved for the diagnosis of PTSD (Table 10.5). Many psychiatric patients with a history of trauma do not strictly have PTSD because their traumas have become integrated into the totality of their personality organization, and they often cannot elaborate the traumatic event(s) (van der Kolk, 1988). The specifics of the event are inaccessible in the trauma memories, and these patients do not have persistent re-experiencing of the traumatic event through vivid flashbacks or dreams. However, without knowing the specifics of the event, these patients can still re-experience the traumatic event through dreams, play, or sensory and affective schema related to the trauma and meet criterion B for PTSD. The traumatic event is persistently re-experienced by demonstrating intense psychological distress or physiological reactivity at exposure to internal or external cues that symbolize or resemble an aspect of the traumatic event. These reactions appear to come "out of the blue" similar to the symptoms of panic attacks. In EI patients, the symbolism is displaced to agents in the physical environment, and panic attacks are reframed as "environmental reactions".

DSM-IV criterion C for PTSD is the persistent avoidance of stimuli associated with the trauma and numbing of general responsiveness. The denial of psychological factors is consistent with criterion C-3 — inability to recall an important aspect of the trauma. There are four other characteristics which describe the feelings of isolation and avoidance behaviors of some EI patients: markedly diminished interest or participation in significant activities, feeling of detachment or estrangement from others, restricted range of affect, sense of foreshortened future.

DSM-IV criterion D requires at least two persistent symptoms of increased arousal selected from sleep difficulties (insomnia), irritability or outbursts of anger, difficulty concentrating, hypervigilance, or exaggerated startle response. Some subset of these symptoms are experienced by most EI patients, and difficulty concentrating is virtually a universal complaint among EI patients. While self-report may suggest that these symptoms were absent prior to the onset of EI (Fiedler et al., 1992), careful examination of history and psychophysiologic parameters often suggests that these symptoms were pre-morbid (Staudenmayer and Selner, 1995).

The remaining criteria of PTSD pertain to duration and disturbance of functioning and are met by severely affected EI patients. PTSD may also occur with delayed onset, defined to be at least 6 months after the stressor. Very often, current life stressors create emotions and physical sensations from the stress response which can evoke re-experiencing symptoms and numbing which have been dormant throughout the intervening period. Assessment of underlying trauma that may be displaced to alleged environmental exposure should be part of the psychological evaluation of every EI patient manifesting symptoms of PTSD.

Experimental models of PTSD

The emotional depletion, immobilization, and lack of insight to escape their predicament so often seen in EI patients are consistent with the clinical descriptions of traumatization by Janet in the 19th century (van der Hart and Friedman, 1989). The effects have been hypothesized to be similar to the experimental effects (passivity, numbing and lack of arousal, and inability to learn subsequent escape tasks) and resultant biological correlates such as catecholamine depletion observed under conditions of inescapable shock (Yehuda and Antelman, 1993). The model has been anthropomorphized as "learned helplessness" and suggested as a model of depression (Maier and Seligman, 1976). In humans, the "experiments" do not always end with the subject reduced to immobilized despair as seen in severe depression. Some people find ways to cope with their feeling of helplessness, and rather than passivity show signs of arousal and hypervigilance, or even outright hostility and aggression. Such arousal and explosive effects are more characteristic of

Table 10.6 Criteria for Models of PTSD

1. Even very brief stressors should be capable of inducing biological and behavioral sequelae of PTSD.
2. The stressor should be capable of producing the PTSD-like sequelae in a dose-dependent manner.
3. The stressor should produce biological alterations that persist over time or become more pronounced with the passage of time.
4. The stressor should induce biobehavioral alterations that have the potential for bidirectional expression.
5. Interindividual variability in response to a stressor should be present either as a function of experience (e.g., prior stress history and post stressor adaptations), or genetics, or an interaction of the two.

Source: Adapted from Yehuda, R. and Antelman, S.M., *Biol. Psychiatry*, 33, 479–486, 1993.

PTSD than of depression. Yehuda and Antelman (1993) presented five criteria for evaluating the relevance of individual stress paradigms to the phenomenology of PTSD (Table 10.6).

Learned helplessness

Learned helplessness is a laboratory phenomenon observed when animals are exposed to uncontrollable trauma induced by inescapable electrical shock. The inescapable shock or learned helplessness model (IS-LH) was initially presented as a model of depression by Seligman (1975). In its initial formulation the behavioral influence was apparent: "Helplessness is the psychological state that frequently results when events are uncontrollable" (Seligman, 1975). Shortly thereafter, the model was extended to humans and reformulated in a cognitive frame; terms such as "perceived uncontrollability" appeared in place of terms such as "attempted action to control" (Abramson et al., 1978). In addition, the individual's attribution of loss of control or uncontrollability became central to the theory (Peterson and Seligman, 1984; Seligman and Nolen-Hoeksema, 1987.) The essential feature for the etiology of learned helplessness is not the experience of trauma, but rather a sense of not having control over the trauma.

The relationship between depression and anxiety in the learned helplessness model is that anxiety resulting from feelings of loss of control underlies depression, rather than anxiety being a symptom of depression. Clinically, this suggests that uncovering the basis of the anxiety, learning to cope with it, and regaining a sense of control are necessary to overcome the depression. The establishment of a feeling of control is usually one of the ultimate goals in self-regulation therapy (e.g., biofeedback) and cognitive-behavioral therapy (Shapiro et al., 1996).

Southwick et al. (1992) and van der Kolk (1987) have hypothesized IS-LH as a model of PTSD. Yehuda and Antelman (1993) suggest that IS-LH does not meet the criteria of PTSD (Table 10.6) based on the following logical analysis. The first criterion requires that onset of PTSD can occur in one, brief trial. The IS-LH paradigm requires numerous repeated trials (60 to 180) and does not account for the possibility of developing symptoms after exposure to a single, very brief acute stressor, indicating that conditioning is not the most likely learning mechanism. They note that the IS-LH paradigm does not provide a way of evaluating behavioral manifestations in an untrained animal in the presence of a novel exposure to trauma.

The second criterion — that any stressor of sufficient intensity should be capable of producing the PTSD-like sequelae — is not met by the IS-LH paradigm because it only holds for laboratory animals that cannot escape the shock, not for those which can. The third criterion — that PTSD symptoms intensify over time — is not met in the IS-LH paradigm because the effects are not chronic and typically return to normal in less than two days. The fourth criterion — bidirectionality of symptoms — is not observed to occur simultaneously or in close temporal proximity. In the paradigm, there is an initial phase of hyperactivity in which the animal makes efforts to escape ("helplessness"), followed by a phase of passivity in which the animal accepts the shock even though escape is available ("hopelessness"). The IS-LH paradigm does meet the interindividual variability for criterion five.

In summary of the IS-LH paradigm, Yehuda and Antelman point out that controllability or helplessness may influence how PTSD is manifested and should be investigated. They suggest that the paradigm may be relevant to exploring vulnerability factors associated with the development of PTSD. However, the identification of factors that influence such manifestations of the disorder need to be clearly differentiated from those that directly contribute to the induction of core PTSD symptoms.

Time-dependent sensitization

Yehuda and Antelman (1993) hypothesized that time-dependent sensitization (TDS) models the clinical syndrome of PTSD. In TDS, one exposure to a stressor can induce an extremely long-lasting alteration in the subsequent responsiveness of the organism to pharmacological or non-pharmacological stressors, characterized by a sensitization to respond to lower levels of the stressor. This effect progresses with time since the first stressor. The time-dependent effects on behavioral and physiological systems are ubiquitous, consistent with the multisystem complaints of EI patients.

Yehuda and Antelman (1993) described how TDS meets the criteria for PTSD. It meets the first criterion (Table 10.6) because it can be induced by stressful events imposed on the organism only for seconds or, alternatively, by more chronic and severe stressors such as inescapable shock. The second criterion is met because TDS has been shown to be dose-dependent on the degree of psychological stressors measured in terms of corticosterone response levels (Antelman et al., 1992). TDS responds to multiple stressors including physical discomfort, both controllable and uncontrollable, and to psychological, pharmacological, and metabolic agents (Antelman 1988).

TDS meets the third criterion because the consequences both persist for extremely long periods (demonstrated to last at least one month) and grow stronger with increased passage of time, similar to what is observed in chronic or delayed PTSD. The fourth criterion — bipolarity — is met because the effects of TDS can be either excitatory or inhibitory (Antelman et al., 1991). The fifth criterion, interindividual variability, is met by TDS as is the case with PTSD. Yehuda and Antelman conclude that the paradigm of TDS is an animal model of stress that fulfills the necessary criteria for successfully modeling the onset of PTSD in response to trauma.

Hyperosmia or cognitive priming?

Hypersensitivity to sensory input and symptoms is one of the defining features of EI patients. For example, some EI patients claim to have unusual acuity to detect odors at lower thresholds called "hyperosmia". However, controlled psychophysical studies with specific odorants have shown that the odor-detection thresholds of EI patients were not different from control subjects (Doty et al., 1988). Studies employing chemosensory event-related potentials, which reflect the cognitive recognition of a stimulus, have not found lower sensory thresholds for olfactory (hydrogen sulfide, rotten egg smell) or trigeminal (carbon dioxide, an irritant) stimuli, suggesting that perceived enhanced intensity reported by EI patients is cognitively mediated (Hummel et al., 1996; Hummel, 1997). How can the discrepancy between the phenomenology of the patients and the controlled laboratory studies be explained? One avenue of explanation assumes that the phenomenology is accurate; patients do experience a sense of heightened ability to detect odors. But the mechanisms for this apparent acuity need not be those hypothesized by a toxicogenic theory, but instead may reflect odor aversion (Amundsen et al., 1996) or associated fears of harmful effects.

These studies illustrate how neural mechanisms activated by processes such as arousal and selective attention offer a plausible cognitive explanation, without invoking super-sensory acuity as a hypothesis (Heilman et al., 1987; Posner and Petersen, 1990). Hypervigilant patients are biased in the interpretation of perceived objects, sensitized to threatening or fearful stimuli. The behavioral sensitization described in animal studies (e.g., Antelman, 1988) has associated cognitive processes in humans, referred to as "priming", a simple form of nonconscious memory in which words are identified more quickly and at a lower threshold after repetition in an experimental word recognition task.

Similar facilitating effects in reaction time have also been observed in word recognition tasks involving conceptual semantic representation in long-term memory. For example, the recognition reaction time that a string of letters was a word as opposed to a non-word was faster when the target word was preceded by another word which was semantically related (e.g., nurse preceded by doctor) compared to an unrelated pair (e.g., butter preceded by doctor). The observed effects are attributed to spreading activation of related concepts in memory networks (Schvaneveldt and Meyer, 1973). Neuroanatomical studies with PET have corroborated this hypothesis in that focusing attention on concepts in the same category showed a cortical activation pattern closer in anatomical proximity compared to categorically unrelated concepts (Spitzer et al., 1995). The spreading activation observed in word-recognition experiments is consistent with more speculative hypotheses that there are distinct neurological correlates for complex conceptual representations of traumatic experiences. A PET study by Rauch et al. (1996) showed that when people were exposed to individualized, script-driven imagery of their trauma, there was more activation in right hemisphere areas involved with emotional arousal as well as in the right visual association cortex. Concurrently, there was also less activation in Broca's area located in the left hemisphere, suggesting an inhibition to verbalizing the experience.

Anxiety lowers the pain threshold and pain tolerance (Sternbach, 1978). In addition, anxious people are self-conscious, and this apprehensive self-scrutiny amplifies pre-existing symptoms and causes them to notice trivial somatic symptoms that would otherwise be ignored. In addition, the somatic concomitants of anxiety may themselves be mistaken as ancillary symptoms of a medical disorder (Barsky et al., 1988a). Patients prone to somatize may be easily distressed by the symptoms of transient hypocapnia, which may then lead to increased arousal, further hyperventilation, and the establishment of a chronic disorder (Bass and Gardner, 1985).

The mechanism of cognitive amplification of pain symptoms characteristic of somatoform disorders has been demonstrated in several of the physiological systems represented in the multisystem complaints of EI patients. The differential contribution of physiological and psychophysiological mechanisms in the perception of epigastric pain characterized by symptoms of dyspepsia, such as early satiety, fullness, bloating, belching, nausea, and vomiting, were evaluated in a controlled study of patients with functional or idiopathic pain (Mearin et al., 1991). Epigastric pain was induced experimentally by incremental increases in the volume of air in an intragastric bag. Patients reported abdominal discomfort at a much lower intragastric pressure than normal controls. Physiologic gastric relaxation responses induced by a cold-pressor test in which the hand is immersed in ice water were the same in dyspectics and controls. The results showed that symptoms in patients with idiopathic dyspepsia may develop from an altered visceral perception rather than from a primary gastric dysfunction or an efferent reflex pathway, suggesting an afferent sensorial pathway at the brain-gut axis. An altered conscious threshold in the CNS is one possible mechanism, consistent with earlier hypotheses linking dyspepsia with psychological disorders (Talley et al., 1986). However these investigators mention that other mechanisms have not been ruled out, such as an altered threshold of gastric receptors or an altered modulation in the conduction of the sensorial input.

Lower pain threshold during an experimental task with a strain-gauge pain stimulator applied to the fingers was found in patients with a history of childhood sexual/physical abuse presenting with functional symptoms associated with irritable bowel syndrome, gastroesophageal reflux, or noncardiac pain (Scarinci et al., 1994). The lower sensory threshold was explained by response bias to judge less intense stimulation as painful, rather than by a physiological mechanism which could account for enhanced sensation. In an accompanying editorial, Drossman (1994) noted that this study supports prior observations that patients with gastrointestinal disorders have normal cutaneous discriminating capability, and it does not exclude the possibility that abuse history may also amplify pain from visceral sites through physiologic mechanisms which may have been altered by the trauma of the abuse or by adaptations resulting from it.

The orienting response can account for the hyperosmia effects reported by EI patients. Under normal circumstances the arousal and selective attention of these patients may be in a constant flux. The sensations of arousal, brought on by non-environmental triggers, may be misinterpreted as chemical sensitivity reactions. Attentional processes may then be involved in the search for and identification of an offending agent.

In another scenario, selective attention can be primary in the sequence of events. When alerted by a limbic-mediated emotional feeling or an appraisal of exposure to an environmental agent, increased arousal may result, even culminating in panic or hysterical reactions. Such interactive interplay between arousal, selective attention and other higher order cognitive processes can create the phenomenological impression that one is more environmentally sensitive and has a lower olfactory threshold (Engen et al., 1991). EI patients report their hyperosmia is not consistent from exposure to exposure, all allegedly explained by the counterbalancing postulates of adaptation/deadaptation. The orienting response, through the neural mechanisms involved in selective attention and arousal followed by habituation and extinction, offers an alternative psychological explanation.

Cognitive-behavioral theory of sensory amplification focuses on the misperception and amplification of physiological sensations, specifically sensations of autonomic arousal and hyperventilation (Clark, 1988). These misperceptions can lead to complaints of symptoms believed to be associated with severe disease. The cognitive-behavioral theory does not attempt to explain the physiological mechanisms of the triggering sensations, and offers no hypotheses on the etiology of the first attack. The theory highlights the cognitive distortions associated with somatization and the subsequent worry and catastrophizing. The processes proposed are consistent with theories of amplification in somatization disorder (Barsky et al., 1988b; Barsky and Klerman, 1983) and symptom perception (Pennebaker and Watson, 1991).

Conclusion

Panic attacks have identifiable physiologic mechanisms of action and symptom expression which can explain many of the symptoms of acute "environmental reactions" manifested by EI patients. There are numerous psychological and psychophysiological factors to consider in the etiology and exacerbation of panic attacks. Pre-morbid trauma is a common predispositional factor associated with biological abnormalities and PTSD and panic disorder. In the next chapter, Chapter 11, the psychological, psychophysiological, and biological abnormalities associated with war-zone exposure and childhood sexual abuse are reviewed. The limbic system is intricately involved in perception, thinking, and feeling. Internal activation, so-called "top down", of the processes and structures of the limbic system and the interlinked attentional processes of the reticular activating system can explain the cognitive, emotional, and arousal symptoms presented by EI patients, as detailed in Chapter 12.

chapter eleven

Trauma and post-traumatic stress disorder

Trauma is as qualitatively and quantitatively different from stress as post-traumatic stress disorder (PTSD) is different from the stress-response. Reflections about the lasting psychological and biological changes caused by trauma and deprivation dates to the dawn of psychology and psychiatry. A century ago, Pierre Janet taught his students that overwhelming experiences are accompanied by vehement emotions that interfere with thinking and appropriate action (van der Kolk and Saporta, 1991). When describing "combat fatigue" during World War II, Kardiner (1941) observed soldiers who acquired enduring vigilance and sensitivity to external threats. The perceived threat in question reflected the fear of being killed by some unpredictable act of war, but he noted vigilance persisted once the soldiers were removed from harm's way. Biological effects associated with hypervigilance such as symptoms of catecholamine depletion (e.g., masked face, cogwheel rigidity, and tremor) were noted. The attribution of trauma was displaced. With the original trauma inaccessible to consciousness, or denied, other elements in the environment became the source of attributed trauma (Grinker and Speigel, 1945). The horrifying experience of trauma has been described (Solnit and Kris, 1967):

> "Sudden disruptive experience that pierces, violates, or rents the
> stimulus barrier leads to a degree of paralysis and immobilization
> and deprives the ego suddenly of its autonomous function, brings
> about regressive phenomena and severe inhibitions."

Kolb (1987) reviewed the enduring nature of autonomic arousal in Vietnam combat veterans with PTSD. Physiological hyperarousal has repeatedly been documented in response to sounds, symbols, and visual stimuli reminiscent of the trauma. Clinical experience suggests that the increased autonomic arousal is rather nonspecific and may be seen in response to a large variety of stimuli. In fact, habituation can occur to the original traumatic stimulus, while associated events are met with hyperreactivity. This results in a cascading effect of further autonomic arousal, disorganization of cognitive processes, and anxiety (van der Kolk, 1988). These findings suggest PTSD as an alternative explanation to the toxicogenic postulate of "spreading".

Traumatized people typically have an all-or-nothing response to emotional stimuli. They have an impaired capacity to modulate the intensity of their emotional responses including anxiety, anger, or intimacy (van der Kolk, 1987). These findings have implications to the all-or-none reactions of EI patients in that the appraised intensity of these reactions does not follow a dose-response relationship correlating with exposure levels of the alleged offending environmental agents.

Post-traumatic stress disorder research has shown that physiologic changes can account for most post-traumatic symptomatology. By analogy, the symptomatology of EI patients during an appraised reaction can be explained by many of the same mechanisms and physiological responses. As Cannon showed, reactions to relatively minor stimuli, if interpreted as threatening, elicit startle reactions and irritability which interfere with an adequate cognitive assessment of current events and result in fight, flight, or freeze responses. Autonomic arousal may further activate long-term memory pathways, allowing access to inhibited memories of past trauma (Gold and Zornetzer, 1983). Prolonged physiological arousal and homeostatic instability can disrupt attention. Because attention is necessary for the functioning of other cognitive processes, disruption of attention creates mental confusion, interferes with "executive function" and memory, and interferes with assimilation and accommodation of new information (Baddeley, 1993). This may explain the symptoms EI patients call "brain fog".

Persistent hyperarousal is one mechanism hypothesized to explain the avoidance and escape behavior characteristic of people with PTSD. Psychologically, avoidance serves as a defense when coping resources are insufficient. Physiologically, avoidance is associated with downregulation of overstimulated receptors through the inhibitory systems of the stress response (van der Kolk, 1987). Trauma has many origins and affects individuals differently for a host of reasons such as developmental stage, genetic predisposition to psychiatric disorders, temperament, coping skills, social support systems, and prior history of stress and trauma. Scientific studies of trauma and PTSD have mostly involved two populations: (1) veterans with war-zone exposure, and (2) adults with a history of childhood trauma, primarily sexual abuse in females. The organization of this chapter presents the criteria for PTSD, studies of veterans, and studies on patients with histories of childhood abuse. As will be shown, these topics are interwoven and suggest an interactive complexity characteristic of EI patients.

War

The mysteries of "Gulf War syndrome" have offered a fortuitous occasion for scientific study of EI. Epidemiological studies failed to identify increased incidence of disease in Gulf War veterans. Controlled physiological studies consistently showed symptoms of stress and PTSD in these veterans. The Presidential Advisory Committee on Gulf War Veterans' Illnesses has concluded (Table 11.1) that veterans likely suffered illnesses resulting from duty in the Gulf, but current scientific evidence does not support a causal link between symptoms and environmental risk factors, and stress is likely to be an important contributing factor (Lashof, 1996).

EI advocates apply the same toxicogenic theory to explain "Gulf War syndrome". While chronic fatigue and chemical sensitivities have historical antecedents, so too does the "Gulf War syndrome", most likely as ancient as the practice of war itself. In a historical review, Hyams et al. (1996) chronicled nonspecific war syndromes (Table 11.2) reported in medical journals beginning with the U.S. Civil War. These various labels were defined in terms of similar multi-system complaints including shortness of breath, palpitations, chest

Table 11.1 Conclusions of the Presidential Advisory Committee
on Gulf War Veterans' Illnesses

1. Among the subset of the Gulf War veteran population examined in the ongoing clinical and research programs, many veterans have illnesses likely to be connected to their service in the Gulf.
2. Current scientific evidence does not support a causal link between the symptoms and illnesses reported today by Gulf War veterans and exposures while in the Gulf region to the following environmental risk factors: pesticides, chemical warfare agents, biological warfare agents, vaccines, pyridostigmine bromide, infectious diseases, depleted uranium, oil-well fires and smoke, and petroleum products.
3. Stress is known to affect the brain, immune system, cardiovascular system, and various hormonal responses. Stress manifests in diverse ways and is likely to be an important contributing factor to the broad range of physical and psychological illnesses currently being reported by Gulf War veterans.

Source: Lashof, J.C., *Report of the Presidential Advisory Committee on Gulf War Veterans' Illnesses*, U.S. Government Printing Office, Washington, D.C., 1996.

pain, fatigue, headache, diarrhea, dizziness, disturbed sleep, and other somatic symptoms. These symptoms were not restricted to combat soldiers, but also occurred in civilian populations. There was no consistent sign of physiological disease, and most patients appeared to be in fair overall health. Nevertheless, they became disabled.

Epidemiology

Two recent cohort retrospective studies (the data are for time periods before the cohort was selected) on veterans of the Persian Gulf War presented findings consistent with these historical observations. The first reported that during the 2 years after the war, there was no excess of unexplained hospitalizations among Americans who remained on active duty after serving in the war (Gray et al., 1996). The second study followed 695,516 deployed and 746,291 non-deployed veterans from approximately 2-1/2 years post-war. Deployed veterans had a significantly higher postwar mortality rate, but most of the increase was due to automobile accidents rather than disease, a pattern similar to that of veterans of previous wars (Kang and Bullman, 1996).

Table 11.2 Multi-System Symptom Phenomena
after U.S. Wars

War	Label
U.S. Civil War	"Irritable heart"
World War I	"Soldier's heart"
World War II	"Battle fatigue"
Vietnam	"Post-traumatic stress disorder"
Persian Gulf War	"Gulf War syndrome"

Source: Adapted from Hyams, K.C. et al., *Ann. Intern. Med.*, 125(5), 398–405, 1996.

Additional prospective follow-up studies of Persian Gulf War veterans are underway (e.g., Schnurr et al., 1993). The unexplained illnesses among Persian Gulf War veterans in one National Guard unit, according to a report from the Centers for Disease Control and Prevention, remain medically unexplained to date (Kizer et al., 1995).

Psychophysiology

Cohort controlled studies (the data are for time periods after the cohort is selected) have shown that combat veterans diagnosed with PTSD manifest significantly greater long-term effects of sympathetic nervous system arousal, e.g., higher resting heart rate, elevated systolic blood pressure, and increased conditioned physiological and emotional startle responses (Blanchard et al., 1982; Pitman et al., 1987). Veterans suffering from war-related PTSD also show increased startle responses as measured by electromyographic eyeblink response and heart rate response. They also show prolonged latency of recovery from startle as measured by skin conductance responses, an abnormality that appears to be specific to PTSD in comparison to psychiatric disorders such as depression (Butler et al., 1990; Morgan et al., 1995; Orr et al., 1995; Shalev et al., 1992).

Cross-sectional studies have reported time-delayed symptoms of PTSD after catastrophic events including war and natural disasters. A progressive study was conducted with a group of National Guard reservists immediately after they participated in Operation Desert Storm (Southwick et al., 1995a). Over a 2-year follow-up period, PTSD symptoms of hyperarousal showed a significant increase with time. The investigators concluded that it may take time for the consequences of traumatic exposure to become apparent, consistent with the time-dependent sensitization model. This suggests that PTSD be considered in the assessment of patients complaining of "Gulf War syndrome"; however, there is another psychological explanation that warrants consideration. The time delay in the reported onset of symptoms is also seen in EI patients who retroactively identify a triggering agent. Often, the iatrogenic effects of suggestion and unsubstantiated treatment along with social contagion from advocacy groups are clearly apparent. Thus, there are at least two psychological explanations for the non-specific symptoms of "Gulf War syndrome": (1) PTSD time-dependent sensitization resulted from traumatizing events during the Persian Gulf War, or (2) iatrogenic influence and social contagion exacerbated existing PTSD and somatization in suggestible individuals.

War-zone experience may lead to increased symptomatology — PTSD symptomatology and multiple physical health complaints. Such increased symptom report was found in the Vietnam Experience Study by the Centers for Disease Control and Prevention which involved 2490 male Army veterans who served in Vietnam (Barrett and Mannino, 1996). Other studies of Vietnam veterans have questioned the link between an increase in general symptomatology and war-zone exposure in women (Kulka et al., 1990).

A case-controlled study on a sample of 912 military personnel, 653 of whom were deployed in the Persian Gulf, stratified by race and gender, failed to find a significant increase of risk in women for PTSD resulting from war-zone exposure (Sutker et al., 1995a). Deployed troops underwent psychological debriefing within a year after returning from the Gulf and, compared to non-deployed troops, generally reported more symptoms of depression and anxiety regardless of race or gender. Female Persian Gulf War veterans did not report greater symptoms of PTSD than did their male counterparts. Although women endorsed significantly more symptoms of physical discomfort and somatic concerns, including headaches, lack of energy, and upset stomach, regardless of war-zone duty, this tendency was not increased by war-zone exposure. In discussing the limitations of this study, the investigators note that "the questions of multiple chemical sensitivity and the direct impact of war-zone exposure on physical functioning require additional investigation."

Cognitive dysfunction

Dysregulation of the stress system to chronic, physiologic, and biochemical basal levels of hyperarousal or hypoarousal has been suggested as a theory for the etiology of certain disorders (Chrousos and Gold, 1992; Sapolsky, 1992). Of particular relevance to EI are disorders defined by CNS symptoms of impairment in attention and concentration, and forgetfulness and memory. Cognitive dysfunction measured by neurobehavioral performance tasks has been documented clinically in heterogeneous groups with histories of trauma (Wolfe and Charney, 1991) and in diverse psychiatric disorders including affective and anxiety disorders (Cohen and O'Donnell, 1993a).

The sequelae of war trauma from combat and imprisonment as reflected in PTSD have also been associated with general cognitive dysfunction (Sutker et al., 1991). As research progresses, specific types of cognitive dysfunction may be uniquely linked to different types of disorders and underlying pathogenesis. Certain types of traumatic experiences in prisoners of war have been selectively linked to deficits in attention (executive conceptual processes such as planning, and memory) (Sutker et al., 1995b). Memory impairment may also have a selective or context-specific quality. For example, one study suggests that memory dysfunction in PTSD may involve the intermingling of past experience with current experiences and that the cognitive deficits are limited to those memories and propositions that are not contextually delineated with respect to past and present (Yehuda et al., 1995). Cognitive dysfunction is consistent with neurobiological hyperarousal (Cohen and O'Donnell, 1993b) which has been linked to the etiology and maintenance of PTSD (Barrett et al., 1996; van der Kolk, 1987) and associated severe personality disorder (Teicher et al., 1994; van der Kolk et al., 1996). Clinically, all of these manifestations have been observed in one or more EI patients.

Pre-war risk factors

The psychological condition of soldiers before going to war is a risk factor for acquiring PTSD after. In one case-controlled study of 38 Vietnam veterans with combat-related PTSD, significantly more (29%) self-reported a history of childhood physical or sexual abuse than did the 28 controls (7%) without PTSD but with other medical problems (Bremner et al., 1993). This study also assessed 12 other precombat experiences unrelated to PTSD, a negative finding consistent with studies of veterans from earlier wars which did not assess abuse histories (Kimberly et al., 1995).

Sexual or physical abuse in adulthood may also have significant psychiatric sequelae in certain individuals (Norris,1992), although prior abuse histories have usually not been evaluated as confounders in these studies. Rape is strongly associated with PTSD, with an 80% incidence of PTSD in rape victims reported in one study (Breslau et al., 1991). In another study of rape victims seen in an emergency room, development of PTSD was more likely to develop after the rape in the women with a history of sexual abuse than those without it (Resnick et al., 1995). This finding supports the hypothesis that individuals who are unable to resolve or integrate earlier traumatic experience are sensitized to repeated trauma and are therefore at greater risk for the triggering of stress symptomatology when faced with events similar to the original stressor (Christenson et al., 1981).

The Persian Gulf War has generated several well-controlled studies of combat-related PTSD that consider premilitary histories of abuse, psychiatric disorders, and gender differences. In the assessment of war-zone related symptomatology, the patients often are without prior psychiatric history and are at an age when one might see the first episode of schizophrenia, mania, depression, or panic disorder. The war-zone stress might be considered a situation that merely hastened the onset of a psychiatric illness that would have declared itself eventually, with or without a war (Friedman et al., 1993). Some 40,000 women (6% of the total U.S. forces) were deployed to the Persian Gulf theater. One stratified analysis of sick-call data found that women made significantly more psychiatric visits than men (Hines, 1993).

In a cross-sectional study conducted by Engel et al. (1993) with volunteers from the First Cavalry Division Mental Health Clinic at Fort Hood, TX, 297 veterans (28 or 9.4% female) of Desert Storm provided self-report data on precombat sexual and physical abuse, precombat psychiatric problems, sociodemographics, combat exposure, and PTSD symptomatology. Precombat abuse correlated with psychiatric history. Gender comparisons revealed no significant differences in age, ethnicity, marital status, years of education, rank, and precombat psychiatric history. Men reported significantly higher levels of combat exposure. In the sample of 28 women, 34% reported precombat sexual abuse, compared to 4% of the 269 men. Analysis of covariance revealed that gender significantly modified the impact of precombat abuse on combat-related and other PTSD symptomatology after adjusting for precombat psychiatric history and level of combat exposure. Of note, precombat abused male veterans had only slightly higher adjusted mean scores on the PTSD symptomatology scale than their nonabused counterparts. However, females describing precombat abuse reported much greater PTSD symptomatology than did females denying precombat abuse. Engel et al. concluded that female veterans with precombat abuse histories may experience greater PTSD symptomatology in response to combat trauma than do female veterans without abuse history.

It remains unclear whether PTSD symptomatology was initiated at the time of the initial abuse and exacerbated by combat experience or if the initial abuse created a predisposition or vulnerability to subsequent trauma. The first explanation — early onset of PTSD — is supported by studies that have shown that PTSD has a chronic, relapsing course exacerbated by stimuli resembling the original traumatic event (Engdahl et al., 1991). Other studies suggest that indirect exposure or re-exposure to more abstract cues which are cognitively mediated and symbolize aspects of the original experiences can trigger symptom reactions in certain individuals. Wolfe et al. (1992) studied a group of female veterans who had experienced moderate war-zone exposure during Vietnam but were not involved in the Persian Gulf War. Shortly before Desert Storm these veterans completed a modified self-report scale on PTSD symptomatology (Mississippi Scale for Combat-Related Post-Traumatic Stress Disorder) (Keane et al., 1988) on the basis of which they were classified as manifesting symptoms of PTSD (n = 50) or not (n = 26). They also completed the SCL-90-R, a self-report measure of a broad range of psychological symptoms reflecting general distress. Both measures were completed again at the conclusion of Desert Storm, and the results showed a significant increase for the PTSD group in all three dimensions of PTSD symptomatology (re-experiencing, avoidance/numbing, and hyperarousal) as well as general psychological symptomatology on the SCL-90-R, particularly on the dimensions of somatization, obsessive-compulsive, and hostility. Wolfe et al. concluded that the indirect forms of stressor exposure (e.g., media) have important implications to susceptible individuals with a history of PTSD.

A second explanation is that the initial abuse heightened vulnerability for onset of PTSD with combat exposure, a sensitization effect. A third explanation is that there was an association of precombat abuse with another characteristic that is also associated with greater PTSD symptomatology after combat. And, finally, the most likely explanation is a combination of these (Engel et al., 1993). Despite the limitations of their study, Wolfe et al. concluded that their study, along with the one by Bremner and colleagues (1993), provided important empirical evidence emphasizing the need for more research on precombat abuse as a combat-related PTSD risk factor.

War-associated PTSD is not limited to combat soldiers. The deprivation and depravation inflicted on civilians can also create PTSD. Twenty-two holocaust survivors with PTSD studied by Yehuda and colleagues had significantly lower 24-hour urinary cortisol excretion compared to holocaust survivors without PTSD (Yehuda et al., 1995b). The results demonstrate that it is not trauma *per se* that is the critical factor in creating biological abnormalities, but rather the individual's reaction to it with PTSD. This study also demonstrated how long the effects of PTSD may persist — in these subjects, five decades after the traumatizing experiences occurred.

Pre-morbid psychological factors were also the most significant variable for the onset of PTSD in firefighters responding to a natural disaster in Australia (McFarlane, 1988). Contrary to expectation, the intensity of exposure, the perceived threat, and the losses sustained in the disaster were not predictors of PTSD. Introversion, neuroticism, and a past history and family history of psychiatric disorder were the best predictors of susceptibility to PTSD.

Adults selected in an epidemiologic study of U.S. citizens who met diagnostic criteria for PTSD were more likely to have co-morbid diagnoses of schizophrenia, panic disorder, or somatization, but in most of their cases the traumatic events occurred in childhood (Davidson et al., 1991). Given the heightened susceptibility to PTSD following childhood abuse, it would seem reasonable to expect an increase in the risk of PTSD in combat veterans with such a history (Friedman et al., 1994).

Yehuda et al. (1995a,b,c) suggested that low basal cortisol levels reflect the presence of a current and chronic post-traumatic stress syndrome. A biological marker for the effects of PTSD has implications to test a hypothesis implied by the toxicogenic theory. EI patients believe their symptoms are caused by chemical or other environmental sensitivities. If low cortisol levels are associated with chemical sensitivities, the effect is expected to be present in all EI patients irrespective of PTSD or history of childhood abuse. However, if the effects are associated with trauma rather than chemicals, differences are expected between EI patients with and without PTSD and history of childhood abuse. EI advocates might not accept the test of this hypothesis. They may not argue with the assumption that PTSD is due to the events which induced the multiple chemical sensitivities. Yet, they might argue that "chemical" PTSD manifests only in EI patients who have a psychological predisposition created by childhood abuse, but only in those individuals susceptible to having adverse, time-delayed PTSD symptoms potentiated by chemical sensitivities.

Childhood trauma and adult sequelae

I apply the terms traumatic child abuse or deprivation only to terrorizing experiences that actually were, or were perceived to be, life-threatening. The long-term sequelae of childhood abuse can have profound effects on mood and affect, cognitive processing, beliefs and attitudes, personality development, social functioning, and certain medical disorders and health-risk behavior (Briere, 1992a; Benjamin and Eminson, 1992; DeBellis and Putnam, 1994; Schwarz and Perry, 1994; Silk, 1994; Springs and Friedrich, 1992). There are physiologic and biochemical response dysregulations that resemble a chronic stress response and are associated with a spectrum of neuropsychiatric disturbances and psychopathological consequences as well as problems of living (Rutter, 1984). Childhood abuse can also predispose the individual to greater susceptibility to traumatic reactions in response to trauma occurring later in life such as rape or war-zone experience. In determining whether a complex of psychological symptoms has an etiology in trauma and does in fact represent PTSD, differential diagnoses must be considered, including organic disorders, alcohol/drug side effects and addiction, and alternative psychiatric disorders.

Trauma, irrational beliefs, and psychopathology can interact in complex ways, as summarized by Bieber (1980):

> "Psychopathology is defined as the study of the nature of injury and the reaction of the individual to injury. Injury was defined as anything detrimental or perceived as detrimental to one's optimal state at any point in personal history. Injury to self could include acts of omission such as defective nurturing, inadequate stimulation, inadequate or inappropriate contact, inadequate affection, interest, and so forth. Acts of commission would include parental hostility, overcontrol, seductiveness, and so forth. Expectations of injury that no longer had a basis in reality or that were exaggerated, as in situations where injury may be incurred, derived from irrational beliefs. Such beliefs may be the consequence of actual injury incurred in childhood and later projected into adult life, a time when it may no longer be reasonable to anticipate injury. Irrational beliefs may also result from indoctrination or the beliefs may be linked to wishful thinking. The target of therapy is the extinction of irrational beliefs."

Childhood emotional deprivation

Some of the most sobering historical observations recorded in the annals of psychiatry about the effects of emotional deprivation in children come from two sources. During World War II, after having come to England with her father, Anna Freud (1939–1945) observed children who lost their parents in the bombings and were left unattended in the shelters of London. Deprived suddenly of the emotional nurturing of their mothers and without surrogates available, many of these children died.

Rene Spitz (1945, 1946) made similar observations in an underdeveloped country in Central America where orphans were left unattended in foundling homes and deprived of human contact after being weaned. These infants had emotional, social, cognitive, and motor deficits. Most alarming was the high incidence of mortality (29.6%) in a sample of 91 infants up to 3 years of age who were followed for a 3-year period. This infant mortality rate was actually lower than that reported in foundling homes for infants in their first year of life in other studies reviewed by Spitz. For example, one large foundling home in Germany studied at the beginning of the century had a mortality rate in the first year of life of 71.5%. In a 1915 study of ten asylums in U.S. cities, the death rates of infants admitted during their first year of life varied from 31.7 to 75% by the end of their second year. Many of these fatalities were due to identifiable but incurable diseases. In 1940, when conditions in the U. S. improved and the general population mortality rate dropped below 10%, the institutional rates where still higher.

Spitz suggested that the effects of deprivation also caused physical illness, noting that "apart from the inadequate psychic and physical development, all these children showed a seriously decreased resistance to disease and an appalling mortality." The surviving children presented a new problem: "Institutionalized children practically without exception

Table 11.3 Sequelae of Childhood Physical Abuse Observed in Infants

Feeding difficulties
Proneness toward negative affects: sadness, distress, and anger
Shallowness in affective expression, with social and affective
  withdrawal
Weak attachment to their mother and caregivers, with diminished
  separation anxiety and stranger anxiety
Attention characteristics include frozen watchfulness — hypervigilance
  with intense scanning of the environment accompanied by motor
  passivity and immobility
Mistrustful of adults
Assaultive to their peers and caregivers in daycare
Central nervous system impairment

Source: Adapted from Green, A.H., in *Comprehensive Textbook of Psychiatry*, Vol. 2, 5th ed., Kaplan, H.I. and Sadek, B.I., Eds., Williams & Wilkins, Baltimore, MD, 1989, pp. 1962–1970.

developed subsequent psychiatric disturbances and became asocial, delinquent, feeble-minded, psychotic, or problem children." Most sobering, in the follow-up study, Spitz (1946) reported that the damage inflicted on the infants in the foundling home in their first year of life from "being deprived of maternal care, maternal stimulation, and maternal love, as well as by their being completely isolated, is irreparable." Even when placed in a more favorable environment on the ward of older children, where nurses constantly accompanied the children, the process of deterioration proved to be progressive.

Endocrinology mechanisms can explain some of the observations made by Anna Freud and Rene Spitz over 50 years ago. In one study of 11 prepuberal children subjected to a psychologically abusive environment with sexual abuse in six cases, failure to thrive was correlated with failure to secrete growth hormone (Albanese, et al. 1994). Encouragingly, the situation was reversible during hospitalization treatment.

Childhood physical abuse

The first report of child abuse as a clinical phenomenon that had an impact on the medical community was in the *Battered Child Syndrome* by Kempe et al. (1962). The focus of that paper was on physical signs, including bone fractures, hematomas, and soft-tissue injuries, that often resulted in disabilities or mortality. It is now estimated that 10 to 15% of all childhood accidents treated in emergency rooms are the result of physical abuse. Knutson (1988) and Green (1989) have written comprehensive reviews of the factors and issues involved in childhood physical abuse. The sequelae of physical child abuse for infants (Table 11.3) and school-age children (Table 11.4) are abstracted from Green's review. The importance of these characteristics with respect to EI patients is that many of them appear in their histories, and some continue into adulthood.

Childhood sexual abuse

"*Soul murder* is my dramatic designation for a certain category of traumatic experience: instances of repetitive and chronic overstimulation, alternating with emotional deprivation, that are deliberately brought about by another individual." (Shengold, 1989)

Table 11.4 Sequelae of Physical Child Abuse Observed in School-Age Children

Cognitive and developmental impairment
Post-traumatic stress disorder
Amplification, intensification of symptoms during exposure to events that symbolize
    or resemble the abusive situation
Pathological relationships
Lack of trust, hypervigilance, masochistic behavior
Primitive defense mechanism
Denial, projection, and splitting
Impaired impulse control
Poor self-concept and depression
Masochistic and self-destructive behavior
Suicide attempts, gestures, and threats
Difficulties in school adjustment
Central nervous system impairment

Source: Adapted from Green, A.H., in *Comprehensive Textbook of Psychiatry*, Vol. 2, 5th ed., Kaplan, H.I. and Sadek, B.L., Eds., Williams & Wilkins, Baltimore, MD, 1989, pp. 1962–1970.

"Soul murder" is a concept originating in literature, defined as taking away a person's reason for living and killing the instinct for love (Shengold, 1989). It is associated with brainwashing, to the point that the victim incorporates the beliefs of the tormentor and may even become like him/her. Of particular interest are the personality effects of infant sexual abuse when it occurs during the child's egocentric stage of development, when all causal attributions are to the self (Piaget, 1970). The child is left blaming itself for the pain and terror it experiences. This misattribution serves to maintain the perpetrator's secret, since revealing it implicates the self of the victim. With respect to the EI patients who have experienced soul murder, their denial takes the form of projecting the debilitating effects onto the physical environment, denying the etiology in their person environment.

Father-daughter incest is a form of traumatic childhood sexual abuse that is particularly disturbing because it occurs within the context of the child's main source of security and socialization, the family. Most incest victims must cope with multiple aspects of the experience:

1. Physical and psychological trauma in the form of the actual sexual experiences, including violation of one's body
2. Extended periods of apprehension, guilt, shame, and fear between sexual contacts (Lewis, 1995)
3. The loss of a trusted relationship with an emotionally significant person

If the abuse is perpetrated by a caregiving parent and the child lacks strong supportive mediating relationships with the other parent or others outside the nuclear family, development is compromised (Cole and Putnam, 1992).

In the years following publication of the *Battered Child Syndrome*, the concept of maltreatment was broadened to include emotional deprivation and neglect and sexual abuse (Steele, 1970, 1976). Because of the social, cultural, and psychiatric attitudes which prevailed when the current adult population in North America were children, psychological and sexual abuse were not usually considered much less assessed. For this and other reasons, much of the current literature on childhood sexual abuse relies on the histories of the victims, often inhibited from conscious memory until adulthood (Briere and Conte, 1993).

In addition to the complexity of problems associated with reliability and veracity of recollection, there are also methodological problems associated with operational definition and sampling (Briere, 1992b). It is difficult to assess the prevalence of a history of childhood abuse in the adult general population because of differences in choice of samples (Haugaard and Emery, 1989), various definitions of what constitutes abuse (Wyatt and Peters, 1986), unwillingness of individuals to disclose past episodes of abuse (Russell, 1983; Siegel et al., 1987), and inhibited memories of abuse.

One recent nationwide survey of adults reported a frequency of attempted or actual penetration in childhood of 14.6% for women and 9.5% for men (Finkelhor et al., 1990). With the enhanced awareness of abuse by clinicians, more studies have reported the effects of abuse on psychiatric patients (Beck and van der Kolk, 1987; Brown and Anderson, 1991; Bryer et al., 1987; Carmen et al., 1984; Jacobson and Richardson, 1987; Pribor and Dinwiddie, 1992), asymptomatic adults seen in a general medical practice (Briere and Runtz, 1988; Felitti, 1991), and patients seen in subspecialty clinics for disorders such as gastrointestinal disorders (Drossman et al., 1990) and paradoxical vocal cord dysfunction masking as asthma (Brown et al., 1988; Christopher et al., 1983; Freedman et al., 1991; Selner et al., 1987).

The incidence of father-daughter incest is far more common than previously realized. The incidence of father-daughter incest has recently been estimated to be about 1.5% (Finkelhor et al., 1990). Sexual abuse has become a leading concern in the treatment of psychiatric disorders and mental health research, uncovered from the cloak of societal secrecy which was partially draped by traditional psychoanalytic theory and only lifted in the past 25 years (Simon, 1992a; Shapiro, 1993).

Early childhood trauma resulting from chronic sexual and physical abuse often, if not always, disrupts personality and personality development, which can result in complex defenses to keep these memories from consciousness. Psychological mechanisms which have been hypothesized to explain this phenomenon include repression, dissociation, denial, and somatization.

In children, unusual somatic or psychosomatic symptoms and behavior problems have been identified as presenting features of concurrent abuse (Hunter et al., 1985; Gilbert, 1988). PTSD symptoms are present in many abused children and can persist into adult life, becoming chronic and integrated into the personality and manifesting as psychiatric illness (Browne and Finkelhor, 1986; Gelinas, 1983; Herman et al., 1986; Krystal, 1978). For example, in a comparison study of 60 women with somatization disorder and 31 women with major affective disorders, the percentage of childhood sexual abuse was significantly greater in the somatization group (Morrison, 1989). Childhood sexual abuse reported in adulthood has also been associated with dissociative symptoms and somatization disorder, although the causal relation among these factors remains to be determined (Pribor et al., 1993). There is an increased incidence of childhood sexual trauma in hypochondriacal patients (Barsky et al., 1994).

Ill-defined, multi-system symptoms of EI patients may be related to past abuse scenarios, as demonstrated in studies focusing on one of these systems (Brier and Runtz, 1988). For example, in studies of the genitourinary system, the prevalence of childhood sexual abuse (broadly defined) is higher (64%) in women with chronic pelvic pain than in women with specific gynecologic conditions (23%) (Walker et al., 1988). The women with chronic pain also had a significantly higher number of other medically unexplained somatic symptoms (Walker et al., 1995).

Table 11.5   Adult Sequelae of Father-Daughter Incest

Mistrust
Poor self-image
Depression
Hysterical symptoms and character traits
Social withdrawal and impaired peer relations
Poor school performance and cognitive impairment
Substance abuse
Delayed post-traumatic stress disorder
Adult sexual relationships are impaired by incest-related fear and guilt
Increased sexual arousal
Sexual avoidance and inhibition

Source: Adapted from Green, A.H., in *Comprehensive Textbook of Psychiatry*, Vol. 2, 5th ed., Kaplan, H.I. and Sadek, B.I., Eds., Williams & Wilkins, Baltimore, MD, 1989, pp. 1962–1970.

Symptoms in the gastrointestinal system are expressed by patients with irritable bowel syndrome, a stress-related disorder (Drossman et al., 1988, 1990). Compared with patients diagnosed as having inflammatory bowel disease, patients with irritable bowel syndrome had a significantly higher rate of severe lifetime sexual trauma (32 vs. 0%), severe childhood sexual abuse (11 vs. 0%), and any lifetime sexual victimization (54 vs. 5%) (Walker et al., 1993). In another study in a gastroenterology clinic, childhood sexual/physical abuse was significantly more prevalent among patients with irritable bowel syndrome (82%), patients with gastroesophageal reflux (92%), and patients with idiopathic chest pain (27%) (Scarinci et al., 1994). Abused patients in that study also reported a significantly greater number of pain syndromes unrelated to gastrointestinal disorders. Neuromuscular pain symptoms associated with fibromyalgia have also been associated with psychosocial factors (Bradley et al., 1992).

Somatization was one of three factors (the others being dissociation and affect dysregulation) identified as sequelae of trauma in the DSM-IV field trial for PTSD (van der Kolk et al., 1996). That case-controlled study was carried out in four hospitals, involving 395 traumatized patients receiving treatment who were compared to a group of 125 non-treatment-seeking controls who had also been exposed to traumatic experiences. For a comprehensive review of somatoform disorders as sequelae of sexual abuse in childhood, the reader is referred to Loewenstein (1990).

Women incest survivors may have their adult sexuality negatively affected in a variety of ways (Table 11.5), including psychological aspects of intimacy and the physical aspects of comfort and pleasure (Courtois, 1979; Herman, 1981), as well as disturbances in sexual behavior and gender role (Scarinci et al. 1994). Two contrasting adaptive styles have been observed in adults. One seeks mastery through active repetition of the trauma through promiscuous sexual behavior, while the other copes by avoiding sexual stimuli (Green, 1989). In patients with dissociative identity disorder, there is often the paradoxical manifestation of both types of behavior. Based on clinical experience, many female EI patients with a history of childhood abuse report sexual dysfunction and express sexual displeasure. Those with co-morbid PTSD often report having flashbacks during sexual relations.

A case of sexual abuse and EI

The following case is one of many treated successfully in our clinic after the patient had been subjected to unsubstantiated EI practices. A 35-year-old white female presented with intense headaches and gastrointestinal symptoms dating back to early childhood. She reported that whenever she smelled cooked foods containing tomatoes, she developed a headache, an upset stomach, and nausea. She had been evaluated by a clinical ecologist for headaches by the method of sublingual provocation food testing. She questioned the validity of the food testing when she was declared "allergic" to everything for which she was tested and indicated that symptoms induced by provocation testing were not discriminated from symptoms already present before the testing session had begun. Stress or psychological factors were not considered as a possible mechanism for her illness.

In our evaluation, the patient revealed traumatic experiences in childhood. Her father was physically violent and sexually abusive with all six of his children. She was the oldest daughter and reported being subjected to sexual abuse not only by her father but also by his friends, for money. In the home, she was also assaulted by her brothers, one of whom was institutionalized as a violent sex offender when a teenager and is still incarcerated as a middle-aged adult; the other was imprisoned as an adult for money fraud. She recalled that a friend of her father, a policeman, would often take her and her brothers to the local jail to show them what could happen to them if they did not comply with sexual demands. This left her with fear and resentment of authority figures. To compound the trauma, she reported rape by a policeman when she was living alone as an adult, away from her family. Also devastating for her was the knowledge that her mother was aware of the abuse but had failed to protect her. During therapy she confronted her mother about her memories and her mother denied them. Over 3 years of psychotherapy, many of these issues and traumas were successfully addressed and their impact resolved. The patient sued her father, at a time when there were legal hurdles as to the court's jurisdiction across state lines and statute of limitations; she noted that this was the first time in her life that authority figures had supported her. The lawsuit was settled out of court. Before the father died of cancer, she was able to face him and establish that he no longer intimidated her or had control over her.

The patient's siblings entered counseling with independent therapists in another state and have acknowledged and corroborated the patient's history. The oldest brother related an explicit account of how he, at age seven, witnessed our patient as a 4-year-old being sexually assaulted by her father on the family kitchen table, which was now in the patient's home. After the rape, he recalls his mother cleaning blood from the table and washing the patient's clothes. The patient had not recalled the event until it was disclosed by her older brother. She subsequently took an ax to the table and burned it.

In recalling the events of her childhood, she realized that spaghetti sauce was her father's favorite food and the staple of her parents' household. The odor of cooked tomatoes was a sensory stimulus for the anxiety associated with these memories that she had inhibited from consciousness. The headache and stomach pain represent the somatization of anxiety. Treatment consisted of biofeedback assisted relaxation training in conjunction with supportive and cognitive therapy. She was able to overcome her aversion to the smell of cooked tomatoes and now is able to enjoy Italian food. This case represents a remarkable recovery over a 3-year period of psychotherapy. The significant social factors included an unconditionally supportive and loving husband and siblings who came forward to corroborate her story. Currently, she is a highly functioning professional and appears to be succeeding in attaining a normal and happy family life with her husband and children. In her mind, the terror of the past has been put to rest, although she continues to have occasional exacerbations of PTSD symptoms when unresolved issues with her family arise, e.g., the imprisonment of her abusive brother who denied wrongdoing even though he swindled innocent investors, including the patient.

This case illustrates how someone raised in an abusive, traumatizing family environment can keep the original trauma out of conscious awareness only to have it surface as somatization. The iatrogenic influence of a clinical ecologist was demonstrated by the use of unsubstantiated sublingual provocation testing results to attribute symptoms falsely to food sensitivity. This case demonstrates that EI patients are not doomed to a life of avoidance of foods and chemicals falsely believed to be the cause of symptoms and misery.

Neurobiologic effects of childhood sexual abuse

There is increasing evidence that the emotional and psychophysiologic sequelae associated with severe stress and trauma have neuroendocrine biological markers (Chrousos and Gold, 1992). Associated neuroendocrine receptors which affect the HPA axis, e.g., glucocorticoid receptors, have implicated circulating corticosteroid concentrations (Yehuda et al., 1991). Elevated levels of catecholamines (Kosten et al., 1987; Mason et al., 1988; Yehuda et al., 1992) and endogenous opioids (van der Kolk et al., 1989) have been reported. Chronic neuroendocrine activation was also observed in a group of women with PTSD and history of childhood sexual abuse (Limieux and Coe, 1995).

In three companion studies, DeBellis and colleagues studied a group of 13-year-old girls with a prior history of sexual abuse (defined as genital contact) and concurrent symptoms of psychopathology including depression and suicidal ideation. Measurements of catecholamines, neuroendocrine parameters, and antinuclear antibodies (ANA) were taken 5 years on average after disclosure of the abuse to social services. Group comparisons were made to non-abused, demographically matched control subjects. In the first study, DeBellis et al. (1994b) found significantly higher total catecholamine functional activity in the abused girls. Total catecholamine synthesis was measured as the sum of NE, EPI, DA, and their metabolites, and no special dietary restrictions were placed on the subjects. The additive yet non-specific effects of catecholamine elevations were noted to be similar to the psychobiology of both adult and childhood depressive disorders and the effects of PTSD.

In the second study, DeBellis et al. (1994a) found neuroendocrine abnormality consisting of alterations in the HPA axis, independent of co-morbid history of depression. Mean basal ACTH plasma levels and mean CRH-stimulated secretions of ACTH were significantly lower in abused girls than in non-abused controls. The blunted ACTH response to CRH suggests the possibility of excessive endogenous CRH hypersecretion. These responses in the abused girls were similar to those observed in depressed patients. However, unlike depressed patients, mean basal plasma cortisol, mean CRH-stimulated plasma cortisol secretions, and 24-hour urinary free-cortisol levels were not different for the groups. These results show that sexually abused girls, on average of 5 years after disclosure of abuse, manifest dysregulatory disorder of the HPA axis, associated with hyporesponsiveness of the pituitary to exogenous CRH and normal overall cortisol secretion. DeBellis et al. suggested that the attenuated plasma ACTH, with corresponding robust plasma cortisol responses to CRH stimulation, indicate hyperresponsiveness of the adrenal cortices to ACTH.

In an editorial accompanying the publication of this study, Rosler (1994) made the following observations. The results imply the existence of a normally functional negative feedback control mechanism in which pituitary secretion of ACTH is reduced to guarantee a normal cortisol output. This compensatory mechanism would be operative on a chronic basis in patients with a history of childhood sexual abuse so that less ACTH should be necessary to maintain normal cortisol secretion, a situation that would also imply compensatory hyperresponsiveness of the adrenal cortex as suggested by DeBellis and colleagues.

Other research has shown lowered basal urinary-free cortisol levels associated with PTSD in veterans (Mason et al., 1990) and elderly holocaust survivors (Yehuda et al., 1995b). In interpreting these studies, Charney et al. (1993) suggested that certain central inhibitory mechanisms suppressing CRH and ACTH function may be increased in chronic PTSD, resulting in decreased basal cortisol level. The suppression or inhibition of the HPA axis responses may reflect the phenomenological symptoms of numbing as a coping mechanism observed clinically in PTSD (van der Kolk and Saporta, 1991). Further studies may substantiate and enhance this hypothesis.

DeBellis and associates also commented on the differences they previously reported in a cohort study of sexually abused girls from the same population, studied within one year of disclosure of abuse, who showed elevated morning plasma cortisol levels (Putnam et al., 1991). The investigators speculated about hypothetical negative feedback control mechanisms to account for the differences in short- and long-term effects on cortisol levels in these young girls. It is significant that depressed adults show continued elevated 24-hour cortisol excretion, and aging has been postulated to account for the failure to adapt (Jacobs et al., 1984).

DeBellis and associates have demonstrated that childhood sexual abuse is not only a traumatic experience that causes psychiatric morbidity, but one that is associated with clear and sustained changes in the dynamics of the HPA axis. These investigators also caution that the age of the individual, the severity of the abuse, and genetic predisposition may contribute to the degree of psychopathology and HPA axis dysregulation.

The findings of elevated catecholamines associated with differences in cortisol levels were commented on earlier by Rose (1980) who speculated that the hormonal differences were related to cognitive appraisal of events. Two events were compared, one representing perceived threat of harm, the other challenge with an achievement reward. Both types of events had associated elevations in catecholamine levels. However, the hormonal response was different for the two events, with elevations in cortisol level associated with perceived threat or harm, but not with challenge.

These cognitive formulations may help explain the effects observed in the young girls studied by DeBellis and associates. When the girls disclosed their abuse to social services, they were removed from the situation. The effects observed about one year later — elevated basal plasma cortisol levels — may reflect residue of the trauma they underwent while in the abusive home or the fear that they might be returned. Much like the combat veterans who maintained vigilance and fear of the environment after returning from war, the girls may have continued to anticipate harm. Five years after removal from the abusive situation, the vigilance remained as reflected by the elevated catecholamine levels, but the threat of eminent harm seemed to have adapted or resolved as reflected in the normalization of basal plasma cortisol levels.

In the third companion study with the same sexually abused adolescent study group, DeBellis and colleagues postulated abnormalities in thyroid functioning and plasma antinuclear antibody tests (DeBellis et al., 1996). An antinuclear antibody (ANA) refers to an antibody against various proteins or nucleotides within the cell nucleus. It is hypothesized that elevated ANA titers reflect a failure to inhibit autoreactive lymphocytes through hypofunctioning of the immune cells that actively suppress autoimmunity (Agius and Arnason, 1991). In reviewing prior clinical studies, DeBellis et al. noted that some studies have reported a higher incidence of antinuclear antibodies in adults with depression (Maes et al., 1991, 1993). Their results showed that thyroid function tests and plasma antinuclear antibody titers did not differ between sexually abused and matched control adolescent girls. However, it was noted that the control group was biased to positive reactions because of the inclusion of African-American girls, who have a higher prevalence of autoimmune phenomena (Steinberg, 1992). DeBellis et al. then employed a second control group of 22 relatively older healthy white female volunteers, ages 20 to 58 years, who showed a significantly lower incidence of positive ANA levels (even though older age is associated with greater risk) compared to the sexually abused girls in the study group. The investigators concluded that this is an exploratory investigation suggesting that sexually abused girls may show evidence of an alteration in normal immune homeostatic function.

Conclusion

When there is a documented, high-dose chemical exposure, the event may meet DSM-IV diagnostic criteria for PTSD. I treated a maintenance worker who experienced near-suffocation in response to all-day exposure to a solvent (toluene) used to remove paint from a fan while alone in a closed electrical room without ventilation. Fortunately, he was working down the hall from the hospital emergency room and staff helped him when he stumbled out into the hall. He was stabilized but did not recover fully. He experienced the onset of asthma, which was treated, and PTSD, which was not recognized immediately. He had flashbacks to the episode and experienced anxiety whenever he passed by the electrical room in which he was exposed. He developed intolerance of odors to solvents and similar smelling chemicals. He continued to work but was unable to wear a respirator because it impeded his breathing and triggered PTSD symptoms, including terror of suffocation. He was enraged by the callous disregard of his condition by his co-workers and the failure of his physicians to recognize the severity of his condition. With long-term psychotherapy, the anger has dissipated but the sadness over his loss of function in work and recreation and his financial difficulties remain. Systematic desensitization helped him overcome his odor aversions, but it has not been successful in overcoming his fear of suffocation when wearing a respirator.

In contrast to this case of industrial asthma and PTSD, EI patients do not meet the event criteria for PTSD on the basis of their alleged chemical exposures. The criterion that "the person's response involves intense fear, hopelessness, or horror" does not apply to EI patients who retroactively make the attribution to a chemical exposure that was considered benign at the time of occurrence. The Presidential Advisory Committee found no accounts of acute chemical exposure complaints from soldiers during the Persian Gulf War. This precludes attribution of PTSD to any since discovered chemical exposure event, e.g., the weapons depot destroyed by explosion at Khamisiyah, Iraq. An event deemed traumatic *ex post facto* does not appear to meet the DSM-IV event criteria for PTSD. Effects of chronic low-level chemical exposure, even if they could be substantiated in EI patients, would not meet the event criteria for PTSD. EI patients who have identifiable underlying psychological trauma which could explain legitimate PTSD symptoms either do not report it or deny association to current symptoms. The relevance of any psychological trauma often is lost to both physician and patient in the current clinical picture dominated by the belief of a toxic origin of EI.

chapter twelve

The limbic system and trauma

The limbic system is the focus of toxicogenic mechanistic hypotheses such as "olfactory limbic kindling" or "partial limbic kindling" (Ashford and Miller, 1991; Bell et al., 1992). EI advocates suggest limbic kindling as one pathophysiologic mechanism to account for panic symptoms (Bell et al., 1992). Functionally, the effect observed is "seizure like" and is to be triggered by inhaled chemicals, the odor of which is transported from chemoreceptors in the nose via the olfactory bulb to the limbic system, where a variety of affective responses are kindled. "Partial limbic kindling" hypothesizes that odor elicits behavioral and emotional symptoms analogous to cortical electrical stimulation eliciting emotional responses independent of psychological factors.

This chapter reviews the structure and function of the limbic system and illustrates how it is interconnected with the stress-response. Studies of electrical stimulation elicited emotions, and emotions in turn activated the limbic system. The reticular activating system is also described because of its essential role in attention and vigilance. Physiological studies of patients with post-traumatic stress disorder (PTSD) associated with trauma are presented to illustrate (1) how they have different state responses to startle and recall of trauma, and (2) how their basal levels of stress physiology are dysregulated from homeostasis. Also addressed are the structural abnormalities in the hippocampus measured by magnetic resonance imaging (MRI) and the functional abnormalities in the brain measured by brain waves (EEG). Related studies with EI patients are also presented.

The limbic system

The limbic system is integral to the stress-response. Studies of the mechanisms of emotion and its psychophysiological correlates invariably implicate brain regions traditionally described as components of the limbic system. EI patient symptoms mediated by the limbic system are non-specific with respect to origin and can be induced by thinking and feeling as readily as by an environmental agent. Individual differences are noted with respect to susceptibility to limbic activation, associated with beliefs and personality traits.

The limbic system (Figure 12.1) is not a specific anatomical region, but rather a collection of neuroanatomical regions adjacent to the cortex, forming a "rim" around it which includes the anterior thalamus, hippocampus, amygdala, and the prefrontal and temporal lobe regions of the cortex (Nauta and Feirtag, 1986). The sensory receptors,

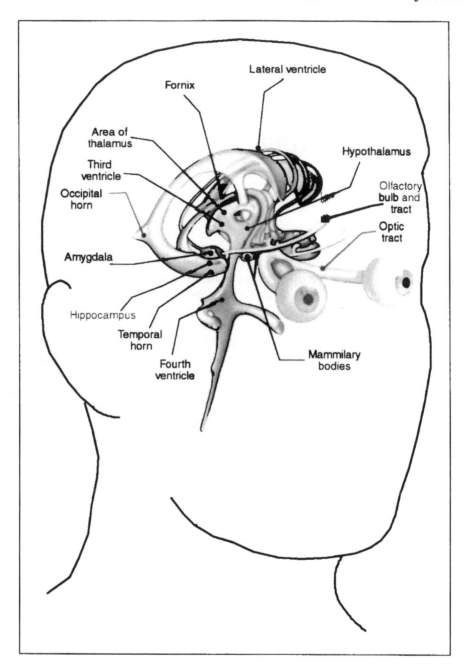

Figure 12.1 Illustration of the limbic system. (Adapted from Fincher, J., *The Brain: Mystery of Matter and Mind*, Torstar Books, New York, 1984.)

particularly the olfactory bulbs and tracts, are intimately connected with it (Getchell et al., 1991; Livingston, 1978). The hypothalamus is also included, although some consider it a separate system (Swanson, 1983). Brain areas at every level of the neural axis, from the cerebral cortex to the spinal cord, are interconnected with the limbic system through the ascending and descending pathways of the hypothalamus (Isaacson, 1982). Thus, the limbic system is a complex of brain structures interconnected by a variety of neural networks with both afferent and efferent pathways to other brain regions and peripheral systems that regulate neuronal catecholamines, autonomic arousal, endocrine hormones, and immune competence (Chrousos, 1995; Isaacson, 1982; Livingston and Hornykiewicz, 1978). It is difficult to imagine a neurobiological process which is not linked with the limbic system. This has led some to criticize the inclusion of so many brain areas in the limbic system (Brodal, 1982, quoted by LeDoux, 1987):

> "The limbic system appears to be on its way to including all brain regions and functions. As this process continues, the value of the term as a useful concept is correspondingly reduced ... it is the author's opinion that the use of the term limbic system should be abandoned."

Regardless of the position taken on the boundaries of the limbic system, it remains useful to consider individual limbic structures when seeking the neural substrates of specific emotional functions (LeDoux, 1987).

Electrical stimulation of emotion

Historically, the limbic system has been described as the "visceral brain", initially thought to be the anatomical storehouse of emotions (MacLean, 1949, 1952, 1955). Early research on electrical stimulation of the limbic system in animals showed that stimulation of certain regions in the thalamus and hippocampus could produce the behavioral effects of normal pain and fear (Delgado et al., 1954). Penfield explored most of the cortical areas with electrical stimulation in the course of neurosurgery on human patients with focal epilepsy who were conscious, having received only local anesthesia (Penfield, 1956; Penfield and Jasper, 1954; Penfield and Perot, 1963). Feelings of fear, sorrow, loneliness, or disgust were elicited by stimulation of the anterior and inferior temporal cortex of the limbic system, but not other areas of the cortex and neocortex. Reports of the hallucinatory effect of temporal lobe stimulation are particularly relevant to episodic memory recall of feelings and emotions, which patients described as vivid re-enactment (Penfield and Roberts, 1959). Similar memory effects were localized in the hippocampus, which had been characterized as the "mechanism which reactivates the record" (Roberts, 1961).

Subsequent research found that memory for emotions is not so localized but is distributed widely throughout regions of the brain. During recall, this information is reintegrated or reconstructed. Limbic structures may have the capability to evoke experiential phenomena because of their afferent and efferent connections to widespread areas of the cortex and neocortex (Squire, 1987). Penfield's studies were criticized for lack of specificity of location, i.e., stimulation of different sites produced the same emotional response. Also, the results were not reliable, i.e., stimulation of the same site at different times in the same patient did not reproduce the same emotional response. The stimulation effects did not generalize to the population but were limited to less than 10% of the surgery patients. Penfield's initial interpretations that the responses were veridical reproductions of past experience and that the neural connections permitting access to them were located near the stimulating electrode were incorrect (Squire, 1987).

Nonetheless, fear and anxiety were commonly reported following direct electrical stimulation of the medial temporal region of the limbic system which includes the hippocampus and amygdala (Gloor et al., 1982; Halgren et al., 1978). In these studies, a response was associated with widespread electrical effects radiating from the stimulation site. The biasing effect of cognition was demonstrated in a single-case study in which continued interviewing of the patient during surgery showed that the responses were sometimes clearly associated with the thoughts of the patient at the moment of stimulation (Mahl et al., 1964). Other studies about the mental experiences elicited by electrical stimulation showed that it is difficult to distinguish among memory, dreams, fabricated reconstructions, and fantasies (Squire, 1987). In addition, no association was found between any anatomical site and a specific experience such as the feeling of fear, the sensation of an odor, or complex visual hallucinations (Horowitz et al., 1968).

Individual differences in emotions learned through experience would be expected. The mental phenomenon elicited depended more on the individual personality and expectation characteristics of the patient (LeDoux, 1996; Halgren et al., 1978). Numerous theories suggest that the limbic system plays a critical role in the inhibition of traumatic memory through processes of amnesia, repression, and dissociation. Neurophysiologically, the amygdala has been metaphorically called the hub in a wheel of fear representing other parts of the cortex and components of the midbrain, collectively referred to as the "emotional brain" (LeDoux, 1996).

Emotions activate the limbic system

"Do thoughts and emotions activate the limbic system?" In a brain-imaging study using positron emission tomography (PET) in healthy, normal adult women, the induction of transitory emotions of happiness and sadness were associated with significant and unique changes in regional cerebral blood flow in the limbic system (George et al., 1995). However, the causal inferences are limited because the regions that demonstrate changes in blood flow may be an epiphenomenon, perhaps related to attention or associated memories and not directly involved in emotion (Robinson, 1995). This caution highlights the interconnections among emotions, memory, and cognition.

The interactions of these factors are studied experimentally in humans using uncontrollable tasks or unsolvable problems. The physiological effects on humans are well documented and include passivity, cognitive deficits, and emotional change. A PET study measuring the regional cerebral blood flow clarified the role of specific subcortical neuronal activity in the limbic system induced by unsolvable anagram tasks (Schneider et al., 1996). Subjects without psychiatric, neurological, and general medical disorders performed both solvable and unsolvable anagrams while in a PET scanner and also rated their perceived level of negative affect which turned out to be greater during the unsolvable problems. In post-experiment interview, all subjects attributed their failure to solve the unsolvable anagrams to their own inability (self-attribution of failure). Of 27 cerebral regions evaluated, the main finding differentiating the anagram conditions was that regional cerebral blood flow increased in the mammillary bodies and the amygdala and decreased in the hippocampus for unsolvable problems. In conclusion, the authors suggested that the activation of the amygdala seems related to the negative affect, while the deactivation of the hippocampus may relate to the passivity and cognitive retardation. These two limbic structures are implicated in assigning affective importance to stimuli, integrating cognition and emotion, problem solving, and memory (LeDoux, 1995; Squire, 1993; Squire et al., 1993).

Studies like this not only demonstrate the confounding roles that cognition and emotion play in activation of the limbic system, but also highlight the nonspecificity of limbic activation. Nonspecificity underlies the precautions expressed concerning speculations

Table 12.1 Physiological Correlates of the Orienting Response

1. Cessation of motion and activity
2. Increased muscle tension at rest
3. Autonomic nervous system activation:
    a. Alterations in heart rate
    b. Alterations in respiration
    c. Pupillary dilation
    d. Peripheral vasoconstriction
    e. Activation of electrodermal response (palmar sweating)
4. Lowered thresholds to respond to sensory stimuli

Source: Adapted from Lynn, R., *Attention Arousal and the Orientation Reaction*, Pergamon Press, Oxford, 1966.

about brain-imaging technologies as indicating causal effects of low-level chemical exposure on neuronal systems. Given these findings, appraisal of a reaction attributed to a perceived exposure event is sufficient to activate the limbic system and associated pathways of the human stress-response.

Reticular activating system

The reticular formation, which is made up of a network of neural fibers that runs throughout the center of the brain stem, interacts with the limbic system, particularly the hypothalamus. The reticular formation mediates the processes of arousal and sensory activation referred to as the reticular activating system. It also modulates autonomic nervous system (ANS) activity, particularly arousal of sympathetic end organs, including muscular tension, pupillary dilation, cardiovascular response, galvanic skin response, and glandular secretions (Lindsley, 1960). Centrally, it controls overall arousal as well as specific attentional processes associated with selective activation of specific areas of the cortex associated with vigilance, orienting response, and signal detection (Magoun, 1963; Moruzzi and Magoun, 1949).

When an organism is faced with a novel or significant stimulus, it aligns its eyes, ears, or nose to the stimulus and reacts with an orienting response (Sokolov, 1963), which has several physiological correlates (Table 12.1). The mechanisms of the orienting response include the mesencephalic reticular formation (Moruzzi and Magoun, 1949), part of the reticular activating system which can affect the excitability of neurons to their natural stimulation. In non-human primate studies, this enhancing effect is most apparent when the subject is not alert, but is either drowsy or not paying attention. When the subject is alert or oriented, reticular stimulation has little additional effect in that it does not enhance perception. The implications of this to EI will be discussed shortly with respect to olfactory sensory threshold.

Selective attention differs from arousal in that some members of a given class of stimuli, e.g., odors, are treated preferentially by the organism. At the electrophysiological level, this principle has been shown by the selective enhancement of the response of sensory neurons to their appropriate stimuli when that stimulus has some biological significance to the animal's behavior. Sensory thresholds have been demonstrated to be affected by levels of arousal in the visual system, the auditory system, and the somatosensory system.

Anatomically, selective attention is mediated through neural systems in the parietal cortex which integrate sensory information, modulated by nonsensory stimuli generated from the limbic system. The limbic system projects strongly to the parietal cortex, and these connections may provide an anatomical substrate by which motivational activity (e.g., biological needs, psychological set, long-term goals) can influence stimulus processing (Heilman and Watson, 1977). Experimentally, a learned effect may be demonstrated by a neutral orienting stimulus that precedes the target stimulus (Posner and Petersen, 1990). For example, when a primate attends to a location, events occurring at that location are responded to more rapidly (Eriksen and Hoffman, 1972; Posner, 1988), give rise to enhanced scalp-evoked potential (Mangoun and Hillyard, 1987), and can be reported at a lower threshold (Bashinski and Bachrach, 1984; Downing, 1988).

In addition to motivational factors, novelty and the interest of the stimuli also enhance selective attention. This was first observed in the cat auditory cortex where neurons responded better to specific stimuli such as the squeaking of a toy mouse (Hubel et al., 1959). The orienting response is not necessarily constant across replications of a stimulus. Repeated presentation of a stimulus can lead to a reduced cortical response called "habituation". If the stimulus becomes totally insignificant and the organism ceases to respond to it completely, it is called "extinction". A failure of the selective attention process may also lead to inappropriate distraction, phenomenologically characterized as "inability to concentrate". Selective attention, arousal, habituation, and extinction are closely related and interdependent processes in the orienting response (Sokolov, 1963).

Cognitive processes are intricately mediated and modulated by, and in turn can affect, the limbic and reticular activating systems. Thinking probably results from the momentary pattern of stimulation of many different parts of the nervous system at the same time, involving most importantly the cerebral cortex, the thalamus, the limbic system, and the upper reticular formation of the brain stem. The stimulated areas of the limbic system, thalamus, and reticular formation perhaps determine the general nature of the thought, giving it such qualities as pleasure, displeasure, pain, fear, anxiety, comfort, crude modalities of sensation, localization to gross areas of the body, and other general characteristics.

The cerebral cortex contains elaborate memory representations for knowledge, affect and emotion, sensation, and motor action. Contextual information gives meaning to memory networks. The processes of storage and retrieval of knowledge in memory and the higher order cognitive processes of reasoning and productive thinking are affected by the brain systems below the cerebral cortex. The pervasive involvement of limbic activation in human social, cognitive, and emotional processes makes it difficult, if not impossible, to identify experimentally variables or manipulations which uniquely and specifically affect the components of these anatomical regions (Joseph, 1992).

Disruptive effects of trauma

In 1987, van der Kolk and Greenberg (1987) proposed the bold and provocative theory that repeated traumatization, particularly child abuse, may disrupt normal development of the brain and thus may contribute to psychiatric sequelae. This theory was consistent with the theory concurrently proposed by Kolb to explain PTSD in veterans (Kolb, 1987). The core postulate is that excessive stimulation of the CNS at the time of the trauma may result in permanent changes in brain systems that affect the cognitive processes which regulate learning, habituation, and stimulus discrimination.

This neurogenic theory stands in contrast to the misconception commonly held by the public that the effects of early abuse are completely psychologic in nature. The defensive expression by patients — "they think its all in my head" — has very different implications given current developments in neuroscience. Be that as it may, van der Kolk and Greenberg postulated a specific mechanism of the limbic system — amygdaloid kindling — to explain neurological abnormalities with behavioral manifestations of impulsivity, aggression, and promiscuous sexual activity in adults who had been abused as children. In general, studies of traumatized patients have shown a sensitization of several biological systems including those of the stress-response (DeBellis and Putman, 1994; Southwick et al., 1995b). The sensitization of biological and behavioral responses observed in traumatized patients is opposite of the attenuated, adapted, or dysregulated biological and behavioral responses seen in clinically depressed patients. Observations of sensitization have led to theories of the neurogenic hypothesis of trauma which include animal models of kindling (Post, 1992) and time-dependent sensitization (Antelman and Yehuda, 1995; Yehuda and Antelman, 1993).

The pathophysiology of trauma may involve dysfunction of several brain structures, particularly the amygdala and the hippocampus in the limbic system. Also affected are the neurotransmitter systems, particularly the noradrenergic systems, and the HPA-axis neurochemical systems which represent two of the three major pathways of the stress-response (Charney et al., 1993), described in detail in Chapter 10. In short, all of the systems, mechanisms, and pathways involved in the stress-response may be affected by trauma. According to Cannon's fight-or-flight model, acutely severe psychological trauma activates the stress-response for the purpose of adaptation. Normally, with resolution of the stress or trauma, the systems return to homeostasis. In contrast, Kolb suggested that acute PTSD responses do not adapt in a state response fashion, but instead evolve into maladaptive neurophysiological and neurobiological sequelae which become traits (Southwick et al., 1994).

The amygdala has been linked to fearful responses which are characteristically associated with sensory memory (LeDoux, 1996). Numerous studies have shown that simple sensory memories such as tastes and odors which have prior associations with a traumatic event may serve to trigger PTSD responses (van der Kolk, 1994). Charney and colleagues proposed a neural model (Figure 12.2) in which the amygdala is the structural basis for memory of trauma and related behaviors (Charney et al, 1993). Neural mechanisms of fear conditioning, extinction and sensitization involve certain brain regions and neurochemical systems and contribute to the persistence of traumatic memories and associated symptoms. The amygdala plays a necessary role in processes involved in traumatic memories because of its extensive connections to all of the sensory memory systems in the cortex. The interconnection of these pathways with the amygdala may help explain how neutral stimuli come to be associated with fear, anger, pain, and other negative emotions and experiences associated with trauma (Pincus and Tucker, 1985). For example, the central nucleus of the amygdala plays a critical role in the fear-potentiated startle response because it has direct neuronal projections to one of the brain-stem nuclei necessary for startle (Rosen et al., 1991).

The amygdala is also one of the brain structures most susceptible to kindling as demonstrated in animal studies, which results in long-term changes in neural excitability that can manifest as disruptions in behavior (Adamec, 1978) and, in the extreme, result in seizure activity (Post et al., 1984). In these animal studies, the triggers of limbic kindling responses were restricted to electrical stimulation or drugs. van der Kolk hypothesized that chronic traumatic child abuse may lead to limbic kindling and to associated neurological abnormalities, which can lead to inappropriate affective expression and behavior (van der Kolk and Greenberg, 1987).

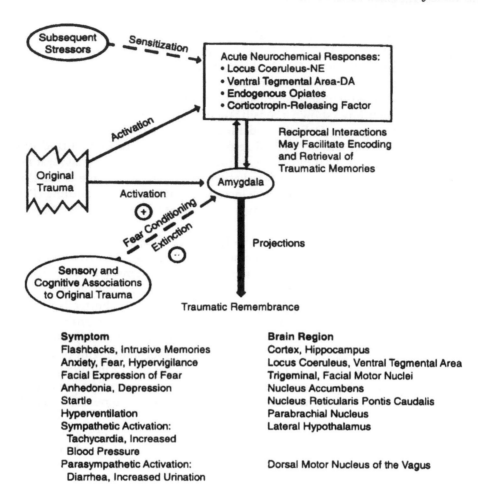

Figure 12.2   The role of the amygdala in memories of trauma. (From Charney, D.S. et al., *Arch. Gen. Psychiatry*, 50, 294–305, 1993. With permission.)

Limbic system symptoms

Teicher and colleagues presented data from the self-report questionnaire with 33 items called the Limbic System Checklist (LSCL-33) to corroborate the hypothesis that psychopathology sequelae seen in adults are related to traumatic abuse in childhood (Teicher et al., 1993). Subjects rated the lifetime frequency with which they experienced LSCL-33 symptoms on a four-point scale from never to often. Validation of the symptoms was done with a positive control group of patients with temporal lobe epilepsy who scored high on the LSCL-33, as expected. Criterion-related concurrent validity was evaluated by having psychiatric subjects complete two self-report measures with overlapping domains of inquiry: the Dissociative Experience Scale (DES) (Bernstein and Putnam, 1986) and the Hopkins Symptom Checklist (SCL-90-R) (Derogatis, 1983).

Not surprisingly, the LSCL-33 correlated very highly with the DES as both scales have some similar content items. The results are consistent with research showing increased symptom report on the DES in psychiatric patients with a history of childhood physical and sexual abuse (Chu and Dill, 1990). The LSCL-33 also correlated with the SCL-90-R subscales of somatization and psychoticism. The SCL-90-R has been employed on EI patients in several independent studies, all of which found significant elevations in the somatization scale compared to control groups (Black et al., 1993; Selner and Staudenmayer, 1992b; Simon et al., 1990).

In Teicher's study, history of physical and sexual abuse was assessed using a self-report measure, the Life Experience Questionnaire (LEQ) (Bryer et al., 1987). The LSCL-33 and the LEQ were administered to 253 outpatients, 58% female, who presented for psychiatric evaluation. Overall, 56% reported a history of some abuse during their lifetime. Based on self-report, patients were placed in four categories as follows: no abuse, 44%; physical abuse only, 30%; sexual abuse (genital contact) only, 10%; and both physical and sexual abuse, 16%. Comparison of LSCL-33 scores for the four groups was statistically significant. Relative to non-abused controls, the magnitude of effect was progressively greater for physical abuse (38% greater), sexual abuse (49% greater), and both physical and sexual abuse (113% greater).

Teicher and colleagues were careful to point out the limitations of these findings. First, the LSCL-33 is a new self-report instrument and validation of limbic system dysfunction with physiologic and biochemical markers needs to be established. Second, these findings are correlational, and no claims as to a cause-effect relationship could be established.

PTSD, limbic abnormalities, and trauma

Pitman (1997) reviewed the various explanations that have been suggested for the association of abuse, limbic system abnormality, and post-traumatic stress disorder. One hypothesis is that limbic system dysfunction makes the child more susceptible to being abused (Davies, 1979). An extension of this hypothesis is that limbic system dysfunction may be a hereditary disturbance passed from generation to generation. Irrespective of etiology, such predisposition causes the child to act in a way that invites abusive behavior on the part of parents or others. There is one aspect of Teicher's findings that makes this hypothesis less plausible. Traumatic events produce disintegrating effects proportional to their intensity, duration, and repetition. It appears that the effects of trauma, such as substantiated toxicologic effects of chemicals, follow a duration-dose-response curve. In another study, Silk et al. (1995) found a correlation between symptoms of borderline personality disorder and severity of retrospectively reported experiences of sexual abuse in childhood. Why the likelihood of being abused should follow such a growth curve is difficult to explain.

The strongest causal model for the effects of trauma suggests that the traumatic event independently caused both the biological abnormality and the PTSD. PTSD may represent the most extreme example of the ability of external events to induce lasting brain alterations analogous to electrical or pharmacological stimulation of the limbic system; this could explain symptoms attributed to "partial limbic kindling" in the human brain (Pitman et al., 1993). Other models allow for sequential effects among the three constructs, e.g., trauma initiates PTSD which in turn causes biologic abnormalities, or trauma initiates biologic abnormalities which then cause PTSD. With respect to EI patients, the commonality of these models is that trauma came first. The unfolding of PTSD symptomatology and biological abnormalities come with aging, additional traumatic or stressful events, and depletion of coping resources.

The intensity of the initial physiological response (vehement emotion) to a traumatic experience, combined with cognitive interpretation, is probably the most significant predictor of long-term outcome in PTSD (Figley, 1985). Its severity has been correlated with the initial degree of hyperarousal and dissociative reactions, developmental maturity, unfamiliarity with the environment, severity of interpersonal loss, length of exposure to the trauma, degree of life threat and personal injury, and participation in abusive violence (van der Kolk and van der Hart, 1989). All of these factors are characteristics associated with the overstimulation that usually occurs with trauma.

What is unique to PTSD patients and distinct from other anxiety disorders is that they describe experiences of "reliving" the traumatic experience during panic attacks or flash-backs, what psychologists call "flashbulb memories". The sudden recall of these vivid emotional memories do not fit well with current semantic theories of memory, especially if they are recalled after a lengthy period during which they were inhibited. Nonetheless, they are supported by physiological evidence when memories are experimentally evoked in PTSD patients. The procedure developed by Pitman et al. (1987) involves a script-driven imagery technique in which subjects imagine personal experiences as vividly as possible while physiologic parameters (e.g., heart rate, galvanic skin response, muscle tension, or brain imaging) are measured. Exaggerated responses to noise startle and evoked images have been replicated in veterans with PTSD. Orr et al. (1997) demonstrated noradrenergic activation specific to imaging traumatic experiences in women with PTSD and history of childhood sexual abuse. Similar effects were observed for memories that were continuous since the trauma and those recovered after a period of inhibition. Cognitive effects include dysregulation of attention with decreased brain metabolism observed by PET imaging studies (Southwick et al., 1997).

Hippocampal abnormalities

Anatomical changes in hippocampal volume measured by magnetic resonance imaging (MRI) have been identified in patients with Cushing's syndrome, characterized by a genetic predisposition to secrete excessive amounts of cortisol (Starkman et al., 1992). The mechanisms involved in elevated endogenous levels of cortisol are also affected by the stress-response which suggested that PTSD patients have similar anatomical signs and behavioral symptoms seen in Cushing's syndrome. Bremner et al. (1995a) found that the average right hippocampal volume of 26 combat veterans with PTSD was 8% smaller compared to that of 22 combat veterans without PTSD. No differences in volume were found in the left hippocampus or in the adjacent brain regions (caudate and temporal lobe) on either side, suggesting the effect was anatomically specific.

A specific effect associated with the pathophysiology of the hippocampus is short-term memory deficit (Zola-Morgan and Squire, 1990), which is also one of the common central nervous system complaints presented by EI patients. Smaller right hippocampal volume was associated with deficits in short-term memory in these PTSD patients. Similar deficits were observed in Cushing's syndrome patients (Starkman et al., 1992), in medical patients following neurosurgery, and among epileptics (Lencz et al., 1992). The link to memory difficulty is consistent with clinical reports from combat veterans who demonstrate alterations in memory, including flashbacks, nightmares, intrusive memories, and amnesia for traumatic experiences. Independent neuropsychological studies of Persian Gulf War veterans have shown significant attention and memory performance deficits in the veterans with PTSD compared to those without PTSD (Vasterling et al., 1998).

The finding for decreased hippocampal volume in combat-related PTSD was observed in a second study by Gurvits et al. (1996), but the 26% total decrease was much larger and manifested bilaterally. This difference from the earlier study was attributed to refined MRI technology and the greater extent of hippocampal structure measured. Neuropsychological data did not show significant differences between PTSD patients and controls on verbal memory, as had been found in the earlier study, but there were significant differences on tests of attention and concentration.

The effects in the studies by Bremner et al. (1995a) and Gurvits et al. (1996) were also found to be specific sequelae of trauma. Control subjects were matched for alcohol use. Depression was ruled out because there were no differences found in hippocampal volume between PTSD patients with and without a lifetime of major depression or dysthymia. This has implications for the treatment of EI patients with antidepressant medications in that these medications can exacerbate some of the symptoms of PTSD. A large number of EI patients report noncompliance or resistance to antidepressant medications (LeRoy et al., 1996). PTSD should be explored in a comprehensive assessment of these difficult to manage EI patients.

The findings of Bremner et al. also have implications for certain theoretical arguments for the validity of the toxicogenic theory. In effect, EI patients are said to be unlike "typical psychiatric patients" because they have a lower incidence of dual diagnosis for alcohol/ drug abuse (Bell, 1994b). First of all, the trauma paradigm offers an alternative to biological psychiatry's theoretical bias to over-diagnose depression, which then becomes the basis for the characterization of the "typical" psychiatric patient. Given the multitude and complexity of psychiatric disorders in the DSM-IV and the heterogeneity of psychiatric conditions observed in EI patients, the term "typical" psychiatric patient is undefined and fosters the allusion that such a population exists. Secondly, there is also a logical error in the argument. Given the conditional statement, "If one has a psychiatric condition, then one is likely to have an alcohol/drug problem," it is logical fallacy to conclude that given the absence of an alcohol/drug problem, one does not have a psychiatric condition (Staudenmayer, 1975).

The adult effects seen with war veterans are often similar to those seen in studies of adults and adolescents who have a history of traumatic childhood abuse. While there has been no systematic comparison of these populations, studies from various sources show striking similarities.

Bremner and his colleagues conducted MRI studies in 17 women who had suffered traumatic sexual abuse in childhood, and found significant 12% smaller left hippocampal volume compared to non-abused controls, controlling for alcohol, age, and education (Bremner et al., 1997b). These abused women also showed greater decrements in performance on a test of verbal memory (Bremner et al., 1995b), similar to the deficits found in combat-related PTSD by these investigators (Bremner et al., 1995a).

In a second controlled MRI study of 21 women with histories of traumatic childhood abuse, Stein and colleagues found a significant 5% smaller left hippocampal volume, but failed to find a significant deficit in memory in the group of abused women (Stein et al., 1997). Scale scores of dissociation correlated significantly with decreased left hippocampal volume, supporting the hypothesis that hippocampal abnormality may directly mediate many aspects of dissociation (Krystal et al., 1995).

These MRI imaging studies of hippocampal volume in patients with PTSD related to combat or childhood sexual abuse consistently showed anatomical abnormalities, although the magnitude and the location of the decreases varied. The neuropsychological task performance data are also inconsistent. These inconsistencies may reflect differences in MRI techniques, variability in neuropsychological task performance, specificity associated with the type of trauma and its onset in developmental stages, and/or possible other factors which may be identified in future studies. While these studies are suggestive, the demonstration of a causal relation between trauma and structural abnormalities requires further study and demonstration.

In the context of EI, it may be informative to conduct structural and functional brain-imaging studies on EI patients, stratifying them according to histories of abuse, PTSD symptomatology, depression, and/or other psychological characteristics. EI patients could be compared to psychiatric control subjects who have similar characteristics, patients with identifiable toxicogenic disease (e.g., industrial asthma), and asymptomatic controls.

EEG abnormalities

Non-specific electroencephalographic (EEG) abnormalities have been reported in studies of physically maltreated children (Green et al., 1981) and incest victims (Davies, 1979), although no causal attribution to abuse was hypothesized, allowing for the etiologic possibility that neurophysiologic abnormalities increased the risk of the child being abused.

Panic attacks experimentally induced by lactate infusion show associated state changes in EEG parameters including decrease in voltage in alpha frequency with associated increase in voltage in beta frequency (Fink et al., 1971; Knott et al., 1981) or increase in slow wave delta activity (Knott et al., 1981; Lapierre et al., 1984). These EEG abnormalities are typical for anxiety patients in general and are consistent with effects associated with hyperarousal (Lader, 1975; Lindsley, 1951, 1960). A normal state response to induced hyperventilation during a clinical EEG is slow wave theta frequency.

In a companion study to that reporting limbic system symptoms assessed with the LSCL-33, Teicher's group conducted comprehensive assessment for neurologic abnormalities using neuroimaging, neuropsychological testing, and electrophysiological measurements in a subgroup of 104 adolescent patients, 51% male (Ito et al., 1993). The sample was stratified on the basis of history of childhood abuse: none, psychological, or physical and sexual (genital contact). Of the subjects, 55% were classified in the physical and sexual abuse group, and of these, there were significantly more females (68.4%) than males. There were no differences between abused and non-abused patients in the prevalence of abnormal neurological exams or abnormal neuropsychological tests. Brain-imaging studies with computed tomography (CT) or magnetic resonance imaging (MRI) did not show statistically significant differences between the groups, although there was a trend in the direction of more abnormalities in the abused group.

Electrophysiologic studies were conducted to identify the presence of paroxysmal events, asymmetries, or regions of focal slowing by conventional EEG or brain electrical activity mapping, which includes EEG, quantitative EEG spectral analysis (QEEG), auditory evoked potentials, and visual evoked potentials. There were no differences observed between abused and non-abused patients in the prevalence of evoked potential abnormalities. There were, however, significant differences between the groups on combined EEG abnormalities, which were observed 3.58-fold more frequently in patients with a history of any abuse — psychological, physical, or sexual. Spectral EEG analysis showed the most dramatic differences between the non-abused patients (10% abnormalities) and the severe physical/sexual abuse subgroup (47.8% abnormalities).

These findings raise many questions about the specific effects of certain brain regions, especially the prefrontal cortex, and the neurotransmitters that innervate them. For example, dopamine projections to the prefrontal cortex are specifically activated by mild stress (Kalivas and Duffy, 1989). It has been suggested that repeated activation may adversely affect the development of this brain region, noting that it is the last region of the central nervous system to reach maturity in early adolescence (Ito et al., 1993; Teicher et al., 1997). The most sobering conclusion suggested is the possibility that early abuse may produce enduring neurobiological abnormalities.

In a case-control study presented in Chapter 4, Staudenmayer and Selner (1990) studied 58 EI patients with quantitative EEG under conditions of eyes-closed relaxation, reflecting stable baseline patterns. A group of 55 medical patients treated in a private allergy practice without significant psychological symptoms was used as asymptomatic controls. Also, a group of 89 individuals with heterogeneous psychological disorders receiving psychological treatment in private practice, but without beliefs about their symptoms being associated with EI, was used as a positive control group.

The EI patients showed abnormalities on several EEG parameters derived from spectral patterns. A spectral pattern is the x,y plot of brain-wave frequency by electrical energy or amplitude of the signal. A spectral pattern may have one or more significant peaks which define the dominant frequency. A homeostatic frequency is about 10 Hz. Higher or lower dominant frequencies are correlated with attentional difficulties, with difficulty focusing attention associated with lower frequencies and acute focus or vigilance associated with higher frequencies. The results of the study showed that the QEEG spectral pattern classification of EI patients had a similar distribution to the group of patients with psychological disorders. The distribution for each group was significantly different from that of the asymptomatic control group, which had fewer subjects classified into abnormal EEG spectral patterns. EI patients showed two kinds of attentional abnormalities. Spectral patterns in one subgroup showed no identifiable dominant peak and are called desynchrony, which reflects cortical activation associated with rapid attentional scanning or hypervigilance, what patients describe as "racing mind". Spectral patterns in another subgroup showed distinct peaks in higher frequencies (11 or 12 Hz) which reflects focused vigilance.

In a companion study also presented in Chapter 4, a subset of these EI patients (20 females and 10 males) underwent psychological treatment after initial evaluation (Staudenmayer et al., 1993b). They were statistically compared to a group of control patients (25 females and 11 males) taken from those with psychologic disorders who also underwent therapy. These EI patients self-reported a high incidence of childhood abuse. In the study, childhood sexual abuse was defined conservatively to include only penetration, and physical abuse was defined as perceived to be life-threatening to the child. Significant differences were found between the groups of women labeled "universal reactors". The comparison percentages are relative to 100% of women in each group. Of the 20 EI women, 50% reported physical abuse compared to 12% of the controls; 60% of the EI women reported sexual abuse compared to 25% of controls. Combined sexual and physical abuse was reported by 40% of the EI women and 8% of controls. Of all the EI women, 50% reported incest compared to 16% of controls. Sexual abuse involving multiple perpetrators was reported by 25% of EI women and 0% of controls.

The extremely high self-reported incidence of incest (50%) among EI patients who underwent psychotherapy indicates that they represent a population distinct from the general population. The high incidence reported in this study also suggests that the prevalence of childhood sexual abuse among EI women may be higher than that found in a general outpatient psychiatric population, or even in some inpatient psychiatric populations. In one study of 68 female psychiatric inpatients, the incidence of reported abuse was as follows: physical abuse in 12%, sexual abuse (defined as forced contact) in 21%, and both types of abuse in 33% (Pribor and Dinwiddie, 1992). In another study of 100 psychiatric male and female inpatients, a history of sexual abuse (defined as sexual contact) was reported by 18% (Jacobson and Herald, 1990).

These data support the theory that intrafamilial abuse in childhood produces complex post-traumatic syndromes characterized by disturbances of affect, self-image, and effective coping; dissociative disorders; and, particularly, somatization. The severity of the psychiatric disturbances observed in some of these EI patients was associated with histories of early childhood traumas that were processed in therapy. This clinical observation, while limited, is consistent with the field trials of severe stress disorders for DSM-IV which showed that number and severity of these disturbances was proportional to the duration of trauma and the age of the child at onset (van der Kolk et al., 1992).

Developmental stage and vulnerability

"Dreams never come true, and broken hearts keep beating."
(Survivor of iatrogenic EI, coping with dissociation)

The adult psychopathologic sequelae of trauma seem to depend on how severe and chronic the traumatic experiences were and the age of the victim at trauma onset. A growing body of evidence suggests that the developing brain organizes in response to the pattern, intensity, and nature of sensory, perceptual, and affective experiences during childhood development when the brain is undergoing critical sensitive periods of differentiation. During this time, stressor-activated neurotransmitters and hormones can play major roles in neurogenisis, migration, synaptogenesis, and neurochemical differentiation (Schwarz and Perry, 1994). While no longitudinal study has been reported linking childhood sexual abuse with adult sequelae, similarities form cross-sectional studies of adults abused as children show disturbances of self-esteem, mood and affect, social and interpersonal behavior, and impulse control, as well as deviations from physiological and biological homeostasis.

If the abuse occurs before the age of approximately 2 years, the hippocampus has not yet matured anatomically. If the trauma causes chronic hypersecretion of glucocorticoids, which seems plausible in the context of life-threatening events (real or perceived), the HPA-axis does not yet have adequate negative feedback systems available to inhibit hypersecretion along its pathways. Furthermore, the hypersecretion of cortisol may not only damage the existing glucocorticoid receptors in the hippocampus and other cortical regions, but may also somehow disturb their anatomical development and maturation. With ongoing, chronic abuse, the damage to the child's stress-response mechanisms may become indelible. This seems especially likely given damage to the negative feedback systems which regulate inhibition of secretions along the HPA axis.

It is generally proposed that the earlier the physical or sexual trauma the greater its impact on the concept of self and social functioning. The age at which sexual abuse occurs has been reported to have different effects on limbic symptoms. In one study, patients sexually abused before age 18 had significantly more symptoms than control subjects, while patients sexually abused after age 18 were not different from non-abused controls (Teicher et al., 1993). This suggests that age tempers the effects of abuse, and there is evidence that with advancing age humans develop more effective coping mechanisms for abuse.

Compared to other personality disorders, borderline personality disorders are uniquely associated with abuse which occurs before age six. (Herman, 1981, 1992; Herman et al., 1989; Ogata et al., 1990; van der Kolk, 1987). Dissociation of experience is often seen in the most severe personality disorders and related psychopathology (Ross, 1989). It is possible that damage to the hippocampus, whether directly or through the functions it mediates, may play a role in the etiology of dissociation and dissociative disorders. It may be more than coincidence that the majority of patients diagnosed with dissociative identity disorder (DID) have reported onset of traumatic sexual and physical abuse between the ages of 3 and 5 (Coons et al., 1989; Ross et al., 1991).

Complete memories of a traumatic event usually are not recalled before these ages, and, typically, when the most hysterical of patients are screened, the memories are soma-tosensory. Are these residual memories of fear and pain which are mediated by the amygdala left intact, while semantic memory of the event is disrupted by glucocorticoid damage to the hippocampus? Does the hippocampus also mediate the cognitive defenses that inhibit recall of trauma and are these memories disinhibited with aging as the hippoc-ampus is progressively damaged from prolonged, chronic stress-responses? If these hy-potheses are confirmed, can these effects be ameliorated by any manner of therapeutics?

Conclusion

The fundamental difference between toxicogenic and psychogenic theories is the role of environmental agents vs. the role of psychologic and psychophysiologic factors. The toxicogenic theory postulates that environmental agents are necessary but not sufficient to explain EI, requiring additional predispositional and susceptibility considerations. Ac-cording to the psychogenic theory, psychological factors are necessary and sufficient to explain EI.

A presupposition of the psychogenic theory is that PTSD onset occurs before the alleged onset of biological abnormalities attributed to chemical sensitivities by EI patients or their clinical ecologists. This implies that trauma rather than chemical sensitization is the necessary factor. Both theories postulate that biological abnormalities associated with PTSD are sufficient to explain symptoms in EI patients. Both theories postulate the same mediating biological mechanisms to explain the etiology and maintenance of multi-system complaints.

I contend that there is supportive evidence that psychological factors are necessary and sufficient to explain the symptoms and characteristics of EI patients. I also contend that the psychogenic theory has concurrent validity from other research programs studying the psychology, psychophysiology, and biology of psychiatric disorders. Psychological factors such as learning, suggestion, and placebo have been discussed in previous chapters. This chapter focused on limbic system symptoms as they are associated with trauma, PTSD, and limbic system abnormalities. Since most of the recent research on emotions and the limbic system has been with patients with trauma history and/or PTSD, the findings are specifi-cally relevant to EI patients with similar characteristics.

chapter thirteen

Personality disorders

In 18 years of clinical inpatient and outpatient experience with several hundred EI patients, I have come to believe that personality disorders should be the focus of diagnosis and treatment. I have also come to appreciate the association between trauma and personality disruption. EI patients demonstrate the heterogeneity of symptomatology and psychopathology associated with trauma.

Affective and anxiety disorders encompass a wide spectrum of human problems. General anxiety disorder entails many of the basic characteristics of all the anxiety disorders: apprehension, worry, psychophysiological arousal, motor tension, vigilance, and attentional disruption. Phobias are highlighted by unrealistic fear of real or imagined threat. Phobias allow the expression of the feelings of anxiety, but the underlying fears are displaced onto the phobic object. Obsession is characterized by more focused, repetitive, and intrusive thoughts. Repetitive overt behavior to alleviate anxiety is the hallmark of compulsion. Panic and PTSD are characterized by more dramatic and intense emotional and physiological reactions. PTSD also has specific requirements for the identification of a traumatic experience deemed to be life-threatening. But sometimes that knowledge is not accessible to awareness, and the traumatizing event is far removed from any possibly harmful chemicals in the environment. When anxiety is converted to somatic and pain symptoms, there are somatoform disorders for classification. The multi-system complaints of EI patients are consistent with criteria for a number of disorders in the affective, anxiety, dissociative, somatoform, and delusional disorder classifications. Depression is commonly seen in EI patients, but I have chosen to discuss it in the context of personality disorders because I wish to dispel the notion that EI patients are "simply depressed".

Studies comparing lifetime co-occurrence of several neurotic syndromes found that symptoms consistent with more than one type of disorder appear to be the rule rather than the exception (Andrews et al., 1990). For example, depression and co-morbidity among the anxiety disorders were observed by structured interview in 126 patients presenting at an anxiety disorder research clinic (Barlow et al., 1986). Of particular relevance to EI patients is the strong relation between obsessive-compulsive disorder and obsessive-compulsive personality disorder and depression reported in another study (Mavissakalian and Barlow, 1981). Many EI patients express intense issues around control, or the fear of losing control, and perfectionism. In three independent samples of EI patients, each obtained through

Table 13.1   Characteristics Suggestive of Mood Disorder

1. Depressed mood: loss of interest in hobbies, sex, or pleasure; anxiety
2. Sleep disturbance: difficulty falling asleep; early morning awakening; hypersomnia
3. Fatigue and lack of energy, daily and often most of the day
4. Change in appetite and associated weight change
5. Diminished ability to think, concentrate, or focus attention
6. Heightened irritability
7. Recurrent thoughts of death or suicide
8. Negative view of oneself, the world, and the future: feelings of worthlessness; feelings of excessive or inappropriate guilt
9. Cognitive distortions: arbitrary inference (reacting, then looking for an environmental cause); selective abstraction (able to travel on vacation, but not to work); overgeneralization, catastrophizing; magnification of symptoms (somatization); minimization of consequences (*la belle indifference*)
10. Selective attention to negative events: seldom talk about times when they feel good
11. Learned helplessness: passivity; retarded learning; lack of aggressiveness and competitiveness; weight loss and undereating; an expression of sympathy elicits uncontrollable tears; feeling helpless or hopeless about ever getting better

different referral sources, the obsessive-compulsive scale scores on the Derogatis symptom report instruments — the SCL-90R and the Brief Symptom Inventory — were all significantly higher compared to control groups (Black et al., 1993; Selner and Staudenmayer, 1992b; Simon et al., 1990).

Another consideration is that increasing frequency of somatic symptoms has been correlated with increasing distress, disability, and maladaptive illness behavior (Katon et al., 1991). Anxiety and depression are more commonly seen in patients with somatization than has been previously recognized (Katon et al., 1982a,b; Rix et al., 1984; Katon, 1986; Stewart, 1990). In short, there is greater recognition that a wide range of physical symptoms may be associated with anxiety and depressive disorders. In two longitudinal studies of older adults, negative affect predicted increased self-report of physical symptoms 6 months later, indicating that mood states are consistent predictors of later physical symptom reports (Leventhal et al., 1996). These findings do not support EI advocates' argument that somatic symptoms are toxicodynamic effects and should therefore be removed as criteria for the diagnoses of depression and anxiety in EI patients (Davidoff, 1992a). In this chapter, the multitude of Axis I disorders seen in EI patients will be presented in the context of Axis II personality disorders.

Depression as a symptom

Mood disorders are differentiated on the basis of severity, chronicity, and persistence and on associated mood features such as mania or hypomania and cyclical periods. The differential diagnoses are often difficult to make because of the subjective nature of the criteria. There are some illustrative signs and symptoms (Table 13.1) which suggest further assessment for specific mood disorders in EI patients.

The cardinal manifestations of melancholic depression are hyperarousal and redirection of energy. Both the HPA-axis and the sympathetic system appear chronically activated in depression (Gold et al., 1988a,b). The normal arousal seen in the state stress-response becomes dysphoric hyperarousal and anxiety when it remains chronic. Focused vigilance and attentional absorption are disrupted, replaced by rapid scanning and hypervigilance, which are often associated with anxiety disorders, eating disorders, or sleep disorders.

In some EI patients, attention is focused obsessively on depressive ideas (e.g., chemical sensitivities) adversely influencing the ability of the individual to remember and to solve practical, everyday, pertinent problems. Assertiveness is often transformed into excessive cautiousness, phobias, and nervousness which may result in passivity. Alternatively, the feelings of loss of control may result in displaced aggression and hostility, at times bordering on psychotic rage.

Symptoms of low energy, fatigue, hypersomnia, hyperphagia and associated weight gain, rejection sensitivity, and leaden paralysis are characteristic of atypical depression. This form of depression is considered atypical because the symptoms are opposite those expected from a CRH hyperarousal model of depression. CRH hyposecretion has been observed in patients with atypical depression and Cushing's disease (Kling et al., 1991). Preliminary results suggest that patients with atypical depression have a less dysfunctional LC-NE system compared to patients with other forms of depression (Asnis et al., 1995).

Depression can be co-morbid with virtually any other of the anxiety, somatoform, and personality disorders (Ariety, 1982; Oldham et al., 1995). Depression is associated with adverse life-events such as difficulties in personal relations, sexual dysfunction, occupational and academic problems, or what in general is referred to as "problems of living". When the patient realizes the pattern of living has failed, there may be the added feeling that he or she is unable to change it, accompanied by a state of helplessness and hopelessness. The limitations of the patient, determined by rigidity and adherence to an inflexible pattern of life, do not provide effective alternatives — i.e., there are no existing or accessible cognitive structures for effective coping. The only remaining escape from reality is to create an illusion that justifies the hopelessness and to withdraw physically from reality through avoidance behaviors, as EI patients do.

Albert Ellis, founder of rational emotive therapy, attaches primary importance to irrational beliefs in the development of depression (Ellis and Harper, 1961). Working from a cognitive theory, he hypothesized that depression occurs when a particular situation triggers an irrational belief. It is the belief which is hypothesized to cause the person to over-react emotionally to the situation. The implication to EI patients is that distorted beliefs about the etiology of the phenomenon or belief in unsubstantiated methods to treat it contribute to their depression. Ellis' theory also suggests that iatrogenic reinforcement of belief in EI by clinical ecologists directly contributes to the patients's depression.

Historically, 19th-century German psychiatrist Kraeplin (1921) used the term "depressive temperament" which was characterized by persistent gloominess, joylessness, anxiety, and a predominantly depressed, despondent, and despairing mood. Patients with this temperament were described as serious, burdened, guilt-ridden, self-reproaching, self-denying, unable to relax, dependent, burdened, and lacking in self-confidence. Compulsive traits were noted including orderliness, striving for achievement, devotion to duty, conscientiousness, and scrupulousness. For depression that is pervasive over time and across most situations and circumstances, it has been suggested that a DSM Axis II diagnosis be created — depressive personality disorder (Phillips et al., 1990), which is not a diagnosis in DSM-IV but one that is recognized in the standard international nosologic system, ICD-10.

Personality

"Personality" is a term used to define the highest organizational category in theories about the makeup of a person. Other terms have also been employed as labels for this all-encompassing construct, e.g., the "self" (Baker and Baker, 1987; Kohut, 1971) or the colloquial "character". A comprehensive definition (Barnett, 1980) is

> "Character [personality] is described as a structural phenomenon whose design and formal attributes determine much of the mental life of the individual. Character [personality] may be viewed as a behavioral and cognitive reflection of the organization of experience in a person's life. Character [personality] is a template, formed by the impact of historical experience, which functions to determine the organization of ongoing experience, its perception and interpretation, the designs of expectation and anticipation, and the behavioral tendencies and interpersonal operations of the individual."

Personality is made up of the perceptual and cognitive processes that describe both the background, trait characteristic of the individual and the phasic, physiological state changes associated with the stress-response. Every day, present world experiences and precepts derive their personalized meaning from these systems as the individual organizes and plans interactive behavior with the world. Well-adjusted individuals express their personality through effective coping styles. Maladjusted individuals express their personality disorders through ineffective psychological defenses. Depression, anxieties, phobias, and somatization are often the surface disorders which reflect the output of personality disorders (Hudziak et al., 1996). When that is the case, these clinical syndromes have quite different meanings depending on the which personality disorder they are an expression of. To make matters more complex, a host of psychological defense mechanisms are employed in various combinations and hierarchies by different individuals.

Based on the increased recognition in the past 30 years of horrifying effects of childhood sexual and physical abuse, as well as the debilitating effects of a lack of emotional attachment to parental objects, numerous theorists are postulating that physical depravity and emotional deprivation are the primary etiological causes of psychopathology, especially that which is associated with disruption of personality (Bowlby, 1971, 1975, 1977, 1980; Braun, 1984; Herman, 1981, 1992; Kluft, 1985; Miller, 1984; Putnam, 1989a,b; Ross, 1989; van der Kolk, 1987). These theories share the psychodynamic view that the origins of general anxiety neurosis and its various manifestations are caused by fears of separation/abandonment, going "crazy", and/or suffocation or being killed. They share the presupposition that these are unconscious processes which can project frightening thoughts, images, sensations, and affects into conscious awareness. A principle difference from classical psychoanalytic thinking is that the basis of these negative affects are not fantasies, at least not without foundation in prior experience. These fears may well be amplified or generalized, leading to unrealistic overreactions in the present, adult world of the individual. In that sense, the fears stem from inaccurate perceptions or memories and may even take on the characteristics of a delusion. However, it is unlikely that they are primary fantasies in the sense that the occurrence of the prior childhood trauma was imagined. Rather, the fantasy may be thought of as the expectation that the trauma can recur in the

Table 13.2   Illustration of Defense Mechanisms

Narcissistic defenses

Projection	Projection identification	Primitive idealization
Splitting	Denial	Distortion

Immature defenses

Acting out	Blocking	Hypochondriasis
Introjection	Passive-aggressive behavior	Projection
Regression	Regression	Schizoid fantasy
Somatization		

Neurotic defenses

Controlling	Displacement	Dissociation
Externalization	Inhibition	Intellectualization
Isolation	Rationalization	Reaction formation
Repression	Sexualization	

Mature defenses

Altruism	Anticipation	Asceticism
Humor	Sublimation	Suppression

Source: Adapted from Wong, N., in *Comprehensive Textbook of Psychiatry*, Vol. I, 5th ed., Kaplan, H.I. and Sadock, B.J., Eds., Williams & Wilkins, Baltimore, MD, 1989, pp. 356–403.

present world with the same intensity and effects it had in the past. This then leads to the key question, "What are these fantasies about?" Are they the result of the instinctual psychosexual desires or hostility the child feels for the parent, as Freud proposed? Or, are they the result of the real trauma of abuse the child experienced at the hands of a love object? Whether false appraisals of an unsafe physical environment are motivated by actual or fantasized abuse does not impact the psychogenic theory of EI, although it may help identify the personality structure.

Psychological defense mechanisms

Psychological defense mechanisms are mental processes used to resolve or avoid conflicts and to mitigate anxiety and depression. They can be adaptive as well as pathological. Because these mechanisms give the patient an illusion of having resolved a problem, they are resistant to change. Abandoning a defense, even one that does not solve the problem, is likely to increase conscious distress unless the defense is replaced by an effective coping style. There is an emotional price to pay when moving from false defenses to effective coping strategies. No consensus exists as to the number of psychological defenses or how they should be classified, but an illustrative list is presented in Table 13.2. The use of any particular defense or group of defenses, as well as the effectiveness of their employment, depends on the abilities and personality of the patient.

EI patients often use their particular combination of somatization together with denial of psychological conflict and projection of attribution to environmental causes. Both Freud (1936) and Jung (1967) described these defenses, particularly the projection of problems onto external factors. Jung (1971) detailed such a patient:

> "What comes from inside seems to him morbid and suspect. He always reduces his thoughts and feelings to objective causes, to influences emanating from objects, quite unperturbed by the most

glaring violations of logic. Once he can get back to tangible reality in any form, he can breathe again. In this respect he is surprisingly credulous. He will unhesitatingly connect a psychogenic symptom with a drop in the barometer, while on the other hand the existence of a psychic conflict seems to him morbid imagination."

Personality disorders

A personality disorder is an enduring pattern of inner experience and behavior that deviates markedly from the expectations of the individual's culture, is pervasive and inflexible, has an onset in adolescence or early adulthood, is stable over time, and leads to distress or impairment (American Psychiatric Association, 1994).

Etiologically, personality disorders have been characterized as patterns of intrapsychic defenses and coping processes to manage the effects of insurmountable injustice (Perry and Vaillant, 1989). From the patient's perspective, a personality disorder is "a way of making a painful truce with people one can neither live with nor live without; or it is the modes by which one copes with unbearable people." From the perspective of loved ones, individuals with personality disorders have a peculiar capacity to get "under their skin" and cause distress. Individuals with personality disorders have trouble working and loving. Problematic responses are elicited by interpersonal conflict. There is a vicious cycle in which already precarious interpersonal relationships are made worse by the person's manner of adapting.

Patients with personality disorders have maladaptive and inflexible responses to stress and do not cope well with the demands of the outside world. Instead, they attempt to make the world adapt to them. Lacking empathy with others, they consistently fail to see themselves as others see them. Their interpersonal relationships remain perpetually troubled and entangled. Many of these patients tend to describe the past as irrelevant, externalize their difficulties, and focus on their immediate troubles. Such a presentation makes it particularly difficult for the doctor to establish an enduring pattern that originates in adolescence or early adulthood. Patients with personality disorders are often irritating and consequently are more scorned than studied and treated.

Historically, the study of personality and personality disorders has been an attempt to identify and define a parsimonious classification system. The fields of psychology and psychiatry have approached this task differently, each having contributed to unraveling a piece of the puzzle. Reservations about losing sight of the big picture were also expressed.

One concern is that the understanding inherent in a global concept of anxiety and a depressive personality may be lost in the subclassifications. Sixty years ago, Horney (1937) anticipated these concerns:

> "When speaking of neuroses, I shall refer to character neuroses, that is, conditions in which — though the symptomatic picture may be exactly like that of a situation neurosis — the main disturbance lies in the formations of the character. They are the result of an insidious chronic process, starting as a rule in childhood and involving greater or lesser parts of the personality in a greater or lesser intensity. Seen from the surface a character neurosis, too, may result from an actual situation conflict, but a carefully collected history of the person may show that difficult character traits were present long before any

> confusing situation arose, that the momentary predicament is itself to a large extent due to previously existing personal difficulties, and furthermore that the person reacts neurotically to a life situation which for the average healthy individual does not imply any conflict at all. The situation merely reveals the presence of a neurosis which may have existed for some time."

Horney (1937) went on to minimize the importance of the specific symptoms in understanding the dynamics and etiology of the neurosis because they are ever-changing:

> "We are not so much interested in the symptomatic picture of the neurosis. Our interest lies predominantly in the character disturbances themselves, because deformations of the personality are the ever-recurring picture in neuroses, whereas symptoms in the clinical sense may vary or be entirely lacking."

She also commented on the implications to treatment:

> "With greater knowledge of the structure of neuroses and with the realization that the cure of a symptom does not necessarily mean the cure of a neurosis, psychoanalysts in general have shifted their interest and given more attention to character deformations than to symptoms. Speaking figuratively we may say that the neurotic symptoms are not the volcano itself but rather its eruptions, while the pathogenic conflict, like the volcano, is hidden deep down in the individual, unknown to himself."

While specific symptoms in the multi-system complaints of EI patients may be important to evaluate alternative medical and psychiatric diagnoses, these may be less important to understanding the underlying psychological problems and personality disorders.

DSM-IV Axis II classification

DSM-IV lists ten personality disorders, grouped into three clusters based on similarities in descriptive traits: odd-eccentric, dramatic-emotional-erratic, and anxious-fearful (Table 13.3). DSM-IV guidelines allow for the diagnosis of more than one Axis II personality disorder in an individual. Also, there is a category for disorders of personality functioning that do not meet criteria for any specific personality disorder — personality disorder not otherwise specified. Passive-aggressive personality characteristics are one example in this category.

Self-management of explosive affect

Barnett described certain acting-out traits, impulsive behavior, and explosion of affect, characteristic of dramatic personality structure (Barnett, 1966, 1968, 1980). Explosion of affect is a cognitive operation in which a forceful ejection or voiding of apprehended experience takes place before adequate structuring or thoughtful integration of experience can occur. Barnett's analysis captures the interpersonal interactions between some EI patients and the spouse who finds him/herself in a defensive and subservient role.

Table 13.3   Abbreviated Character Traits of Personality Disorders

A. *Odd, eccentric*

Paranoid:	Distrust and suspiciousness such that others' motives are interpreted as malevolent
Schizoid:	Detachment from social relationships and a restricted range of emotional expression
Schizotypal:	Acute discomfort in close relationships, cognitive or perceptual distortions, and eccentricities of behavior

B. *Dramatic, emotional, erratic*

Antisocial:	Disregard for, and violation of, the rights of others
Borderline:	Instability in interpersonal relationships, self-image, and affects and marked impulsivity
Histrionic:	Excessive emotionality and attention seeking
Narcissistic:	Grandiosity, need for admiration, and lack of empathy

C. *Anxious, fearful*

Avoidant:	Social inhibition, feelings of inadequacy, and hypersensitivity to negative evaluation
Dependent:	Submissive and clinging behavior related to an excessive need to be taken care of
Obsessive-compulsive:	Preoccupation with orderliness, perfectionism, and control

Source: Diagnostic and Statistical Manual of Mental Disorders, 4th ed., American Psychiatric Association, Washington, D.C., 1994. With permission.

Explosion of affect also benefits EI patients internally in the sense that comprehension of incongruent precepts remains imprecise and therefore avoids a challenge to their unsubstantiated attributions for EI. The defense mechanism of denial also serves to preserve the attitude of innocence, and, by attributing the explosive affect to EI, the patient disclaims intention and responsibility for his actions. If further prone to displace rage, the patient can feel indignant and display a posture of self-righteousness. The emotional consequences on others, usually the spouse or significant other, are written off to the unfortunate circumstance of being involved with someone suffering from EI. Armed with this projective externalization, the EI patient feels free to violate the basic interchanges of socially acceptable interpersonal relations without the slightest appreciation or concern for their meaning or impact.

Some EI patients have florid characteristics of histrionic or obsessional personality disorders that may manifest in their emotional reaction to their doctor. Initially, they may seek consultation with the hope that the physician will confirm and validate their EI belief and perhaps heighten their expectation for improvement by recommending a specific treatment, even one whose effectiveness has not been established. They may present with exaggerated flattery about how this doctor can finally help them, when others in past medical contacts are described with terms such as "just did not understand EI" or with harsher disparagements. However, this adoring, idealizing attitude can change abruptly. A disconfirmation of their EI belief or a recommendation to seek psychological care will often elicit incredulity, cynicism, and contempt and termination of treatment. Impulsive reactions such as running away in the service of denial are typical defenses. This enables them to avoid understanding what they have experienced and leaves their belief about an environmental cause for EI intact. On the other hand, the physicians or therapists who confirm their EI beliefs can do no wrong in their eyes. A patient with firm conviction in EI has to face his/her own gullibility, self-deception, or outright self-delusion to discredit a clinical ecologist. The associated feelings of embarrassment, loss of face, or humiliation are extremely anxiety provoking for patients with personality disorders. This helps explain the reluctance of EI patients to talk about their experiences at the hands of clinical ecologists once they give up their EI beliefs. It may also contribute to their reluctance to take action against doctors who subjected them to unsubstantiated diagnoses and treatments.

Self-management of implosive affect

Implosion of affect is a cognitive operation in which affect is forced inward on the cognitive processes and disorganizes them (Barnett, 1980). For example, incongruent information (e.g., unsupportive medical opinion and negative experimental evidence) is left simply as raw data without being integrated into the knowledge system. Somewhat paradoxically, the patient may be able to talk about this information without insight as to implications it has to his/her conviction about an environmental cause for EI. The obsessional personality represents the most characteristic example of the phenomenon of implosion of affect. It has been suggested that the etiology of implosion of affect arises from a climate of ambiguity in childhood, where explicit avowals of love and concern covered implicit rejection or hostility (Barnett, 1980), what has been called the "double-bind" (Bateson et al., 1978). The adult sequelae are that the obsessional person has learned to fear the implicit in interpersonal situations. The implosion of the affective experience, while creating and maintaining systems of control and innocence, has the effect of disintegrating cognitive organization and the meaning of social interactions. The self-reported history of these individuals is often dominated by analytic thinking and intellectualization, and their speech is excessively literal and legalistic.

EI patients who have obsessive traits are fluid in arguing the latest toxicogenic theory, often presenting with the emotional fervor characteristic of faith in dogma. They can eloquently argue the elaborate explanations for negative medical findings, often parroting what they have been told by a clinical ecologist or learned from advocacy publications. During initial history taking, they usually have extensive, well-documented summaries for their complaints and the circumstances under which they react. They are usually quite eager to share them with the doctor. Spending extra time with these patients is necessary to establish a working doctor-patient relationship. When undergoing controlled chamber challenges, negative results may never be conclusive to them; they can be the masters of the "yes, but ..." rebuttal.

Patients with obsessional personality traits usually do not present aberrant, as hysterical patients tend to, and their mental status evaluation typically reflects an individual who presents well oriented, is cooperative, appropriately dressed and groomed, and normal in demeanor, speech, and language. However, they often show flat, non-responsive, or guarded affect without insight into the psychological significance of their symptoms. Typically, these patients deny present or past psychological conflicts that may be contributing to their current symptoms, including their symptoms of depression and anxiety. Review of medical records may reveal no prior psychiatric evaluation or treatment although there may be entries by physicians indicating concern about stress, depression, or anxiety.

The paranoid disorders are closely related to the obsessional disorders in that both utilize implosion of affect to disorganize ongoing inference-making in interpersonal situations (Barnett, 1980). Individuals with both types try to avoid knowing about their emotional conflicts because of the central damage it would cause to their sense of self and their very precarious self-esteem. But, while the obsessional person substitutes static, clichéd, and stereotyped judgments for ongoing inferences, the paranoid personality substitutes inferences of an arbitrary sort based on existing suppositions of a threatening and hostile interpersonal environment. The similarities in obsessional and paranoid cognition account for their tendency to overlap clinically. Paranoid phenomena frequently appear in obsessional patients, and most individuals with paranoid traits utilize obsessional defenses. Psychotic, paranoid presentation in EI patients is atypical. One explanation is that the attribution of psychological conflicts onto the physical environment is not consistent with the paranoid patient's inherent assumption of a hostile interpersonal environment. In that respect, belief in EI does not serve their particular personality disorder. Another explanation is that paranoid EI patients lack the trust to subject themselves to objective medical and psychological assessment. Finally, patients with personality disorders often have approach-avoidance conflicts. While they have a fundamental fear of trust in interpersonal relationships, they also experience the need to establish a trusting relationship.

A case of folie a deux

A mechanic was involved in an alleged industrial exposure to elemental mercury from a broken industrial thermometer. The day of the accident, he had an altercation with his supervisor because he did not get the sympathy he expected. He acted out his frustration by throwing his tools and walking off the job. That night he experienced symptoms of anxiety. His wife took charge of finding an explanation and treatment for his condition. A few phone calls gained entry to the clinical ecology network, which led him to Colorado Springs for dental amalgam removal, to Dallas for sauna depuration, and to other clinical ecologists who provided additional unsubstantiated treatments. The patient adapted the life style and rituals of a severely affected EI patient, living in a cabin in the mountains, dependent on his wife.

Inpatient psychiatric records indicated this man harbored anger toward his wife. On one occasion, he held a knife to her throat for several minutes. He had a collection of guns, which she hid from him, but he could easily find them. He had several Axis I diagnoses including depression and anxiety. He also was diagnosed with dependent personality disorder.

I was asked to conduct an independent psychological examination on behalf of the defendant in a civil action. I was informed that this man was extremely environmentally sensitive and given a list of accommodations he required, including environmental exposure restrictions. He arrived for his appointment accompanied by his wife who spoke for him, indicating she had power of attorney. After some posturing and a call to their attorney, she agreed to have me conduct the evaluation in a conference room the patient selected. His wife waited outside. He acted disoriented to person, place, and time and was unable to repeat my name after I said it several times. This was a remarkable change in mental status from minutes before when he selected the interview room. He was dressed in all white cotton clothing, wore a dust mask, and carried an oxygen canister with a canula inserted under the mask. He wore darkened glasses to accommodate his "light sensitivity". His appearance was unkempt. He slouched in his chair, looking down to avoid eye contact. He started with a barrage of profanity directed toward me, reflecting his anger about his condition and paranoia about the legal process. Suddenly oriented, he demanded to take the "tests" he was sent for. I indicated that I would begin with a history, and he cooperated by giving me his symptoms and attributions. Spontaneously, he became teary and described the accident. Then he offered how he would get "crazy" and beat his wife, even threatening her with a knife. He attributed all his hostile actions to chemical sensitivity. He stated that he kept guns because his former employers were out to kill him. As the interview progressed, he used my name and showed less hostility toward me, talking calmly and logically about some of the good relationships he had had with his co-workers. After some time, he requested a break.

During the break, he talked to his wife and she seemed infuriated. They approached me together, she demanding that the interview be moved to another office. Since he had been showing tolerance of the meeting room, I suggested we continue there. She began shouting profanities at me, which instigated an attack on me by her husband. He raised the oxygen tank over his head to throw it at me, but threw it on the ground next to me and pushed me against the wall. The commotion had attracted other staff, who intervened. The patient stormed out of the building screaming he was going to get a gun to kill me. (Subsequent psychiatric records indicated he had a gun in his car that day.) At this point, the police were called. Instead of leaving, the couple sat outside my office asking me to complete the interview. He no longer showed any hostility after his wife instructed him to calm down. The police came and removed them from the premises.

Subsequently, she accused me of having incited her husband to violence. He was admitted to a psychiatric hospital for alleged complications resulting from the event and additional chemical exposure from the building. During that hospitalization, his guns were put in the custody of his wife, who refused to tell the doctors where she put them. After I submitted my evaluation, they filed a complaint against me which was dismissed. This case represents the interaction of two people, each with their own form of insanity, reinforcing each other in a clear case of malingering.

Communicated emotional illness between two people is called *folie a deux*, meaning double insanity. In EI, the psychopathology of *folie a deux* is characterized by an induced disorder in which the overvalued ideas are transferred from one person to another and mutually reinforced. Typically, both persons have been intimately associated for a long time and live in relative social isolation. The dyad usually consists of a dominant, controlling person who established the belief and a secondary person who is less intelligent, more gullible and suggestive, and passive and has low self-esteem. The submissive person is compliant and uncritically obedient to an idea or the influence of the dominant party. The motive for such submissiveness lies in a dependent personality, characterized by fear of abandonment and a need to be taken care of. The dependent partner may also harbor deep-seated resentment and hostility toward the controlling party. There is some question as to whether patients who are the passive partners are truly delusional rather than highly impressionable, because frequently there is merely passive acceptance of the delusional beliefs of the more dominant person until they are separated, and the distorted belief may remit spontaneously.

In the above case, the husband acquired the EI belief through his wife and the clinical ecology network. While his bizarre presentation and behavior appear psychotic, they were not. His expressions of psychotic behavior were context dependent, incited, and controlled by his wife. When alone with him in the interview, he began to reveal his feelings. His wife was controlling the situation and him. She held his power of attorney, which fostered his dependence, as did the alleged EI. The husband expressed his resentment toward her through physical violence and threats to kill her. But they stayed together, united in the legal action.

Dimensional models of personality

Dimensional models of personality are empirically generated by statistical procedures called "principal component analysis" (e.g., factor analysis). The factors are based on the correlations among numerous adjectives and descriptors rated by large samples of patients or subjects. From this often vast array of data, a few factors are abstracted and are interpreted to represent personality constructs or dimensions. Over the years, different numbers of factors have been suggested, and I will illustrated this approach with a recent model by Cloninger which has seven factors because two of these seem apropos to EI patients (Cloninger et al., 1993).

In Cloninger's model, temperament is defined in terms of automatic, pre-conceptual responses to perceptual stimuli, presumably reflecting heritable biases in informational processing which manifest early in life. Character is defined to reflect acquired self-concepts. There are four dimensions of temperament: novelty seeking, harm avoidance, reward dependence, and persistence; and three dimensions of character: self-directedness, cooperativeness, and self-transcendence. Novelty seeking is a bias toward frequent exploratory activity in response to novelty, impulsive decision-making, extravagance in approach to cues of reward, and quick loss of temper and active avoidance of frustration. Harm avoidance is a bias of failing to inhibit certain behaviors such as pessimistic worry in anticipation of future problems, passive avoidant behaviors such as fear of uncertainty and shyness of strangers, and rapid fatigability. Reward dependence largely reflects dependency and need for approval. Persistence was measured in terms of perseverance despite frustration and fatigue.

Two of these seven dimensions characterize EI patients — harm avoidance and low self-directedness. The temperament dimension of harm avoidance is consistent with EI patients' beliefs about the harmful effects of the environment and the avoidance rituals many practice. The character dimension of self-directedness refers to self-determination and willpower, or the ability of an individual to control, regulate, and adapt behavior to fit the situation in accordance with goals and values. Individuals lacking self-directedness have low self-esteem, blame others for their problems, feel uncertain of their identity or purpose, and are often reactive, dependent, and resourceless. In the sample of 136 psychiatric inpatients used to develop this instrument, lack of self-directedness alone predicted the presence of any DSM-IV personality disorder (Svrakic et al., 1993).

Limitations of categorical diagnoses

The heterogeneity of symptoms among EI patients raises perplexing problems for diagnosing according to categories of disorders defined by symptom clusters and demographic characteristics. Many EI patients do not to fit neatly into any one major DSM-IV diagnostic category (e.g., affective disorders, anxiety disorders, or somatoform disorders), much less a specific disorder within one of these categories. Their polysymptomatic presentations combined with their behavioral manifestations range over a multitude of diagnostic categories in the DSM-IV on both Axis I and Axis II. Phobic avoidance, psychophysiologic abnormalities, attentional and memory difficulties, perceptual distortions of body sensations, amplification of pain sensation, dramatic panic-like reactions to perceived chemical exposure, and unsubstantiated beliefs are some of the more obvious characteristics.

This is more than an academic exercise in the classification of a disorder. In a medical model, diagnosis has implications for treatment and prognosis. For example, a diagnosis of clinical depression implies consideration of antidepressant medications and/or cognitive therapy (Beck, 1991; Antonuccio et al., 1995). Clinically, the combination of the two is often beneficial in that medications manage symptoms and facilitate effective cognitive therapy. By contrast, somatoform disorders are considered among the most difficult to treat and generally have a poor prognosis (Quill, 1985; Ford, 1983, 1985). The greater potential for successful treatment of depression as well as standard treatment modalities may unwittingly bias diagnosis in its favor. There may well be a tacit bias among many psychiatrists to diagnose only conditions for which there is a clinical treatment, most often pharmacologic intervention for Axis I diagnoses. Ross (1995) criticized this bias of biological psychiatrists, particularly as it applies to difficult to manage patients with personality disorders.

In the context of litigation and compensation hearings, plaintiff attorneys argue about the specific criteria of any one psychiatric diagnosis with the motive of discrediting the witness. The DSMs have allowed them to argue about details while clouding the big picture. For example, the onset of EI patients' multi-system complaints are attributed to an alleged exposure event which occurred after age 30 based on patient self-report history. Does the plaintiff have enough, or the right kind of, symptoms for the diagnosis of somatization disorder and is the onset before or after age 30? EI advocates would like to argue that late onset precludes somatization disorder as a diagnosis, even though their own studies show EI patients report problems that typically begin in their 20s (Bell et al., 1995). Any expert witness would be in a difficult situation if he had to defend a clinical classification system were it presented to be definitive. The editors of DSM-IV addressed this dilemma and included the following reservation (American Psychiatric Association, 1994):

> "In most situations, the clinical diagnosis of a DSM-IV mental disorder is not sufficient to establish the existence for legal purposes of a 'mental disorder', 'mental disability', 'mental disease', or 'mental defect'. In determining whether an individual meets a specified legal standard (e.g., for competence, criminal responsibility, or disability), additional information is usually required beyond that contained in the DSM-IV diagnosis. This might include information about the individual's functional impairments and how these impairments affect the particular abilities in question. It is precisely because impairments, abilities, and disabilities vary widely within each diagnostic category that assignment of a particular diagnosis does not imply a specific level of impairment or disability."

Attorneys would also like jurors to be perplexed by the multitude of symptoms that do not fit any one diagnostic category and come to the conclusion that there must be nothing psychologically wrong with the plaintiff. In anticipation of this misuse, the editors of the DSM-IV noted (American Psychiatric Association, 1994):

> "The specific diagnostic criteria included in DSM-IV are meant to serve as guidelines to be informed by clinical judgment and are not meant to be used in a cookbook fashion. For example, the exercise of clinical judgment may justify giving a certain diagnosis to an individual even though the clinical presentation falls just short of meeting the full criteria for the diagnosis as long as the symptoms that are present are persistent and severe."

During assessment, some EI patients deny any symptoms pre-morbid to the alleged environmental incident, which typically occurs in middle age. Nevertheless, the limits of self-report do not preclude consideration of personality traits associated with personality disorders (American Psychiatric Association, 1994):

> "It should be recognized that the traits of a personality disorder that appear in childhood will often not persist unchanged into adult life. ... Although, by definition, a personality disorder requires an onset no later than early adulthood, individuals may not come to clinical attention until relatively late in life. A personality disorder may be exacerbated following the loss of significant supporting persons (e.g., a spouse) or previously stabilizing social situations (e.g., a job). However, the development of a change in personality in middle adulthood or later life warrants a thorough evaluation to determine the possible presence of a personality change due to a general medical condition or an unrecognized substance related disorder."

It is incorrect to suggest that the specific symptoms of EI acquired in adulthood are those necessarily observable at an earlier age. While self-report often reflects denial of pre-morbid psychological symptoms, psychological testing with personality assessment instruments or structure clinical interviews can often identify long-term conditions suggestive of Axis II diagnoses (Millon and Davis, 1996). There is growing evidence that personality disorders are preceded by childhood emotional and behavioral disturbances (Bernstein et al., 1996).

Dimensional diagnostic approaches

> "Should researchers and clinicians continue to act as if diagnostic categories were actual 'entities' by choosing one category and excluding another, or should they attend more closely to presenting coexisting problems and the overall pattern of psychopathology?" (Barlow et al., 1986)

This question is apropos to the assessment of EI patients. A fundamental question is whether classification should be categorical, as in DSM-IV, or dimensional. The dimensional approach is more consistent with psychotherapy in that psychiatrists and psychologists treat symptoms and complaints, not diagnoses. With the marked overlap in symptomatology it is tempting to consider more dimensional analyses for anxiety disorders (Barlow et al., 1986). Barlow noted that almost all of the anxiety disorders, except possibly simple or specific phobia, have the four features of generalized anxiety disorder (motor tension, autonomic hyperactivity, apprehensive expectation, and vigilance and scanning) (Barlow, 1985). The behavioral and cognitive dimensions common to all the anxiety disorders include distorted cognitive processes, somatic manifestation, and avoidance (Barlow et al., 1986). For EI patients, a dimension could be added to reflect strength and degree of irrational belief about appraisal and attribution, the continuum spanning from uncommitted curiosity at one end, to overvalued ideas in the middle, to delusional conviction at the other extreme.

Barlow and colleagues have suggested a rapprochement that would allow using DSM-IV symptom clusters as a dimensional model. If one can forgo the notion that these

Table 13.4 Dimensional Model for Axis I Diagnoses

Attention disorder:	Difficulty concentrating ("brain fog")
Panic:	Dramatic "EI reactions"
Agoraphobia:	Withdrawal behavior into "safe house", geographical isolation
Simple phobia:	Aversion to specific environmental agents
Obsessive-compulsive:	Preoccupation with EI beliefs and rituals
PTSD:	Hypersensitivity and symptoms of arousal and numbing
General anxiety:	Diffuse, pervasive feelings of apprehension or worry
Somatization disorder:	Multi-system complaints, onset before age 30
Undifferentiated somatoform:	Fewer symptoms, late onset
Hypochondriasis:	Excessive worry about EI with insight
Conversion:	Medically unsubstantiated motor symptoms, e.g., pseudoseizure
Delusional:	Overvalued ideas without insight
Depression:	Vegetative signs, hopelessness, self-destructive ideation
Dissociation:	Amnesia, derealism, and depersonalization
Malingering:	Symptoms are intentionally produced or feigned for compensation
Factitious:	Symptoms are intentionally produced for primary gain

diagnostic classifications are distinct clinical syndromes, then identifying certain disorders seems more useful than a strictly dimensional analysis. DSM-IV could work as a dimensional model for EI patients were it to permit multiple diagnoses within a category of disorders, including mental disorders due to a general medical condition, substance-induced disorders, psychotic disorders, mood disorders, anxiety disorders, and somatoform disorders. Currently, using the decision trees for differential diagnosis (DSM-IV, Appendix A), the diagnoses within any one of these categories are mutually exclusive. At minimum, a complex EI patient would have to be evaluated on symptoms associated with several Axis I diagnoses (Table 13.4).

Discussion

The classification of multi-system, general malaise symptoms is problematic for a psychological nosology. In a system such as the DSM-IV, which relies primarily on symptoms, statistical sensitivity is often high, meaning that there is an over-inclusion of individuals into any one of the categories. Symptoms are not specific to any particular classification, and the same symptom may be in several symptom clusters which define the different diagnoses. The hierarchical-selection priority among the diagnoses seems somewhat arbitrary as applied to EI patients. Dimensional models attempt to overcome this by focusing on general characteristics seen in a variety of disorders. The disadvantage in this approach is that etiologically different functions and mechanisms leading to similar symptoms may be lumped together, whereas treatment may be most effective when targeted at the underlying mechanisms. The clinical benefit of the dimensional approach is that each of the problems identified by the dimensions is addressed, as there is no exclusion through diagnostic priority. A balance between these conceptual systems remains to be achieved.

Severely affected EI patients tend to be heterogenous with respect to the psychopathology among them. Furthermore, any one individual presenting with multi-system complaints is difficult to treat as a unique anxiety disorder, a phobia, PTSD, depression, somatization, a personality disorder, etc. Treatment of the typical EI patient involves addressing more than one psychological condition, and too often "all of the above".

The psychological complexity and heterogeneity seen in EI patients warrant consideration of arguments for a general neurotic syndrome (Tyrer, 1985, 1990), which echo Horney's (1937) forewarning.

The devastating effects of abuse on personality disruption and resulting pathological sequelae were illustrated in a newspaper editorial commenting on the case of Susan Smith, a South Carolina mother with a history of ongoing parental incest who drowned her two young children (Santhuff, 1994):

> "Child abuse of any kind causes a child's psyche to register a level of pain it cannot tolerate, but also cannot erase. The hurt, grief, and rage get buried deeply below consciousness — only to leak out sooner or later, against self or others, in small toxic doses or massive eruptions."

The Susan Smith case is only one of many cases of infanticide that are reported. In 1994, 538 children in the U.S. were reported killed by their parents (Chavez, 1995). Those who survive the horror of childhood abuse are left with an indelible mark. Some repeat the cycle and displace their rage onto others, including their own children. Oliver (1993) summarized a series of renowned criminal cases of intergenerational transmission of child abuse in Britain, documented by Scotland Yard, which span psychiatry, psychology, pediatrics, public health, anthropology, social work, sociology, and teaching:

> "Omission, confusion, irrationality, distortion, and — most bizarre of all — idealization in accounts of cruel, rejecting, or neglectful grandparents by parents with rearing difficulties and/or ill-treated children are universal findings."

But, the disruptions of personality do not always end in homicide. Others project their frustration, anger, and rage onto the physical environment, as is the case with some EI patients who have suffered such ignominy. In doing so, some re-enact the frustrating interactions they had with their perpetrator(s). Examples of such behavior include withdrawal into "safe houses" or becoming obsessed with eradicating a food or household product. They may also manifest a psychological defense called "reaction formation" in which one's own feelings — in particular, hostile feelings — are attributed to another. Another way of saying this is, "What I perceive you feel toward me or are doing to me is really what I unconsciously feel for you or would like to do to you." For example, there are too many legal cases in which the plaintiffs displace their vengeful rage onto innocent purveyors of common food or household products that are harmless to all but the plaintiffs.

chapter fourteen

Iatrogenic illness: exploitation and harm

I wish to illustrate two points in this chapter — first, the processes by which iatrogenic EI is induced and embraced by vulnerable patients, and, second, how it is exploited by unscrupulous physicians and EI advocates. To begin, I revisit the power of suggestion and how it can be harnessed in service of pseudoscience and malevolence. I shall describe the characteristics of cult movements and leaders and followers in cults, all of which characterize the EI phenomenon. I present the indictment that the symptoms of EI patients are exacerbated by clinical ecologist practices contributing to patient morbidity and mortality.

A medical cult

Cults are subsets of larger movements referred to as charismatic groups (Galanter, 1982). There are common characteristics to cults shared by anarchistic groups and new-age religions (Conway and Siegleman, 1978), zealous self-help groups (Galanter, 1990), traditional religious affiliations (Booth, 1991), and medical subcultures (Brodsky, 1983). Appel (1983) abstracted the characteristics common to different types of cults (Table 14.1). The EI movement exemplifies the medical cult (Staudenmayer and Selner, 1987):

> "The ecology belief system usually is deeply entrenched and its logic
> well developed by intricate rationalizations and indoctrination. So-
> cial factors feed on the primary and secondary gain of the victim.
> 'True believers' are more than willing to present their testimonials,
> seeking and affording mutual assurance. The social and psychologi-
> cal dynamics of the cult apply. In addition, a plethora of 'health
> publications' provide the authority of print, while an impulsive
> media, eager for news, often is duped by unsubstantiated and unsci-
> entific claims of so-called ecology authorities."

Characteristics commonly associated with medical cults were listed in a U.S. Department of Health Education and Welfare (H.E.W., 1978) analysis of laetrile as an unsubstantiated cancer treatment. Allergist/immunologist Ellis (1986) summarized them as they apply to EI (Table 14.2).

Table 14.1   Features of Cults

Attributes
Authoritarian structure
Regimentation of followers
Renunciation of the world
Belief that adherents alone are gifted with the truth

Derived characteristics
An attitude of moral superiority
A contempt for secular [scientific] laws
Rigidity of thought
Diminution of regard for the individual

Degree of control
Totalistic cults control the total environment of the individual.
Devotees sever all ties with the past and the outside world.
Devotees give up independent thought or action.
Most time is spent on rituals, confessions, and group testimonials.
Martyr role is reinforced, marked for persecution by outsiders.

Indoctrination
The individual is isolated, malnourished, and sleep deprived,
    causing symptoms of cognitive dysfunction.

Advocates
The indoctrinated become teachers.
They need to be larger than life, embarked on an all-important
    mission.
Evil is projected onto scapegoats who become its symbols.
Absolved of responsibility, antisocial feelings are legitimized.
Initial expression of skepticism is overcome by realization there is
    no other choice.
Breakdown is a public manifestation of being special.
Self-delusion is an occupational hazard.

Source: Adapted from Appel, W., *Cults in America: Programmed for Paradise,*
Holt, Rinehart, and Winston, New York, 1983.

In the most extreme forms of mental coercion (so-called "brainwashing"), when techniques of torture and psychic terror are employed with prisoners of war or under conditions of oppression, there is ample evidence that beliefs can be molded (Lifton, 1956; Schein, 1956). Less brutal but equally effective psychological techniques can be used to manipulate the need for affiliation and fear of abandonment (Conway and Siegelman, 1978; Feldmann and Johnson, 1995). In the charismatic group, the forces of group cohesiveness, shared belief, and altered consciousness operate to compel behavioral conformity and modulate affect without overt coercion (Galanter, 1990). Group cohesiveness is defined as the product of all the forces that act on members to keep them engaged in a group. Patient testimonials, enumerating anecdotal experiences with their chemical sensitivities, are commonly shared among members of an EI advocacy group or inpatients in an environmental care facility. Narratives are intrinsically appealing in a way that lectures on abstract subjects and technical documents are not (Epstein, 1994), as evidenced by the popularity of daytime shows such as Oprah Winfrey. Experiential anecdotes appeal to the emotions and facilitate conviction in a commonly shared belief.

Table 14.2 Common Characteristics of Medical Cults

1. The proponents don the mantle of science while at the same time traducing the reputable scientists of their day.
2. They claim that prejudice of organized medicine hinders their efforts.
3. They cite examples of physicians and scientists of the past who were forced to fight the rigid dogma of their day.
4. They rely heavily on testimonials and anecdotes as evidence that their remedy is safe and effective.
5. They do not use regular channels of communications, such as journals, for reporting scientific information, but rely instead on the mass media and word of mouth.
6. Their chief supporters are not people trained or experienced in treating the disease or in scientific methodology.
7. They offer a simplistic theory for causation of disease.
8. Their remedy is easy and pleasant, compared with the frightening therapies wielded by orthodox physicians.
9. They claim the mode of administration of a drug and the method of treatment can be learned only from them.

Source: Adapted from Ellis, E., *Buffalo Phys.*, 19(5), 23–28, 1986.

Basic cognitive processing research showed that including anecdotes increases the persuasiveness of messages (Kahneman and Tversky, 1973). In medical and psychiatric practice, the case presentation is also a common mode of communicating ideas and findings. In scientific contexts such examples are often used to illustrate mechanisms and principles that have been independently supported with research evidence. They are also employed to describe phenomenon or syndromes which remain enigmas. However, testimonials may be no more than hearsay and in and of themselves do not constitute evidence to validate the existence of a toxicogenic illness, or any illness for that matter, much less explain it.

The compulsive behavior and obsessive thinking associated with counterphobic ritual taxes the limited attentional processing capacity of cognition. Hypervigilance and narrow focus of attention for dangerous contaminants in the environment and monitoring of bodily functions consumes attentional processing capacity, leaving little left to direct to other activities (Easterbrook, 1959). Such preoccupation disrupts effective coping and problem solving (Folkman and Lazarus, 1988). With the limited attention allocated to anxiety-related thoughts, there is little left to process information from other sources or to process ideas involving executive function and memory recall (Baddeley, 1993; Broadbent, 1971). This may partially explain why some EI patients cannot attend to other responsibilities in their daily lives. Anxiety-related thoughts impede problem-solving coping because they are irrelevant or counterproductive for performance (Lazarus, 1966). It is not surprising that a lack of concentration or difficulty thinking is the primary symptom reported by so many EI patients who are obsessively preoccupied with perceived threats from the environment.

Cognitive difficulties and other multi-system complaints may be exacerbated or induced by classic brainwashing techniques such as malnutrition and sleep deprivation. Fasting can induce a state response of fatigue and other general malaise symptoms. Iatrogenic malnutrition has been reported in EI patients who believed in universal food sensitivities (Selner and Staudenmayer, 1986). Sleep deprivation may result indirectly from anxiety-induced insomnia from worrying about EI. Iatrogenic sleep deprivation results directly from adherence to unsubstantiated purification rituals throughout a day which allows for only a few hours of sleep.

Quacks

> "Under the influence of my college and medical school education, I had been led to the supposition that reason, logic, and common-sense interpretation of the obvious are the prevalent modes of human thought in America. ... But, what I discovered in interactions with the proponents of 'alternative' health care were some different thought processes. They had, to me, a remarkably alien perception of science and the world." (Fitzgerald, 1983)

The *American Heritage Dictionary* defines "quack" as an untrained person who pretends to be a physician and dispenses medical advice and treatment; synonyms include "charlatan" and "imposter". The irony is that EI advocates, particularly clinical ecologists or "environmental physicians" who practice unsubstantiated procedures, are mostly physicians board certified in several specialties including internal medicine, allergy/immunology, surgery, and occupational medicine. Whether they measure up to the definition of physician as envisioned by Hippocrates is debatable. Most of the devoted EI advocates with whom I am familiar profess to suffer or to have suffered from EI themselves. The motivational issues of EI patients also apply to them and need to be considered when assessing the degree of conviction and involvement they have with EI. This is not to say that all EI advocates engage in the practices of clinical ecology, and certainly not all are deliberate frauds. The emotionally charged nature of the subject, the lack of definition, and the allusion to real environmental concerns understandably may cloud critical judgment of well-intentioned, otherwise level-headed and rational professionals and academicians.

EI advocates fall into three categories: misinformed, deluded, or dishonest. The misinformed endorse principles of environmental intolerance that they do not understand, often relying on analogy to well-documented models of biologic or toxic effects with which they are familiar. Typically, they are well-meaning individuals who perhaps can be educated and rehabilitated.

Deluded EI advocates view themselves as crusaders who are capable of redefining science and medicine. Theirs is a delusional mission but certainly grandiose ("I can save the world from toxic pollution") and/or persecutory ("the medical/industrial complex is out to get us"). Experience confirms these deluded EI advocates as being beyond rehabilitation. They demonstrate a "don't bother me with the facts" attitude that stifles legitimate debate. They flatly refuse to acknowledge compelling evidence contrary to their views. They are unwilling to acknowledge theoretical rationale which is proven to be implausible. They cling to hypothesized mechanisms that are disproved. They never stop preaching illogical reasoning that has been totally discredited. Retraction of false ideas and refuted hypotheses, gracefully allowed in the course of scientific progression, is not part of this evangelistic movement. They refuse to accept that self-report is not sufficient to validate their theories. They deflect argument from refuted ideas by further speculation and the introduction of new labels for discredited ideas in the hope of salvaging the hard-core postulates of the toxicogenic theory and their own credibility. While many deluded EI advocates deny using unsubstantiated clinical ecology procedures, they refer EI patients within a closed network. Once inside, patients become indoctrinated with a bankrupt belief system that often leaves them helpless and hopeless.

Intellectually dishonest EI advocates are malevolent. They are motivated by more than belief in their overvalued ideas and delusions. They have no scruples. They lie repeatedly for personal gain. When clinical ecology was discredited, they simply called themselves "environmental physicians" practicing environmental medicine while maintaining the discredited methods. Their self proclamation as "researchers, scientists, and environmental clinicians" is a transparent veneer that dissolves under the scrutiny of testimony under oath. While professing to be champions of environmental victims, they show blatant disregard for the dependency they construct, often abandoning the patient when his resources are exhausted without guilt or remorse. Psychiatrists refer to these physicians as antisocial or psychopathic personalities (Barrett and Herbert, 1994). They should be held accountable for their predatory behavior.

Patient vulnerability

> "Erasing history by cultivating denial is essential to the brainwashing that is an inevitable part of psychic murder, resulting all too often in what Nietzsche called the worst form of slavery: That of the slave who has lost the knowledge of being a slave." (Shengold, 1989)

Cults generally attract vulnerable, depressed individuals described as lonely, rejected, and sad (Levine and Salter, 1976), those who have inadequate, borderline, or antisocial characteristics (Etemad, 1978), or those who have borderline/narcissistic personality organization and excessive familial enmeshment (Markowitz, 1983). It is difficult to determine the extent to which cult activities actually cause these symptoms, and it is more likely that the personality traits are pre-morbid and motivate entry into the cult. The cult replaces some missing element for its members and thus provides structure. The missing element may be the need for affiliation which was not fulfilled in the family of origin; it may be a mythical cure which engenders hope for recovery from an "incurable disease" — real, such as cancer, or imagined; it may be an "answer" for unexplained psychopathology; or it may serve to blame the secular world.

Once indoctrinated, the teachings and practices of the cult become the center of life for the devotee. Treatment centers are revered as shrines, reflected in the following statement made by an EI patient in adulation of a clinical ecology environmental care unit, "You'll believe it when you see it." What there is to see is no mystery, and this patient's adoration reflects the experience of religious conversion (Frank and Frank, 1991):

> "The period before religious conversion is one of severe demoralization. This may be so intense that the person becomes confused and attributes inner experiences to the outer world, entering a state of transient psychosis. ... The person in such a state longs to submit to an all-powerful, benevolent figure who can give absolution and restore order to his or her assumptive world. At the moment of conversion the convert feels closer to God and confident of divine favor. This experience is intensely emotional and may be followed by a sense of inner joy or peace."

Individuals initially skeptical about EI may eventually succumb to peer pressure that encourages adherence to a common belief. Pressure to conformity and the human need for affiliation influence that accommodation process. Once concepts are incorporated into the conceptual networks, feedback loops preserve and enhance the beliefs, and incompatible ideas or alternative explanations are displaced.

The personality structure of many cult members resembles the definition of self pathology (Feldmann and Johnson, 1995; Kohut, 1971, 1977; Kohut and Wolf, 1978). For an individual who has a sense of damaged self, a pathological relationship or attachment is viewed as preferable to the emptiness and isolation that he would otherwise experience. Membership in a cult provides a sense of identity or belonging for those personalities whose underlying sense of identity is flawed, who are said to manifest self pathology. The relatively close-knit nature of most cults, for example, may offer an attractive alternative to the fears of rejection and self-doubts that mark a damaged or vulnerable self. To maintain affiliation, members of a cult will come to embrace a common set of beliefs, no matter how unrealistic in the context of the outside world. In some instances the cult will also encourage a degree of conformity among its members. In this way the individual becomes surrounded by a group of like-minded members, all of whom reflect similar views and feelings and who confirm the reality of each member's self. Commitment to the group's beliefs implies a sense of importance and uniqueness, something to which individuals with low-self esteem are vulnerable, especially those with dependent personality disorder. The severely disrupted personality (including schizoid, paranoid, antisocial, and sadomasochistic tendencies) is more likely to find cult membership a source of strength that offsets the vulnerability of the self (Feldmann and Johnson, 1995). Extrication from the cult is made very difficult because the damaged self risks losing the one belief system and group affiliation that maintains its sense of cohesion, which is often maintained at any cost.

Devotion to the guru

> "But rather than question the nature of their own beliefs, those who 'know' that occult practices work see science as too dogmatic."
> (Hyman, 1993)

The members of a cult generally surround a central figure who influences his followers in a direction that deviates strongly from that of the dominant culture (Deutsch, 1980). Gurus are generally charming and affable and appear caring. In my experience, this has been true for most of the advocates of EI, with the exception of some extreme zealots. Gurus are well versed in rhetoric that can be mistaken for reason and understanding by the initiate or by those in search of such a message. A seemingly essential asset of an alternative medicine practitioner is to meet the emotional needs of his patients very well (Hewer, 1983). Quack gurus profess confidence in their ability to cure all ailments, or at least the current one on their agenda, and often have an evangelical fervor. Genuine concern for patients appears to be compatible with irregular life styles, financial unscrupulousness, and even psychosis (Deutsch, 1980).

Gurus cloak themselves in an atmosphere of certainty which complements the needs of those in search of a medical guru who can provide an answer for their feeling miserable. Beyond trust and hope every patient is looking for patience, understanding, empathy, concern, and tolerance in the doctors' dealings with them (Dominian, 1983). When the medical guru meets the potential devotee, there is a synergism that creates an optimal context for the placebo effect, regardless of what particular assessment procedure or treatment is employed or how absurd it seems.

Cult members offer more than devotion to their guru. Money is paramount. The practices employed by clinical ecologists are not cheap. While professing poverty and high overhead, the profit and income clinical ecologists reveal in testimony under oath in legal cases is astounding, enviable by many chief executive officers of Fortune 500 companies. Equally disturbing is the extent to which some EI patients will go to pay for their treatment. They will liquidate their assets, borrow from loved ones, sue for divorce if their spouse resists, demand maintenance for full disability, and even mortgage the future of their children. Devotees have no regard for the imposition on family and friends, whom they often proselytize or shun. They seem more concerned with rejection from the cult, feeling humiliated in front of their "brothers and sisters" if they cannot pay. When the money runs out, some EI patients have reported working some type of service labor in exchange for treatments, such as cleaning the residential quarters affiliated with a clinical ecology environmental care facility. This resembles the characterization of some religious cults where cult members are expected to promote or support the cult by begging, preaching, making and selling objects, etc. These activities strengthen adherence to the cult's doctrines, foster submission to the leader, cement ties among the cult's members, and accentuate the separateness between them and the people in the mainstream of society. Members distinguish themselves from their neighbors by behaviors that, while not antisocial, often violate the implicit values or behavioral standards of the community (Frank and Frank, 1991).

In the 1980s several fundamental Christian evangelists were exposed for their schemes and scams. One in particular was convicted for misappropriating large amounts of money from his flock and was sent to prison. One might have expected his followers to be outraged and to dismantle his church. Yet, the movement faltered little. The resilience of a belief system, once entrenched, was illustrated by Hyman (1993) in a historical account of "trance-channeling":

> "The spiritualist movement was launched in 1848 when Margaret and Kate Fox claimed they were able to communicate with the dead. Through a series of rapping noises, the 'spirits from beyond' gave advice, made predictions, and consoled a loved one. The Fox sisters performed in large arenas and charged clients for the opportunity to communicate with spirits. Soon after the Fox sisters began performing, thousands of mediums around the world claimed similar abilities. Years later, Maggie Fox admitted that she and her sister had been perpetrating a hoax. But this had no effect on committed believers."

This seemingly defies rational behavior. How are we to explain this? Once a set of beliefs is established and serves the purpose of reducing anxiety, the beliefs seem to take on a life of their own. Even after psychological treatment, some EI patients cling to some aspect of their belief about chemical sensitivities, even though they complain about the "craziness" they were exposed to by clinical ecologists. A false prophet does not invalidate a religion; instead, a new prophet is anointed who will perpetuate it.

One explanation for this behavior is group affiliation and attachment needs. A hardcore postulate of psychodynamic theory is that early childhood experiences with the parental figure are critical to character development. That bond can be healthy and nurturing or trauma inducing. The devastating effects resulting from lack of appropriate attachment during infancy were demonstrated experimentally by Harlow in the surrogate mother studies with monkeys (Harlow, 1958; Harlow and Zimmerman, 1959). Infant monkeys deprived of the bodily contact of their mother and given a wire mannequin in her place became neurotic, and some even died. Other monkeys given a wire mannequin covered with terry cloth fared much better, although not as well as those who had actual contact with their mothers. The emotional damage of inadequate attachment resulting in separation anxiety has also been demonstrated in humans (Bowlby, 1960).

These effects are relevant to EI in that unfulfilled attachment needs can be displaced to influence the interpersonal interactions that contribute to the placebo effect. For example, patients with dependency needs resulting from separation anxiety are more vulnerable in seeking and maintaining social affiliation. Some affiliation, even when perceived to be harmful from a third-party perspective, is preferable to no affiliation, be it with an individual such as an exploiting doctor or a group of similarly afflicted EI patients. The group pressure among individuals in a cult, be that a religious cult or a medical cult, is toward conformity with a consensus of belief based on common presuppositions. The individual susceptible to fear of rejection, abandonment, or banishment is more vulnerable to group pressure. Such an affiliation does not meet the requirements of a nurturing bond which has the wellness of the individual as the one and only goal. Often, the harmful effects of poor childhood experiences are re-enacted in the cult environment out of the individual's need to resolve attachment needs.

Bias and expectation invalidate provocation challenges

The "Hawthorne effect" and the "Rosenthal effect" are psychology terms referring to experimenter expectations that significantly influence the outcome of a study; that is, experimenters obtain from their subjects — human or animal — the data they want and/or expect to obtain (Rosenthal, 1963). In a series of experiments, Rosenthal demonstrated that he could influence the outcome of animal learning studies and desirability ratings of photographs of human faces by biasing the expectations of student experimenters. For example, in one study the rats were all from the same genetic strain, but some of the experimenters were told the animals were from a dull strain, while others where told the rats were from a bright strain. The performance results were consistent with the expectations of the experimenters.

Experimenter expectation can affect the observation and interpretation of dependent variables. With EI patients, the dependent variables are most often subjective symptom complaints which are much more amenable to bias than are objective signs or laboratory results. In general, studies showed that biased experimenters tended to make more and larger recording and computational errors in the direction of their hypotheses in unblinded experiments. The experimenter's bias may be transmitted to the staff, and their attitudes, expectations, biases, and conflicts, or harmony with the subject may influence placebo-control effects (Shapiro, 1964). Human beings can engage in highly effective and influential unprogrammed and unintended communication with one another. Reinforcement, overtly by a verbal comment or covertly by body language, can influence symptom reports. Individuals who are particularly in need of affection and acceptance are vulnerable to such influence. The subject's expectations are influenced by the contextual circumstances of the testing and the actions and attitudes of others in that context, be they experimenters or other subjects (Wolf, 1959):

> "Investigators have often been naive in failing to recognize that patients, like dogs and children, are likely to know what is in the atmosphere without our telling them and even when we try desperately to conceal our attitudes."

Modeling refers to the extent to which the experimenter's own experiences with an exposure determines the effect reported by the subject. Personal experiences of the experimenter must not be shared with subjects. In ECUs run by clinical ecologists, it is not uncommon for the staff to be sufferers of EI. Such influence can heighten the apprehension of subjects who have fearful expectations about low-level exposure and thereby set off psychophysiologic disturbances. The effects of modeling can also come from other patients, especially those who have firm conviction about the effects of exposure. Conducting provocation challenges in groups is contraindicated. Modeling is a common procedure used in the indoctrination of novitiates into a cult. Modeling is also part of the dynamics of support groups where individuals beliefs and experiences are molded to conform to the group stereotypes.

When running a series of challenges, the results should be kept from both subject and experimenter until all the challenges are completed or until a reliable conclusion can be reached. The interpretation of early results may lead to premature conclusions which can bias the observation and interpretation of the results of later challenges (Tversky and Kahneman, 1971). The bias of a "self-fulfilling prophecy" on a small sample, the first few challenges, can inadvertently break the blindness of the double-blind procedure for later challenges.

Having a vested interest in the confirmation of a hypothesis, especially among experimenters who share the belief in EI, can bias to confirmation or positive results. Unconscious enthusiasm and solicitude, or outright faith and conviction of belief on the part of the staff may unduly influence symptom experience and report. This bias was noted by Canadian investigators who visited the environmental care units run during the mid 1980s by William Rea in Dallas and the late Theron Randolph in Chicago (Orme, 1994). Rea (1994) claimed his air-borne challenges were conducted under double-blind, placebo-controlled conditions, even though ambient dose levels (above detectable odor threshold) were delivered without a masking odor; this makes the challenges with a putative active agent easily discriminable. The Canadian investigators reported a major problem with diagnosis, noting that only four out of 2000 of Rea's patients had tested negative for environmental sensitivity, and those four were found to have malignancies during their assessment. They concluded that (McCourtie, 1990): "Dr. Rea used no appropriate controls and the patients were assumed to have environmental hypersensitivity mainly by being referred to the unit."

The implications for the need of double-blind, placebo-controlled studies in pharmacologic research, where the stakes and consequences are high, are obvious. The same can be said for the assessment of environmental intolerances with provocation challenges. As a note of caution, great care must be taken to ensure true blindness in a research study. In many intended double-blind studies, true blindness may not occur (Margraf et al., 1991).

Doctor's expectation and treatment outcome

Physician attitude toward a treatment and enthusiasm for anticipated outcome can also influence the patient's heightened expectation and response to a placebo treatment (Rosenthal and Frank, 1956; Fisher and Greenberg, 1989). Jewett (1992a) has presented a candid and remarkable clinical account of his own experiences both as an EI patient and as an orthopaedic physician practicing some of the methods of clinical ecology. At a particularly difficult period in his life, his life-long allergies became worse. He suffered from what he thought was EI and was treated by a clinical ecologist and ultimately spent 5 weeks in the Dallas ECU, labeled a "universal reactor". Jewett believed in the dietary treatments he recommended to his EI patients. Patients saw him as one of them. When his double-blind study (see Chapter 4) failed to confirm his hypothesis about methods to diagnose food sensitivity (Jewett et al., 1990), Jewett questioned the validity of the treatments implied by the clinical ecology theory. Though ostracized by the clinical ecology community when the negative findings were presented, patients still came to him for treatment. Despite his own doubts, he administered the same written material for the diets he had given to patients earlier when he had firmly believed in it. But now, the therapeutic effect of the diets was noticeably less. Ultimately he abandoned the diet treatment, shifted to a counseling model, helped his patients with problems of living, and reported good outcomes. His own symptoms became manageable with traditional medical practices.

Counterphobic rituals and obsessive-compulsiveness

The importance of catharsis inherent in counterphobic rituals may contribute to the placebo effect (Shapiro, 1964):

> "The history of medicine is full of reports about procedures or substances which have important symbolic meaning. The list includes emetics, cathartics, enemas, purges, stomachics, sweating, bleeding, leeching, cupping, starvation, and dehydration. ... Methods of depletion and expulsion, manipulation of internal body wastes and vital fluids may relieve symptoms by symbolically expelling bad thoughts and conflictual ego-alien impulses. However, the same expression and relief of symptoms may occur when the patient is able to express verbally conflictual and guilt-ridden thoughts and feelings in the free, nonjudging and accepting atmosphere of the doctor's office. In other words, the alimentary and other primitive methods of catharsis of the past are no longer appropriate for our culture and have been superseded by a more intellectual and appropriate verbal catharsis. The fundamental mechanism, however, may be quite similar."

Despite the optimism for the effective use of communication expressed in this summary, many EI patients regress with counterphobic rituals. Those who have acquired obsessive conviction about their beliefs may also manifest compulsiveness about avoidance and purification procedures. Individuals with obsessive-compulsive personality traits are most susceptible to acquire counterphobic rituals. When presenting their history, they are methodical and precise in laying out the most minute details of their problem. When alternative possibilities such as stress or anxiety are broached, they invariably return to discussing EI. Typically elaborate coping procedures are the rule. Medical records may be very lengthy. Exhaustive self-generated chronologies have often been compiled. The following case illustrates this type of ritualistic obsessive-compulsive behavior.

Selner and I jointly interviewed a 35-year-old woman in New Mexico who alleged multiple chemical sensitivity. She indicated she required megadoses of vitamin C every 30 minutes. To control her symptoms we were required to move from the referring physician's office (where she could not tolerate the odors of office furnishings) to a courtyard adjacent to the medical building, where she complained of symptoms (she detected roofers working with asphalt on a distant building). We moved to the sanctuary of her apartment across the valley by auto without further complaints of symptoms. We expected stark furnishings, but upon entering found a typically furnished apartment including electronic devices and gas appliances. The exception was an unusual, elevated platform upon which rested a cotton-cloth covered mattress with an oxygen tank and a conspicuously large timer nearby. The woman immediately set the timer for 30 minutes, and when the alarm sounded proceeded to take a large dose of vitamin C, expressing immediate, euphoric relief. The interview apparently distracted her enough that she neglected to reset the timer and did not take vitamin C during the ensuing two-and-a-half hours. She showed no adverse effects and her mood seemed elevated during our interview. She thanked us for our visit when we departed. The behaviors we observed in her apartment were inconsistent with her expressed ideation, and the environment inside her apartment was not markedly different from others for which she expressed intolerance.

Some insight about this kind of ritualistic behavior may be gleaned from research on the psychology of obsessives (Rachman and Hodgson, 1980):

> "Their cognitive style is preoccupied with achieving certainty. They believe they should be perfectly competent (some in everything). They are afraid of the consequences of criticism or disapproval and avoid it. They fear severe punishment for their mistakes and imperfections."

EI patients with obsessive traits often believe that meticulous adherence to ritual will prevent the dangerous exposures just waiting around the corner. Some air out letters and printed materials by hanging them from a clothesline. Reading materials are often placed in a glass "reading box" which supposedly protects against offgassing of harmful print and paper related toxin. Appliances, telephones, and even walls may be covered in aluminum foil to protect against alleged contamination from offgassing chemicals or radiation of electromagnetic forces.

State revokes license of mercury amalgam guru

In 1995, the Colorado State Attorney General's office representing the State Board of Dental Examiners charged Hal A. Huggins, D.D.S, a Colorado Springs dentist and self-proclaimed world leader in the treatment of alleged mercury toxicity caused by amalgam fillings, with five violations of the Dental Practice Act. The case (no. DE 95-04) was adjudicated by administrative law judge Nancy Connick, who, on February 29, 1996, recommended that the license of Dr. Huggins be revoked. (The full report may be obtained by writing to Assistant Attorney General Robert N. Spencer, State of Colorado Department of Law, Office of the Attorney General, 1525 Sherman Street, 5th Floor, Denver, CO 80203.) The initial decision was reviewed and adopted by the Colorado State Board of Dental Examiners in May 1996. Dr. Huggins did not appeal. What follows is abstracted from that report.

The first two of the five charges concerned the use of misleading, deceptive, or false advertising and whether Dr. Huggins's professional corporation, the Huggins Center, with approximately 50 employees, practiced outside the scope of dentistry. The remaining charges dealt with the care of eight patients (details presented in the report) who were said

Table 14.3   Illustrative Diseases and Disorders
Allegedly Caused by Dental Amalgams

Neurological
Tremors, seizures, multiple sclerosis (MS), amyotrophic lateral
sclerosis (ALS, or Lou Gehrig's disease), Alzheimer's disease,
emotional disturbances, unexplained depression, anxiety, and
unprovoked suicidal thoughts

Immunological
Systemic lupus erythematosus, scleroderma, and rheumatoid arthritis

Cardiovascular
Unexplained heart pains, high and low blood pressure, tachycardia,
and irregular heart beat

Collagen:
Connective tissue disease, including osteoarthritis

Miscellaneous
Chronic fatigue, "brain fog", digestive problems, and Crohn's disease

Source: Adapted from Connick, N., Case no. DE 95-04, Before the State Board of
Dental Examiners, State of Colorado, 1996.

to have received substandard and grossly negligent dental care, were not referred to
physicians, and were subjected to unnecessary tests and treatments which were without
clinical justification.

Dr. Huggins contends that a multitude of diseases and disorders are caused by
amalgam fillings (Table 14.3). The Huggins Center accepts for treatment patients with just
about any symptoms. The justification offered in Dr. Huggins defense to treat these as a
dentist was that these were all caused by mercury toxicity coming from dental amalgams
located in the oral cavity. In his opinion, general dentistry addresses everything which
affects the health of the oral cavity and everything in the oral cavity which has a systemic
effect.

The Huggins Center widely advertised a toll-free number which persons interested in
the issue of mercury toxicity from amalgams called to obtain information. Patient represen-
tatives, who were essentially sales personnel paid on commission, answered telephone
inquiries (approximately 5000 callers per month). They encouraged callers to purchase and
read Dr. Huggins's book, *It's All in Your Head: The Link Between Mercury Amalgams and
Illness*. They also provided advertising materials including brochures, position papers,
videos, and other materials prepared by the Center outlining Dr. Huggins's theories, and
they encouraged enrollment at the Center for treatment. They referred out-of-area callers
to dentists and physicians who had attended his seminars and presumably shared his
beliefs on amalgam. Those who come to the center were personally met by Dr. Huggins,
who exhibited sympathy and warmth and effectively conveyed to them his strong belief
that the treatment offered at the Center would improve their health. If a patient appeared
reluctant to proceed with the treatment, Dr. Huggins stepped in with a harder sell approach, telling the patient something to the effect of "You want to get well, don't you?" On
several occasions, he told patients that he had multiple sclerosis (MS) and was cured, even
though in fact he has never suffered from this disease.

Individual testimonials of improvement were relied upon. Patients were routinely
videotaped before and after treatment. In one, a young woman diagnosed with MS, within

minutes after having her amalgams removed, felt her muscle control returning and could stand, whereas before she could hardly walk. The administrative law judge remarked on the placebo effect: "It is widely accepted that there is a powerful therapeutic effect from any treatment administered to patients."

The Huggins Center accepted the patient's medical diagnosis (e.g., MS), made a diagnosis of mercury toxicity, and then treated the purported mercury toxicity and medical disease according to a single protocol developed by Dr. Huggins. Central to the treatment was removal of all dental amalgams, metallic crowns, or bridges and replacing these with composites. The removal procedure followed a special sequence said to counteract the adverse effects of certain magnetic forces in the oral cavity which would otherwise be out of balance. The program included five steps:

1. The diagnosis is made with the use of complete body chemistry, urine testing, hair analysis (all conducted in a laboratory in which Dr. Huggins has a financial interest and is run by his son), and a self-report symptom questionnaire, the "mercury toxicity questionnaire".
2. Based on the body chemistry analysis, nutritional supplements are used which are said to enhance the functioning of the cell membranes and increase the body's ability to release toxic metals such as mercury.
3. A dental examination including electrical readings with a meter known as the "ammeter" is used to justify removal of all amalgams, the first step toward recovery from the "ravages of mercury toxicity".
4. Serum compatibility tests (developed by Dr. Huggins himself) are conducted with the stated goal of determining which dental materials are compatible with an individuals immune system.
5. Follow-up analyses of blood and urine are conducted to determine if the chemistries are changing as desired. This program generally lasts 2 weeks (3 weeks for ALS and leukemia patients).

The Huggins Center has approximately 250 patients a year in this in-office program, at an individual cost of about $6000 plus charges for the actual dentistry, generating an annual revenue of approximately $1.5 million. The Center provides numerous non-dental therapies such as massage, sauna, acupressure, and Feldenkrais, which is a movement training program to re-educate muscles that are claimed to have been re-innervated with removal of mercury amalgams.

During the revocation proceedings, Dr. Huggins presented a list of hypotheses and alluded to thousands of references to support his position, none of whom he was able to cite during his testimony. He also sought to portray questions seeking to identify these studies as unreasonable by, for example, indicating that his goal is to treat patients and not to "rattle off" citations in the literature. Dr. Huggins admitted that he cannot prove the link between mercury form dental amalgam and disease but believes that he is entitled to rely on his clinical experience which suggests such a link. In addition, he offered his philosophy: "The absence of proof is not the proof of absence." He contended that further research will validate his theories and treatment modalities and asserted that there must be room for pioneers in the dental field.

An extremely impressive array of expert witnesses testified on behalf of the Dental Board and provided unanimous support for the propositions that dental amalgam has not been shown to have any connection to the diseases identified by Dr. Huggins, and its removal is not an effective treatment. These experts were also able to demonstrate to the court that the diagnostic and treatment methods employed at the Huggins Center were invalid.

Among the hypotheses put forward by Dr. Huggins was how elemental mercury in amalgams is converted to methylmercury in the oral cavity. This alludes to well-characterized toxic effects, particularly the event in Minomata Bay, Japan, where elemental mercury was dumped into the bay and converted in the water to organic or methylmercury. Eating the fish in which the mercury was bioconcentrated resulted in mercury toxicity. The administrative law judge found that the scientific authorities relied on by Dr. Huggins could generally be described as poorly constructed, biased, improperly conducted, unblinded, or irrelevant. Review of the scientific studies led her to conclude that there was no reliable scientific basis to state that mercury methylates in the mouth to any extent.

Dr. Huggins also believed that the body absorbs mercury released from fillings and concentrates it in the central nervous system by two separate routes: first, absorption into the blood stream and, second, mercury in the oral cavity entering directly into the nerves of the oral cavity and being transported directly back up the nerve sheath through the axons of the nerves themselves into the brain, where it causes damage. If Dr. Huggins's theory about axonal transport of mercury were correct, one would expect that mercury would be toxic to the trigeminal ganglia, but there is no evidence of this.

There was also a hypothesis as to how mercury causes autoimmune disease by embedding in cell membranes, giving the cell the appearance of a foreign body and triggering the immune system to destroy the specific cell. Scientific studies were reviewed which found no evidence to support the theory that amalgams affect the immune system.

The following are some of the conclusions arrived at by the administrative law judge. The diagnosis and treatment of mercury toxicity is the practice of medicine, not the practice of dentistry. In addition, the diagnosis of treatment of MS or any other disease (Table 14.3) is the practice of medicine, not dentistry. Dr. Huggins's statements in his publications attributing these diseases and disorders to dental amalgam and indicating that amalgam removal is an effective treatment were judged to be misleading, deceptive, and false.

Dr. Huggins consistently represented that the Huggins Center had 85% success rate, measured by an improvement in patient symptoms and chemistries. He stated it was only the patient's unwillingness to continue adequate nutritional and other follow-up or the patient's falling into the unlucky 15% that prevented success. Because there is no scientific basis for the diagnosis or treatment performed at the Huggins Center, this statement was judged to be misleading, deceptive, and false.

Dr. Huggins claimed his staff were qualified to explore the patient's "current physical, emotional and psychological condition" and that tests identified potential sources of toxic reactions. The administrative law judge deemed these practices outside the scope of dentistry.

With respect to the efficacy of the methods used to diagnose and treat mercury toxicity, Dr. Huggins publications were judged to be misleading, deceptive, and false. The diagnostic methods specifically identified were the mercury toxicity questionnaire, blood chemistries, the lymphocyte viability test, co-oximetry testing, and hair analysis. The treatment methods specifically identified were serum compatibility testing, the bubble operatory (curved walls said to have unique air circulation and control for adverse effects of electromagnetic forces), sequential removal of amalgams, nutritional supplements (Trans-Mix, Eaters Digest, X-IT, and Jogger Juice all distributed or endorsed by Dr. Huggins), intravenous vitamin C, vitamin C flush, insulin, lithium, thyroid, posterior pituitary extract, massage, sauna (said to sweat out mercury lodged in tissue), acupressure for pain relief, and Feldenkrais movement.

In addition, certain theoretical positions influencing treatment were also noted as having been falsely presented, and these warrant comment. According to "retention toxicity", a person excreting low levels of mercury in his urine is retaining mercury in the body and is severely ill. In fact, low excretion indicates a low body burden of mercury. A patient with normal kidney function excretes more mercury the more he has on board. The fact that urine mercury levels indeed increase after amalgam removal is due to the removal process, which causes transient exposure from the grinding and displacement of the mercury, and not to an enhanced ability to excrete mercury.

The Huggins Center scheduled patient treatment based on a "7-, 14-, 21-day immune cycle". In Dr. Huggins' view, day one represents an immune system challenge such as the replacement or removal of amalgam. Thereafter, on the 7th, 14th, and 21st days, the patient experiences flu-like symptoms. On the 21st day, if there is additional immunological challenge, the patient may develop serious autoimmune disease due to the cyclical dying off of the white blood cells. Based on this cycle, patients are scheduled to avoid treatment on the "low defense" days.

Finally, it was the standard procedure at the Huggins Center to have prospective patients sign a financial agreement which committed them to pay for services at the Center before they arrived for treatment. Statements of informed consent were obtained at a later time after the patient arrived at the Center.

In summary, Judge Connick compiled a scholarly treatise delineating the philosophy, allusions, distortions, and outright fraudulent practices associated with the dental amalgam phenomenon. She also presented a thorough review of the scientific aspects of the case. The same psychopathological characteristics identified in this case may be seen in treatment strategies developed for EI patients. Serious illness is not taken seriously. Vulnerable patients who would grasp at any illusion of hope are exploited for profit and blamed for the failure of treatment. Fear is instilled, and false beliefs are created which influence the person's life. Patients are subjected to unwarranted risks from unnecessary procedures. Some treatment modalities have potential harmful side effects. Judge Connick left a forceful impact on Colorado case law. Yet, one harmful effect which is not addressed in this legal decision is the psychological harm which results from false explanations for symptoms, diseases, and disorders.

Harmful effects

The lady in the desert

Consideration of harmful effects brings me full circle to the case of "an episode in the desert" which intrigued Selner and me when we initiated our studies of the EI phenomenon. The patient was a 45-year-old woman who lived a hermit's existence in a small, 7 × 12-foot trailer in the Arizona desert.

Communication with the world was maintained by use of a remote control phone covered with aluminum foil to avoid touching plastic. The patient explained that it was necessary for her to remain in the darkened trailer most of the time because of the adverse effects of ultraviolet rays. However, she appeared deeply tanned. The patient had a letter from a California physician, addressed to the Visually Handicapped Division of the Tucson Health Department, outlining why she required a seeing eye dog to lead her about her darkened trailer. How the dog was going to see to lead her remains a mystery. She described intolerance of many foods and chemicals. Her multi-system complaints were all subjective. Further medical evaluation was not possible, as she could not travel to the medical school for examination because of the power lines and associated magnetic field perturbations which she believed would exacerbate her symptoms. This woman was articulate and disarmingly ready to share her history with the listener in exquisite detail. Her story had been the subject of articles in the lay press.

The patient described multiple medical assessments in which examinations and laboratory tests failed to demonstrate physical pathology. She underwent drug therapies and psychotherapy for years without success. Continuing to search for an explanation for her symptoms, she discovered clinical ecology. A well-known California clinical ecologist evaluated her and diagnosed her as having "classical ecologic disease". Uncontrolled provocation tests with a variety of foods and chemicals were interpreted to be positive. Leukocyte T and B cells were called abnormal and used as corroborating evidence for a diagnosis of total immune disorder syndrome induced by environmental illness. The patient was declared allergic to the twentieth century.

This experience was bewildering and motivated a detailed review of the case. On careful scrutiny of leukocyte subset data, T and B cells appeared to fall within normal limits. The provocation tests were uncontrolled, not replicated, and interpreted on the basis of subjective complaints. There was no indication that this patient's psychiatric history had been considered as a possible contributing factor in the illness. During discussions with the woman, intimate revelations of childhood, family, marital, and social interactions were shared which suggested a heritage of non-nurturing, criticism, rejection, and unrealistic expectations for performance. She manifested low self-esteem, perfectionism, obsessive-compulsive characteristics, and paranoid ideation. The conclusion of the clinical ecologist and accompanying recommendations seemed to represent a self-fulfilling prophecy, reinforcing her belief system which protected her form facing the reality of psychiatric illness (see Selner and Staudenmayer, 1986).

Isolation and ineffective rituals

The indiscriminate use of controversial and unproven diagnostic and treatment techniques can reinforce inappropriate ideation. The danger of grouping patients under an appealing label such as EI is that this all-encompassing phenomenon may obscure detection of more serious medical or psychiatric illness. Adherence to ineffective treatment may create more severe or persistent morbidity than the initial illness.

Isolation is one harmful effect seen in EI patients. Withdrawal into a "chemically safe" house or community may implicitly confirm the patient's EI belief that symptoms result from biologic processes caused by "environmental poisons". Isolation removes the patient from feedback from people who function in the real world, information that could potentially facilitate the restructuring of distorted beliefs. Even among normal college students, it has been demonstrated that social isolation impairs perception of reality and ability to use one's faculties (Asch, 1957). Social withdrawal is also one of the characteristics of patients with post-traumatic stress disorder. Isolation can also create or exacerbate psychological problems such as depression and associated symptoms of greater hopelessness and despair. In interpersonal interactions, depressed people are often more confrontational which drives others away (Coyne, 1976). In one study, depressed individuals reported more anger and hostility during their encounters than non-depressed individuals (Folkman and Lazarus, 1986). Finally, there is also a risk of lethality. Hopelessness has been predictive of suicide in studies of depressed inpatients (Beck et al., 1985) and depressed outpatients (Beck et al., 1990).

Avoidance strategies can be maladaptive if they draw the person's attention away from a problem that needs to be addressed. Another set of avoidance strategies called "escape-avoidance strategies" are even less adaptive and describe efforts to escape through wishful thinking (e.g., wishing that a situation would go away or somehow be over with), eating, drinking, smoking, using drugs or medications, or sleeping (Folkman and Lazarus, 1988). Although these strategies may provide temporary relief, several studies have shown that escape-avoidance is associated with depression and anxiety (Folkman et al., 1986) and somatization (Benner, 1984).

When an EI patient believes that his/her symptoms can only be caused by an environmental exposure, by definition the belief system is closed (Gomez et al., 1996). As a consequence, symptoms can be resolved only by an external intervention that counteracts the effects of the exposure, e.g., isolation, avoidance of odors, oxygen administration, and detoxification rituals such as megadoses of vitamins and supplements, purging enemas, or sauna depuration. This creates a logical trap for the EI patient in that it precludes the possibility of resolution of symptoms by intrapsychic mechanisms. That is, if symptoms resolve, the patient cannot attribute this improvement to his own initiated action because the closed belief system does not permit this. Such a recovery effect would constitute incongruent information which would require restructuring beliefs about chemical sensitivities. It also precludes treatment modalities that emphasize self-control, self-regulation, and internal locus of control, which are some of the therapeutic processes inherent in successful psychotherapies (Strickland, 1978). With EI patients, successful psychotherapy is based on overcoming the limiting effects of the false environmental appraisals and by restructuring the EI beliefs to be more consistent with reality.

The implementation of a coping strategy may serve as a source of stress (Cohen et al., 1986). For example, some coping strategies may be inherently aversive or medically dangerous. Disappointment over a failed coping strategy may exacerbate anxiety and feelings of helplessness and hopelessness. The search for and the failure to find an effective coping strategy may create feelings of immobilization and lead to procrastination, avoidance, or social withdrawal. Implementation of a false defense or an ineffective coping strategy can create a negative feedback loop that depletes adaptive energy, making the individual vulnerable to additional stressful encounters. Increased release of stress hormones and adrenal catecholamines mobilizes blood sugars and fats, the fuel for both mental and physical action. However, in the long term, the sustained release of sugars and fats leads to depletion and fatigue. Mental work can be as exhausting as physical work because of the electro-cortical arousal and the consequent stress-responses (Hamilton and Warburton, 1979). Stress has also been found to have deleterious effects on decision-making processes (Cohen, 1980). These examples highlight that false defenses and ineffective coping strategies may have disruptive and potentially harmful effects.

A frightening experience is not always harmless. A pseudoreaction based on belief of exposure, like a panic reaction, has associated psychophysiologic changes that may potentially contribute to adverse health effects. For example, stress-induced hormonal discharges may create effects mediated in the autonomic nervous system such as increased blood pressure or elevated heart rate, which are risk factors in patients with cardiovascular disease. Heightened nervous system activity may also drive anxiety and depression. In acute, terror responses there is always the risk of induced death, as documented in anthropological accounts of "voodoo" death, which were interpreted in terms of the stress response by Cannon (1957) and reviewed in detail elsewhere (Morse et al., 1991).

Munchausen syndrome by proxy

One particularly distressing manifestation of EI is that of parents who express their psychological conflicts by displacing somatization to their children, holding them hostage to EI. In most cases the perpetrator is the mother, with the father physically or emotionally uninvolved with the children and subservient to the dominant mother. The perpetrators of this type of childhood abuse present themselves as devoted parents, doing everything within their power to care for the children's illnesses, including inordinate demands for medical investigations which invariably show negative findings for underlying disease processes. Their presentations are often credible and mislead relatives, friends, teachers, school officials, and their treating physicians. These interactions seem incredulous and could be regarded not only as a *folie a deux* between the mother and child, but also a *folie a deux* between the mother and the professionals concerned. Yet, it is the child who suffers and is at risk.

This resembles a form of child abuse called Munchausen syndrome by proxy, (Livingston, 1987; Meadow, 1977; Rosenberg, 1987), in which signs and symptoms are falsified, either by physical induction or verbal fabrication. There are numerous manifestations of signs and symptoms which, for the sake of presentation, may be classified into three categories. The first involves premeditated harm to the child, such as suffocation, poisoning, or breaking a limb, which reflects what has been called apparent life-threatening events (Little et al., 1987) and is typically seen in infant victims (Southall et al., 1997). The second includes suffering such as failure to thrive, often through the active withholding of food; allegation of allergy and withholding of food; inflicting pain by slapping or pinching; or disrupting medical treatment. The third includes fabricating illnesses without direct evidence for physical harm, one of the most documented being fictitious epilepsy (Meadow, 1984), although others have been identified including hematemesis (vomiting of blood), hematuria (blood in the urine), hemoptysis (blood in the sputum), urinary tract infection, glycosuria (high glucose level in urine), fever (by altering a temperature chart), vomiting, food allergy, and cancer (Bools et al., 1992). Regardless of the degree of physical harm, there is compelling evidence that these children suffer emotional abuse (Garbarino, 1978) and are at greater risk to develop (1) conduct and emotional disorders, problems related to school (including difficulties in attention/concentration and non-attendance), and antisocial behavior (Bools et al., 1993; Rosenberg, 1987); (2) iatrogenic problems as a result of investigations, medications, and other interventions (Meadows, 1982); and (3) psychological morbidity as adult sequelae (McGuire and Feldman, 1989), including somatoform disorders (Bools et al., 1993.) Siblings of identified cases of abuse have also been found to be at greater risk for Munchausen syndrome by proxy and, in some cases, death (Bools et al., 1992; Southall et al., 1997).

EI is another fabricated condition used by parents, particularly mothers, to express the dynamics of Munchausen syndrome by proxy and has been labeled "multiple chemical sensitivity by proxy" (Robertson, 1994). Unfortunately, labels often do not capture the complexity and heterogeneity seen in psychological conditions, and this condition is no exception. The injuries and suffering inflicted and the illnesses fabricated encompass far more than those initially researched when the term "Munchausen syndrome by proxy" was introduced by Meadow in 1977. The current trend is to avoid the term and to describe the actual abuse identified. This is more in line with differential diagnoses, and the DSM-IV does not have a specific diagnosis for Munchausen syndrome by proxy. Rather, the term "factitious disorder by proxy" is used and considered an experimental category. Nonetheless, several differential DSM-IV diagnoses are applicable. For the child victim of the abuse or neglect, the diagnostic code is 995.5. For the perpetrator, there are two differential diagnoses for factitious disorder by proxy: (1) factitious disorder not otherwise specified

(300.19), or (2) malingering (V65.2). The essential feature of factitious disorder by proxy as given in DSM-IV is

"... deliberate production or feigning of physical or psychological signs or symptoms in another person who is under the individual's care. The type and severity of signs and symptoms are limited only by the medical sophistication and opportunities of the perpetrator. Cases are often characterized by an atypical clinical course in the victim and inconsistent laboratory test results that are at variance with the seeming health of the victim."

Malingering differs from factitious disorder in that the motivation for the symptom production in malingering is an external incentive, whereas in factitious disorder external incentives are absent.

In the context of EI, these are not mutually exclusive, as secondary gain issues in the form of litigation may arise — against a school district for having constructed a "sick building" or failing to accommodate the child's "unique needs" or against a spouse in a custody battle. Such aggressive litigation may stem purely form secondary gains such as financial gain or attention for the parent, or it may represent a defense to deflect authorities from the parent's role in a factitious disorder by proxy.

The motives underlying such abusive and deceptive behavior in the mother are heterogeneous, and to date poorly understood in terms of psychological mechanism. In the first historical account of Munchausen's syndrome by proxy, Meadow (1977) noted that mothers had a history of falsifying their own medical records and treatment and had been labeled as hysterical personalities who also tended to be depressed. Personality disorders have been almost universally diagnosed in studied cases of mothers who abused their children in the form of factitious disorder by proxy (Bools et al., 1994; Famularo et al., 1992; Southall et al., 1997), with most meeting criteria for more than one type of personality disorder. Although no unique and reliable personality profile has been identified, Meadow and colleagues, in a unique long-term follow-up study of perpetrators from their case series, have suggested that there is a high incidence of histrionic and borderline types (Bools et al., 1994), a finding consistent with an extensive literature which they reviewed showing borderline, histrionic/hysterical, and narcissistic features in perpetrators. These mothers often have personal histories of somatization and fabricating their own illness, as well as self-harm and substance misuse (Meadow, 1984; Bools et al., 1994). The psychological factor most commonly reported in studies of mothers is their report of previous (childhood incest or adult rape) and ongoing physical and sexual abuse in their own history (Southall et al., 1987, 1997). This finding is consistent with studies of abusive mothers in general (Famularo et al., 1992).

Gray and Bentovim (1996) identified another psychosocial factor underlying illness induction, namely, parents use the focus on the child's illness as a way of solving major personal, marital, and family problems. In that study, as in others, the mothers also reported histories of childhood sexual, physical, or emotional abuse.

Munchausen syndrome by proxy has been unusually resistant to effective therapeutic intervention (Jones, 1987). Denial by the perpetrator is common, and repeated offenses toward the child after detection often continues. Professional partnerships with perpetrators who have personality traits of deception and falsification are tenuous at best and may be neither safe nor effective in protecting their children (Southall et al., 1997).

Where are the bodies?

Given all the environmental, technological, social, and disease exposure changes that have been imposed on the modern human condition, it is remarkable that people, for the most part, still adapt quite well. EI advocates would have us believe that there is a yet-undefined killer illness. But where are the bodies? To my knowledge, only one account of deaths attributed to EI has been presented, that by William Rea, arguably the preeminent clinical ecologist in the U.S. In a presentation to the California Department of Health, Berkeley (May 5, 1994), Rea stated that he knew of 40 cases of deaths from EI and an additional 20 who committed suicide over the despair of having EI. If these deaths are erroneously attributed to a nonexistent illness, then the implications are unthinkable.

Discussion

EI cannot be dismissed as a harmless fad by which believers fulfill their delusions and the rest of the world is not affected. EI has well-identified, harmful effects to both EI patients and society, including:

1. Iatrogenic belief in EI obviates proper psychological care. Unsubstantiated diagnostic and treatment methods emphasize the permanence of chemical sensitivities and impairment rather than resolution. This can lead to interminable unsubstantiated treatment and social withdrawal. Often, there results a total depletion of the patient's or family's financial resources as the EI patient, and the advocate and doctor are engaged in a process of mutual deception to overcome the fantasized effects of a myriad of mythical toxins. What may at first be seen by the patient as puzzling events during a "reaction" may become fixed beliefs about environmental sensitivity reinforced by the doctor. This can lead to deeper despair and greater sense of loss of control as the patient has lost the stance of puzzlement and the option of challenging the interpretations of environmental attribution. This process is similar to that of induction into a cult. Doctors who engage in this iatrogenic practice may well become the defendants in future lawsuits.

2. Projection of attribution to external agents, foods, chemicals, or electromagnetic forces is similar to other fantasy projections throughout history, ranging from the demonic to political conspiracies. All have served to override the concept of personal responsibility for thoughts, feelings, and actions. "The environment made me do it" has even entered the criminal courts through the famous "Gallo wine defense" for a homicide in Hawaii and the "Twinkie defense" used for the murder of the mayor of San Francisco. Aside from these examples which defy credulity, there are the more tragic cases of vulnerable patients with psychiatric conditions that are not treatable because the element of personal responsibility is missing. Many of these patients are suggestible, especially when the suggestion fulfills their wish to be relieved of the responsibility of having to make their own decisions, and to take responsibility for their anxieties, misperceptions, and misconceptions.

3. EI victim advocacy groups, anarchistic social groups (e.g., the AIDS ACT-UP advocates), radical environmental groups such as Green Peace or even the Sierra Club on occasion (Sierra Club, 1981), and consumer advocacy groups motivated by a sociopolitical agenda may knowingly or unwittingly exploit the vulnerability and suggestibility of EI patients. The wisdom of these movements is often clouded by absorption with their own moral or social agendas.

4. Unfounded allegations of harm alleged by EI patients have emerged in pursuit of compensation for disability, entitlement, and product liability. Unscrupulous trial lawyers have targeted deep-pocket defendants without regard to the implications to society. Virtually every chemical product manufactured has been cloaked in the cloud of innuendo of EI. Irreparable damage may result from destroying the reputation of reputable consumer and industrial products, companies, and individuals. Settlements to avoid litigation have been large. While insurance companies pay these settlements, as well as the legal costs to defend their clients, who will ultimately pay is no mystery. The potential negative effect of removing safe, effective products from the market is difficult to estimate, but removing chlorine, for example, would be catastrophic. There is also the cost of unnecessary services. Numerous state and local municipal agencies are virtual hostages to communities of EI patients who demand "chemically free zones" and corridors in which to travel. Again, who will bear the cost? Juries may well sympathize with the pathetic theater before them, but they may think differently when they come to realize the social injustice this creates for everyone.

5. Social withdrawal can become a self fulfilling prophecy, leading vulnerable individuals to hermetic existences in remote geographical locations. The banishment from society is self-imposed, enforced by overvalued ideas about environmental sensitivities. Communications to the outside world often run through EI networks, with supply lines to clinical ecology marketeers selling medically unsubstantiated products that reinforce false hope and illusions. Banded together in communities, encouraged by mutual reinforcement and processes of mass psychogenic illness, EI patients can create an inescapable prison.

chapter fifteen

Treatment

This chapter suggests treatment approaches for both physician and psychotherapist. Treatment of EI patients, as for other patients with somatization and psychogenic illness (Barsky, 1979), requires a complementary relationship between medical and mental health-care professionals. Where there is no objective medical disease, the healing quality of the doctor-patient relationship becomes primary. Under these circumstances, the healing skill of the doctor lies in bringing out the self-healing potential of the patient. These interpersonal interactions are the essence of psychotherapy. No single approach has been shown to be more effective than the rest as a general approach to a broad spectrum of psychological disorders, which implies that the effectiveness of psychotherapy lies in the essential nonspecific characteristics shared by all of them (Frank and Frank, 1991; Luborsky et al., 1975). This clinical insight was supported by the 1994 *Consumer Reports* survey on the effectiveness of psychotherapies (Seligman, 1995). Some therapeutic changes may result as much from client inferences as from the skill of the therapist. For example, specific therapeutic goals in the patient's rationale or preconceptions about therapy may activate a deductive process that can lead to altered self-perceptions and attitudes that are important to treatment outcome (Kirsch, 1978).

Like other patients with functional somatic symptoms, EI patients cling tenaciously to their beliefs about the cause of their symptoms. They are unique only in that their overvalued ideas are about EI. Somatizers are usually threatened by self-exploration and therefore adamantly preserve representations of a false self for fear of discovering and realizing their own true-self. These patients typically think it ludicrous that they might benefit from psychological or psychiatric intervention.

There are primary and secondary gains in the projections of attribution of harmful appraisals onto the environment. Preoccupation with false beliefs serves to distract attention away from underlying conflicts, often associated with recalling trauma-inducing experiences from the past and the burden of coping with stressors in current life. For individuals with personality disorders, such experiences are often perceived to be incriminating to self. Those with a history of childhood abuse and trauma often resist entertaining the possibility that their personalities have been disrupted by trauma, or that they suffer the physical and psychological sequelae associated with PTSD or other forms of anxiety. Some may be afraid of any label that triggers their primal fear of being or going insane. Surviving the horrors of trauma in childhood can certainly drive one to the edge of insanity, if not beyond. Such severely traumatized patients have a vested interest in keeping their trauma secret, motivated by a perception of the perpetrator(s) that reinforces the fear of being killed if they were to reveal the secret. Many EI patients with a history of trauma resist explanations that suggest their current symptoms represent somatization and can be explained by conversion of intrapsychic conflict through the physiological systems of the stress response (Cohen and Rodriguez, 1995). Furthermore, identification with EI advocacy/victim groups may provide a sense of affiliation, a sense of acceptance and importance, which counteracts feelings of isolation, vulnerability, emptiness, and sadness (Staudenmayer, 1996).

Patients with closed beliefs about the causality of their symptoms have taken Randolph's dedication to heart, which appears in many clinical ecology publications (e.g., Randolph and Moss, 1980) and reads, "This book is dedicated to all patients who have ever been called neurotic, hypochondriac, hysterical, or starved for attention, while actually suffering from environmentally induced illness."

The path of psychotherapy

Psychotherapy should be viewed as a safe haven in which to learn about self-defeating personality characteristics, to face significant fears, to restructure distorted beliefs and overvalued ideas, to self-regulate stress physiology and somatization, to talk about painful feelings, to weigh the facilitating effects of psychotropic medication, to deal with unresolved feelings about childhood trauma, and to convert unhelpful defenses into effective coping styles. These are noble sentiments and admirable long-term objectives. But can they be achieved with the most guarded and difficult EI patients?

The therapeutic approach that I have found to be effective in the treatment of EI is somewhat hierarchical, beginning with education and assurance by the physician, followed by supportive psychotherapy, self-regulation of symptoms, and cognitive-behavioral treatment, culminating in insight-oriented psychotherapy. The specific techniques include explanation of stress-response physiology, self-regulation treatment employing biofeedback, non-critical confrontation of the unsubstantiated beliefs, and interpretation of the projection of anxiety. Whenever relevant, explanation of an alternative etiology is explored, an etiology which originates in disruptions of personality occurring during childhood development often as a consequence of childhood abuse and deprivation. Throughout what may be a lengthy therapy, it is important to demonstrate understanding and empathy for the painful and distressing effects these individuals feel. Patience is paramount because healing often requires waiting for the patient to find an exit out of their individual possible world of false beliefs and overvalued ideas. Entering into a world resembling reality is often threatening to an individual who has relied on psychological defenses to protect against a personal reality of trauma and inability to cope. A therapist can only guide these patients along this precarious and painful journey, accepting that ultimately their destination is their own choosing. They cannot be rescued from this journey, nor be transported to the real world by the proclamation that their EI centered beliefs are scientifically unproved.

Knowing the answer is only the beginning for the mentor. The patient/student needs to learn to incorporate the integrated conceptual networks that represent reality into their own knowledge. This entails correcting not only the false propositions, but also the associated emotional and behavioral memory networks that often manifest as panic reactions but are appraised as chemical reactions. The journey in search of self and reality is fraught with pitfalls, hardship, and suffering. For some EI patients, the journey is too threatening to undertake, and projecting their misfortune onto EI serves their need for denial and avoidance. Others may undertake this journey reluctantly and never complete it. Of these, some will return to EI, and some will find escape into another false world of beliefs. But some will persevere and complete the journey to wellness. In one sample of EI patients who underwent psychotherapy, 75% had a successful outcome in the sense that they put aside EI and addressed their psychological issues (Staudenmayer et al., 1993b).

Difficult patients and resistance

"The physician who is unable to provide the time or who is unable to maintain a more flexible role, often compromising his own omnipotent or narcissistic needs, should carefully consider not accepting the commitment to care for such patients." (Drossman, 1978)

The physician often finds that certain patients are poorly responsive to usual methods of treatment, especially those individuals with somatic symptoms in whom psychosocial factors are a major, if not the sole, determinant of the illness. To care for such patients one must understand their illness as the product of mutually interacting biologic, psychologic, and sociologic determinants (Drossman, 1978).

The interaction between a physician or therapist and an EI patient who has a history of childhood trauma often is characterized by attitudes and fears that come from another source in the past, having nothing to do with the immediate interaction. In the context in which the unresolved trauma occurred, the family is usually in overt denial or has actively threatened the child not to reveal the intrafamilial abuse. For the patient to risk revealing the abuse in treatment, the physician/therapist needs to be open, non-judgmental, and supportive. The language used in discussing abuse is also critical, and recent guidelines for a structured interview for childhood trauma include the use of objective descriptions of acts rather than terms such as abuse, rape, or trauma (Fink et al., 1995). My physician colleague, John C. Selner, asked, "Has anyone ever made any mistakes with you?" as an opening statement in exploring childhood abuse. Given an opportunity, it has been reassuring to see how many patients are motivated to seek help.

The responses many abused patients expect from others based on their childhood experiences are disbelief, rejection, judgment, withdrawal, or banishment (Saporta and Gans, 1995). If any such perceived hostility is inferred from the language or actions of the physician, the patient is unlikely to risk revealing and exploring aspects of his underlying anxiety and trauma. Instead, the patient is more likely to revert to false defenses, which in the case of EI patients is the repetition of rationalizations about overvalued ideas of environmental sensitivities. If confronted about that logic, the patient is likely to go elsewhere for treatment.

By now it should be obvious to the reader that treating an EI patient can be frustrating for the physician. Patients are generally reluctant to accept psychiatric referral, and if they do, it often backfires, as noted by Ducatman (1993), an occupational medicine physician:

"The use of antidepressant medication and psychotherapy as 'rational approaches' to the therapy is controversial. The success of these approaches has been dismal. Any practitioner not wishing to see EI patients need merely recommend early pharmacologic treatment for depression and psychiatric referral, before clinical trust is established and other possible clinical needs ruled out. The patient will never return. This advice may be 'rational' if the solution to the physician's immediate dilemma of what to do with EI patients is to chase them from the office. Actually helping them is more complex. Healing may require frequent revisitation of disease beliefs and suggestions for coping with symptoms, without the aid of psychiatrists or drugs." (Ducatman, 1993)

Similar circumstances are often encountered with patients who have a variety of functional illnesses accompanied by somatization such as chronic pain, fibromyalgia, or chronic fatigue syndrome. Kleinman (1982, 1986) reported similar experiences with neurasthenia patients in China. His patients were receiving polypharmacy with traditional herbs, sedatives, and antianxiety agents (benzodiazepines), along with vitamins, tonics, and pain medicines. These patients have high utilization rates for medical services and a history of frequent changes in caregivers and medical regimens. The similarity in the pattern of care in two very different cultures indicates a common frustration on the part of both practitioners and patients over chronic problems that fail to respond to specific therapeutic interventions. This frustration is as much a challenge to the physician's coping skills as it is to the patient's.

Kleinman (1982) observed that when his Chinese neurasthenia patients improved, their illness behavior did not. Although psychotropic drugs may have significantly benefited symptoms, they frequently did not have a major effect on important psychosocial aspects of the illness. Such cases demonstrate that medical treatment for chronic conditions without significant psychosocial intervention exerts only a limited effect on the overall illness. Kleinman's observations, as well as Ducatman's editorial, should alert us to the limitations of orthomolecular psychiatry and therapeutic techniques (e.g., behavior modification, systems theory of family therapy) that have in common a relative disregard of a person's life history and a preoccupation with the current situation (Rutter, 1994). Nevertheless, in some articles that identify alternative psychiatric diagnoses in EI patients, antidepressant medications are suggested as the recommended treatment modality, although no clinical data on treatment, adherence, or follow-up are presented (Simon, 1992b; Black et al., 1990).

My medical colleagues have attempted to introduce antidepressant medication into the therapeutic regimen of some EI patients who manifested clinical depression. Some of these patients report intolerable side effects attributed to the medication, as would be expected from their belief about the harmful effects of any chemicals. Paradoxically, these complaints often do not occur immediately after taking the medications, as expected. Rather, the complaints are not registered until several days or even weeks after initiating a trial of antidepressant medication. Could these symptoms be delayed or cumulative toxicologic effects? Probably not. Rather, the complaints correspond to the delayed onset of therapeutic effects characteristic of these medications (Hyman and Nestler, 1996). It seems as though the patients are unconsciously telling us not to take away their symptoms of depression because to do so would also require them to alter their beliefs about an environmental etiology which would render the psychological defense projection ineffective.

Transference and countertransference

In reaction to the patient's transference and resistant behavior, the doctor may displace feelings toward the patient. This is called "countertransference" (Wachtel, 1982). Physicians rated patients with somatization as more frustrating, help-rejecting, and demanding (Barsky et al., 1991). Consider the paradoxes of working with EI patients. For the traditional physician, who usually follows his training in objective science and focuses on truth in diagnosis, the irrational elements of healing are incongruent to his/her knowledge and may create conflict. Patients with somatization overuse medical services. They typically have views of their symptoms very different from the views of their physicians (Lin et al., 1991). They are more dissatisfied with their physicians (Barsky et al., 1991). Somatoform disorders involve a unique healing paradox. For the doctor the dilemma is how to relieve symptoms without removing the sick role. For the patient the dilemma is how to function without being well.

Hateful and dangerous patients

Hateful patients are those with whom the physician or psychotherapist has an occasional personality clash. These are the patients most doctors dread, and they are the ones most likely to be labeled pejoratively as crocks (Lipsitt, 1970). The insatiable dependency or domineering and controlling manipulations of hateful patients leads to behaviors that group them into four stereotypes: dependent clingers, entitled demanders, manipulative help-rejecters, and self-destructive deniers (Groves, 1978).

Clinical experience suggests that psychiatric patients who have dissociative identity disorder (DID) co-morbid with borderline-narcissistic personality structure (Adler, 1981) have great hostility and the most fantastic accounts of victimization. They are the most likely to profess satanic, ritualistic abuse involving sexual orgies, human sacrifice, participatory murder, and other horrifying atrocities, all without corroborative evidence. EI patients who have these types of psychological characteristics may also present fantastic accounts of environmental conspiracies. These types of DID patients also tend to have the most elaborate entourage of "alter personalities". These personalities include a "dark side" which comprises punishing, destructive, alter personalities reflecting their childhood aggressors. In a dissociative state, such alter personalities identify with the aggressors and act out the associated aggressive behavior against themselves or others. These patients are also potentially dangerous to the therapist when they act out a malignant transference of their perpetrator with their therapist (Vickery, 1992).

In dangerous patients with a history of childhood trauma, abusive interactions are unconsciously perpetuated for a number of reasons. Most relevant, the patient's active perpetration of abuse against the doctor or therapist is a defense against the feelings of powerlessness and rage that characterized the experience of childhood trauma; the action serves to inhibit these original feelings out of consciousness (Saporta and Gans, 1995). The displaced rage toward the doctor or "those poisoning the environment" is a form of resistance. By displacing the abusive or "bad parent" onto the doctor, the illusion of the aggressor as a "good parent" remains. Psychiatrists call this preserving an ambivalent perception of the perpetrator. Patients who have these internal dynamics can often be identified by their idolization of the abusive parent.

In clinical practice, there are several identifiable clues that some form of malevolent transference has occurred with a difficult patient. These include resistance to treatment, attempts to dominate the therapist and control the therapy, or efforts to entrap the therapist in a boundary violation. The perceived harm from the object of transference (i.e., the therapist) can become the obsessional focus of their lives and amplified to delusional proportions which motivate retribution, all rationalized and enacted as a crusade against "injustice". The aggression often takes on a vindictive, destructive form in which the therapist is entrapped in situations which later are reframed into boundary violations by crafty lawyers in civil actions or by the patient portraying victimization in the star-chamber environment of regulatory boards.

Table 15.1   Five Stages of the Loss-of-Control Model

Stage	Symptoms
1. Fear of losing control (feeling in control, but focused on maintaining it)	Afraid of embarrassment or humiliation Obsessive thoughts Narrowly focused attention Compulsive activity High achievement
2. Helplessness (the sense of having control is lost, but helplessness is offset by active attempts to regain control)	Perfectionism Indecision and procrastination Approach-avoidance conflict Agitated depression Irrational avoidance behavior
3. Hopelessness (despair that control is lost and cannot be regained)	Atypical, vegetative depression Acute, generalized anxiety Inability to focus attention or concentrate Mind racing Numbing
4. Hallucinations and psychotic-like behavior (psychological defenses against going insane)	Exacerbation of paranoia Social withdrawal Delusions and overvalued ideas Overt manifestation of personality disorders
5. Hysterical reenactment of being killed (physical sensations of a near-death experience, e.g., choking, suffocation, loss of consciousness)	Dissociation of pain and suffering Living in a perpetual state of terror Severe dissociative disorders Severe personality disorders Preoccupation with death as a last defense Suicidal ideation and enactments, often not genuine

This kind of displaced hostility illustrates a psychological defense called projective identification and is often observed in child-abuse victims who externalize unwanted self-images (Saporta and Gans, 1995). The victim identifies with the childhood perpetrator and has internalized those abusive characteristics. The devalued false self is externalized and projected onto the doctor, who is now actively victimized by the formerly passive victim. To avoid conscious awareness of their own malevolence and to justify their persecutory actions, a host of false accusations may be rationalized (Klein, 1989). In fact, these patients re-enact their own childhood experiences of "blaming the victim", sometimes unconsciously but at other times consciously and sadistically assuming the role of an omnipotent parent with narcissistic delusions of being God.

A depth model for selecting an intervention strategy

Assessing an EI patient's attitude about control, or rather, loss of control, is often helpful in identifying the degree of psychopathology with which one is working. For purposes of illustration, I suggest five progressive stages to capture the feelings of loss of control and the associated depth of psychopathology (Table 15.1). The stage to which an individual progresses in this depth model is hypothesized to be dependent on the severity and chronicity of the underlying trauma, counterbalanced by the resilience of the individual and utilization of his effective coping skills.

Table 15.2   Treatment Strategies for Difficult Somatizers

1. Establish the significance of psychosocial factors in the patient's illness — use the symptom-oriented, nondirective interview.
2. Maintain an unbiased interest; communicate an air of non-judgmental concern.
3. Take a complete history and perform a physical examination.
4. "Don't just do something, stand there!" Have the confidence not to order unnecessary diagnostic studies to rule out organic disease.
5. Do not attempt to reassure by stating that the problem is emotional.
6. Accept the symptoms. Respect the adaptive value of the illness.
7. Set up regular visits to develop the therapeutic relationship.
8. Be alert for new developments. Alternative medical diagnoses can develop in the patient who cries wolf.
9. Treatment may be prolonged; ending a therapeutic relationship can be a difficult task.
10. Be aware of personal attitudes [countertransference].

Source: Drossman, D., *Ann. Intern. Med.*, 88, 366–372, 1978. With permission.

Drossman (1978) suggested some therapeutic guidelines for physicians which are directed toward modification of psychosocial and somatic variables (Table 15.2). Weaver (1996) suggested similar approaches in working with EI patients. I have elaborated on these, emphasizing issues in working with EI patients who present with overvalued ideas and illogical thinking that have been acquired, often through iatrogenic influence. Physicians and psychotherapists working with such patients may benefit by having some familiarity with the processes of mind control and indoctrination effects identified in victims of cults (e.g., Morse and Morse, 1987). Following are some guidelines for working with EI patients.

Trust, empathy, and rapport

First and foremost, trust, empathy, and rapport are necessary conditions for any therapy to be effective. Generally, if trust and rapport are not established at the beginning of the intervention, the likelihood of having a successful impact is reduced. One helpful technique is to focus on what the doctor and patient can agree upon, namely, that the pain and suffering are real! It is not imagined, it is not "all in your head". It is the cause of symptoms over which there is disagreement. Minimization of pain and coping with suffering are shared objectives in therapy. The introduction of interventions directed towards achieving relaxation and pain reduction such as biofeedback reinforces the recognition of the pain as real. Discussion of stress psychophysiology which mediates symptoms associated with muscle tension, hyperventilation, anxiety, and autonomic arousal are usually non-threatening topics. An educational approach counteracts the patient's fear that therapy will uncover some deep-seated anxiety or personality disorder, which in fact they may well have. Most of these patients do not believe they are in need of psychiatric care, and some may become hostile if the topic is broached. This aversion often comes from misunderstandings about what psychiatric or psychological care involves and who it may benefit. Individuals who believe that psychotherapy is reserved for the insane avoid it because participating would logically confirm their belief that they are "crazy". Initially the treating physician and the psychotherapist need to convey that their role is to help with the emotional consequences of having symptoms and pain, which may include medications for depression and/or anxiety (Kellner, 1990, 1991).

Logical, scientific argument is usually not effective with patients who have closed beliefs about the appraisal of their symptoms and attribution to environmental sensitivities. Such strong conviction often masks fear, anxiety, and anger. Explanations of logical, medical models to disprove their irrational beliefs are usually interpreted as aiming at domination and control, two constructs which drive their fears and anxiety in the first place.

Psychological assessment is incomplete if focus is placed on the alleged precipitating exposure event involving a possible environmental agent without investigating the psychosocial demands on the patient at the time of the event, as well as the course of life preceding it. EI patients, particularly those with obsessive traits, prefer to restrict discussion of psychological symptoms to those they believe were caused by their perceived environmental sensitivities. It is important to ask them about their personal and social relations, their work, and their fears and anxieties. A history of experiences in the family of origin is important, especially an assessment of childhood abuse and emotional deprivation. Collateral contacts with family members, spouses, or significant others may enhance this type of information. Such contacts can help others to better understand the disorder, and if they are not believers in EI themselves or contributors to the patient's stress or anxiety, to support the psychological treatment.

The treating physician should maintain regular contact with the patient. This shows concern about looking after the medical aspects of the treatment and respect for the patient. Cooperation between physician and psychotherapist decreases the cost and risk of unnecessary medical diagnostic procedures and treatments (Sharpe et al., 1992). It also reduces the effect on the patient of being referred for psychological intervention, the risk of the stigma of it being "all in the head". It seems that physicians sometimes need to be reminded to support their local psychotherapist.

An exposure event as an organizing concept

When a stressful event has taken place, such as a potentially harmful chemical exposure (real or perceived), it becomes a meaningful, accommodating knowledge network. An exposure event can very quickly be elaborated to reinforce overvalued ideas about EI. One strategy for the therapist is to access the distorted belief and show the patient a way to integrate these ideas with a more rational model of the real world. During the process of restructuring a memory network representing a belief about a harmful physical environment, there is the psychological risk for the patient of accessing anxiety-related memory networks with links to inhibited memories of trauma.

Such an interactive therapeutic approach may be frustrating or distasteful to some therapists. The resistance these patients manifest may be frustrating to the physician, especially one who expects these patients to respect and trust the doctor's medical and scientific judgment. Antagonistic confrontation of EI patients usually backfires, often because it triggers a re-enactment of overstimulation and the cognitive processes that were associated with prior stressful interpersonal experiences. It mobilizes their psychological defenses and may overwhelm their capacity to accommodate the incongruent information about psychological explanations for their reactions. Avoiding the traditional allopathic physician is a simple solution for them, but it puts them at risk to turn to pseudoscientists.

In cognitively oriented therapies, the therapist's role involves getting to know the patient's world, to explore it with him. Gentle persuasion through explanation and education is usually more effective than outright challenge of unsubstantiated beliefs. If the patient is open to testing his belief about environmental sensitivities with provocation challenges, it is important to identify assumptions that can be formulated into a testable hypothesis. Methods and procedures must be mutually agreed upon before any testing is initiated. The implications of the different possible outcomes of the challenge study need to be delineated explicitly, so that there will be no hedging once the results are in and no *post hoc* reformulations to preserve the validity of the hypothesis. The assessment of a problem by two scientists with divergent theories is usually done in the context of an amicable, or at least civil, interpersonal relationship. This type of a collegial relationship serves as a good model for the doctor-patient relationship in the assessment of environmental sensitivities.

If the results of provocation challenge study refute the patient's hypothesis about attribution of symptoms, their personal investment in maintaining their belief can still contaminate a rational conclusion. Patients may feel shame about having advocated a false belief. Having a false belief about the cause of symptoms or responding to unsubstantiated practices and treatments associated with EI should not bring discredit to the patient. It does not mean that illness is absent or feigned. It does not imply simple-mindedness or gullibility. Instead it should be framed as a preparedness to be helped, a willingness to trust a caregiver, and a responsiveness to the doctor as a person as well as an agent of healing (Whitehorn, 1959).

Pitfalls

The art of therapy with EI patients involves knowing what facts remain to be discovered and having the patience to either wait for the information to emerge or to gently broach likely topics. Initially, logical analysis based on scientific evidence is not always the most rational therapeutic strategy. Logic and evidence that challenge beliefs about environmental sensitivities need to be introduced slowly. As in all forms of psychotherapy with difficult patients, the art lies in persuasion, leading the patients to discover the insight themselves. The ideal time to offer explanations founded in fact is when the EI patient begins to question his/her own reactions, or the lack thereof, when exposed to a putative harmful agent. Later in therapy, after an alliance has been established, confrontation about false reactions is not as threatening to the patients because they have come to trust that they themselves are not the object of scrutiny or criticism.

Finally, I wish to add a precautionary note on pacing the introduction of new information that is incongruent with the patient's belief. Jumping to conclusions can have detrimental effects, even when correct. The implosion of information may overwhelm the patient's coping capacity and activate psychological defenses. Jumping to conclusions about the psychological implications of certain symptoms, beliefs, affects, or cognition is not without risk. If the therapist's inference is incorrect, the therapeutic alliance can be damaged by undermining the patient's confidence in the therapist. Or, the suggestion may elicit fantasy motivated by the need to please the therapist. This creates another confabulated memory network that will have to be restructured in the subsequent course of therapy. Resistance can be acted out in many forms, depending on the personality of the patient. For example, the patient with a histrionic character disorder may avoid the incongruent information by emotional or impulsive reaction. On the other hand, the patient with an obsessional or paranoid character structure may implode the emotions.

There is always the possible risk that the underlying rage may be untapped and displaced toward the therapist, as was already discussed.

The environmental care unit as a sanctuary

In treating patients who have adopted the sick role, Brody (1980) introduced the concept of "the healing context" and explained it from a socio-cultural perspective. The healing context includes a healer, a setting, and rituals. To overcome the sick role, the patient must submit to the authority of the socially designated healer who may be a doctor, a shaman, an herbalist, etc. It is customary to conduct the healing process in a particular setting, be that a physician's office or hospital, a temple, or a hilltop. Historically in Western societies, the asylum has fulfilled this role. In the sanctuary of the asylum, the patient could play out the sick role behaviors not tolerated elsewhere and also learn new coping skills to facilitate re-entry into society. The raised social consciousness for the abuses within these institutions has contributed to their closing. Today, some asylums still exist, but most are relegated to housing psychotic patients, sociopaths, and the criminally insane which has contributed to the public's perception that asylums are institutions restricted to patients dangerous to themselves or others. Today, the concept of asylum as a mental health facility no longer includes the positive aspects of the sanctuary. What appears to have been lost in this transition is the asylum being a refuge from an unsafe world for frightened individuals.

An environmental care unit (ECU) can provide such sanctuary. The ECU provides an environment perceived to be safe from environmental contamination in which EI patients may better cope with the input of unassimilated percepts of an unsafe world, external or internal. If properly handled, this period of time-out and caring may allow the patient to dispense with unhelpful beliefs and defenses and establish adequate coping mechanisms to return to the outside world (Selner and Staudenmayer, 1986).

On the other hand, if beliefs in unsubstantiated diagnoses and treatments are reinforced, as is the case in the ECU facilities run by clinical ecologists, the hospitalization facilitates avoidance of reality. Withdrawal and isolation may become a perceived way of life necessary for survival in a threatening world, undermining recovery. In the extreme, it can exacerbate hopelessness and despair, leaving suicide as the only exit from uncontrollable misery.

Hope, magic, and the art of healing

In many indigenous cultures, the shaman traditionally treated psychological disorders, as well as truly physical ailments including those for which there was no cure. Folklore tells us that shamans were carefully selected and well trained and were expected to devote their lives to learning the subtle art of healing. They were masters in the use of suggestion, supported by religious ceremony and ritual. Shamans professed to have supernatural powers obtained by communion with metaphysical forces. They maintained their power by demonstrating miraculous healing (successes) or the presence of misfortune and the power of evil forces (failures). People both revered and feared the shamans, who carefully elected themselves into an elite, secret society or guild that transcended tribal and cultural boundaries. What is the secret they shared, the secret they guarded so well and did not let out of their inner circle? Could it be that there is no magic?

In underdeveloped societies that have no adequate medical health-care systems, hope may well be the only avenue of treatment. To take away that hope in such a context is unwarranted. However, when more effective alternatives *are* available, the issue of iatrogenic exploitation and harm is of concern.

While indigenous shamans and recognition of their healing powers have been virtually eradicated in the developed world, the psychological attraction to magical or mystical healing prevails. Howard and Wessely (1993) observed that sufferers from mysterious conditions that lie outside conventional medical practice no longer consider themselves to be oppressed by spirits and demons, but rather by mystery gases, toxins, and viruses. This is particularly visible in the changing nature of mass psychogenic illness (Wessely and Wardle, 1990). When coupled with the cognitions of loss of control underlying depression, anxiety, and phobia, we see individuals vulnerable to exploitation. It is this complex of vulnerabilities that is nurtured in a healing psychotherapy. Those same vulnerabilities are also recognized and exploited by fringe medicine evangelists and charlatans, often masking their ministrations in the cloak of alternative health semantics and counterphobic rituals which create the illusion of scientific credibility.

Psychotherapeutic interventions

The first question a clinician should ask himself when treating an EI patient is, "Is it rational to be logical?" Insight-oriented psychotherapy is contraindicated in many EI patients, at least initially. Often, cognitive-behavioral therapy techniques, stress management, and supportive psychotherapy are the most effective ways to begin treatment. But, even before that, the most effective approach is educational (Barsky et al., 1988). Both physician and mental health provider can help the patient understand the physiology and psychophysiology that underlie their symptoms (Goldberg et al., 1989). Psychophysiological stress profiling can also be utilized as a valuable educational procedure in therapy. Alterations in brain waves associated with attention problems, muscle tension associated with tension headaches, and elevated galvanic skin response often associated with symptoms of autonomic arousal are all objective, physiological measurements. They are usually psychologically non-threatening to the patient because they do not directly challenge their belief about environmental causation of symptoms, nor do they activate defenses against "they think it's all in my head." Often, carefully titrated explanations of learned sensitivity, the placebo-response, and the physiology of the stress response overcomes resistance to consideration of psychogenic factors. Information from individualized provocation challenge studies has been an invaluable educational and therapeutic tool for my colleagues and me. When a patient's own hypothesis of environmental sensitivities is not confirmed under controlled challenge conditions, earlier toxicogenic convictions may soften.

As a therapeutic alliance is established, the bridge to insight-oriented therapy can be crossed with many EI patients, even some who are the most difficult to treat. Clinical experience suggests that these are the patients who reported histories of traumatic and chronic sexual and physical abuse in childhood. For many EI patients who have been able to process their childhood traumas in therapy, the emotionally disruptive sequelae have become manageable. As the psychological and psychophysiological sequelae of the trauma subside, symptoms become more manageable and controllable, more effective coping styles replace false defenses, the overvalued beliefs about EI are displaced by more realistic concepts about health and the environment, and life becomes more functional and productive.

Behavioral, cognitive-behavioral, and self-regulation therapies tend to be short term, goal oriented, and structured to provide reinforcement (Bandura, 1977; Frank and Frank, 1991). Some suggested guidelines to structure the initial interventions with EI patients are

Table 15.3   Guidelines for Short-Term Therapy with EI Patients

1. Confidence in the technique — present a planned rationale that provides an initial structure.
2. Relevance — provide training in practical skills for daily life.
3. Generalization — emphasize independent use of these skills.
4. Success and reinforcement — training mastered in the office should work for real world situations.
5. Self-reliance — improvement is attributed to the patient's increased skillfulness rather than the technique.

presented (Table 15.3). In the context of conducting short-term therapies, it is usually best not to delve into psychodynamic issues because they can distract from learning a self-regulation task or other coping skills. For many patients, some degree of mastery of psychophysiological self-regulation skills is helpful before the introduction of content material into therapy. But it does not always work that way. Patient resistance to relaxation may require discussion of content material, especially the basis of the resistance, before any progress can be made in learning a cognitive-behavioral technique.

Stress management

Stress management and psychophysiologic self-regulation are value-free techniques in regard to the etiology of stress physiology and symptoms. They usually do not trigger the fears many patients have about exploring unconscious, dynamic motivation. They do not imply the social stigma unfortunately associated with psychological disorders and are less likely to trigger the primal fear of going (being) "crazy" in the more severely affected EI patients. Any form of intervention that purports to treat a psychological symptom without overt focus on "real psychological mechanisms" has appeal to patients in denial. In that respect, stress management techniques are less likely to threaten intrapsychic conflicts and do not challenge the psychological defenses.

Some EI patients find themselves unable to relax, even in informal situations or when with loved ones. When engaged in learning a relaxation technique, some express a great deal of frustration because they just cannot relax. They do not lack insight in the sense that they acknowledge that tension and anxiety contribute to their symptoms. "Why can't I relax?" they may ask the therapist as well as themselves. The answer is usually found in a deeper level of analysis. One way to broach the issue is to ask, "What are you afraid will happen to you if you let go?"

Patients who rationalize and justify their anxiety by projecting the cause unto a chemically unsafe world can be asked if they experience the same apprehension about relaxation when they are in the context of an environment they perceive as safe. This is an indirect confrontation of the likelihood that they experience the same fears and anxieties in all contexts, regardless of ambient exposure levels. If the patient professes to feeling relaxed in his "safe house" it gives the therapist an opening to have him practice and learn relaxation techniques in that place, later to be generalized to other contexts perceived to be harmful.

According to the fight-or-flight model of Cannon (1927, 1929), peripheral nervous system changes occur principally in service of central processes, to promote adaptive redirection of energy. For example, muscle bracing is often associated with focused concentration because rigidity serves to maintain vigilance, an evolutionary adaptive response. Clinically, the patient may complain of muscle tension but not be aware of the associated vigilance. Psychophysiologic self-regulation therapy assisted with biofeedback may follow two general strategies. According to one strategy, electromyographic (EMG) biofeedback of the target muscle is employed to reduce the tension directly. General muscle relaxation may be the objective, and the training is conducted with EMG biofeedback from certain muscle groups typically indicative of whole body relaxation (e.g., trapezius). According to the second strategy, electroencephalographic (EEG) biofeedback may be employed to reduce muscle tension indirectly by broadening the focus of attention and reducing attentional vigilance.

The cardiovascular pathway of the stress response provides another example. Heart rate, blood pressure, and respiration usually increase to maintain greater cardiac output to accommodate heightened arousal. Through vasoconstriction, blood is shunted away from the peripheral vascular beds to increase flow to the large muscles and the brain, supplying oxygen and nutrients required for efficient action. The skin of the hands and feet will cool as a consequence of peripheral vasoconstriction. This is the rationale for monitoring skin temperature of the fingers or toes in psychophysiologic stress profiling and biofeedback for cardiovascular disorders.

Clinically, I have found that self-regulation and biofeedback training is often an effective first step in the treatment of EI patients. By recognizing partial successes or accomplishments in regulating the body and symptoms with biofeedback and by experiencing small degrees of pleasure from the sense of accomplishment, patients with certain personality traits may also be helped to re-evaluate their feelings of inadequacy and incompetence.

Behavior modification: systematic desensitization

Validation for the application of conditioning techniques to human fear come from animal studies that demonstrate extinction (Rachman, 1990). It is both a quantitative and qualitative leap in logic to apply animal conditional models to human learning. Since the displacement of behavior theory as a philosophy of psychology, cognition has become an integral part of behavioral theories of psychotherapy. One behavior modification technique — systematic desensitization — has survived this transition, although the postulated mechanisms for its effectiveness are now different from those initially postulated by behavior theory (Kazdin, 1977; Wolpe, 1958). The therapeutic effects of specific self-expectations in treatment also challenge such assumptions. This has implications to the Pavlovian model of the placebo effect, because desensitization in humans may be more cognitive than in animals. In fact, several research studies have found placebo interventions to be as effective in reducing anxiety as therapies supposedly based on specific conditioning mechanisms such as systematic desensitization (Holroyd, 1976; Kirsch, 1978; Kirsch and Henry, 1977; Kirsch et al., 1983).

Wolpe's (1958) application of systematic desensitization to overcome simple phobias, such as animal phobias, was particularly effective when the fears were acquired in response to a frightening experience with the animal. However, in applying the technique, the approach to the phobic stimulus does not need to be progressive or systematic. Those aspects of the procedure are structural in the sense that order and the precision of a ritual give it a greater semblance of credibility. Wolpe's technique succeeds because its ritual persuades patients to remain in actual or imagined contact with the phobic situation long enough for the fear to subside (Frank and Frank, 1991). The success of this and other therapies for phobias depends to a considerable degree on their persuasiveness, to which the scientific rationale, valid or not, contributes.

In contrast to specific phobias, when the avoidance behavior is more profound as in agoraphobia, and the physiologic reactions can reach panic proportions, desensitization techniques are much less effective, if of any benefit at all. Agoraphobia with panic is associated with remarkable impairment in the quality of life. Residual symptoms after the acute attacks are significant, persistent, and long term (Katon, 1986). The chronic and polysymptomatic complaints are also characteristic of EI patients whose environmental phobiga resembles agoraphobia.

Guglielmi et al. (1994) reported on the use of intensive behavioral desensitization treatment of phobic avoidance in three EI patients. By the end of treatment, all three patients had acquired significant relaxation skills and were able to sustain prolonged exposure to a wide array of noxious chemicals without demonstrating physiological or symptomatic activation. However, these effects were sustained in only one of the three patients upon follow-up. This patient was the one of the three who was not under the care of a clinical ecologist:

> "A behavioral rehabilitation program cannot produce long-term change when the patient is surrounded by friends, relatives, and respected professionals who provide emotional support and 'scientific' legitimacy for continuing disability."

This study highlights Rutter's (1984) caution that behavior modification is focused on the symptoms without regard to the person's life situation.

EI case-report involving desensitization

Selner and I reported the employment of systematic desensitization as part of a comprehensive treatment program in an EI case presentation (Selner and Staudenmayer, 1992b). Upon beginning therapy, the patient professed intolerance for his luxurious home, his cars, the company of his wife, and any form of travel. He lived alone in an old, downtown motel room, the only environment he perceived as safe. Among the many chemical agents to which he reported intolerance, he expressed particular sensitivity to the odor of rubbing alcohol. Systematic desensitization was employed only after months of twice weekly psychotherapy in which a therapeutic alliance was established. The patient mastered self-regulation skills using biofeedback for symptom regulation before the systematic desensitization procedure was initiated. Restructuring of the EI belief started underway by educating the patient to the shortcomings of the unsubstantiated theories and practices of clinical ecology. He had undergone removal and replacement of all his dental amalgams, but now he accepted that the heavy metal was not affecting the magnetic polarity in his brain. Cognitive therapy techniques were employed to identify and analyze cognitive distortions and inconsistencies in his logic explaining unreliable reactions to environmental agents. Antidepressant medication was prescribed for symptoms of extreme fatigue and attentional problems associated with depression, with good therapeutic response. With all of these concurrent interventions, systematic sensitization may well have been the ritual that effectively gave him the explanation acceptable to him as to why he was no longer reacting to the odor of alcohol. According to Frank and Frank (1991):

> "A not infrequently and easily overlooked function of therapeutic rituals is to provide a face-saving way for a patient to abandon a

symptom or complaint without admitting that it was trivial or produced for some ulterior motive."

The tolerance for the odor of alcohol generalized to his other perceived chemical sensitivities. Shortly thereafter, he was traveling throughout Europe. Unfortunately, he had a recurrence of his EI several years later, after the publication of our 1992 report. He was involved in an automobile accident in which his wife was severely injured. Initially, she was in critical condition from internal injuries which she survived, but she had a residue of debilitating chronic pain. The anxiety of almost losing his wife followed by her disability led him to project his anxiety onto the environment again. A year after the accident, he was referred for additional psychotherapy. He was unaware that he was experiencing symptoms of anxiety and depression. He had again tried a variety of unsubstantiated treatments, including exotic teas and supplements, none of which worked for him. With the aid of cognitive-behavioral approaches, he came to more realistic appraisals and attributions of his symptoms. After several months of treatment, he appeared less anxious and recognized that his alleged environmental reactions were inconsistent. Whereas he complained of symptoms with trivial chemical detections, he could attend and tolerate large dinners, concerts, and other public gatherings where he had a multitude of chemical encounters. His wife had also recovered to the degree that she could travel again. Her trauma made her less sympathetic, and she resisted his requests to accommodate his "odor sensitivity". It is possible that her becoming less accommodating of his somatization had something to do with his recovery.

Cognitive-behavioral therapy

The cognitive-behavioral approach was initially conceived as a middle ground between the psychoanalytic theories heavily biased to unconscious instinctual motivation and the stimulus-response theories on which behavior-modification was based (Beck, 1967, 1976, 1991; Meichenbaum, 1977). A practical guide for using cognitive-behavioral techniques with patients who have functional somatic symptoms is available (Sharpe et al., 1992). A postulate commonly shared by cognitive-behavior therapists and cognitive therapists from psychodynamic schools is that affect, sensory reactions, and motor responses are associated with beliefs operant in the interpretation of an experience. The differences among cognitive theories centers on what the operational beliefs might be and whether they are conscious or unconscious. A distinctive postulate of the cognitive-behavioral school is that conscious thought alone is sufficient to account for what motivates and determines affect and behavior. Little credence has been given to unconscious motivation and unconscious processes.

As with all therapies, there are pros and cons for the cognitive-behavioral approach, both as a theory and as a therapeutic technique. A fundamental presupposition, one shared with psychoanalysis, is that the patient is open to rational discussion of their cognitive distortions. The patient must also demonstrate the capacity for insight, at least to the extent that he/she recognizes that the cognitive distortions lead to irrational beliefs and anxiety. For the population of EI patients for whom these presuppositions hold, the technique of identifying and analyzing cognitive distortions is an effective therapeutic intervention. Two recent studies have shown positive outcome in using cognitive-behavioral therapy with chronic fatigue syndrome patients (Deale et al., 1997) and "electric hypersensitivity" (Andersson, et al., 1996). These approaches have also been effective in the treatment of patients with functional somatic pain symptoms such as chronic pain (Benjamin, 1989) and irritable bowel syndrome (Guthrie et al., 1991), and phobia and hypochondria (Warwick and Marks, 1988).

Table 15.4   Common
Cognitive Distortions

Dichotomous thinking
Selective abstraction
Arbitrary inference
Circular logic
Overgeneralization
Undergeneralization
Catastrophizing
Decatastrophizing
Misattributed causality
Rationalization of lies

Source: Adapted from Beck,
A.T., *Cognitive Therapy and the
Emotional Disorders*, Interna-
tional Universities Press,
New York, 1976.

Early childhood experiences have frequently been implicated as causal factors in the development of adult psychopathology. Beck (1967) noted that negative memory networks have their origins in the child's early experiences and interactions with his/her parents. Despite this, cognitive-behavioral therapy typically does not include assessment of child-hood experiences and abuse and does not incorporate it into treatment (Brewin et al., 1993). This avoidance may be partly motivated by controversy over the reliability of childhood memories or it may reflect a longer standing behavioral bias to ignore events other than those in the immediate environment. The issue of the accuracy of a patient's memories of childhood experiences has implications to determining the meaning of his negative beliefs about self and others, whether or not the beliefs are, in fact, irrational, as is assumed by principles of rational-emotive therapy and cognitive therapy. Brewin noted that what are often construed as distorted or irrational beliefs may appear in a very different light in the context of the patient's actual experiences (Brewin and Andrews, 1992). There seemingly is the paradox that beliefs may be irrational in the present world, but they represent an accurate assessment of a past world.

Cognitive distortions

Cognitive-behavior therapy is particularly effective with depressed patients who are not aware of how their hopelessness is related to pessimistic appraisals resulting from faulty logic (Beck, 1976, 1991). Some of the most common cognitive distortions initially reported by Beck (1976; Table 15.4) may be observed in EI patients. I have compiled examples from clinical experience which illustrate these cognitive distortions.

Dichotomous thinking

Issues and people are appraised in terms of extreme, black or white categories. In therapy, the idea of grey is introduced. Probability of truth statements may be presented as an alternative to the absolutes of true and false and may be illustrated by the inherent uncertainty in scientific research. The modulating effects of context on truth judgments is presented as an alternative to having universal postulates that transcend all contexts. For example, chemicals are neither necessarily harmful nor safe, but may be either depending on the dose and manner of exposure. The one-molecule exposure theory of toxicity represents an absolute category for which there is no evidence.

Selective abstraction

Elements are taken out of context. By the process of selective attention, salient features that define an integrated event are denied or not perceived. Meaning is abstracted from incomplete data, and a feature or part of the concept may stand for the whole. Partial truths may result because the specific context of a proposition is not determined. A clinical consequence is a distortion of memory and judgment. In the extreme case of overvalued ideas, histories and experiences may be confabulated, distorting recollection of actual events. Elaborating the context and complexity of a situation may be helpful in therapy.

Arbitrary inference

This distortion refers to drawing specific conclusions without supportive evidence (confabulation). Speculations are only remotely related to the supportive data. Arguments are characterized by overgeneralization and misleading inferences. Clinically, those who use arbitrary inference often have anxiety about making their theories meaningful and world cohesive. In therapy, the emphasis is on the necessity for correct logic and valid evidence to support any argument or hypothesis, no matter how self evident the postulates appear to the patient.

Circular logic

The premise and the conclusion in deductive reasoning are the same. The effects of circular reasoning can extend to subsequent logical arguments in which unsubstantiated conclusions from a prior argument are accepted as established fact. Circular logic is characteristic of a closed belief system: "Don't bother me with the facts; they only get in the way." In therapy, the patient is directed to the facts based on the model of reality of the therapist.

Circular logic may be seen in the context of mass psychogenic illness, for which a belief emerges based on ever-growing numbers of testimonials. In the Waterside Mall Complex "sick building" case involving EPA employees in Washington, D.C., the local union newsletter kept a running count of how many newspaper articles appeared over the months leading to filing grievances and lawsuits. Most of these media articles represented plaintiff testimonials and interviews with advocates. When the numbers grew large enough, they were presented by the advocates as overwhelming evidence, implying that it was independent evidence, to warrant presenting their concerns to members of the U.S. Congress.

Overgeneralization

Facts or conclusions are extended beyond the context in which they are valid. This is similar to a category error in which valid facts in one context are presented in another context in which they are irrelevant. Overgeneralization is similar to the cognitive distortion of arbitrary inference but is different in the sense that the scientific fact alluded to is not distorted, but is true only in a limited context. Limiting the scope of the argument to the relevant context is the approach in therapy. In EI advocate writings, it is common practice to extrapolate well-established toxicologic effects due to high-dose exposure to low levels. Overgeneralization of the limits of scientific inquiry ("nothing is really known") is used as a cognitive distortion to justify continued use of unsubstantiated diagnostic and treatment methods.

Undergeneralization

Undergeneralization means not drawing appropriate conclusions from the available information. Examples with respect to EI include proponents ignoring the vast body of scientific literature or patients ignoring their own experiences with unsubstantiated treatments which have failed to help them. Undergeneralization may also be observed in the service of the defense mechanism of denial. EI patients may minimize the personal factor in the sick role or dismiss pre-existing psychopathology as irrelevant to their present condition.

Catastrophizing

When dire consequences are inappropriately inferred, usually from trivial events, it is called "catastrophizing". Excessive worry about illness is characteristic of hypochondria. Worrying about something that may happen in the future serves to avoid awareness of something that has happened in the past, such as traumatic experiences, or coping with present day stressful situations. In EI patients, there is a tendency to exaggerate, overplay, or dramatize the assumed causal relationship of the environment to their illness. In therapy, presenting examples of well-documented cases of hysterical reactions associated with mass psychogenic illness may help to minimize the effect of catastrophizing.

Decatastrophizing

Important warnings are not accorded appropriate weight. Alternative diagnoses are often overlooked in the treatment of EI patients, e.g., a terminal illness, such as cancer, or a psychiatric disorder that can be readily treated, such as clinical depression. Decatastrophizing may undermine a basic health principle that early detection is one of the best preventive medicine techniques.

Misattributed causality

EI patients and advocates maintain a belief about possible environmental causes for symptoms despite evidence to the contrary. In therapy, it often helps to work on a testable hypothesis such that the patient will relinquish the belief if the resulting evidence is disconfirming. The availability of provocation challenge studies using an environmental chamber has been an invaluable tool for me in demonstrating the unreliability of the patient's alleged "chemical reactions".

Rationalization of lies

The attitude that the end justifies the means may be a rationalization to distort facts or to report them imprecisely. Clinically, this may be seen in zealots who take on excessive responsibility, e.g., self-proclaimed guardians of the planet. In a more sinister context, physicians and therapists conducting independent medical/psychological evaluations of EI patients in the course of litigation are open to unsubstantiated allegations of ethical misconduct.

Restructuring beliefs

In restructuring beliefs, the objective is to unassimilate previously assimilated information in memory. This can create anxiety. The release of that anxiety must be carefully titrated in psychotherapy. Implosion or flooding techniques used by some exit counselors (so-called "cult busters") can elicit resistance or overwhelm the defense system and drive it into psychosis if the patient is not properly prepared and a willing participant. In an experimental protocol with volunteer control subjects, it was found that previously established perceptual interpretations are adhered to more tenaciously under conditions of threat (Moffitt and Stagner, 1956).

From a cognitive viewpoint, interpretations are like hypotheses that produce insight when the patient discovers evidence to support them. This can be viewed as a reworking and re-evaluation of a context, often the past, that leads to the discovery of new facts. Taking into account the differences in contextual effects of the present and past, new relationships and meanings are assigned to previously known facts. The reassessment of their meaning in a new context can serve to restructure the memory networks of an overvalued idea. This results in new postulates and presuppositions which develop into complex networks called attitudes, beliefs, and values (Rokeach, 1968).

In cognitive-behavior therapy, the technique of restructuring is employed to reorganize the memory networks representing false defenses and to elaborate the conceptualizations of reality. Therapeutic techniques such as interpretation and confrontation direct the attention of the patient and therapist toward new information that challenge the validity of the presuppositions, assumptions, and postulates on which false beliefs are founded. The same techniques are used to confirm the validity of the therapist's alternative theory, formulated as a model of reality. The sensory, emotional, cognitive, and action memories in a memory network are closely linked or integrated. If all goes well, a change in one type of memory facilitates awareness and change in the others. In patients with a history of trauma, restructuring may also disarm some of the emotional impact of the anxiety of exploring the secrets of their trauma. As a new conceptual system evolves, it offers some semblance of safety and security to counteract the destructive primal fears of separation or abandonment, being killed, or "going insane". Psychological defenses are false defenses by definition because they do not solve the problem or protect from trauma. Instead, they give the illusion that it did not happen or rationalize that the pain and suffering inflicted were justified.

While an effective technique in some aspects of therapy with EI patients, the analysis of cognitive distortions and the restructuring of beliefs do not always work. For example, consider the numerous cognitive distortions of EI patients who believe they have been seriously affected by an alleged exposure in a "sick building". No amount of engineering data about HVAC (heating, ventilating, air-conditioning) systems and no amount of industrial hygiene sampling data can convince them they were not harmed. When challenged with suspected agents under controlled conditions, false responses (symptoms with placebos or no symptoms with the suspected agent) are not necessarily interpreted in line with the obvious conclusion. If rational analysis were an effective cure, toxicologists would make the best therapists for EI patients. The logical techniques of cognitive-behavior therapy often fail to address primary and secondary gain motivation. The insight gained from logical analysis is not synonymous with change, even though it may be an indispensable tool for change. Fundamental changes occur in therapy only when there is a substantial change in irrational beliefs (Bieber, 1980). Rational therapy is pointless with patients who are malingering or who have factitious disorder.

Cognitive therapy includes a panoply of techniques, all aimed at facilitating the patient's perception and acceptance of reality. One could argue that all forms of psychotherapy are cognitive at some level. What makes cognitive therapy unique is its emphasis on a rational world of reference, the real world. In describing and identifying the different kinds of cognitive therapies, as in all therapies, it is the techniques that differentiate them. With regard to somatization, cognitive therapy assists the patient to make more realistic attributions for somatic sensations. The rational aspect of cognitive therapy is often facilitated by behavioral/psychophysiologic relaxation techniques such as deep breathing, muscle relaxation, and meditation or attention training to facilitate a calmer state of mind which facilitates cognitive processing. Education about basic physiology and psychophysi-

Table 15.5   Principles of Psychodynamic Psychotherapy

1.   During treatment, unconscious (inhibited) material becomes conscious.
2.   The mobilization of unconscious material is achieved mainly through the interpretation of information presented during free association and the patient's emotional interpersonal experience (transference).
3.   The patient shows resistance against recognizing unconscious content.
4.   Through the transference relation with the therapist, resistance is overcome and childhood experiences are re-enacted to be better understood with insight and resolved.

Source: Adapted from Alexander, F., *Am. J. Psychiatry*, 120, 440–448, 1963.

ology is also part of cognitive therapy. For example, panic patients often complain of chest pains and are relieved when informed that the sensations are really mediated through skeletal muscles rather than the heart. Preventricular contractions, which often send these patients to the emergency room, lose their life-threatening implications when they are explained as a common symptom of an anxiety response, rather than an occult event induced by environmental exposure. Cognitive/educational therapy also may be used with somatizing patients to make them aware of the physiological symptoms of anxiety and how cognitive distortions can amplify symptoms and foster hypochondriacal beliefs.

For many years, the Japanese have practiced Morita therapy, which incorporates supportive therapy and cognitive techniques to address the psychological symptoms of intractable disease as well as somatization and the imagined fears about disease (Miura and Usa, 1970; Reynolds, 1980). Morita therapy practices include facing and accepting reality for what it is or was, refocusing attention, and choosing purpose and mobilizing action to achieve it (Ishiyama, 1990). The therapeutic techniques of cognitive therapy and Morita therapy have many similarities, although they evolved in different cultures, seemingly independently. Clinically, my colleagues and I have used supportive and cognitive techniques with many EI patients to facilitate acceptance and mastery of the psychological etiology of their symptoms.

Psychodynamic psychology

A review of the many psychodynamic or ego psychology theories that have evolved over the past 100 years is beyond the scope of this presentation. Rather, I wish to highlight some of the presuppositions about therapy held in common by all these different and often diverse formulations. Franz Alexander (1963), the psychoanalyst best known for his work with psychosomatic illness, shortly before his death summarized what he considered to be the essential psychodynamic principles (Table 15.5).

Psychodynamic psychotherapy emphasizes early experience in the formulation of personality. Emotional problems later in life are often related to childhood experiences which are not part of conscious knowledge. The basic techniques used in psychodynamic psychotherapy are interpretation, confrontation, and clarification. Interpretation refers to identifying something the patient is not aware of because of a lack of conscious awareness or lack of insight. Confrontation means bringing up an issue or problem the patient has been consciously avoiding. Clarification means recapitulating, elaborating, or summarizing insights achieved by either patient or therapist. The reaction one can anticipate with any one patient depends on the subtleties of transference and countertransference interactions associated with different personality types (Masterson and Klein, 1989).

The one significant change from psychoanalytic therapy advocated by Alexander was a departure from the neutral role of the therapist. Alexander took the position shared by other schools of psychology that the therapist should be flexible and adaptive to accommodate the idiosyncratic needs of each patient. His challenge to the classical and invariant neutral role to be played by the therapist was met with considerable alarm among traditional psychoanalysts. Alexander's change of attitude was based on extensive study of the behavioral interactions between therapist and patient, and findings noted seem particularly relevant (Matarazzo, 1971). One, psychotherapists are, first and foremost, individual human beings and only secondarily members of a particular class called psychotherapists. And two, Alexander's research showed that, in fact, neither patient nor psychoanalyst behaved as he and other theorists and practitioners had believed they did. Matarazzo (1971) summarized the implications of these findings:

> "Those psychotherapists who heretofore have seemed to make a fetish of such concepts as 'the process' of psychotherapy or who have exalted the 'transference' relationship to a position of almost religious preeminence will learn, to their surprise I believe, that such mystical phenomena are probably little more that what occurs in most, if not all, social interactions; and that the same general laws and principles which are relevant to the study of other behavior (e.g., parent-child, peer-peer, teacher-student, experimenter-subject, and other important human pairings) also apply to the study of psychotherapy."

Working with EI patients is really no different that interacting with anybody else. The mystery and mystique associated with psychodynamic psychotherapy should not intimidate the physician, much less the patient. Over the years, I have come to realize that the problems of EI patients are much more common and much simpler to understand than are their overvalued ideas and the toxicogenic theory.

The shift of emphasis away from instinct and drive theory toward cognition in the development of the dynamics of personality is characteristic of the post-Freudian theorists, whose writings began appearing regularly after World War II. Prior to that, in the schools of classical psychoanalysis in the Freudian tradition, cognition was not considered as a source of conscious or unconscious motivation. In a large part of the psychiatric, psychoanalytic, and psychological literature of that period, concepts were considered static, purely intellectual entities, separate from human emotions and unimportant in psychodynamic studies. According to Arieti (1980), recognition of the importance of mental representation in ego psychology is summarized as:

> "If I want to discover my unconscious, I must do so in a cognitive way. Introspection and self-awareness require openness to one's memory. For what I have kept out of consciousness is predominantly cognitive in nature; i.e., I repress ideas, attitudes, mental dispositions, and in most cases the emotions derived from them."

Barnett (1978) noted that the central task of cognitive therapy is as that of cognitive repair:

> "Cognitive repair implies the correction of the distortions and imbal-
> ances of the person's systems of knowing and meaning, and the
> repair of those dysfunctional or nonfunctioning mental processes
> and operations which maintain a system of innocence and prevent
> integration of meaning with experience and action. Without such a
> concept as an integral part of theories of therapy, there is too much
> danger that new experience cannot occur, that insight cannot be
> integrated into experience, that the patient will continue the dys-
> functional patterns of the past, and that therapy will become futile
> and even interminable."

Cognitive repair is presented as a process involved with cognition and oriented by personality (Barnett, 1978, 1980). With EI patients, denial of psychogenic factors is a significant obstacle to insight-oriented therapy. This precaution is raised not to invalidate the psychodynamic approach to therapy; quite the contrary, as it is often required for resolution of somatization (Hickie et al., 1997). In my treatment approach, I have incorporated the psychodynamic principle that insight into underlying personality formation is necessary for complete resolution and understanding of the etiology of psychological disorders in EI patients. My intent is to emphasize that certain steps need to be taken before insight-oriented therapy may be utilized effectively when working with patients holding overvalued ideas.

Psychoanalysis is undergoing a paradigm shift with the recognition by some more traditional psychoanalysts that childhood abuse has adult psychiatric sequelae. Childhood trauma has been incorporated into psychodynamic therapy (Miller, 1984; Shengold, 1989; Stone, 1989). The approach to taking a history for possible experiences of trauma and sexual abuse in childhood presented by Saporta and Gans (1995) is an excellent model to incorporate in the assessment of EI patients.

Treatment of personality disorders

Should an EI patient become a good candidate for insight therapy, Barnett offered some helpful therapeutic guidelines for treating certain types of personality disorders (Barnett, 1978, 1980). When explosion of affect is more dominant, therapy is directed to identification and analysis of the acting out. This pertains particularly to the DSM-IV cluster B personality disorders: antisocial, borderline, histrionic, and narcissistic (American Psychiatric Association, 1994). Individuals with these disorders often appear dramatic, emotional, or erratic. Acting out is a form of venting that serves to prevent incongruent information from reaching conscious awareness, such that neurotic defenses are maintained. Another way of saying this is that acting out unconsciously sabotages any possibility of change through the cognitive restructuring of false conceptual networks. Patients with these kinds of personality disorders tend to overvalue the affectivity and the apprehension of experience. They believe that what they feel must be right. But what they feel are reactions precipitated by perceptions out of context in the present world. Explosive characters often use the cognitive distortion of overgeneralization in the service of denial. The overreaction serves to discard the initial precept entirely. The recommended therapeutic strategy is to contain the affect. Focus is directed to analyzing the emotional experience associated with the explosion of affect or compulsive activity and to relate the experience in the context of the present to the experience and fantasies associated with the context of the past. The cognitive emphasis is to clarify how emotional experiences in the past contributed to certain cognitive accommodations resulting in memory networks that define an overvalued idea.

Individuals prone to the implosion of affect are more likely to have DSM-IV cluster C personality disorders: avoidant, dependent, and obsessive-compulsive personality disorders. Such individuals may be more difficult to treat successfully with psychodynamic therapy because the very nature of the therapy is so intellectualized. There is the risk these patients might master the caricaturization of real insight. Repair of the cognitive faults of the implosive personality types requires analysis of their need to maintain innocence, to avoid knowing and comprehension of experience. The patient's dynamics involve a need to obscure and hide through deviousness, inhibition, and denial. For therapy to be successful, insight about these defenses must be demonstrated by the patient. In cases with a history of trauma, the therapeutic task becomes more difficult because the psychological defenses serve to inhibit memories of trauma from consciousness. A cognitive-behavioral approach which analyzes the cognitive distortions is recommended as a tactic in the early phase of therapy, with the caution to be alert for intellectualization. Identifying the context in which interpersonal inference making is disturbed provides clues to the historical and developmental origins of the defensive systems to preserve the illusion of innocence. The stereotypic labeling the patient uses in place of these inferences must be recognized as such and related to the context of the patient's life experiences.

Psychotropic medications

Pharmacotherapy with EI patients may be indicated if clear differential diagnoses are established for which medications are the preferred treatment. Also, stabilization of certain symptoms such as depressed affect, explosive hysteria, panic attacks, insomnia, attentional difficulties, and fatigue often renders patients more amenable to psychotherapeutic intervention and facilitates progress in psychotherapy. The neurophysiologic mechanisms by which these drugs enhance psychotherapy include their effects on vigilance, attention, and concentration; memory storage and retrieval; and promotion of an adequate emotional arousal or dampening overarousal and anxiety (Herrmann, 1982). However, the propensity for cognitive distortion seen in patients with depression and anxiety may be more effectively addressed with cognitive therapy (Beck, 1976, 1991; Beck et al., 1979, 1985a).

The critical question in regard to the use of psychotropic medications with EI patients is, "Can irrational beliefs about environmental appraisals be medicated so they restructure to rational beliefs?" Regrettably, clinical experience with EI patients indicates that the answer seems to be "no". It is unlikely that any psychotropic medication alone can directly resolve incongruencies and bring about assimilation in conceptual networks, especially if primary or secondary gain is involved. With the exception of certain types of delusions associated with schizophrenia and other psychotic disorders, it seems unlikely that medications alone can alter a belief system.

Goldberg (1996) has suggested that there may be a tacit bias among many psychiatrists to diagnose only conditions for which there is a clinical treatment, most often depression and anxiety. Existing assessment procedures are biased to categorize multi-system somatic complaints as symptoms of depression and anxiety (Goldberg and Bridges, 1991). This bias may be enhanced because somatoform disorders are considered among the most difficult to treat and generally have poor prognosis (Quill, 1985; Ford, 1983, 1985). Selective serotonin reuptake inhibitors (SSRIs; e.g., Prozac, Zoloft, Paxil) have not been effective in treating chronic fatigue syndrome (Hickie and Wilson, 1994; Vercoulen et al., 1996) or fibromyalgia (Wolfe et al., 1994). Negative health perception is characteristic of patients with a history of childhood sexual abuse. In a review of seven studies from demographically diverse populations, controlling for depression did not remove patients' self perceptions of their health (Golding et al., 1997). This means that high levels of depression do not account for patients' negative perceptions of their health.

I see the role of medication in the treatment of EI patients as one to create an emotional state in which they can participate in effective psychotherapy. The clinical art with EI patients is convincing them that they could benefit from a trial of medication, and that they do not have sensitivities to these medications, as many have been led to believe. Taking psychotropic medications represents an incongruence in the overvalued idea about the harmful effects of any chemical agent. Further, if the patient has an underlying primal fear of being "crazy" or a secondary fear of being seen as "crazy", taking medications confirms that fear. If the patient takes the medication and experiences symptom relief, that information is incongruent in that it challenges the validity of the patient's overvalued idea about EI. If the patient's belief system is closed, information which threatens the fundamental postulates of the toxicogenic theory is avoided, and nonadherence to a schedule of psychotropic medications achieves that end. However, in patients with an open belief system, a therapeutic response may confirm the alternative psychogenic theory which they may be ready to accept because it puts them on a path to wellness.

Discussion

Behavior therapies have the characteristic that they are directed to a well-defined goal, usually a specific symptom relief. Inferences about one's capacity for control are based on mastery of skills in self-regulation training. In behavior therapies such as systematic desensitization or skills training, the tasks are usually structured in a hierarchical fashion so the patient can master them gradually and achieve success. Through the mastery of structured situations, patients may infer that they have acquired new skills or made personal gains that are readily transferable to real life. In some cases, self-regulation demonstrates that the patient can have some sense of control, even though it is limited. For EI patients, this is a great improvement over feeling hopeless about living in an unsafe environment. However, in other cases such inferences or conclusions may not be accurate, or the context of the mastery is more limited than the patient's expectations. Structured mastery of techniques in the clinic may produce self-attributed abilities that are overgeneralization and not transferable to the real world.

In the course of long-term cognitive therapy, the therapist's model of reality becomes progressively reinforced with evidence of improved functioning by the patient. As the patient creates his own reality, it contains positive memory representations of greater self-worth and self-esteem. The appraisal of other individuals is more accurate and contact with them becomes less threatening. With the learning of coping skills, stressors and negative life events are less disruptive. With more experiences of success, a sense of control displaces feelings of helplessness, hopelessness, and despair. With less uncertainty and a greater sense of control of their lives, patients usually report less anxiety, improved focus of attention, good memory, and some positive emotions. These behavioral changes provide feedback to the cognitive structures to strengthen the patient's restructured possible world of reality.

For traumatized patients, especially those with a history of childhood abuse and emotional deprivation, it is important to offer a safe environment in therapy for them to reconstruct deeply wounding and humiliating experiences for which they may continue to feel in some way responsible (Brewin et al., 1993). The more the memory of the trauma can be re-evaluated and contained in the context of the past, the less frightening are the effects of past traumas in the context of the present. With reduced fear and terror, the more credible is the prospect of giving up the false defenses. One sign of a successful therapy is that the elaboration and effectiveness of beliefs about reality are at the expense of the patient's original distortions and overvalued ideas. The sometimes unexpected lack of resistance that may be encountered by the therapist during this process should not be surprising, as an effective coping mechanism serves the patient better than a false defense. Ultimately, the test of any therapy is whether the patient can overcome the fears of the past and maintain effective coping behaviors in daily life.

As a psychologist, the first question I am frequently asked is, "How do you treat EI?" My answer: "I do not treat EI; I treat the symptoms, concerns, behaviors, and underlying dynamics of the patient." The second question I am often asked is, "What kind of therapy would you recommend for EI patients?" The person asking the question usually anticipates one specific approach such as biofeedback, behavior therapy, cognitive therapy, antidepressant medications, etc. My answer is usually something to the effect:

> "Any technique that facilitates rapport, gives the patient confidence
> that he may be helped, addresses the underlying personality issues
> that manifest as somatization, and especially offers the opportunity
> for the patient to explore their childhood experiences which are so
> often laden with traumatic experiences of abuse and deprivation."

Anything less will most likely not help the more severely disturbed EI patients who have erroneously projected the cause of their suffering onto the physical environment. Even in a successful therapy, the ideal outcome of complete wellness is rarely, if ever, fully achieved with difficult patients. It is hoped that the approach I have outlined provides some useful heuristics for setting therapeutic goals, planning intervention strategies, and applying behavioral and cognitive methods. These suggestions are no more than that and should not be taken as definitive. After all, clinical experience has taught us that no two therapists practice alike and that no particular form of psychotherapy is generally more effective than any other. Regardless of the kind of therapy practiced, the common objective is to restructure the knowledge of the patient toward a more positive concept of self, a more realistic appraisal of the physical environment, and a less frightening interpretation of the person environment.

Last, it may be a theoretical debate as to when to terminate therapy, and, as in all good arguments, there are valid conflicting positions. With the advent of managed health care and the associated mandate to shift away from long-term therapies in mental health services, the argument may well be out of the hands of the academicians and clinicians (Barsky and Borus, 1995). As a note of caution to policymakers and health-care managers, limiting psychological services to EI patients with personality disorders may not save money in the long run. Somatization of underlying conflicts can lead to numerous, often expensive diagnostic procedures and also increase the number of primary care physician contacts (Smith et al., 1986). For such patients, preventive psychotherapeutic maintenance may serve all parties concerned.

chapter sixteen

Politics

Americans have increasingly come to regard such conditions as alcoholism, drug abuse, impulsive crime, and random homicidal violence as illnesses requiring treatment rather than as manifestations of wickedness. This humanitarian development follows from the recognition that constitutional vulnerabilities may contribute to such behaviors and that some social deviants are trying to deal with the same internal conflicts and external stressors that confront persons who are consensually regarded as mentally ill. The only difference is that the offenders' efforts to cope take the form of socially destructive behavior (Frank and Frank, 1991).

Treatment based on the biomedical disease model is not likely to be effective in the psychologically motivated aspects of EI. The disease model implies that the EI sufferer is the blameless victim of such impersonal forces as foods, chemicals, fungal and viral infections, or energy force fields. The doctor's responsibility is to provide a remedy. The patient's responsibility is to adhere to the treatment. To be effective, psychological treatment should emphasize that responsibility lies within the individual. Without that fundamental assumption, the process of experienced self-control and self-efficacy is taken out of the treatment milieu. Invoking belief in an incurable illness may temporarily relieve the EI patient from self-deprecating feelings of guilt, shame, and inadequacy. Such self-deception and iatrogenic reinforcement only serves to create the illusion of a solution. But these painful feelings usually do not go away, and the pathway to a psychotherapeutic relationship is often blocked by distorted beliefs associated with EI.

Victims and a herd of sacred cows

It is not politically correct to blame the victim. But, before being blinded by political expediency, it may serve to consider the definition of victim. The *American Heritage Dictionary* lists several definitions, two of which seem most relevant. First, a victim is one who is harmed by or made to suffer from an act, circumstance, agency, or condition — e.g., a victim of war. Second, a victim is defined as a person who is tricked, swindled, or taken advantage of — e.g., the victim of a cruel hoax. It seems that the first definition is one for which political correctness applies, and it is also the definition by which advocates of EI define themselves. The second definition represents advocates and patients according to the psychogenic theory. I believe that adherence to the first definition adversely biases decisions about disability and entitlement and mistakenly reinforces a false concept. I also believe that the second meaning of victim is not receiving enough attention from the medical, scientific, and political communities.

Blaming social ills

A driving force behind the EI movement is the subset of vocal, self-proclaimed environmental victims eager to present their testimonials to the world. Individuals with hostility that is displaced from intrapsychic conflicts are likely to seek destructive, revolutionary outlets to satisfy their basic human need for self-assertive activity and the expression of displaced anger. Their influence on society is garnered from a coalition of social factors external to the distortions of belief within their psyche, but interactive with it.

The projection of intrafamilial responsibility onto the faults of society has been observed and questioned in the area of child abuse. Extrafamilial issues such as poverty, poor housing, major health problems, drug use, permissive sexual mores, etc. have all been held up as culprits. Oliver (1993) criticized research suggesting that social factors account for intergenerational transmission of child abuse as being "plausible, popular, convincing, and *wrong*." The lesson to be garnered from Oliver's frank conclusion is that low-level chemical sensitivity is another factor that can be added to the list of environmental factors that are *wrong* in the explanation of the problem of antisocial behavior resulting from suppressed hostility.

While in some instances the label of mental illness seems prejudicial, in others it may be too forgiving. The EI patients who seem to attract the most attention from the media are often those who are most bizarre in their public presentation and lifestyle. It is these self-defined "victims", shrouded in militant piety, who have created a negative atmosphere for all EI patients, most of whom are not militant and are more concerned with becoming well.

Sociologist Emil Durkheim (1952) described the attitude of entitlement and health in his commentary on neurasthenia in the 19th century:

> "Now today neurasthenia is rather considered a mark of distinction
> than a weakness. In our refined societies, enamored of things intel-
> lectual, nervous members constitute almost a *nobility*."

If a diagnosis of EI based on what patients think is wrong with them were to be legitimized, the economic cost of entitlement programs could be huge. If EI were to be imposed politically as a *bona fide* diagnosis based on subjective complaints and attributions, then anyone claiming disability from it would be entitled to a subsidy under rights guaranteed by such programs as workers' compensation and Social Security.

Reasonable accommodations?

Lobbyists for EI demand equal rights that are granted to the physically handicapped, as evidenced by efforts to include EI as a disability under the Americans with Disabilities Act, which went into effect in 1992. The act defines a disabled person as someone with a physical or mental impairment that substantially limits a major life activity. But, without explicit definitions of these terms, the Equal Employment Opportunity Commission says the act could be applied too broadly (Sharpe, 1995). The limits on a major life activity include mental processes such as thinking, concentrating, and interacting with other people. The Commission has issued a compliance manual further defining an "impairment" as being a physiological disorder affecting at least one body system or a mental or psychological disorder. The implication is clearly that psychological factors cannot be premorbid to the physiological impairment. In fact, the Commission explicitly stated that physical characteristics, personality traits, and environmental, cultural, and economic disadvantages are not impairments. Some EI advocates argue that all psychological impairments are secondary to physiological abnormalities caused by chemical sensitivities. The EI advocacy group, MCS Referral & Resources, offers pamphlets on how to file complaints against institutions for alleged noncompliance with the Americans with Disabilities Act (Chemical Injury Litigation Project, 1994).

Mary Lamielle, appointed to the 600-member President's Committee of Employment of People with Disabilities, founded and directed the National Center for Environmental Health Strategies (NCEHS), of Voorhees, NJ, an EI advocacy organization founded in 1986 and with a current membership said to number several thousand. How many other members of the President's Committee represent constituencies who profess to have disabilities based on a phenomenon with no medical or scientific credibility?

The Department of Housing and Urban Development (HUD) recognized EI as a handicap under the Fair Housing Act in 1992, and individuals claiming to be so afflicted are entitled to reasonable accommodation in housing. With an initial grant of $837,000, HUD sponsored a special housing project for EI patients in San Rafael, CA. Ecology House, a two-story building with 11 one-bedroom apartments, was constructed according to chemically clean guidelines worked out with advocates in advance. Special exclusions included no carpets, no "noxious" chemicals, no petrochemicals, no particleboard containing formaldehyde (the cabinets and cupboards were steel), no herbicides or pesticides, and no air fresheners. To avoid paints and stains, the interior surfaces where finished with plaster. Resin-free fiberglass was used for insulation, covered with a foil vapor barrier to reduce the release of gases to the interior. Ventilation fans were placed on the roof to draw air out of the attic space and the exterior walls. Personal hygiene products were restricted.

Because several hundred volunteers applied for residency, the maximum 22 inhabitants where selected by lottery. This random selection process makes it possible to generalize the effects of this residential experiment beyond the subjects selected. In the Fall of 1994, within one month of occupancy, the residents deemed the house unfit to live in. They also demanded further assistance and additional compensation for having been "subjected to such a terrible experience" by the agency.

This seemingly rational solution attempted by HUD was doomed before it began. Problem-solving strategies that are applied to a situation in which no solution is possible are likely to generate frustration and anger, regardless of the quality of the problem-solving activity (Folkman and Lazarus, 1988). The Ecology House saga illustrates another aspect of failed coping — namely, if escape-avoidance fails, the individuals use confrontation to rally for greater social support. The EI advocates, so deeply vested in their illness, would have been faced with insurmountable conflict and anxiety were they to be free of their somatization and appraisal of chemical sensitivities. Nothing is ever enough to appease a misguided victim, especially when the wrong solution is tried.

I have treated several EI patients who, as recommended by clinical ecologists, had purchased special residential trailers constructed with aluminum and porcelain-lined steel. After investing significant amounts of money, usually about $50,000, they found that they could not live in this custom-built environment, either. What EI patients need more than money and special housing is guidance. Compromises and political accommodations cannot cure psychological disorders. Rather, they reinforce unhelpful beliefs.

Special accommodations for the medical evaluation of patients involved in civil law suits are demanded by EI advocates. For example, a lengthy list of accommodations to be met by the physician's office are spelled out in a letter I received (Table 16.1). The justification is presented in a footnote to the letter:

Table 16.1 Contents of Letter from EI Advocate Stating
Accommodations Required to Evaluate an EI Patient

1. No current use of any petrochemical pesticides within the last 3 months at minimum.

2. No history of building treatment with chlordane at any time.

3. No history of use of dioxin-containing material on the building grounds.

4. No history of spill of pesticides or other petrochemical agents in the building. If spill history is present, please specify time frame, location, complete list of chemicals and their amounts, and detailed description of clean-up methods and follow-up building evaluation.

5. A non-smoking environment.

6. No "air fresheners" in clinic area or in restrooms because of their petrochemical content.

7. No use of storage of carbonless copy paper in, or around, the area of the patient.

8. No recent use of petrochemical germicidal agents on open surfaces in the clinic environment or for sterilizing items which will be used on or near the patient.

9. No use of cleaning agents containing phenol, formaldehyde, glutaraldehyde, or other petrochemicals or potent irritants (e.g., ammonia, chlorine-containing bleaches and cleaning products).

10. No recent minor renovations (within 3 to 6 months) or major renovations (within 1 to 2 years). This includes painting, carpeting, synthetic fabrics with stain-resistant or other coatings, and the use of sealants, adhesives, or any other products containing petrochemicals. If renovations have been done recently, the office should be equipped with an activated charcoal filter system with adequate capacity for all areas sharing a common air source.

11. On the day of the patient visit, staff in the patient's area should refrain from use of perfumes, colognes, hair spray, or other scented personal products (because these are typically petrochemicals) and should not wear recently (within 2 months) drycleaned garments or any clothes treated with fabric softener during the last laundry/drying cycle.

12. If the patient's symptoms increase after entry into the clinic, this indicates that there is a probable aggravating exposure. If this occurs, the patient must be allowed to leave promptly and not be kept inside awaiting further tests, procedures, etc. Unless the clinic is then able to identify and adequately correct the exposure, which may require a filter system as described above, the evaluation should be conducted in the patient's home.

13. If the clinic is unable to achieve the controls described above, the evaluation should be done in the patient's home, using staff precautions as described in #8 and #11 above, and no germicidal-treated items which may leave any residue.

Source: Letter from Grace Ziem, M.D., Occupational and Environmental Health, Baltimore, MD, January 6, 1997.

> "Reasonable accommodation for handicaps is required (under the Americans with Disabilities Act) by facilities used by the public, including medical settings. The patient's condition is severe enough to be considered a handicap under this Act since it interferes significantly with one or more major life activity."

This EI advocate's definition of reasonable is subject to consideration. But, even if these accommodations could be met, it would not ensure that an examination could be completed because the patient may feel symptoms which would necessitate leaving the building. How many of these conditions are met by the offices of EI plaintiffs' treating physicians is of interest.

Victims: a neglected population?

Scientific investigations and mental health services are generally focused on managing the more severe psychological disorders. Treatment of these patients is undoubtedly necessary, and research to alleviate their suffering deserves support. But, much of the research money is allocated to the biological psychiatry paradigm, even though the validity of the orthomolecular paradigm is seriously in question (Fisher and Greenberg, 1989; Ross and Pam, 1995). This priority has left many demoralized people without the attention they deserve. Their psychologically troubling problems have a vague, general feeling of malaise, usually and correctly deemed insufficient to warrant pharmacological intervention. Feeling neglected, yet suffering, many of these people can only get attention from the medical system if they escalate their problems. Unfortunately, this makes them likely to receive psychotropic medications (and the necessary diagnoses to justify such prescriptions), which they neither want nor necessarily need. Rejecting medication and the stigma of psychiatric treatment, people tend to turn elsewhere for an answer.

If the projection of causality is away from the self, as it is with EI patients, these people can easily fall into the role of victim. The politics of victimization are a hallmark of the American culture, and numerous movements seeking followers offer absolution of self-responsibility as a caveat (Sykes, 1992). The illness excuse is one avenue by which demoralized people suffering from somatization can achieve both institutional attention and avoidance of psychological issues. There are several pathways by which these vulnerable people find their way to the answer offered by EI. Some undertake a crusade with obsessional intensity, "researching the literature" of alternative health care practices until they suddenly, miraculously come upon an accommodating belief system. Based on personal experience, there is usually as much valid scientific information in one such pamphlet as in all of them combined. Others, who stumble along in life wondering why they do not feel well, naively follow the advice of quack physicians who always have another answer for their ills. Irrespective of how they find the path, many of these patients follow it to its source, where they enter the shrines erected in the name of a false ideology.

In the case of EI, there exist clinical ecology referral networks, now linked by computerized worldwide information systems, that channel many EI patients into special inpatient environmental care facilities, if they can afford it. In these facilities, the belief in EI is reinforced, and the patient often becomes a "true believer". All of the symptoms and problems of living are then attributed to the physical environment by fabricated explanations. Such overvalued ideas fulfill the patient's search for an explanation external to self and temporarily may relieve the anxiety generated by the incongruence of unexplained symptoms. But, the phenomenon does not end there. The afflicted still seek a solution to their suffering. Desperate people are vulnerable to manipulation and exploitation. More familiar accounts of cancer patients seeking miraculous cures in clinics in the underdeveloped countries have become headlines in tabloid media, as illustrated by the death of actor Steve McQueen.

Clinical ecologists offer a host of treatments that have not been proven effective and have been summarily rejected by the scientific medical community. Treatment is offered to purge the poisons from the body, correct immunologic abnormalities, or realign magnetic polarity. Other techniques and procedures that are valid when applied in the appropriate context, are employed to treat these unrecognized phenomena. For example, biofeedback has been incorporated as a treatment by clinical ecologists to realign polarity, said to have been misaligned by electromagnetic forces (EMF). The misapplication of legitimate procedures may even extend to aligning the EMF forces of two people who are said to set each other off if they come into contact.

A rational analysis of these procedures leaves many scientists with the feeling of incredibility, and many lay people with incredulity. Both may feel that something so bizarre is "either too crazy to really be happening" or, if it is, it is contained within the delusional worlds of the patients. Both of these conclusions are wrong.

During the settling of America, covered-wagon "snake oil" salesmen were notorious for selling their health elixirs. To help establish a market in a town, an accomplice would visit to create the perception of a mythical crisis or ailment. Once the need was established, the "savior" would arrive with the magic elixir. The afflicted would partake, and all would revert to the state they were in before the crisis. These exploiting profiteers would move on to the next town, to which the ailment mysteriously spread. Today in the U.S., the unsubstantiated elixirs are administered in established centers, and the "spreading" involves illusions of fantasized poisons.

Politics of intimidation and the law

One perplexing question often asked by spouses and parents who have seen their loved ones fall into the hands of clinical ecologists is, "Why doesn't the medical community do something about these practices and practitioners?" First of all, today there is no lack of documentation about the unsubstantiated practices associated with EI, something which could not be said 15 years ago. The defense of naiveté, at least within the health-care community, has no credibility today. Several medical associations have written position papers critical of these practices. The silence on the part of many traditional physicians may be motivated by several factors, not the least of which is the fear of being sued. Federal Trade Commission rulings dating to the mid-1970s have discouraged physicians from criticizing other physicians even in the face of overt fraud. This is the result of threats of litigation based on allegations of restraint of trade and attempts to limit competition — charges easily alleged but very difficult and costly to defend against. In the practice of quackery, threat of litigation for the freedom to express their pseudoscientific views is the refuge of the scoundrel. This is yet another example of how the law is suffocating America which could be added to the many illustrations in the provocative book by Philip Howard, *The Death of Common Sense* (1994).

In 1984, Earon S. Davis, J.D., a former director of the Human Ecology Action League (HEAL), presented a blueprint for directions for legal action on behalf of patients said to be suffering from EI (Table 16.2). This presentation was made at the annual meeting of clinical ecologists, the American Academy of Environmental Medicine, in Chicago. Today, all of these actions have come to pass. Davis (1985) justified this victim advocacy:

> "One of the reasons this illness is so devastating is that, without the
> proper coverages, compensations, and assistance listed above [Table
> 16.2], many individuals probably will never get well. And, to wage
> a legal battle each time one applies for an 'entitlement' is outra-
> geous."

Unrealistic product safety regulations

There are two principles that currently undermine any realistic assessment of the potential health hazards of low-level environmental exposure in the U.S. The first is called "no threshold", which argues that if large amounts of any chemical agent are found to be harmful, there must not exist a minute amount that could be safe. This conclusion is based

Table 16.2 The Legal Side of Ecologic Illness

1. Securing reimbursement for medical costs from health care insurers and workers' compensation payments
2. Securing coverage under Medicaid or Medicare and various state and federal assistance programs for poor and aged victims
3. Securing financial assistance for the rehabilitation of living space
4. Securing proper income tax deductions for expenses associated with ecological illness, especially excess costs of remodeling, changing of heating systems, and organic foods
5. Establishing disability under private disability policies and under Social Security and other federal programs
6. Securing fair treatment under federal and state protections for the disabled.
7. Securing the services of a state vocational rehabilitation agency
8. Preventing employers from improperly dismissing or demoting them
9. Securing and maintaining a "safe" work environment
10. Securing just compensation from the companies and individuals responsible for chemical exposures which caused the disabling illness
11. Contesting a divorce and/or child custody matter in which the spouse tries to use the patient's ecological illness as "proof" of severe emotional disturbance
12. Fighting discriminatory practices in colleges and universities, including law and medical schools, which threaten their ability to complete school and to enter into further specialty training
13. Securing "safe" environments and foods in public and private schools where ecologically ill children are involved
14. Securing safe environments and foods in prisons, mental hospitals, hospitals, and other public and private institutions
15. And, sometimes, even using ecological illness as a possible defense or mitigation in a criminal action

Source: Adapted from Davis, E.S., *Ecol. Illness Law Rep.*, 2(6), 4–10, 1985.

purely on assumption, derived by linear extrapolation of effects from high to low doses. These simulation models assume a static host and ignore research on cellular repair and replacement capacity (Ames and Gold, 1995).

To illustrate the absurdity of the no-threshold principle, one can argue that drinking a glass of water with your meal is dangerous because ingesting large amounts of water can kill you, a possibility demonstrated in an unusual case of child abuse. Based on the 1994 National Academy of Science/National Research Council report, *Science and Judgement in Risk Assessment,* there is some indication that the unintended consequences of regulatory legislation may be addressed. The report noted that default options in the risk assessment process and the specific data selected can have a major impact on the estimated level of risk. These more scientific attitudes may have had some impact on Congress in 1996, when regulatory reform legislation was passed with respect to some products. Specifically, the Food Quality Protection Act, which was signed into law, states that pesticides are no longer subject to the Delaney clause. In the 1958 Food, Drug, and Cosmetic Act, a section that has come to be known as the Delaney clause was included which reads, "No additive shall be deemed to be safe if it is found to induce cancer when ingested by man or animal, or if it is found, after tests which are appropriate for the evaluation of the safety of food additives, to induce cancer in man or animals." There was no allowance for the astounding improvement of measurement technology of today in this legislation, written almost 40 years ago. In the last decade alone, chemical detection methodology has improved from parts per million to parts per quadrillion. A part per quadrillion is equivalent to the thickness of one dime in a stack of dimes, back to back, reaching from the Earth to the moon (Robert Vogt, personal communication). The new standard is "reasonable certainty of no harm."

Aside from technological considerations, there are also some suggestive psychological issues that warrant attention. Congressman James Delaney, prime sponsor of the legislation that bears his name, had not been concerned about such subtle distinctions. An avowed anti-fluoridationist, he regarded carcinogens as "stealthy, sinister saboteurs of life" (Whelan, 1993). Such attitudes can have a defining impact on the political climate which impacts funding and regulatory policy.

Regulating trivial risks impedes effective risk management (Breyer, 1993). The Delaney clause served to foster public pressure for a safe environment, regardless of consideration of real-life economics or need for the product. This legislation and the political pressure which it facilitated has undermined the FDA's ability to conduct sound science and to present guidelines that reflect a more realistic concept of risk to humans.

The second principle, referred to as the "false link cancer risk", is that no food or chemical shall be deemed safe if it is found to induce cancer in an animal model, regardless of induction dose and with total disregard for comparative physiology and body mass. Federal regulations have mandated an inflexible stand that bans ingestion exposure to agents, whether the risk is great or not.

Similar restrictions exist for the Environmental Protection Agency (EPA) scientists at the Human Effects Research Laboratory (HERL), located on the campus of the University of North Carolina. According to regulatory guidelines, no chemical compound that has been associated with high-dose cancer research with animals may be employed in human research. The prohibition holds even if the dose administered in a provocation challenge was so low that the subject would be exposed to it as a matter of course from inhaling ambient air walking around the campus. As one illustrative example, the biological effects of exposure to side-stream tobacco smoke could not be studied under experimental conditions because of the link between tobacco smoke and cancer. The fact that the proposed project was with children who live in homes with smokers and are chronically exposed to side-stream tobacco smoke anyway could not sway the argument. As another example, the alleged effects of low-level exposure to formaldehyde cannot be studied because formaldehyde is on the list of carcinogens, even though challenge levels would be no greater than those found in most homes. The absurdity of this exaggerated risk in research extends to ethics, where the possibility of exposure to humans is dictated by regulatory EPA guidelines resulting from the irrationality of the Delaney clause. Once again, common sense has been subjugated to misguided government regulation, with the added threat of ethics violations for scientists who would challenge the regulation. This represents yet another example of epicyclical, self-serving ideation.

"More research"

The argument against publicly funded research on EI by those who would allocate the money to "real science" is understandable but not practical. However, it is fair to argue that any research must follow scientific protocol with plausible and testable hypotheses. EI poses special problems for policy development and for regulatory agencies (Gori, 1996). The EPA is responsible for protecting public health and environmental quality from the adverse consequences of pollution. The EPA is charged with enforcing more than a dozen major environmental laws in the U.S., including the Clean Air Act; the Federal Insecticide, Fungicide, and Rodenticide Act; the Resource Conservation and Recovery Act; the Safe

Drinking Water Act; the Toxic Substances Control Act; and the Comprehensive Environmental Response, Compensation, and Liability Act (Superfund). Under this scenario, decisions about whether regulatory actions are warranted to safeguard public health would be based, at least in part, on an assessment of EI-related health risks. Dyer and Sexton (1996), EPA employees expressing their personal opinions, argued that high-priority research on EI is distinguishable by four attributes:

1. Results are valuable for risk-related decisions.
2. Findings significantly advance scientific knowledge and understanding.
3. The hypothesis being tested is biologically plausible.
4. The hypothesis being tested is readily testable.

The EI advocate argument for more elaborate testing, involving sequestration in "environmentally safe" hospital units (an ECU by any other name) is made without a shred of even preliminary pilot data to justify such a costly enterprise (Ashford and Miller, 1989, 1991, 1997). The "new" theoretical rationale and mechanistic hypotheses are no more than semantic relabeling of the postulates of clinical ecology. Only additional testimonials from self-labeled "canaries" (Berkson, 1994) and self-report surveys of EI patients are offered as evidence for the need of such a facility. Were such a unit to be established, by definition there is no assurance it could meet the purity requirements of the hard-core postulates of the toxicogenic theory. Standards and levels for environmental exposure set by scientists with no bias to protect following established toxicologic exposure criteria should be sufficient but probably would be judged insufficient. Unfortunately, if past experience is any indication, the ultimate judgement of purity would depend on patients' *post hoc* subjective appraisals of acceptability.

Conspiracy theories

"Conspiracy to hide the truth" is a cry that may be heard in alleged toxic spill cases and class-action suits contaminated by mass psychogenic illness processes such as contagion or media escalation. We live in a time in the Western world when there is understandable concern for preservation of the environment and the elimination of pollution. It is easy to see how clinical ecology can ally its interest with well-intended, pro-environment movements. In my own experience, talking to a journalist about the Love Canal incident, I mentioned Whelan's review of scientific studies (1985), which showed that there were no harmful health effects traceable to the alleged harmful chemical agent at the site, dioxin. The incidence of health abnormalities among the residents of Love Canal were within expected epidemiological prevalence parameters. Furthermore, people have moved back into the same houses that had been declared uninhabitable, houses that sold at market price. Her response to my presentation was immediate: "Obviously a coverup by the government." When I suggested she review the evidence, she remarked, "That wasn't necessary," and terminated the discussion. This journalist was not a radical environmentalist and had a very superficial understanding of EI. Yet, she had written factual articles about environmental contamination in the Southwest, the coverups by responsible parties, and their prosecution. This experience resulted in justifiable suspicion and distrust, but all too often this caution can be inadvertently or deliberately used to bolster the credibility for EI among the lay public.

The conspiracy theories of EI advocates have recently been portrayed in Radetsky's book, *Allergic to the Twentieth Century* (1997), implicating a host of conspirators, including the chemical and drug industries, mainstream or traditional physicians, the military, and Federal intelligence agencies. Yes, there may well be a conspiracy around EI, but it may involve those least suspect by a naive public, the accusers.

Media

EI (or MCS) is an acronym that is becoming part of the colloquial semantics of our culture. When a concept becomes part of the language, it can take on an air of validity and have an influence on an unsuspecting public. How did this come to be? The press and television media are enamored with the audience fascination found in sensational victim testimonials. The coverage given toxicity incidents often reflects classic toxic terrorist tactics (Whelan, 1985). Examples of this may be found in litigation regarding the harmful effects of dioxin and particularly Agent Orange (Franklin, 1994), radon, air pollution, indoor air in so-called "sick buildings", pesticides, herbicides, formaldehyde, carpet, paints and lacquers, solvents, and other household products that need only come to the attention of plaintiffs.

The media effect is heightened when Hollywood celebrities become activists. One memorable example is the apple scare over the chemical alar, used on apples to lengthen shelf life and enhance the cosmetic appearance of the fruit. The allegation of poisoning was initiated in 1989 by the radical environmental group, the Natural Resources Defense Council, which claimed that children were developing cancer from the alar. The charge was led by Meryl Streep, an Oscar-winning actress. Ed Bradley of CBS-TV's *60 Minutes* aired a program implying the allegations were factual. This mass hysteria introduced by environmental radicals and promoted by the media and a Hollywood icon led mothers to halt school buses to extricate apples from their children's lunch bags as if they were grabbing an apple planted by the wicked witch in *Sleeping Beauty*. Whelan has presented a detailed account of the media involvement, the substantial financial loss to the fruit growers' industries, and subsequent legal action by them against CBS (Whelan, 1993). Recently, the same CBS reporter and program took up the advocates' cause in the Gulf War syndrome, airing unfounded toxicological speculations and suggesting government coverup and conspiracy. None has been substantiated (Lashof, 1996).

Manufacturing industry

The manufacturing industry has been placed in an untenable position by unfounded law suits and threat of further litigation. Some manufacturing corporations risk bankruptcy from litigation or settlements, as evidenced by the recent demise of Dow-Corning which filed for bankruptcy in May 1995 because of its involvement with silicone breast implants. Angell (1996), expressing an initial inclination to find fault, reviewed the evidence as the scientist she is and concluded there was no evidence linking silicone breast implants to disease.

Politically, Dow-Corning was accused of hiding damning animal data from the Federal Food and Drug Administration, which allegedly linked implants to disease. Both the FDA and the Justice Department cleared the company of any wrong-doing or withholding data. The company was besieged with claims of rheumatoid and immune disease, for which there was at least a plausible hypothesis and mechanism, and some cases of scleroderma were reported in women with silicone breast implants (Bridges et al., 1993). However, several epidemiological studies have found no association between silicone breast implants and rheumatalogic disease (e.g., Gabriel et al., 1994). There were also claims of neurological problems characterized by symptoms of memory loss, fatigue, and other cognitive impairment which came to be called "silicone encephalopathy". In a review of cases involved in litigation, Rosenberg (1996a) found there was no neurological evidence of abnormality, much less a neurologically plausible hypothesis for such an effect. Despite a multibillion-dollar class-action settlement, individual plaintiffs maintain the option of suing the remaining companies associated with Dow-Corning under the existing tort laws. As of this writing, settlement negotiations are in progress.

To some extent, the manufactures may have unwittingly contributed to their own predicament. Within the industry, decisions about research on possible low-level chemical effects of products are sometimes made by attorneys concerned with liabilities, rather than by the scientific advisory boards. There is fear that any effects found to be caused by low-level exposure to chemicals, no matter how small or harmless, would be misconstrued to support allegations of EI. By analogy, the risk of medical malpractice claims has seriously affected how physicians are practicing, and sometimes *not* practicing in the U.S. Similar concerns have created a justifiable paranoia in U.S. industry. Will this curtail the orderly development of new safe products such as antibiotics for the growing number of resistant organisms?

When it comes to EI, many traditional, scientifically well-founded physicians appear to be immobilized. Some do not intervene, either because they cannot fathom the depths of the psychological issues they are confronting or they just do not know how to deal with EI patients. Some avoid the issue by invoking the prerogative of the specialist, and send the difficult EI patients elsewhere. Most physicians know what is going on but fear the personal and legal risks associated with their involvement.

Some well-intended, negotiation-minded politicians and policymakers are naively trying to appease both sides, hoping the controversy over EI will "just go away". However, there can be no compromise between systems founded on science and pseudoscience because the fundamental motives and objectives are incompatible, an issue I shall revisit in the following chapter.

Insurance industry

After some egregious abuses of insurance benefits by clinical ecologists, the insurance industry has curtailed payment for their unsubstantiated diagnostic and treatment practices. Prescriptions written for the construction of "ecologically safe houses", special saunas, organic diets, etc. certainly warrant scrutiny. But the pendulum should not swing too far in the other direction. Proper medical assessment of chemical sensitivities may require controlled chamber challenges (IPCS, 1996; Selner, 1989; 1996).

Successful psychological intervention with most EI patients is not accomplished by short-term psychotherapy, much less pharmacotherapy alone. Rather, a variety of short-term therapies may be used initially as strategies in an overall comprehensive, long-term plan. With the increased pressure from managed care insurers to curtail long-term psychological treatment in favor of short-term psychotherapy and/or medication, EI patients are at risk of not receiving the psychological treatment they require. Such indiscriminate curtailment of psychological services reflects poor risk-management philosophy. Once the belief in a physical cause is entrenched, these patients feel no expense should be spared to substantiate their illness. CAT scans and MRIs to view anatomical structure and PET, SPECT, and functional MRI to assess cerebral function are costly procedures to be employed as diagnostic methods with patients whose problems are psychological. The cost of unnecessary diagnostic procedures conducted with patients suffering from somatoform disorders is significantly greater than the cost of psychological treatment, even if it involves long-term psychotherapy. Failure to offer appropriate treatment to EI patients can only worsens their psychological condition. Furthermore, they are vulnerable to iatrogenic therapies which only escalate the problem for the patient and the potential cost of disability for society.

Radical politics

In November of 1990, at the annual meeting of the American College of Allergy and Immunology in San Francisco, John C. Selner and I conducted a workshop on environmental illness. The primary purpose of the meeting was to emphasize that EI patients were, indeed, seriously ill. They had a right to anticipate and receive appropriate medical care and should not be abandoned to pseudoscience because of inconvenience or economic considerations. One theme of the workshop was the psychological treatment of EI patients and how they can be helped. Leaving patients with the notion that "it was all in their head" was described as being very destructive to these patients. Presentation of an orderly approach to working with EI patients followed. This workshop was misconstrued by EI advocates as focused on "declaring EI victims crazy". The workshop was disrupted after the lunch break when EI advocates bolstered by members of the radical AIDS group Act Up stormed into the meeting room, media television cameras not far behind. For those unfamiliar with Act Up, among its claims to notoriety is the desecration of St. Patrick's Cathedral in New York City.

In the workshop, one self-proclaimed EI victim, who had been sitting passively in a wheelchair with oxygen cannula and dust mask in place all morning, promptly rose to her feet, threw off her mask and cannula, and launched into a litany of complaints for the media cameras. A local clinical ecologist unsuccessfully tried to take control of the podium microphone at which I was standing. Other insurgents attempted to take the speakers' materials. After some tense moments, security reinforced by California State Patrol appeared and cleared the room and the rest of the hotel of the demonstrators. In subsequent years, the same workshops at the annual meetings of the American College of Allergy and Immunology were conducted under strict security precautions.

The alliance formed between the radicals of EI and Act Up seemed ironic to me at the time. "Chemical AIDS" was a label used during the late 1980s in an attempt to link EI to the terribly devastating immunologic disease, AIDS. I believe greater recognition of AIDS as a societal problem, anarchistic tactics aside, was not helped by identification with EI.

Discussion

Kleinman (1986) made some perceptive political and social observations about the diagnosis of neurasthenia in a Chinese culture that had no other forms of dealing with the emotional ailments and demoralization of its people. In China, neurasthenia bore the blame for misfortune and misery. Generally, it freed the sick person from taking responsibility for his actions, and scientific explanations may not always serve the common good of society. In the period from 1958 to 1962, Pavlovian classical conditioning was introduced to explain neurasthenia under the influence of Russian psychology. But it had little lasting effect on the meaning of neurasthenia for the people, though it may well have controlled, at least temporarily, its misuse and abuse. Ultimately, the state may have come to accept that neurasthenia serves certain indirectly useful personal and social functions for Chinese society. Chronic illness behavior offered workers some leverage over a centralized economic system that otherwise did not offer many opportunities for changing jobs, getting out of very difficult work situations and stresses, resolving lengthy family separations, and getting time off with compensation.

The great advantage of neurasthenia as a diagnostic category is its flexibility. Here its very ambiguity, which makes it scientifically suspect, is a clinical virtue. It is easy to use and it covers the range of possible complaints. These multiple, even contradictory meanings allow for the needed flexibility, slippage, and multiple reinterpretations essential in the day-to-day care of patients with a chronic disorder of greatly uncertain treatment response, course, and outcome. It maps a phenomenon in the world, sanctions a wide range of possible interventions for it, and is understood and accepted in the popular culture. It may be the best available way of dealing with somatization, which collapses biomedical categories and makes a hash of reductionism and mind-body dualism. Even if treatment practice changes (e.g., psychotropic medication appears to be more acceptable in China today), the diagnosis of neurasthenia will not (Kleinman, 1986).

Through advocacy effort, EI has received increased cultural and social acceptance by the general public. In a diverse culture such as that of the U.S., entrenched in the right to self-expression and freedom of belief, it is difficult to establish what is a socially sanctioned belief and what is considered to be aberrant thinking. One question to consider is, "To what extent are the practices and speculations of a medical subculture a culturally sanctioned behavior under the First Amendment?" In many if not most EI cases, the patient has been exposed to inappropriate medical treatment by physicians and other health-care professionals who instill or reinforce the patient's beliefs. In general, somatizers are more suggestible and EI patients have been shown to have a predisposition to a personality trait characterized as lack of intuition and insight about their condition. These people turn to outside sources to make decisions and are therefore more suggestible and vulnerable to external influences and opinions from acquaintances and the media and to iatrogenic influence and exploitation. How much is the constitutional right to freedom of expression to be counterbalanced by exploitation and induced harmful effects to patients?

Finally, somatization is not typically motivated by malingering. However, even where there is insight, it cannot readily be expressed in an open fashion as this removes the social sanctioning of the illness discourse and labels the patient a malingerer. Avenues of rapprochement with the employer and the legal and workers' compensation systems must be left open to allow for implementation of the goals attained in psychotherapy.

The irony reflected in EI is that humanity may not have evolved in its ability to overcome the evil that has been represented in the beliefs and actions of some of its sociopaths throughout time (Carus, 1974). Society has failed to heal the sequelae of its own brutality, creating false attributions, unfounded expectations, and unrealistic demands for restitution in the victims. Historically, there have been numerous examples of deviant beliefs and movements appearing when society fails to address certain fundamental social problems (Kephart, 1976) or has erroneous models which camouflage them. Freud's theories of the seduction hypothesis, and later its rejection, unwittingly influenced society's views about the effects of childhood trauma on adult behavior. The legal world, the medical community, the mental health professions, the sexologists, and the sociologists for decades minimized or ignored the actuality of incest (Simon, 1992a).

Unable to face the injustices stemming from personal trauma and terror, some victims project their misery onto false icons such as a metaphysical devil or, in the context of EI, a physical environment perceived to be unsafe. As victims, the aspect they have least insight about and are most reluctant to accept is that they are victims of a hoax.

chapter seventeen

Future
directions

Testimonials and subjective impression are the empirical foundations of the toxicogenic theory. Speculations about underlying physiological mechanisms are based on fallacious and implausible rationale, yet continue unabated. Allusion to credible science and distortion of findings are commonly seen in publications advocating EI, in which valid scientific studies are referenced but often misinterpreted, reinterpreted, or used out of context. The distortions of speculation that build a particular theory are referenced with every subsequent writing, magnifying the distortion and creating the illusion of inductive evidence. The toxicogenic theory re-emerges under new names, but the postulates remain the same. Testability of hypotheses is professed, but upon inspection of the reframed postulates, the basal state of the patient is indeterminable. An allusion repeated often enough creates an illusion.

Inconsistent findings from carefully controlled scientific studies (negative results) are criticized as being woefully inadequate in research design and conducted improperly, usually with a phrase such as "too numerous to comment on" in editorials and letters to the editor (e.g., Ziem and Davidoff, 1992). Beware the exploiting profiteer masquerading as a concerned scientist and caring physician! The results of double-blind, placebo-controlled studies which provide experimental refutation of the controversial methods practiced by clinical ecologists are dismissed as "merely generating controversy". Investigators who report negative findings are deemed biased (Davidoff and Fogarty, 1994). Credible scientists who once believed in the possibility of low-level multiple chemical sensitivity (MCS) and their treatment by unproven techniques and who conducted well-controlled studies in collaboration with clinical ecologists are treated as defectors when the results are negative. Accusations of conspiracy with chemical manufacturers, insurance companies, or the workers' compensation administration are either explicitly stated or covertly insinuated. Researchers who have reported results from well-controlled studies that refute the postulates of the toxicogenic theory have been falsely accused of ethical misconduct by EI advocates (Daniell, 1996). Exoneration by ethical review committees is ignored by the accusers, and instead additional unsubstantiated charges may be filed.

These self-anointed critics have an uncanny ability to rephrase another's statements in a way that bears some small resemblance to the original. Their editorials are also used to support their own speculations and to report testimonial data in the form of surveys. Despite protest from legitimate institutions of medicine and science, proponents insist on repeating willful distortions.

TILT

An example of these assertions is the "yet to be proven mechanism or theory of disease" which Miller (1997) labeled toxicant-induced loss of tolerance (TILT). TILT is an all encompassing theory of disease which subsumes chemical sensitivity as a consequence of it. What this means is that the limiting definition of MCS proposed by Cullen which has generated studies of negative findings is now deemed insufficient, not because of the negative evidence, but because it is not comprehensive enough to account for onset and subsequent triggering of symptoms. Miller (1997) offers two reasons:

> "Although chemical sensitivity certainly sounds like an inconve-
> nient problem to have, the words fail to convey the potentially
> disabling nature of the condition and its postulated origins in a toxic
> exposure. ... Chemical sensitivity suggests that those afflicted be-
> come intolerant of chemical exposures only when, in fact, caffeine,
> alcohol beverages, various drugs, and foods reportedly trigger symp-
> toms in these individuals once the process has been initiated."

Toxicant-induced loss of tolerance is the same theory proposed by Randolph and other clinical ecologists (Randolph, 1962; Rea, 1992, 1994, 1995, 1996). Any and all environmental agents, not just chemicals, can induce and subsequently trigger not only multi-system symptoms but diseases (Miller, 1997):

> "A consequent of viewing TILT as a theory of disease would be a
> shift in perspective from chemical sensitivity as a syndrome to chemi-
> cal sensitivity, now TILT, as a class of disorders parallel to infectious
> diseases or immunologic diseases."

The specific postulates are masking or unmasking, the same as adaptation/deadaptation. Masking involves three interrelated components which may interfere with the outcome of low-level chemical challenges: acclimatization, apposition, and addiction. Miller argues that these components may be controlled only by testing patients under the controlled condition offered by a hospital environmental care unit (ECU). Inspection of the require-ments of fasting to put a subject in a "deadapted" state reveals that once again the theories are not testable. Four to seven days of acclimatization are proposed, with equal time periods between challenges to wash out the effects of a reaction. If errors of appraisal-exposure occur during double-blind, placebo-controlled challenges (i.e., false positive response to placebo or false negative to active agent), the theory has a ready answer: the subjects were in fact not in an optimal state of adaptation/deadaptation. More disturbing to me is that the potential effects of malnutrition on both the dependent measurements and the status of the patient are not addressed. Psychological effects of isolation in an ECU are also not addressed but are left to experimental design with a control group of volunteers.

Scientific honesty

In 1919 and 1920, Karl Popper proposed a series of principles that became the foundations of the study of the growth of knowledge in the philosophy of science (Popper, 1968). The paradigm of normal science adopted in this book is guided by Popper's philosophy of science and its refinements by Lakatos (1970). The following seven principles are relevant to the discussion of EI (Popper, 1968):

1. It is easy to obtain confirmations, or verifications, for nearly every theory — if we look for confirmations.

2. Confirmations should count only if they are the result of risky predictions; that is to say, if unenlightened by the theory in question, we should have expected an event which was incompatible with the theory — an event which would have refuted the theory.

3. Every "good" scientific theory is a prohibition; it forbids certain things to happen. The more a theory forbids, the better it is.

4. A theory which is not refutable by any conceivable event is non-scientific. Irrefutability is not a virtue of a theory (as people often think) but a vice.

5. Every genuine test of a theory is an attempt to falsify or to refute it. Testability is refutability, but there are degrees of testability. Some theories are more testable, more exposed to refutation, than others; they take, as it were, greater risks.

6. Confirming evidence should not count except when it is the result of a genuine test of the theory; and this means that it can be presented as a serious but unsuccessful attempt to falsify the theory. This is "corroborating evidence."

7. Some genuinely testable theories, when found to be false, are still upheld by their admirers — for example, by introducing *ad hoc* some auxiliary assumption or by re-interpreting the theory *ad hoc* in such a way that it escapes refutation. Such a procedure is always possible, but it rescues the theory from refutation only at the price of destroying, or at least lowering, its scientific status. [Popper later described such a rescuing operation as a "conventionalist twist" or a "conventionalist stratagem"].

These principles may be summed up by saying that the criterion of the scientific status of a theory is its ability to be falsified, or refuted, or tested. The specific methodology to decide about the plausibility of a theory requires familiarity with concepts and terms peculiar to philosophy of science. The interested reader is referred to Appendix A.

Popper's principles have implications for scientific honesty, which Feyerabend (1975) summarized as follows:

1. Belief may be a regrettably unavoidable biological weakness to be kept under the control of criticism, but commitment is for Popper an outright crime.

2. Scientific honesty consists of specifying, in advance, an experiment such that if the result contradicts the theory, the theory has to be given up. Falsification demands that once a proposition is disproved, there must be no prevarication: The proposition must be unconditionally rejected.

3. The most admired scientific theories simply fail to forbid any observable state of affairs (i.e., no basis of refutation). Tenacity of a theory against empirical evidence would then be an argument for rather than against regarding it as scientific. Irrefutability would become a hallmark of science. Only those theories which forbid certain "observable" states of affairs, and therefore may be falsified or rejected, are scientific.

4. To be scientific, a theory must have an empirical basis.

5. Apologists will argue for holding onto a theory unless it is conclusively disproved, claiming the discrepancies which are asserted to exist between the experimental results and the theory are only apparent, and they will disappear with the advance of our understanding. A scientific research program based on falsification must deplore the fact that stubborn theoreticians frequently challenge experimental

verdicts and have them reversed in a political or legal forum rather than scientific forum. There is no place in science for such successful appeals.

The personal factor in disposing of contradiction

> "Little attention was paid to the experiments, the evidence being set aside in the hope that it would one day turn out to be wrong."
> (Polanyi, 1958)

Henle (1974) summarized how information is shaped by attitudes, hypotheses, or convictions and how it affects progressive or degenerative problemshifts in research programs. Productive thinking, defined as the creative processes by which new questions arise, affects the evolution of theories and research programs. In a progressive research program, contradictions in observations, evidence, and theory lead to formulations of new, testable hypotheses. In a degenerative research program, personal conviction biases the interpretation of evidence and obscures contradictions which stagnates a research program. Henle described five cognitive distortions used to avoid acknowledgment or recognition of contradiction in a degenerative research program. To these I have added a sixth which is particularly relevant to EI advocates.

1.   It is easy to ignore facts that contradict one's view if one can question their source, the conditions under which they were obtained, and the like

EI advocates attack physicians critical of EI on the grounds that they do not wish to jeopardize their referral sources from chemical manufacturers or insurance companies (Ziem and Davidoff, 1992). Two recent examples of personal attacks on reputable scientists deserve mention. First, EI advocates attempted to impugn the negative findings on immune parameters obtained by allergist and immunologist Terr (1986, 1987, 1989) on the grounds that EI patients were evaluated in the context of workers' compensation disability litigation. What is ignored in these attacks is that similar results have been reported in controlled studies of patients not involved in workers' compensation or civil litigation. The National Research Council of the National Academy of Sciences published a book on EI, *Biologic Markers of Immunotoxicology*, which contained a libelous passage criticizing Terr. EI advocates petitioned the American College of Physicians to withdraw its 1989 position statement critical of clinical ecology (of which Terr was the principal author) on the grounds that Terr's work had now been "discredited by no less than the National Academy of Sciences." The American College of Physicians denied this request and, along with Terr, protested that the National Research Council's criticism of Terr was inappropriate. Following review by the National Research Council's Subcommittee on Immunotoxicology, an erratum was issued with the attack on Terr deleted (Orme, 1994).

In the second case, Simon, Daniell, and colleagues were accused of ethical misconduct after publishing negative findings which showed no differences in immune parameters between a group of EI patients and a control group of chronic pain patients (Simon et al., 1993). The controversy centered on having split samples (serum from the same patient is sent in twice with different identification) assayed by a immunology laboratory in California of dubious credibility but commonly used by clinical ecologists. The assay results were found to be unreliable, a fact not presented in the original study, but mentioned in a conference presentation by Simon. EI advocates, through MCS Referral & Resources

(Donnay, 1994a), filed a complaint with the Office of Research Integrity, Department of Health and Human Services, alleging that the study should not have been published because of the unreliability of the California laboratory. A second complaint filed by the same accusers alleged improprieties in the reporting of the psychological data from the same study (Donnay, 1994b).

Daniell has discussed these issues and events at scientific conferences in the hopes that, in the future, such naked disregard for travesties of the academic process and attempts to stifle freedom of expression of intellectual ideas are not tolerated. He notes that the institutional procedures for responding to allegations of misconduct generally include no safeguards against abuse of the process by accusers (Daniell, 1996):

> "Depending on the institution, and depending on the stage of the institutional review process, accused investigators may have no or limited entitlement to the constitutional rights that are specified for judicial proceedings. Investigators' records can be seized without notice and without proof of sufficient cause; and investigators can be denied rights of due process, expedient review, legal representation, rules of evidence, confidentiality, or protection against multiple jeopardy. Scientists who are accused of misconduct can suffer ongoing or permanent damage to their professional reputations, particularly while 'gagged' from public discussion during the institutional review, and even if ultimately exonerated."

Simon, Daniell, and colleagues were cleared on all charges of misconduct in their research. One would hope that research oversight agencies and ethical review boards come to recognize the political agenda of EI advocates who instigated this process and the dubious credibility of their accusations. One would also hope that the risk of harm to the reputations of scientists and the personal pain and suffering of false accusations be considered by these agencies before unleashing the bureaucratic processes of an investigation. What I find particularly revealing is that EI advocates continue to rely on interpretations of immunologic data obtained from laboratories which lack credibility in the scientific community.

2. Contradictory facts may be recognized, but their relevance to the issue questioned

EI patients who have retreated to the confines of forests, deserts, or enclosed "safe houses" are considered to be too extreme, not representative, by EI advocates who would distance themselves from the clinical ecologists who foster such isolation. They argue that "real" EI sufferers are not so extreme in their beliefs and behaviors. This is a transparent attempt to establish a politically correct posture. It should be interpreted within the context of prior advocacy.

I have one memorable example of a "sick building" case I was involved with (Bahura et al. v. S.E.W. Investors et al.; summarized in Appendix B). The plaintiffs were Federal EPA workers at the Waterside Mall Office complex in the Washington, D.C., area who were alleging some extremely bizarre symptoms which EI advocates had diagnosed as a form of chemical encephalopathy. The symptoms were attributed to contaminated indoor air, particularly chemicals emanating from carpet. The facts of the case showed no objective scientific evidence to support these claims. One reason the jury felt the plaintiffs were entitled to some compensation was that the workers were victimized by the belief that they had suffered chemical sensitivities. However, "belief of harm" was not part of the theory of harm argued by plaintiff attorneys. One particular strategy employed by the plaintiff attorneys warrants comment. During preliminary hearings and depositions, some plaintiffs presented with typical paraphernalia (mask, oxygen cannula, wheel chair, walking cane, water bottles, etc.) and disrupted the proceedings periodically, claiming ill effects due to a chemical reaction. During the trial, these plaintiffs were restricted from the courtroom, at the request of their own attorneys. What motivated the attorneys is not clear. I suspect they did not want the jury to see this unusual behavior. After all, their contention was that such behavior resulted from EI rather than any premorbid psychiatric disorder.

3. Interpreting contradictory evidence in line with one's own hypothesis

Environmental illness advocates sometimes maintain that psychological symptoms are caused by chemical sensitivity brought on after an alleged exposure event; however, studies of EI patients have shown that the symptoms of somatization, depression, and anxiety, as well as personality disorders, are often pre-morbid to any alleged toxic event. To preclude the possibility that psychological factors are necessary and sufficient to explain the symptoms, the interaction explanation suggests that psychological vulnerabilities are predisposing factors for EI.

4. One particular way of reinterpreting contradictory evidence is by means of the *ad hoc* hypothesis

When subjects do not respond consistently to double-blind, placebo-controlled provocation challenges, EI advocates are quick to explain negative findings with added postulates about conditional effects of previous experience. For example, according to the adaptation/deadaptation theory, reaction to low-level exposure can only be detected by the person if he has been "detoxified" in a clean environment for several days or weeks before provocation challenge. What is glossed over is that there may also be false positives observed (reactions to placebos), something that should not occur under these "most purified" of environmental conditions.

5. Identifying facts with a theory that raises controversy in its own right

In the clinical cases of EI patients we began reporting in 1986, there was an inordinately high incidence of self-reported childhood physical and sexual abuse. During a conference presentation, Meggs suggested that the childhood sexual abuse data were not credible, but reflected false memories either fantasized by the patients or implanted by me (Bill Meggs, panel discussion, AOEC Conference, Washington, D.C., 1991). This was an understandable emotional reaction, in light of the atmosphere of incredulity surrounding the "false memory syndrome". What Meggs failed to mention is that many of these patients had continuous memories of their abuse since it occurred, not "repressed" memories. I defer to the scientific processes of verification and replications of reports of childhood abuse which have begun in independent clinics (Kehrl, 1995). Furthermore, the interaction version of the toxicogenic theory now accepts the childhood abuse data but proposes that childhood abuse is a predisposing factor for acquiring toxicologically mediated EI.

6. We are the exception to the rule

I have added this sixth cognitive distortion to Henle's list. A negative finding is not questioned but accepted as true. Despite it, however, a postulate is still held out to be true because there is a unique subpopulation postulated which is the exception to the rule. This cognitive distortion was demonstrated in Rosenthal's study of induced experimenter bias in which rats from the same genetic strain were characterized as dull or bright to different groups of graduate students who evaluated their performance in identical tasks. As expected, the performance ratings correlated with the characterization of the animals' innate intelligence. Rosenthal reported an observation which shows how resilient beliefs can be. When experimenters who had run "dull" rats were told that their subjects were really not dull at all, their uniform reaction was, "How very interesting that you took in those other experimenters. Our rat, however, was obviously really dull" (Rosenthal, 1963).

Adversaries

EI advocates suggest that the difference between traditional medicine and their diagnosis and treatment procedures is simply a matter of approach. In fact, EI is much more than a simple matter of two research programs in science with incompatible postulates at their respective cores. Scientific medicine and toxicology have met the methodological criteria of a scientific research program; clinical ecology and the toxicogenic theory have not. The issue is not a simple intellectual disagreement among scientific "ladies and gentlemen". Rather, this represents an intellectual war between the rational progression of scientific knowledge based on methodological evidence, on the one hand, and the unfounded speculation and unsubstantiated methodology of pseudoscience on the other. Historically, this can be construed as a repetition of the differences between scientific and metaphysical investigation or the difference between scientific civilization and anarchy. This fundamental incompatibility makes it impossible to have meaningful dialogue between a scientifically based community and EI advocates, much less approach problems of common interest in a cooperative enterprise.

The opposing factions in the controversy over EI appear to have taken positions which stagnate development of a solution. Most position statements from the reputable medical and scientific community place the burden of justification squarely on the shoulders of the EI advocates, as exemplified by this quote from the statement by the American Medical Association, Council of Scientific Affairs (1992):

> "Those who support a new test, procedure, or treatment must prove, by appropriate controlled trials, that it is effective for the purposes for which it is used. The burden should not be shifted to opponents to prove that a new test or therapy is invalid."

On the other hand, advocates of the clinical ecology movement argue "prove us wrong".

Practically, these extreme positions have created a standoff. In science, such a standoff usually forces the advocates of a new position to establish credibility by accepted standards or to go away. In the complex social, personal, and political milieu in which EI is promoted, the standoff has quite another effect; it perpetuates pseudoscience. In non-science arenas such as the media, the courts, and legislatures, the disagreement is being portrayed as a showdown among scientific equals.

Historically, the position of justification (that theories are provable) was linked to empiricism, which meant accumulating data which support the theory. Keeping in mind that experimental results cannot prove a theory, data nevertheless can demonstrate the existence or non-existence of hypothesized events logically implied by the theories of a research program. The clinical ecology movement has not presented experimental evidence meeting credible peer review which could be construed as inductive justification for a hypothesis or postulate of their theories. For so long, EI advocates have been able to foster the illusion that they represented the position of authority in regards to environmental illness. Those who opposed their views were label skeptics and their arguments ignored. Offering no competently peer-reviewed research data to substantiate their theory, EI advocates left it up to the disbelievers to prove them wrong. In essence, they would have the positions of the established and revolutionary paradigm reversed, with legitimate science cloaked in the mantle of heresy.

The clinical ecology movement's "prove us wrong" position is not consistent with Popper's algorithm of conjecture and refutation. Popper was referring to what should personally motivate each scientist in formulating experiments about their hypotheses and theories. By designing careful experiments which refuted hypotheses of well-conceived theories, the theories are refined and enriched. Popper was not advocating the "we conjecture and you refute" position of the EI advocates. However, psychogenic theory has not offered a testable conjecture or hypothesis, one which if shown to be without empirical corroboration would imply the refutation of the underlying theory. Selner's (1991) critical review of Ashford and Miller's *Chemical Exposures: Low Levels and High Stakes* points out that the authors unwittingly reveal the critical flaw in their clinical ecology theories: "They are not testable." If the theories are not testable, then no rational, scientific research program is applicable.

EI is without scientific foundation and is a pseudoscience that has been characterized as "practice by hypothesis" (Gots, 1993b). While in agreement with the direction of Dr. Gots' characterization, I contend that the term "hypothesis" is too generous and does not go far enough when used in the context of the toxicogenic theory. The *American Heritage Dictionary* defines "hypothesis" as a tentative explanation that accounts for a set of facts and can be tested by further investigation. Where are the facts? Have the EI advocates put forward hypotheses that are testable? I believe a better characterization of EI is "practice by postulates". The *American Heritage Dictionary* defines postulate as something assumed without proof as being self-evident. EI advocates hold up the hard-core postulates of the toxicogenic theory to be both true and self-evident. Clinical ecologists continue to employ unsubstantiated practices in the face of scientific evidence that has shown them to be bogus.

Paradigm shifts

Kuhn (1970), in *The Structure of Scientific Revolutions*, suggested that "normal science" is a time-relative term used to describe a theoretical and methodological framework called a "paradigm". Normal refers to what society adopts at a particular period, making normal science subject to social consensus. Kuhn attempted to restrict his definition of the social meaning of a scientific paradigm to the members of a scientific community. In retrospect, it seems unrealistic that he thought he could exclude the socio-political members from the structure of the community when he made it relativistic. With respect to EI, the "community" is made up of advocates and victims who share the unsubstantiated beliefs of the harmful effects of low-level chemical sensitivities. According to Kuhn, a paradigm is characterized by a group of proponents who are kept busy by addressing unanswered questions and applying the paradigm. A scientific revolution occurs when a new paradigm displaces the old one. But that displacement process is not logical or absolute, but instead is relativistic as it depends on social consensus rather than scientific fact alone.

Kuhn describes the situation when an entrenched paradigm is faced with embarrassing data it cannot explain. The area of the anomaly is extensively investigated, and new postulates and theories are proposed to accommodate the inconsistent findings with the intent of protecting the core postulates and preserving the validity of the paradigm.

In a degenerative research program, so characteristic of pseudoscience, speculation replaces investigation. Typically, these elaborations and accommodations are unsuccessful. Nevertheless the prevailing, but deficient, paradigm survives. The proponents of the paradigm go on with business as usual, ignoring the contradictory evidence. The revolution takes hold when others break with the paradigm and develop an alternative, new paradigm which can account for the incongruent results that could not be assimilated by the existing theories of the old paradigm. The struggle between the paradigms may go on for years, and its final resolution, if it comes, is based not on logical analysis or irrefutable empirical results, but rather on the mortality of the proponents of the older paradigm. The older paradigm is typically not relinquished by those who developed and worked in it. These advocates refuse to understand the new paradigm, while at the same time arguing against its validity, arguments unfounded in evidence. According to the rational analysis offered by Kuhn, more members of the community come to realize the superiority of the new paradigm and adopt it, abandoning the old paradigm. A state of normal science prevails until the next round begins.

From this vignette, it should be clear that EI advocates are reversing the roles of normal science and scientific revolution. EI advocates would have the toxicogenic theory be the prevailing paradigm of normal science, determined by the social milieu of advocates, victims, exploiting physicians, opportunistic trial lawyers, confused policymakers, and constituent-minded politicians. Traditional scientists who have ample data to discredit the EI paradigm are cast in the role of "revolutionaries". This role reversal has historical irony. Were the toxicogenic theory to become the paradigm of normal science, the new paradigm which would then have to displace it is in fact the one that has been progressively developed in traditional science since the Renaissance.

This example highlights a criticism Lakatos (1970) expressed about Kuhn's description of the evolution of science, i.e., social relativism. Lakatos rejected the idea that scientific anarchists can affect the progression in science, as suggested by Kuhn. EI advocates may not affect progression or degeneration among the scientific community, and in that context Lakatos seems to be correct. But Kuhn has expanded the concept of paradigm beyond the meaning of research program by opening up membership in the "community" structure of science to non-scientists, with the only initiation requirement seemingly being a hypothesis or postulate. In the extreme, the paradigm of science would be determined by sociocultural forces alone (Bloor, 1976). In that context, Kuhn's analysis may be closer to the truth with respect to EI. For society, there is always the risk that the advocates of a degenerative research program may get control of the prevailing paradigm through nonscientific means.

Lakatos summarized the differences between Popper and Kuhn on the topic of scientific paradigms, research programs, and paradigm shifts (Lakatos, 1970):

> "Kuhn objects to the entire Popperian research programme, and he
> excludes any possibility of a rational reconstruction of the growth of
> science. In a succinct comparison of Hume, Carnap, and Popper,
> Watkins points out that the growth of science is inductive and irra-
> tional according to Hume, inductive and rational according to Carnap,
> non-inductive and rational according to Popper. Watkins' compari-
> son may be extended by adding that it is non-inductive and irratio-
> nal according to Kuhn. In Kuhn's view there can be no logic, but only

psychology of discovery. For instance, according to Kuhn's conception, anomalies, inconsistencies always abound in science, but in normal periods the dominant paradigm secures a pattern of growth which is eventually overthrown by a crisis. There is no particular rational cause for the appearance of a Kuhnian crisis. Crisis is a psychological concept; it is a contagious panic. Then a new paradigm emerges, incommensurable with its predecessor. There are no rational standards for their comparison. Each paradigm contains its own standards. The crisis sweeps away not only the old theories and rules but also the standards which made us respect them. The new paradigm brings a totally new rationality. There are no super-paradigmatic standards. The change is a bandwagon effect. Thus, in Kuhn's view, scientific revolution is irrational, a matter for mob psychology."

The reasons proposed for the toxicogenic theory are either foreign or unconvincing to the scientific way of thinking, or they rely on evidence and allude to theories within the existing paradigm of science. I believe the toxicogenic theory meets Lakatos's definition of being non-inductive and irrational. Furthermore, the movement behind EI is not characterized by scientific truth but rather by faith and conviction, self-serving interests, and the abuse of power and politics in a benevolent and permissive society. The multi-system symptoms of EI patients can be explained by known mechanisms and processes in medicine and psychology. EI advocates have no basis on which to claim toxicogenic theory should be the prevailing paradigm of normal science. The argument that such a shift would enhance our understanding of science and nature is a myth.

Endless rhetoric

> "'Absence of evidence is not the evidence of absence.' — The greatest pseudoscientific argument of all time!" (Rosenberg, 1996b)

Given the lack of evidence and implausibility of speculative theory, the only basis on which EI advocates perpetuate the toxicogenic theory is rhetoric. One presupposition of such rhetoric is that the discovery of evidence will come to pass in the future when the correct mechanism is hypothesized and technology becomes sophisticated enough to detect the hypothesized effects. But how much advancement in the technology of chemical detection is required? How much is enough? We can now measure the body load of a chemical to one part per quadrillion. The argument for a toxicologically harmful effect from such a minuscule amount of matter would be analogous to identifying that one dime lost in the lifetime of a 100-year-old woman who has made one billion dollars ($1,000,000,000) during every year of her life.

Another line of rhetoric is that "it is not logically impossible" which is not to be confused with the argument that "it is logically possible". This is more than a play on words, because there are different implications as to who has the burden of proof. The argument for something that is not logically impossible can be made for the existence of aliens, despite the fact that all the presented evidence of sightings or "close encounters" have failed to produce the object. This is particularly relevant to litigation, where the admissibility criteria of expert testimony is *probable*, defined as "within a reasonable degree of medical, scientific, or psychological certainty."

A fundamental tenet of science is that negative evidence cannot prove the null hypothesis that there is no effect. That is to say that failure to find an experimental effect does not mean there is no effect. There are sound, defensible reasons for this principle in science. Shortcomings of methodology or research design should not prematurely close off speculation and hypotheses about a phenomenon. Scientific principles to ensure that premature rejection of a theory does not occur should not be confused with sociopolitical or legal principles motivated by concern to protect and uphold freedom of expression and dispel unfair restraint of the individual. The principles are similar, but their context is different. What they share in common is that both may be subject to distortion and abuse. The blurring of boundaries between these contexts has formed the basis of many EI advocate arguments. In the sociopolitical context, there are EI patients who claim victimization at the hands of an establishment conspiring to harm them with environmental chemicals and pollution, the Gulf War syndrome for example. In the context of science, EI advocates contend that further research is necessary to demonstrate the illusive but harmful effects of low-level chemical exposures. What is disturbing is that the political agenda is influencing decisions in the context of science. At the time of this writing, these dynamics are being played out in several contexts including "sick building syndrome" and the Gulf War syndrome.

Expressing dismay at such behavior is minimally, if at all, effective in generating change in the context of the greater society. Negative evidence, although influential in the scientific world, may have little impact in a sociopolitical world. The number of studies which have demonstrated negative effects in the assessment of clinical ecology theory and practice are numerous and ever growing. Nevertheless, the failure to find supportive evidence has not impeded toxicogenic speculation. More alarmingly, the number of EI patients seems to be growing.

Model for a progressive research program

A more effective scientific strategy is one that avoids the trap of the null hypothesis, i.e., it is not possible to prove that something does not exist. That strategy entails a research program with testable hypotheses and research design which generates empirical evidence with unambiguous implications about the meaning of the results. There can be no compromise on methodological rigor and reliable and valid measurement in the accumulation of experimental data. Nor can there be any compromise on scientific honesty in the formulation of theory. These are deemed necessary conditions in the evolution of scientific knowledge.

Careful consideration should be given to alternative medical, toxicological, and psychological diagnoses in individuals who present with multi-system symptoms attributed to EI. To repeat, there exist well-documented diseases whose etiology and exacerbation are attributable to environmental toxins that exert their effects through measurable neurologic, allergic, endocrine, dermatologic, respiratory, cardiac, and other physiological mechanisms. These disorders have objective signs and identifiable biological markers. Their symptoms are elicited consistently by exposure. They have specific symptoms that are replicable. And the magnitude of the symptoms follows the dynamics of dose-duration response. Traditional medical science would do well to establish research programs to study possible effects of low-level chemical exposure on physiological systems and mechanisms of disease or illness. This should be done not only for humanitarian motives, but also because in the evolution of scientific knowledge false theories are not laid to rest by refuting evidence alone. Rather, false theories must be supplanted by more powerful research programs which demonstrate a progressive problem shift. What follows is a proposed model to study EI in which the toxicogenic and the psychogenic theories make different predictions about outcome in the same experiment.

Objective measurement and biomarkers

Miller (1997) observed that there is "no single biomaker" common to a variety of diseases such as cholera, AIDS, or shingles. To argue that there is "no single biomaker" is a truism, not to be confused with disease processes with unique biomarkers. "No single biomarker" is not the same as saying "not a single biomarker in any physiologic system has been identified for EI."

In addition to subjective appraisal, objective physiologic, behavioral, and biologic measurements can be incorporated in the experimental design. For purposes of illustration of the model, the hypothesized effects of a putative toxic agent is represented by a binary hypothetical dependent variable that shows a response or not. Under the null hypothesis, there are no expected significant differences.

The ideal dependent variable that could serve as a biological marker is one that is sensitive to a reliable response, defined by a well-characterized biological mechanism, and specific to provocation challenge with a chemical agent. The ideal measure is also unaffected by placebo, or at least less so than when exposed to the chemical agent. Finally, the measure should be distinguishable from psychophysiological and biochemical effects associated with long-term, pre-morbid psychopathology or state stress-responses. The history of medicine is replete with examples of such markers in well-defined infectious diseases that have been validated by adherence to Koch's postulates of identification, isolation, infection, and retrieval. However, in the poorly defined environmental milieu of EI, where low-level chemical sensitivities to a multitude of undefined chemicals and chemical mixtures from indoor and outdoor sources are at issue, the identification of reliable and valid biological markers remains to be determined (Straight and Vogt, 1997).

Certain features of laboratory methods used to measure biologic markers in epidemiologic studies can be assessed statistically, e.g., variability inherent to measurement and human error in laboratory analysis and indices of central tendency of the distribution of scores in different populations (Vineis et al., 1993). Experimental design can be employed to identify and remove effects due to individual differences by the use of between-subject control groups, as is done in case-control and cohort-control studies. The effects of depression and anxiety on a dependent variable, for example, can be estimated by using positive control groups of individuals with these conditions, either independent or co-morbid. However, epidemiological studies are most effective when a well-defined target effect is being studied. Such population designs, when applied to an ephemeral phenomenon such as EI, may be more difficult to interpret and only add to the confusion.

Another experimental approach is to use individuals as their own controls — the single case study design (Kazdin, 1976, 1978). For example, Jewett and colleagues used this approach in their double-blind study of subcutaneous provocation testing with yes-no appraisal of symptom responses (Jewett et al., 1990). With this design, the inherent variability of a dependent variable that is measured on an ordinal scale can also be determined under various conditions of exposure to a putative environmental agent and to placebo. The variance of the dependent variable under conditions of placebo response may be used as an estimate of its inherent variability when employed as a biological marker. Ideally, the effect of a chemical on the dependent variable would be significantly above and beyond that seen under placebo-control, as is illustrated in pharmacologic clinical-trials research. A dependent variable that shows statistically reliable differences

Table 17.1 Examples of Dependent Measures

1. Hypothalamic-pituitary-adrenal axis — neurotransmitters, immune parameters, and neurohumoral modulators (e.g., catecholamines, histamine, cortisol)
2. Central nervous system changes in activation (e.g., neuroimaging such as QEEG, PET, SPECT, functional MRI)
3. Autonomic nervous system arousal (e.g., heart rate, respiration, blood pressure, peripheral blood flow)
4. Neuropsychological/neurobehavioral task performance (e.g., speed and accuracy of cognitive and motor functioning)
5. Objective signs of physiologic impairment (e.g., pulmonary functions, rhinometry)
6. Laboratory assay (e.g., immunomodulators)
7. Perception/appraisal (e.g., subjective symptom ratings)

between a chemical provocation and a placebo within one individual would suggest a toxicologic effect. If such a variable produced reliable effects in a series of single case studies, it would warrant further study in a larger, population-controlled epidemiologic study.

In practice, the actual measures used as the dependent variables to test a specific toxicologic hypothesis may be selected from a menu that reflects different levels of analysis of behavioral, physiological, and biochemical responses (Table 17.1).

Predicted outcomes for competing theories

The appraisal-exposure contingency table discussed in Chapter 5 is reproduced in Figure 17.1. The pattern of provocation challenge results expected according to the predictions of the two explanations of the toxicogenic theory (toxicologic and interaction) and the cognitive/belief explanation of the psychogenic theory are presented in Figure 17.2. The Pavlovian classical conditioning explanation for the psychogenic theory is not considered because the blinded procedure does not include detectable sensory stimuli which could serve as conditioned stimuli.

True positive

If the subject correctly appraises and reports a reaction to the presence of the non-perceptible chemical agent, the response is classified as a true positive (TP). All three of the explanations presented predict that a TP response should occur.

EI advocates who postulate that adaptation/deadaptation must be considered (Ashford and Miller, 1991; Miller, 1997) predict a TP response when the patient is "deadapted" but not if the patient is "adapted" (symptoms are masked). These contingencies are determined *ex post facto* depending on the response of the patient, a circular logic and a scientifically unacceptable explanation. To overcome this circularity, it has been suggested that a patient be induced into a state of "deadaptation" through prolonged isolation from any putative toxicologic agents in an ECU. In our clinical experience with EI patients who were tested with double-blind, placebo-controlled chamber challenges while inpatients in our ECU at Presbyterian Hospital in Denver between 1980 and 1985, subjects did not demonstrate a TP response to putative food or chemical agent consistently even though they had been in the ECU long enough to be "deadapted" (Selner and Staudenmayer, 1986).

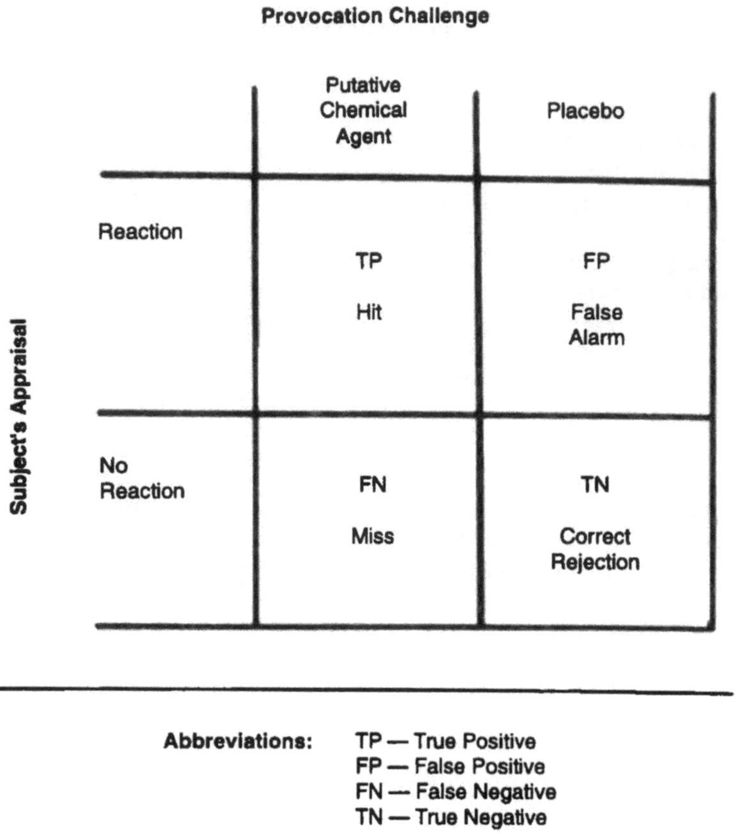

Provocation Challenge

	Putative Chemical Agent	Placebo
Reaction	TP Hit	FP False Alarm
No Reaction	FN Miss	TN Correct Rejection

(Subject's Appraisal — vertical axis label)

Abbreviations: TP — True Positive
                  FP — False Positive
                  FN — False Negative
                  TN — True Negative

Figure 17.1 Classification of responses from double-blind, placebo-controlled provocation challenges. (From Staudenmayer, H., *Regul. Toxicol. Pharmacol.*, 24, S96–S110, 1996. With permission.)

From a practical point of view the requirement for prolonged isolation is inconsistent with typical observations and the histories provided by EI patients. When a patient complains that he is in a particular exposure environment and a constellation of symptoms occur immediately upon detection of an odor or suspicion of the presence of an environmental agent, it is hard to argue for the "adaptation/deadaptation" scenario. If it were operative, ideally only a reaction which occurred when the patient was hovering at equilibrium between the stimulatory and withdrawal conditions would satisfy pre-challenge requirements (Selner, 1996). If a patient were in fact to be deadapted for seven days in an ECU, the first provocation challenge (even if it were a placebo and the patient reacted) could upset this equilibrium so that an additional 7-day deadaptation period would be required, leading to an endless spiral which precludes testing.

For practical reasons, the existence of the TP event should be verified with an open challenge. There is no point in initiating a series of time-consuming and costly double-blind, placebo-controlled provocation challenges when the subject does not demonstrate the symptoms when openly exposed to the putative agent. In the context of plaintiff litigation, it seems inconsistent to argue that EI patients are permanently impaired in their natural environment yet cannot reliably demonstrate the alleged toxic effects in a different (and presumably less safe) environment. From a toxicologic perspective, if the subject cannot demonstrate an adverse effect, what is the problem?

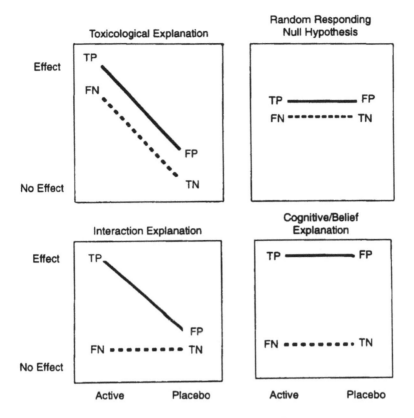

Figure 17.2 Appraisal of reaction (effect or no effect) and challenge agent (active or placebo) and predicted effects on a hypothetical dependent variable for the null hypothesis and competing explanations of EI. (From Staudenmayer, H., *Regul. Toxicol. Pharmacol.*, 24, S96–S110, 1996. With permission.)

False negative

If the subject does not appraise a reaction during provocation challenge with a putative toxic agent and shows no changes in objective measurements, the response is classified a false negative (FN). Based on that challenge, it can be concluded that there are no effects attributable to that dose of the putative chemical agent, and the strict toxicologic explanation is refuted.

The interactive explanation of the toxicogenic theory accounts for the lack of effect in the dependent variable with mitigating postulates, e.g., the subject is in a positive state of mind (or has adapted) such that hypothesized psychologic or psychophysiologic factors that are said to interact with the effects of the chemical agent have not lowered the sensitivity threshold sufficiently. That being the case, there should also be no FPs. Furthermore, were the patient to be sequestered and deadapted in a clean environment such as an ECU, the sensitivity would be heightened and therefore the FN should not occur.

The cognitive/belief explanation predicts no effect in the dependent variable. Under conditions of provocation challenge, the biological mechanism of the stress-response are assumed to be cognitively mediated and should not react when there is an appraisal of no reaction, as is the case in the FN contingency.

True negative

When the subject does not report an effect to placebo, the response is classified a true negative (TN). The TN response, like the TP discussed above, is not helpful in differentiating the explanations because it is consistent with all three. However, failure to demonstrate a TN means the subject's responses are not specific to the putative chemical agent being tested, or any other chemical agent for that matter. Failure to demonstrate a TN is inconsistent with the predictions of the toxicologic or interactive explanations. For practical reasons, demonstration of tolerance of the placebo is a necessary condition that should be fulfilled in open, unblinded challenges before costly and time-consuming series of double-blind, placebo-controlled provocation challenges are undertaken. A subject who cannot demonstrate tolerance under open conditions has an unrealistic response bias to respond whenever challenged, no matter what the circumstances, and generates mostly TP and FP contingencies, which obviates discrimination of the effects of the agent from the placebo, and the response pattern across trials lacks specificity. With respect to personal injury litigation, it would seem that lack of specificity makes it impossible to present a credible argument for an effect due to an environmental agent, much less to assign liability.

False positive

If the subject appraises a reaction to a placebo, the response is classified a false positive (FP). An observable effect is not expected according to the toxicologic explanation because subjective rating is considered irrelevant.

An observable effect in the dependent variable for the FP is consistent with the cognitive/belief explanation in that the subject has appraised a reaction based on belief of exposure. In that respect, the appraisal of a reaction is necessary and sufficient to explain that the alleged sensitivity to the environmental agent was in fact a placebo effect.

The interaction explanation predicts there is no effect in the dependent variable because there is no environmental agent, a necessary factor to trigger symptoms. Lowering of the sensitivity threshold by psychologic or psychophysiologic factors is necessary, but not sufficient to trigger a chemical sensitivity response to placebo. There is no basis on which to rationalize the response to placebo, unless one argues that there is no such thing as a benign placebo or that the person showed a delayed reaction to an environmental exposure unrelated to testing. There is at least one logical flaw in Miller's argument (Miller, 1997) that there are numerous intersecting sinusoidal time waves representing acclimatization, apposition, and addiction which determine the EI patient's susceptibility which where not considered in our challenge studies (Staudenmayer et al., 1993a). A challenge was initiated only when pre-symptom ratings were absent or minimal. Reactions occurred shortly after exposure, usually within minutes, not randomly over time as suggested by her model of intersecting curves. The argument for the need for prolonged ECU sequestration is not supported by our empirical findings.

Miller (1977) asks the obvious question, "What evidence is there that unmasking patients in an ECU and conducting challenges within a 4- to 7-day window of time is either useful or necessary"? Her answer:

> "Thousands of credible patients and dozens of physicians have attempted this approach. They report that patients' symptoms resolve within a few days after they enter such a facility and that robust symptoms occur when challenges are conducted after several days of avoidance."

The foundations of the latest renaming of the toxicogenic theory, TILT, are exposed: patients' subjective reports and clinical ecologists' pseudoscience.

Conclusion

With all the speculation about possible underlying mechanisms for EI, we cannot lose sight of the fundamental primary question, "Can the harmful effects of low-level chemical sensitivities be demonstrated to exist reliably?" Rationality would dictate that an effect be established before speculation about etiology and mechanism is undertaken. But, never mind; social and political pressures from EI advocates would have the validity of an effect be secondary, arguing that catering to the "disabilities" claimed by EI victims is primary. But, in a scientific forum there can be no tolerance for such pressure. Furthermore, accommodating a false treatment can do harm.

As theoretical postulates of the toxicogenic theory are translated into testable hypotheses following accepted scientific methods, it may come to pass that the phenomenon may not exist outside the collective consciousness of those who believe in it. EI may be no more than a network of advocates and victims who intentionally or unwittingly are more vested in holding on to their beliefs than in discovering truth.

Appendix A

A methodology of scientific research programs

The paradigm of normal science adopted in this book is guided by the philosophy of science, originating with Karl Popper (1968) and refined by Imre Lakatos (1970). The following is a reflection of some of their ideas which may facilitate discussion of the assessment of a research program with respect to EI. The basic problem is how to appraise scientific growth objectively in terms of progressive and degenerating problem shifts in a series of scientific theories. To begin, some terms used in the language of the philosophy of science require definition.

The "hard core" of a research program consists of the fundamental postulates, presuppositions, and assumptions held to be irrefutable by the protagonists of the research program. To maintain the research program, certain methodological decisions are made to preserve the hard core.

A "negative heuristic" refers to a methodological rule about what paths of research to avoid. This heuristic diverts evaluation to auxiliary hypotheses, a protective belt around the hard core which bears the brunt of tests and gets adjusted and re-adjusted, or even replaced, in defense of the hard core.

A research program is successful if the negative heuristic leads to a progressive problem shift, unsuccessful if it leads to a degenerating problem shift. A progressive problem shift is characterized by good experimental results that corroborate the hypotheses in the protective belt and therefore do not challenge the hard core of the research program. Restructuring and refinement of hypotheses are made to elaborate more sophisticated research protocols to better explain known phenomena and predict new ones implied by the hard core postulates. The primary indicator of progress in a research program is the extent to which it successfully explains known facts and evolves new hypotheses that predict new, meaningful outcomes.

Each successive link predicts some new fact, an increase in empirical content. It requires that each step of a research program be consistently content increasing. In the end, each prediction is verified (allows for momentary refutation and re-evaluation of methodology). To avoid a dogmatic position such that a single finding could refute the research program, a less stringent requirement is that the research program display an intermittently progressive shift.

A degenerative problem shift is characterized by stagnation, which occurs when theoretical growth lags behind empirical growth. A theory undergoing a degenerative problem shift gives only *post hoc* explanations of either chance discoveries or of facts anticipated by and discovered in a rival program.

Poor experimental results refute a hypothesis in the protective belt and do not corroborate the postulates in the hard core. The hard core can be protected with a negative heuristic which can take the avenue of proliferating new hypotheses (offensive strategy) or introducing *ad hoc* hypotheses (defensive strategy) that complicate and stagnate the research. This can have the effect of slowing the proliferation of new research implications from the core assumptions. The model will not be able to predict new empirical facts and there will be an overall reduction of assimilated knowledge.

To illustrate a degenerative problem shift, EI proponents imply that chemicals are unsafe because of some 60,000 widely used chemicals only 2% have been comprehensively studied for toxic effects. This implies that we can never exhaustively study all chemicals, and therefore the hard-core postulate that these chemicals cause EI can never be rejected. Furthermore, a second hard-core postulate would preclude the consideration of psychological explanations of EI. Advocates of EI contend that, "It is an axiom of medicine that, when investigating the origins of illness, physiological causes must be ruled out before ascribing psychological etiologies." If this argument were accepted, and taking into account that it is impossible to explore physiological causes exhaustively because of the endless speculations about mechanisms, it would follow that one can never entertain psychogenic explanations of EI.

A research program undergoing a degenerative problem shift may stagnate further until it contains nothing but solemn re-assertions of the original position coupled with a repetition of the successes of rival programs. When a research program ceases to anticipate novel facts, the hard core might have to be abandoned. In science, the bases for abandonment of a research program are logical and empirical. In pseudoscience, the bases are more likely to be socio-political.

A positive heuristic consists of a partially articulated set of suggestions or hints on how to change, how to develop the "refutable" protective belt. The advantage of the positive heuristic of the research program is that it saves the scientist from becoming confused by an ocean of anomalies and overwhelmed by the problem. The positive heuristic sets a program to develop a chain of ever more complicated models simulating reality. By definition, a model is a set of initial conditions which one knows is bound to be replaced during the further development of the program, and one even knows more or less how. A positive heuristic is generally more flexible than a negative heuristic, one reason being that it does not have the burden of protecting the hard-core postulates from refutation. The positive heuristic is directed by the search for truth. The negative heuristic is directed by the conviction that a theory is true.

Appendix B

Court rulings unfavorable to environmental illness[*]

Below are some court decisions excluding expert testimony about diagnostic and treatment methodology by advocates of EI.

1. **Bloomquist v. Wapello County, Iowa, et al.** (No. CL 0174-0687, Mahaska City, Iowa, District Court, August 28, 1990). The judge overturned a $1,000,000 verdict for two employees of a "sick building" ruling that plaintiffs's clinical ecology evidence was "unproven medical speculation which is not accepted by mainstream medicine." This was subsequently reversed on appeal by the Iowa Supreme Court which held that epidemiological evidence was not required (No. 419/90-1371, Iowa Supreme Court, April 21, 1993).

2. **Brandon v. First RepublicBank Group Medical Plan** (No. CA-7-89-002, Northern District, Texas, November 27, 1990). The court ruled that the services of clinical ecologists Drs. William Rea and Alfred Johnson were not medically necessary and therefore not covered under an employee welfare benefit plan.

3. **La-Z-Boy Chair Co. v. Reed** (No. 90-6013, 6th Circuit, June 28, 1991). The U.S. Court of Appeals for the 6th Circuit affirmed the trial court's decision to bar the testimony of plaintiff's clinical ecologist, Fred Furr, that the plaintiff was permanently disabled as a result of exposure to trichloromethane at work. The court held that such testimony was "only a theory which is not generally accepted by the medical profession."

4. **Newman v. Stringfellow** (No. 165994, California Superior Court, Riverside County, January 17, 1991). The trial court ruled that plaintiffs' immune assays, including calla and porphyrin antibody testing performed by Bertram Carnow, were inadmissible, as plaintiffs failed to prove that the testing was "acceptable to at least a substantial minority of the relevant scientific community."

---

[*] Source: With the assistance of Timothy Kapshandy, J.D., of the law firm of Sidley & Austin, Chicago, IL.

5. **Rea v. Aetna Life Insurance Co.** (No. 3-84-0219-H Northern District, Texas, February 25, 1985). The Federal court rejected plaintiffs' attempt to bring a class action on behalf of clinical ecologists and their patients against Aetna and the American Academy of Allergy and Immunology, holding that plaintiffs failed to establish that clinical ecologists and their patients were a "recognizable and identifiable class."

6. **Schickele v. Rhodes** (No. C 451843, Arizona Superior Court, Marcopa County, August 1, 1986). The court excluded the clinical ecology testimony of Alan Levin, M.D., suggesting "immune dysfunction".

7. **Sterling v. Velsicol Chemical Corp.** (855 F.2d 1188, 6th Circuit, 1988). The U.S. Court of Appeals for the 6th Circuit excluded from evidence the clinical ecology testimony of Alan Levin, M.D., as generally unaccepted, based in part on the position papers of the American Academy of Allergy and Immunology and the California Medical Association, and reversed an award of damages for injuries to plaintiff's immune system.

8. **Taylor v. Airport Transport and Warehouse Services, Ltd.** (No. 90/NJ/5076, High Court of Justice, Queen's Bench Division October 24, 1991). This English court rejected the claim of plaintiff's clinical ecologist that her multiple chemical sensitivity was triggered by exposure to chemical fumes in a truck she was driving, holding that "her evidence was in many respects bizarre and unscientific ... [and] unacceptable to the vast majority of doctors."

9. **Carroll v. Litton Systems** (No. B-C-88-253, Western District, North Carolina, October 29, 1990; reversed on other grounds — No. 92-2219, 4th Circuit January 13, 1995). This Federal Court excluded the lymphocyte testing and autoantibody testing of Dr. Alan Broughton as lacking a proper factual basis (i.e., no proper controls, alternative causes not excluded).

10. **Bradley v. Brown** (No. CIV-H85-958, 1994 WL 199827, Northern District, Indiana, May 17, 1994; affirmed — No. 94-2467, 7th Circuit, December 13, 1994). Two Federal courts excluded testimony of two clinical ecologists, Drs. William Rea and Alfred Johnson. The trial court found their methodology anecdotal and speculative. As to the general concept of EI, the court held that scientific knowledge about the etiology of EI has not progressed from hypothesis to knowledge capable of assisting the jury.

11. **Mullenax v. McRae's** (No. 87-13915-D-3130, Miss. Workers' Compensation Commission, March 18, 1993). The Commission denied a multiple chemical sensitivities claim allegedly resulting from workplace exposure to solvents in art supplies. The Commission found that the unorthodox methodology of Dr. William Rea did not establish causal connection. Even if they were to accept this theory that exposure to one chemical can cause multiple chemical sensitivities, there were other legitimate explanations that were not excluded.

12. **Maritime Overseas Corp. v. Ellis.** (No. C14-91-00795-CV, App. 14th Div., Texas, December 31, 1992). The Appellate Court reversed the $12.6 million judgment for plaintiff in the Jones Act case where the plaintiff alleged that a one-time exposure to diazinon caused delayed sundry neurotoxic effects. The Texas Court of Appeals held that the methodology of plaintiff's experts (Drs. Alfred Johnson, F. Waikman, and R. Austin) was speculative and not sufficiently validated to establish a causal connection.

13. *In re* Paoli R. R. Yard PCB Litigation (35F 3rd 717, 3d Circuit, 1994). The 3rd Circuit upheld the exclusion of the causation opinion of Dr. Sherman for those plaintiffs on whom she did not perform the traditional clinical method (i.e., exam, history, etc.), but allowed it for those on whom she did. The court also excluded the immunological testing of Dr. Alan Broughton.

14. **Claar et al. v. Burlington Northern Railroad** (No. 92-35337, 92-35539, U.S. District Court, Montana; 9th Circuit Court of Appeals, July 14, 1994). Six plaintiffs were selected from 27 cases of railroad workers filed under the Federal Employees Liability Act (FELA) suffering from unspecified multiple chemical exposures. U.S. District Court of Montana provided summary judgment for the railroad because experts (Drs. Mark Hines and Richard Nelson) failed to explain the basis of their multiple chemical sensitivities diagnosis and failed to specify which chemicals caused which injury; moreover, neither expert made any effort to rule out other possible causes. Plaintiffs argued that the court had erred in demanding that their experts demonstrate a causal connection between specific injuries and specific chemicals. The appellate court upheld the lower court: "This argument misconceives both the standards for causation under FELA and its relationship to the Federal Rules of Evidence."

15. **Benny v. Shaw Industries, Inc.** (No. 93-685-CIT-T-21(A), Middle District, Florida, 1995). The Court excluded the opinion of Dr. Staninger that plaintiff's EI was caused by carpeting and a "bug bomb" as his methodology was unreliable. The court also excluded as unreliable the testing of Dr. Alan Broughton's laboratory as not the type reasonably relied upon by experts in the field.

16. **Ardith Cavallo v. Star Enterprise et al.** (No. 94-1499-A, Eastern District, Virginia, 1995). Plaintiff claimed that she had chronic respiratory illnesses through exposure to aviation jet fuel (AvJet) while walking across a parking lot of a restaurant about 500 feet away from a distribution facility where a 34,000-gallon spill had taken place. The court concluded that the opinion of the plaintiff's expert, Joseph Bellanti, M.D., was largely based on hypothesis and speculation. In granting summary judgment, the judge stated: "It may well be that the AvJet spill forever 'sensitized' Ms. Cavallo to petroleum vapors and various household chemicals. But the published scientific literature and test results simply do not support that conclusion at this time."

17. **Carlin v. REF Industries et al.** (No. 88-CV-842, Northern District, New York, November 27, 1995). Plaintiff asserted that while working in the printing department she was exposed to isopropanol alcohol, Freon TMC containing methylene chloride, surface cleaners such as Kleanz-Easy, bar solder such as Electroalloy, liquid flux, flux thinner, and flux cored wire, all of which caused various physical disorders including toxic encephalopathy. Judge Daniel Scanlon, Jr., excluded the testimony of Drs. James Miller and Michael B. Lax who had diagnosed multiple chemical sensitivities for lack of reliability of the tests and absence of corroborating physical or neurological findings. Additionally, the Magistrate said he had serious doubts about the scientific validity of the MCS syndrome.

18. **Kristin Phillips v. Velsicol Chemical Corporation** (No. 93-CV-140-J, District of Wyoming, September 19, 1995). Plaintiff was a timpanist with the Hong Kong Philharmonic Orchestra who alleged multiple chemical sensitivities resulting from

exposure to pesticides (chlordane) applied in the concert hall. The court excluded chlorinated pesticide screening tests performed by Dr. Robert K. Simon of Accu-Chem Laboratories because they were scientifically unreliable and not trustworthy and failed to follow established protocol. Dr. William Rea's opinion for the harmful effects of chlordane on the plaintiff by "double-blind" tests at the Environmental Health Center-Dallas were deemed irrelevant for lack of specifically identifying chlordane in the alleged incident in the concert hall.

19. **Summers and Potts v. Missouri Pacific Railroad System** (No. 94-468-P, U.S. District Court, Eastern District, Oklahoma, August 25, 1995). Railroad employees alleged long-term harmful effects as a result of short-term exposure to diesel exhaust fumes. Magistrate Judge Payne ruled that employees' experts' diagnosis was not that of "chemical sensitivities" in the generally accepted sense, but rather synonymous with "multiple chemical sensitivity" the theories of which were not shown to be adequately tested. The court excluded Dr. Alfred Johnson testimony, citing a case prior to Daubert (Viterbo v. Dow Chemical, 826 F2d 420 424, 5th Circuit Court, 1987) in which the Fifth Circuit found "Dr. Johnson's testimony is no more than Viterbo's testimony dressed up and sanctified as the opinion of an expert." The testimony of psychologist Susan Franks, Ph.D., was also excluded, in part because the psychological test had not been validated, and three physicians, including the employees' cardiologist, emphatically stated that there were no long-term effects due to short-term exposure to diesel exhaust.

20. **Bahura et al. v. S.E.W. Investors et al.** (No. 90-CA-10594, District of Columbia Superior Court, November 29, 1995). Judge Rufus King tossed out four of five plaintiff verdicts in a "sick building syndrome" action brought by selected plaintiffs from a pool of 19 Environmental Protection Agency workers at the Waterside Mall Office Complex. Plaintiffs asserted multiple chemical sensitivities and toxic encephalopathy appraised to be caused by building renovations. During the trial, four of the five awards were allowed to comply with a procedural issue raised by the jury who awarded based on the plaintiffs' belief that they were injured. The defendants appealed. The four jury awards were set aside in the subsequent ruling by Judge King on the basis that "no treating physician testified, and no expert evidence linked brain abnormalities with emotional or psychological injury." Plaintiffs' witness, Dr. Iris Bell, even testified that psychological causes could be ruled out. Dr. Bell's testimony on olfactory limbic kindling hypothesizing somatization as an effect of chemical sensitivities was excluded as unreliable. "Dr. Bell conceded the olfactory limbic kindling model is not generally accepted in the fields of psychiatry or neurology, and that low-level exposure to everyday chemicals does not cause permanent injury. ... Dr. Bell's own testimony showed that no published data support the Plaintiffs' theory that low-level exposures cause permanent toxic encephalopathy in human beings."

21. **Conradt v. Mt. Carmel School Fireman's Fund Insurance Commission** (No. 94-2842, Wisc. App. 2nd Dist. Sept. 27, 1995). The Wisconsin Court of Appeals upheld the Labor and Industry Review Commission's denial of plaintiff's claim (based on the opinion of Dr. Theron Randolph) that building materials at the school where she worked had caused her to develop MCS. The appeals court rejected claimant's contention that her treating physicians should be accorded more credibility than the employer's experts.

22. **Kuehm v. Hearnen Air Conditioning** (No. A-4289-93T3, N.J. Super., App. Div., July 13, 1995). Plaintiff brought a "sick building syndrome" case alleging mite and fungal allergies due to a defective ventilation system. The trial court summarily dismissed the case, holding that her experts' speculation about conditions 4 years previous was not competent evidence.

23. **Hundley v. Norfolk & Western Railway Co.** (No. 91C 6127, N.D. Ill. Jan. 31, 1996). The court excluded the opinions of Drs. William Rea and Alfred Johnson that plaintiff's one-time exposure to herbicides at a railyard was the cause of his MCS.

24. **Nancy Rutigliano v. Valley Business Forms, et al.** (No. 90-1432, D.N.J., June 27, 1996). The court excluded the opinions of Elaine Panitz, M.D., and Thaddeus J. Godish, Ph.D., that exposure to carbonless paper had made plaintiff sensitive to formaldehyde. The court noted that Panitz was basically a full-time witness who made her diagnosis after an initial visit, based on self-reported symptoms and history. The court also rejected her reliance on blood tests, done in Dr. Alan Broughton's lab, which she had accepted if supportive but dismissed if negative.

25. **Nethery v. Servicemaster Co.** (No. 92-167(G)(L), Miss. Cir. Ct., Lee Co., Feb. 15, 1996). The trial court excluded the testimony of Drs. Glasgow and Lieberman, holding that MCS is an "unproven theory".

26. **Sanderson v. International Flavors and Fragrances et al.** (No. CV-95-3387, C.D. Cal., Aug. 28, 1996). A federal judge summarily dismissed plaintiff's claim that exposure to perfumes and colognes over an 18-month period has caused her to develop MCS, toxic encephalopathy, and impairment of her sense of smell. The court held that the testimony of Drs. Nachman Brautbar, Gunnar Heuser, Richard Perillo, and Jack Thrasher was not sufficient to establish that her symptoms were caused by defendant's fragrance products. The judge also ruled that the plaintiff had failed to demonstrate that MCS is "good science".

27. **Valentine v. Pioneer Chlor Alkali** (No. CV-S-92-0887-ECR, D. Nev. April 12, 1996). Plaintiffs alleged that they suffered neuropsychological injuries from chlorine gas. The court excluded the testimony of Drs. Kaye Kilburn, Gunnar Heuser, and William Spindell as "novel" and "unsupported by research extraneous to the litigation." Although a study by Kilburn had been published in a peer-reviewed journal, the court distinguished "editorial" peer review from "true peer review" and concluded that Kilburn's study suffered from "very serious flaws".

28. **Kathleen Frank, Gretchen Pusey, and Mary Beaudoin v. State of New York et al.** (No. 95-CV-399, No. Dis. N.Y., July 12, 1997). Plaintiffs alleged MCS resulting from exposure to pesticides while employees of the N.Y. State Department of Taxation and Finance. Plaintiffs claimed that they were not provided adequate accommodations for their "generalized chemical sensitivities" under the Americans with Disabilities Act. The court ruled that the testimony of the medical experts (Drs. Lax, Johanning, Burgess, Schimelman, Erner, Orgel, and Lovejoy) and psychological experts (Drs. Golden, Gold, Lifrak, Calabro, and Horenstein) stating that plaintiffs' disability was attributable to MCS was inadmissible because it "fails to meet the standard of evidentiary reliability established in Daubert." The court also noted that "Plaintiffs' own submissions confirm the notion that an MCS diagnosis is virtually untestable," and concluded that "the controversy surrounding MCS remains to be settled by the methods of science rather than by the methods of litigation."

29. **Linda Carroll v. Marion County Board of Education** (No. 92-C-196, W. Va. Cir., Marion Co., Div. 1, Jan. 24, 1997). Plaintiff was one of 17 families who sued the school district for the high school students' alleged long-term exposure to various pesticides. The testimony of Grace Ziem, M.D., an environmental specialist from Baltimore, that Carroll's son suffered from MCS was excluded. The court found that "MCS does not pass the 'good science' test and that the scientific methodology supporting MCS is somewhat suspect."

30. **Donald and Susan Maxwell v. Sears, Roebuck & Co. et al.** (No. 94-1056, Fla. Cir., Manatee Co., Mar. 3, 1997). The plaintiffs' alleged injury (MCS) was sustained from exposure to pesticides applied to their house. The testimony of Alan Lieberman, M.D., Albert Robbins, D.O., and Susan Franks, Ph.D., was restricted because their "conclusions [were] reached without reliance upon any generally accepted scientific foundation. Specifically, Plaintiffs have come forward with no proof of the bases for these opinions outside of each expert's testimony, each of which is dependent upon the other two." The court excluded reference to multiple chemical sensitivities by witnesses and counsel.

31. **Virginia Gressel et vir. v. Ahern, Macvittie & Hoffman, LTD., et al.** (No. CV 94-13040, Sup. Court Arizona, Maricopa County, Dec. 11, 1997). The plaintiff alleged systemic symptoms attributed to acquired environmental illness from inhalation of carpet fumes which left her disabled. The court excluded the expert testimony of Dr. William Rea, finding that "the procedure used by Dr. Rea [skin testing method] is not generally accepted by the larger scientific community. Essentially Dr. Rea's 'environmental medicine' is the same theory that Dr. Rea espoused as 'clinical ecology'. Even the doctors in this case upon which Dr. Rea relied discredit his methods ... stating the 'barrel effect' or 'total body load' are not generally accepted medical principles."

Glossary

clinical ecology: The original name for the toxicogenic theory. With the rejection of the diagnostic and treatment procedures associated with clinical ecology, advocates dissociated themselves from the name.

clinical ecologist: A practitioner of the methods of clinical ecology and advocate of the toxicogenic theory.

degenerative research program: Empirical evidence does not support a theory. The postulates defend against refutation of the theory. There are no testable hypotheses. Alternative explanations explain the phenomena.

EI: Environmental illness, the term used in this book to include all synonymous labels such as chemical intolerance, multiple chemical sensitivities (MCS), etc.

EI advocate: Someone who promotes the toxicogenic theory through unsubstantiated speculation and denies refuting evidence.

EI patient: Someone who presents with multi-system complaints and attributes them to EI. The basis is not objective but subjective. The belief in EI is on a continuum of conviction.

environmental physician: A new label for a clinical ecologist.

folie a deux: Double insanity, the insanity between two people.

progressive research program: Empirical evidence supports hypotheses. Postulates are rigorously scrutinized by experiments with testable hypotheses. As the theory being tested evolves, there are refinements in the postulates that are in line with the evidence. New facts are predicted and explained.

psychogenic theory: Presupposes that those multi-system complaints not accounted for by alternative medical disorders can be explained by psychological, psychiatric, and/or psychophysiological factors.

toxicogenic theory: Presupposes that low levels of foods, chemicals, or electromagnetic forces (EMF) directly cause and exacerbate multi-system complaints.

true believer: Someone who is absolutely convinced in the truth of EI; EI can explain all of their symptoms and justify their life style and behavior.

universal reactor: A self-reference term used by some EI patients.

unproven: Refers to diagnostic and treatment methods and practices for which there is no scientific evidence; a synonym for unsubstantiated.

unsubstantiated: Refers to diagnostic and treatment methods and practices for which there is no scientific evidence; a synonym for unproven.

References

Abbey, S.E. (1992) Depression, somatization and chronic fatigue syndrome: lessons for the study of multiple chemical sensitivity. In T. Wood (Ed.), *Multiple Chemical Sensitivities and their Relevance to Psychiatric Disorders*, Proceedings of a Workshop, Ottowa, Ontario, December 7, 1992, Laboratory Center for Disease Control, Health Canada, pp. 57–83.

Abbey, S.E. (1993) Somatization, illness attribution and the sociocultural psychiatry of chronic fatigue syndrome, in Bock, G.R. and Whelan, J. (Eds.), *Chronic Fatigue Syndrome* (Ciba Foundation Symposium 173), Chichester: Wiley, pp. 238–261.

Abbey, S.E. and Garfinkel, P.E. (1991) Neurasthenia and chronic fatigue syndrome: the role of culture in the making of a diagnosis, *Am. J. Psychiatry*, 148(12):1638–1646.

Abramson, L.Y., Seligman, M.E.P., and Teasdale, J.D. (1978) Learned helplessness in humans: critique and reformulation, *J. Abnormal Psychol.*, 87(1):49–74.

Adamec, R.E. (1978) Normal and abnormal limbic system mechanisms of emotive biasing, in Livingston, K.E. and Hornykiewicz, O. (Eds.), *Limbic Mechanisms*, New York: Plenum Press, pp. 405–455.

Adamec, R.E. (1990) Does kindling model anything clinically relevant? *Biol. Psychiatry*, 27:249–279.

Ader, R., Ed. (1981) *Psychoneuroimmunology*, New York: Academic Press.

Ader, R. and Cohen, N. (1993) Psychoneuroimmunology: conditioning and stress, *Ann. Rev. Psychol.*, 44:53–85.

Ader, R., Felten, D.L., and Cohen, N., Eds. (1991) *Psychoneuroimmunology*, 2nd ed, New York: Academic Press.

Adler, G. (1981) The borderline-narcissistic personality disorder continuum, *Am. J. Psychiatry*, 138:46–50.

Adler, H.M. and Hammett, V.B.O. (1973) The doctor-patient relationship revisited: an analysis of the placebo effect, *Ann. Intern. Med.*, 78:595–598.

Agius, M.A. and Arnason, B.G.W. (1991) Autoimmune neurological diseases and their potential relevance to psychiatric diseases, in Gorman, J.M. and Kertzner, R.M. (Eds.), *Psychoimmunology Update*, Washington, D.C.: American Psychiatric Press, pp. 9–29.

Albanese, A., Hamill, G., Jones, J., Skuse, D., Mattews, D.R., and Stanhope, R. (1994) Reversibility of physiological growth hormone secretion in children with psychosocial dwarfism, *Clin. Endocrinol.*, 40:687–692.

Alexander, F. (1950) *Psychosomatic Medicine: Its Principles and Applications*, New York: Norton.

Alexander, F. (1963) The dynamics of psychotherapy in the light of learning theory, *Am. J. Psychiatry*, 120:440–448. (Reproduced in Mahrer, A.R. and Pearson L., Eds. (1971) *Creative Developments in Psychotherapy*, Cleveland, OH: Case Western Reserve University Press, pp. 330–344.)

Alexander, L. (1982) Illness maintenance and the new American sick role, in Chrisman, N.J. and Maretzki, T.W., Eds., *Clinically Applied Anthropology*, Boston: Reidel Publishing, pp. 351–367.

Allergy and Environmental Health Association (1994/95) *AEHA Quarterly*, Winter/Spring, XVI(4):5–6.

Altenkirch, H., Hopmann, D., Brockmeier, B., and Walter, G. (1996) Neurological investigations in 23 cases of pyrethroid intoxication reported to the German Federal Health Office, *Neurotoxicology*, 17(3–4):645–652.

Altman, J., Brunner, R.L., and Bayer, S.A. (1973) The hippocampus and behavioral maturation, *Behav. Biol.*, 8:557–596.

American Academy of Allergy and Immunology (1981) Position statement on controversial techniques: report of the Executive Committee, *J. Allergy Immunol.*, 67(5):333–338.

American Academy of Allergy and Immunology (1986) Position statement on clinical ecology: report of the Executive Committee, *J. Allergy Clin. Immunol.*, 78:269–271.

American Academy of Neurology (1996) Report of the Therapeutics and Technology Assessment Subcommittee: assessment of neuropsychological testing in adults, *Neurology*, 47:592–599.

American College of Occupational Medicine (1990) Environmental illness, *J. Occup. Med.*, 32(3):211.

American College of Physicians (1989) Position paper: clinical ecology, *Ann. Intern. Med.*, 111(2):168–178.

American Medical Association (1992) Council on Scientific Affairs report: clinical ecology, *JAMA*, 268(24):3465–3467.

American Psychiatric Association (1987) *Diagnostic and Statistical Manual of Mental Disorders*, 3rd rev. ed. (DSM-III-R), Washington, D.C.: American Psychiatric Press.

American Psychiatric Association (1994) *Diagnostic and Statistical Manual of Mental Disorders*, 4th ed. (DSM-IV), Washington, D.C.: American Psychiatric Press.

Ames, B.N. and Gold, L.S. (1995) The causes and prevention of cancer: the role of the environment, in Bailey, R. (Ed.), *The True State of the Planet*, New York: Free Press, pp. 141–175.

Amundsen, M.A., Hanson, N.P., Bruce, B.K., Lantz, T.D., Schwartz, M.S., and Lukach, B.M. (1996) Odor aversion of multiple chemical sensitivities: recommendation for a name change and description of successful behavioral medicine treatment, *Reg. Toxicol. Pharmacol.*, 24(1):S116–118.

Andersson, B., Berg, M., Arnetz, B.B., Melin, L., Langlet, I., and Liden, S. (1996) A cognitive-behavioral treatment of patients suffering from "electric hypersensitivity", *J. Occup. Environ. Med.*, 38(8):752–758.

Andreasen, N.C. (1994) Changing concepts of schizophrenia and the ahistorical fallacy, *Am. J. Psychiatry*, 151(10):1405–1407.

Andreasen, N.C. (1995) Posttraumatic stress disorder: psychology, biology, and the Manichaean warfare between false dichotomies [editorial], *Am. J. Psychiatry*, 152(7):963–965.

Andrews, G., Stewart, G., Morris-Yates, A., Holt, P., and Henderson, S. (1990) Evidence for a general neurotic syndrome, *Br. J. Psychiatry*, 157:6–12.

Angell, M. (1996) *Science on Trial: The Clash of Medical Evidence and the Law in the Breast Implant Case*, New York: W.W. Norton.

Antelman, S.M. (1988) What are the implications of drug-induced time-dependent sensitization? Conclusions, *Drug Develop. Res.*, 14:1–30.

Antelman, S.M. and Yehuda, R. (1994) Time-dependent change following acute stress: relevance to the chronic and delayed aspects of posttraumatic stress disorder, in Murburg, M.M. (Ed.) *Catecholamine Function in Posttraumatic Stress Disorder: Emerging Concepts*, Washington, D.C.: American Psychiatric Press, pp. 87–98.

Antelman, S.M., Eichler, A.J., Black, C.A., and Kocan, D. (1980) Interchangeability of stress and amphetamine in sensitization, *Science*, 207:329–331.

Antelman, S.M., Kocan, D., Edwards, D.J., and Knopf, S. (1987) A single injection of diazepam induces long-lasting sensitization, *Psychopharmacol. Bull.*, 23:430–434.

Antelman, S.M., Caggiula, A.R., Kocan, D. et al. (1991) One experience with "lower" or "higher" intensity stressors, respectively, enhances or diminishes responsiveness to haloperidol weeks later: implications for understanding drug variability, *Brain Res.*, 566:276–283.

Antelman, S.M., Kocan, D., Knopf, S., Edwards, D.J., and Caggiula, A.R. (1992) One brief exposure to a psychological stressor induces long-lasting, time-dependent sensitization of both the cataleptic and neurochemical responses to haloperidol, *Life Sci.*, 51:261–266.

Antonuccio, D.O., Danton, W.G., and DeNelsky, G.Y. (1995) Psychotherapy versus medication for depression: challenging the conventional wisdom with data, *Prof. Psychol.: Res. Practice*, 26(6):574–585.

Appel, W. (1983) *Cults in America: Programmed for Paradise*, New York: Holt, Rinehart, Winston.

Appley, M.H. and Trumbull, R., Eds. (1986) *Dynamics of Stress: Physiological, Psychological, and Social Perspectives*, New York: Plenum Press.

Archives of Environmental Health (1996) Letters to the Editor and responses in regard to the ICPS 1996 Workshop on multiple chemical sensitivity, *Arch. Environ. Health*, 51(4):338–342.

Arieti, S. (1980) Cognition in psychoanalysis, *J. Am. Acad. Psychoanalysis*, 8(1):3–23.

Arieti, S. (1982) Individual psychotherapy, in Paykel, E.S. (Ed.), *Handbook of Affective Disorders*, New York: Guilford Press, pp. 299–306.

Arnetz, B.B., Berg, M., Anderzen, I., Lundeberg, T., and Haker, E. (1995) A nonconventional approach to the treatment of "environmental illness", *J. Occup. Environ. Med.*, 37(7):838–844.

Aronin, N., Coslovsky, R., and Leeman, S.E. (1986) Substance P and neurotensin: their roles in the regulation of anterior pituitary function, *Ann. Rev. Physiol.*, 48:537–549.

Asch, S.E. (1957) *Experimental Investigation of Group Influence, Symposium on Preventive and Social Psychiatry*, Washington, D.C.: U.S. Government Printing Office, pp. 17–25.

Ashford, N.A. and Miller, C.S. (1989) *Chemical Sensitivity*, a report to the New Jersey State Department of Health.

Ashford, N.A. and Miller, C.S. (1991) *Chemical Exposures: Low Levels and High Stakes*, New York: Van Nostrand-Reinhold.

Ashford, N.A. and Miller, C.S. (1998) *Chemical Exposures: Low Levels and High Stakes*, 2nd ed., New York: Van Nostrand-Reinhold.

Ashford, N.A., Heinzow, B., Lutjen, K., Marouli, C., Molhave, L., Monch, B., Papadopoulos, S., Rest, K., Rosdahl, D., Siskos, P., and Velonakis, E. (1995) *Chemical Sensitivity in Selected European Countries: An Exploratory Study*, Commission of the European Union, DG-V, Athens, Greece: Egonomia, Ltd.

Asnis, G.M., McGinn, L.K., and Sanderson, W.C. (1995) Atypical depression: clinical aspects and noradrenergic function, *Am. J. Psychiatry*, 152:31–36.

Aston-Jones, G., Chiang, C., and Alexinsky, T. (1991a) Discharge of noradrenergic locus coeruleus neurons in behaving rats and monkeys suggests a role in vigilance, *Prog. Brain Res.*, 88:501–520.

Aston-Jones, G. Shipley, M.T., Chouvet, G., Ennis, M., van Bockstaele, E., Pieribone, V., Shiekhattar, R., Akaoka, H., Drolet, G., Astier, B., Charlety, P., Valentino, R.J., and Williams, J.T. (1991b) Afferent regulation of locus coeruleus neurons: anatomy, physiology, and pharmacology, *Prog. Brain Res.*, 88:47–75.

ATSDR (1994a) *Low-Level Exposure to Chemicals and Neurobiologic Sensitivity*, Conference sponsored by the Agency for Toxic Substances and Disease Registry. Baltimore, MD, April 6–7.

ATSDR (1994b) *Panel on Evaluating Individuals Reporting Sensitivities to Multiple Chemicals*, California Department of Health Services, Environmental Health Investigations Branch, Conference sponsored by the Agency for Toxic Substances and Disease Registry, Berkeley, CA, May 5.

Back, K.W. and Gergen, K.J. (1968) The self through the latter span of life, in Gordon, C. and Gergen, K. (Eds.), *The Self in Social Interaction*, New York: Wiley, pp. 101–143.

Baddeley, A.D. (1993) Working memory or working attention, in Baddeley, A.D. and Weiskrantz, L. (Eds.), *Attention: Selection, Awareness and Control*, New York: Oxford University Press.

Bailey, C.H. and Kandel, E.R. (1993) Structural changes accompanying memory storage, *Ann. Rev. Physiol.*, 55:397–426.

Baker, D.B. (1989) Social and organizational factors in office building-associated illness, *Occup. Med. State Art Rev.*, 4(4):607–624.

Baker, H.S. and Baker, M.N. (1987) Heinz Kohut's self psychology: an overview, *Am. J. Psychiatry*, 144(1):1–9.

Ball, S. and Shekhar, A. (1997) Basilar artery response to hyperventilation in panic disorder, *Am. J. Psychiatry*, 154(11):1603–1604.

Ballet, G. (1908) *Neurasthenia*, London: Henry Klimpton.

Bandura, A. (1977) Self-efficacy: toward a unifying theory of behavioral change, *Psychol. Rev.*, 84:191–215.

Bardana, Jr., E.J. and Montanaro, A., Eds. (1997) *Indoor Air Pollution and Health*, New York: Marcel Dekker,.

Barkley, R.A. (1990) *Attention Deficit Hyperactivity Disorder: A Handbook for Diagnosis and Treatment*, New York: Guilford Press.

Barlow, D.H. (1985) The dimensions of anxiety disorder, in Tuma, A.H. and Maser, J.D. (Eds.), *Anxiety and the Anxiety Disorders*, Hillsdale, NJ: Lawrence Erlbaum Assoc., pp. 479–500.

Barlow, D.H. and Craske, M.G. (1988) The phenomenology of panic, in Rachman, S. and Maser, J.D. (Eds.), *Panic: Psychological Perspectives*, Hillsdale, NJ: Lawrence Erlbaum Assoc., pp. 11–35.

Barlow, D.H., DiNardo, P.A., Vermilyea, B.B., Vermilyea, J., and Blanchard, E.B. (1986) Co-morbidity and depression among the anxiety disorders: issues in diagnosis and classification, *J. Nerv. Ment. Dis.*, 174(2):63–72.

Barnes, E.M. (1996–98) *Magnetic Field Effects Still in Twilight Zone*, Forum for Applied Research and Public Policy 1994.

Barnes, P.J. and Adcock, I. (1993) Anti-inflammatory actions of steroids: molecular mechanism, *Trends Pharmacol. Sci.*, 14:436–441.

Barnett, J. (1966) On cognitive disorders in the obsessional, *Contemp. Psychoanalysis*, 2(2):122–134.

Barnett, J. (1968) Cognition, thought and affect in the organization of experience, in Masserman, J.H. (Ed.), *Science and Psychoanalysis: Animal and Human*, Vol. XII, New York: Gurne & Stratton, pp. 237–247.

Barnett, J. (1978) Insight and therapeutic change, *Contemp. Psychoanalysis*, 14:534–544.

Barnett, J. (1980) Cognitive repair in the treatment of the neuroses, *J. Am. Acad. Psychoanalysis*, 8(1): 39–55.

Barrett, D.H. and Mannino, D.M. (1996) Post-traumatic stress disorder and physical health status among U.S. Army veterans serving in Vietnam, 1965–1971, poster presentation, Aspen Environmental Medicine Conference, Aspen, CO, September 4–7.

Barrett, D.H., Green, M.L., Morris, R., Giles, W.H., and Croft, J.B. (1996) Cognitive functioning and posttraumatic stress disorder, *Am. J. Psychiatry*, 153(11):1492–1494.

Barrett, S. (1993) *Unproven "Allergies": An Epidemic of Nonsense*, New York: American Council on Science and Health.

Barrett, S. and Herbert, V. (1994) *The Vitamin Pushers: How the "Health Food" Industry Is Selling America a Bill of Goods*, Amherst, NY: Prometheus Books.

Barsky, A.J. (1979) Patients who amplify bodily sensations, *Ann. Intern. Med.*, 91:63–70.

Barsky, A.J. and Klerman, G.L. (1983) Overview: hypochondriasis, bodily complaints, and somatic styles, *Am. J. Psychiatry*, 140(3):273–283.

Barsky, A.J. and Borus, J.F. (1995) Somatization and medicalization in the era of managed care, *JAMA*, 274(24):1931–1934.

Barsky, A.J., Geringer, E., and Wool, C.A. (1988a) A cognitive-educational treatment for hypochondriasis, *Gen. Hosp. Psychiatry*, 10:322–327.

Barsky, A.J., Goodson, J.D., Lane, R.S., and Cleary, P.D. (1988b) The amplification of somatic symptoms, *Psychosomatic Med.*, 50:510–519.

Barsky, A.J., Wyshak, G., Latham, K.S., and Klerman, G.L. (1991) Hypochondriacal patients, their physicians, and their medical care, *J. Gen. Intern. Med.*, 6:413–419.

Barsky, A.J., Wool, C., Barnett, M.C., and Cleary, P.D. (1994) Histories of childhood trauma in adult hypochondriacal patients, *Am. J. Psychoanalysis*, 151:397–401.

Bartoshuk, L. and Beauchamp, G. (1994) Chemical senses, *Ann. Rev. Psychol.*, 45:419–449.

Bashinski, H.S. and Bachrach, R.T. (1984) Enhancement of perceptual sensitivity as the result of selectively attending to spatial locations, *Perception Psychophysics*, 28:241–48.

Basowitz, H., Korchin, S.J., Oken, D., Goldstein, M.S., and Gussack, H. (1956) Anxiety and performance changes with a minimal dose of epinephrine, *Arch. Neurol. Psychiatry*, 76:98–106.

Bass, C. and Gardner, W.N. (1985) Respiratory and psychiatric abnormalities in chronic symptomatic hyperventilation, *Br. Med. J.*, 290:1387–1390.

Bateman, A., Singh, A., Kral, T., and Solomon, S. (1989) The immune-hypothalamic-pituitary-adrenal axis, *Endocrine Rev.*, 10:92–112.

Bateson, G., Jackson, D.D., Haley, J., and Weakland, J. (1978) Toward a theory of schizophrenia, in Berger, M.M. (Ed.), *Beyond the Double Bind*, New York: Brunner/Mazel, pp. 5–27.

Baust, W., Niemczyk, H., and Vieth, K.J. (1963) The action of blood pressure on the ascending reticular activating system with special reference to adrenaline-induced EEG arousal, *Electroencephalogr. Clin. Neurophysiol.*, 15:63–72.

Baxter, J.D. (1988) Actions of adrenal steroids, in Wyngaarden, J.B. and Smith, Jr., L.H. (Eds.), *Cecil Textbook of Medicine*, 18th ed., Philadelphia: Saunders, pp. 1346–1348.

Beard, G. (1880) *A Practical Treatise on Nervous Exhaustion (Neurasthenia)*, New York: William Wood.

Beard, G. (1881) *American Nervousness*, New York: G.P. Putnam.

Beck, A.T. (1967) *Depression: Clinical, Experimental and Theoretical Aspects*, New York: Harper & Row.

Beck, A.T. (1976) *Cognitive Therapy and the Emotional Disorders*, New York: International Universities Press.

Beck, A.T. (1991) Cognitive therapy: a 30-year retrospective, *Am. Psychol.*, 46(4):368–375.

Beck, J.C. and van der Kolk, B. (1987) Reports of childhood incest and current behavior of chronically hospitalized psychotic women, *Am. J. Psychiatry*, 144:1474–14766.

Beck, A.T., Rush, A.J., Shaw, B.F., and Emery, G. (1979) *Cognitive Therapy of Depression*, New York: Guilford Press.

Beck, A.T., Emery, G., and Greenberg, R.L. (1985a) *Anxiety Disorders and Phobias*, Basic Books, New York.

Beck, A.T., Steer, R.A., Kovacs, M., and Garrison, B. (1985b) Hopelessness and eventual suicide: a 10-year prospective study of patients hospitalized with suicidal ideation, *Am. J. Psychiatry*, 142(5):559–563.

Beck, A.T., Brown, G., Berchick, R.J., Stewart, B.L., and Steer, R.A. (1990) Relationship between hopelessness and ultimate suicide: a replication with psychiatric outpatients, *Am. J. Psychiatry*, 147(2):190–195.

Beecher, H.K. (1955) The powerful placebo, *JAMA*, 159(17):1602–1606.

Beecher, H.K. (1959) *Measurement of the Subjective Response*, New York: Oxford University Press.

Behar, D., Rapaport, J.L., Adams, A.J., Berg, C.J., and Cornblath, M. (1984) Sugar challenge testing with children considered behaviorally "sugar reactive", *J. Nutr. Behav.*, 1:277.

Bell, I.R. (1975) A kinin model of mediation for food and chemical sensitivities: biobehavioral implications, *Ann. Allergy*, 35 (Oct):206–215.

Bell, I.R. (1982) *Clinical Ecology: A New Medical Approach to Environmental Illness*, Bolinas, CA: Common Knowledge Press.

Bell, I.R. (1994a) Somatization disorder: health care costs in the decade of the brain [editorial], *Biol. Psychiatry*, 35:81–83.

Bell, I.R. (1994b) *Neuropsychiatric Aspects of Sensitivity to Low-Level Chemicals: A Neural Sensitization Model*, presented at the Conference on Low-Level Exposure to Chemicals and Neurobiologic Sensitivity, sponsored by Agency for Toxic Substances and Disease Registry (ATSDR), Baltimore, MD, April 6–7, 1994 (proceedings published in *Toxicol. Industr. Health*, 10(4/5):277–312).

Bell, I.R., Miller, C.S., and Schwartz, G.E. (1992) An olfactory-limbic model of multiple chemical sensitivity syndrome: possible relationships to kindling and affective spectrum disorders, *Biol. Psychiatry*, 32:218–242.

Bell, I.R., Schwartz, G.E., Peterson, J.M., and Amend, D. (1993a) Self-reported illness from chemical odors in young adults without clinical syndromes or occupational exposures, *Arch. Environ. Health*, 48:6–13.

Bell, I.R., Schwartz, G.E., Peterson, J.M., Amend, D., and Stini, W.A. (1993b) Possible time-dependent sensitization to xenobiotics: self-reported illness from chemical odors, foods, and opiate drugs in an older adult population, *Arch. Environ. Health*, 48:315–327.

Bell, I.R., Peterson, J.M., and Schwartz, G.E. (1995) Medical histories and psychological profiles of middle-aged women with and without self-reported illness from environmental chemicals, *J. Clin. Psychiatry*, 56(4):151–160.

Bell, I.R., Schwartz, G.E., Baldwin, C.M., and Hardin, E.E. (1996) Neural sensitization and physiological markers in multiple chemical sensitivity, *Reg. Toxicol. Pharmacol.*, 24(1):S39–S47.

Bem, D.J. (1967) Self-perception: an alternative interpretation to cognitive dissonance phenomena, *Psych. Rev.*, 74:183–200.

Benner, P. (1984) *Stress and Satisfaction on the Job. Work Meanings and Coping of Mid-Career Men*, New York: Praeger.

Benes, F.M. and Tamminga, C.A. (1994) Images in neuroscience: cortical GABA-ergic interneurons, *Am. J. Psychiatry*, 151:8.

Benjamin, S. (1989) Psychological treatment of chronic pain: a selective review, *J. Psychosomatic Res.*, 33:121–131.

Benjamin, S. and Eminson, D.M. (1992) Abnormal illness behavior: childhood experiences and long-term consequences, *Int. Rev. Psychiatry*, 4:55–70.

Bennett, J.E. (1990) Searching for the yeast connection [editorial], *New Engl. J. Med.*, 323(25):1766–1767.

Benson, T.E. and Arkins, J.A. (1976) Cytotoxic testing for food allergy: evaluations of reproducibility and correlation, *J. Allergy Clin. Immunol.*, 58(4):471–476.

Bentman, L., Dahlgren, N., and Siesjo, B.K. (1978) Influences of intravenously administered catecholamines on cerebral oxygen consumption in the rat, *Acta Physiol. Scand.*, 104:101–108.

Benton, A.L. (1994) Neuropsychological assessment, *Ann. Rev. Psychol.*, 45:13.

Berkson, J. (1994) *Patient Statement: A Canary's Tale, Proceedings of the Conference on Low-Level Exposure to Chemicals and Neurobiologic Sensitivity*, April 6–7, 1994, Baltimore, MD, sponsored by Agency for Toxic Substances and Disease Registry (ATSDR) (proceedings published in *Toxicol. Industr. Health*, 10(4/5):323–326).

Bernstein, D.P., Cohen, P., Skodol, A., Bezirganian, S., and Brook, J.S. (1996) Childhood antecedents of adolescent personality disorder, *Am. J. Psychiatry*, 153(7):907–913.

Bernstein, E.M. and Putnam, F.W. (1986) Development, reliability and validity of a dissociation scale, *J. Nerv. Ment. Dis.*, 174:727–735.

Bernstein, I.L. (1991) Flavor aversion, in Getchell, T.V., Doty, R.L., Bartoshuk, L.M., and Snow, Jr., J.B. (Eds.), *Smell and Taste in Health and Disease*, New York: Raven Press, pp. 417–428.

Bieber, I. (1980) Psychoanalysis — a cognitive process, *J. Am. Acad. Psychoanalysis*, 8:25–38.

Bienenstock, J. (1991) Summing up: present concepts and future ideas on immunoregulation of asthma, *Eur. Respir. J.*, 4(Suppl. 13):156s–160s.

Binkley, K.E. and Kutcher, S. (1997) Panic response to sodium lactate infusion in patients with multiple chemical sensitivity syndrome, *J. Allergy Clin. Immunol.*, 99(4):570–574.

Black, D.W. (1993) Environmental illness and misdiagnosis — a growing problem, *Reg. Toxicol. Pharmacol.*, 1993, 18(1):23–31.

Black, D.W. (1996) Iatrogenic (physician-induced) hypochondriasis: four patient examples of "chemical sensitivity", *Psychosomatics*, 37(4):390–393.

Black, D.W., Rathe, A., and Goldstein, R.B. (1990) Environmental illness: a controlled study of 26 subjects with "20th century disease", *JAMA*, 264(24):3166–3170.

Black, D.W., Rathe, A., and Goldstein, R.B. (1993) Measures of distress in 26 "environmentally ill" subjects, *Psychosomatics*, 34:131–138.

Blanchard, E.B., Kolb, L.C., Pallmeyer, T.P., and Gerardi, R.J. (1982) A psychophysiological study of post-traumatic stress disorder in Vietnam veterans, *Psychiatric Q.*, 54:220–229.

Blau, G.J. and Katerberg, R. (1982) Toward enhancing research with the social information processing approach to job design, *Acad. Manage. Rev.*, 7:543–550.

Blonz, E.R. (1986) Is there an epidemic of chronic candidiasis in our midst? [commentary], *JAMA*, 256(22):3138–3139.

Bloor, D. (1976) *Knowledge and Social Inquiry*, London: Routledge & Kegan Paul.

Bolla, K.I. (1996) Neurobehavioral performance in multiple chemical sensitivities, *Reg. Toxicol. Pharmacol.*, 24(1):S52–S54.

Bolla, K.I., Schwartz, B.S., Stewart, W., Rignani, J., Agnew, J., and Ford, D.P. (1995) Comparison of neurobehavioral function in workers exposed to a mixture of organic and inorganic lead and in workers exposure to solvents, *Am. J. Industr. Med.*, 27:231–246.

Bolla-Wilson, K.I., Wilson, R.J., and Bleeker, M.L. (1988) Conditioning of physical symptoms after neurotoxic exposure, *J. Occup. Med.*, 30:684–686.

Bools, C.N., Neale, B.A., and Meadow, S.R. (1992) Co-morbidity associated with fabricated illness (Munchausen syndrome by proxy), *Arch. Dis. Childhood*, 67:77–79.

Bools, C.N., Neale, B.A., and Meadow, S.R. (1993) Follow-up of victims of fabricated illness (Munchausen syndrome by proxy), *Arch. Dis. Childhood*, 69:625–630.

Bools, C.N., Neale, B.A., and Meadow, S.R. (1994) Munchausen syndrome by proxy: a study of psychopathology, *Child Abuse Neglect*, 18(9):773–788.

Booth, L. (1991) *When God Becomes a Drug*, New York: Putnam.

Boss, L.P. (1997) Epidemic hysteria: a review of the published literature, *Epidemiol. Rev.*, 19(2):233–243.

Bouwer, C. and Stein, D.J. (1997) Association of panic disorder with a history of traumatic suffocation, *Am. J. Psychiatry*, 154(11):1566–1570.

Bowlby, J. (1960) Separation anxiety, *J. Psychol. Psychiatry*, 1:251–269.

Bowlby, J. (1971) *Attachment and Loss. Vol. I. Attachment*, New York: Basic Books.

Bowlby, J. (1975) *Attachment and Loss. Vol. II. Separation: Anxiety and Anger*, New York: Basic Books.

Bowlby, J. (1977) The making and breaking of affectional bonds: aetiology and psychopathology in the light of attachment theory, *Br. J. Psychiatry*, 130:201–210.

Bowlby, J. (1980) *Attachment and Loss. Vol. III. Loss: Sadness and Depression*, New York: Basic Books,.

Boxer, P.A. (1990) Indoor air quality: a psychosocial perspective, *J. Occup. Med.*, 32(5):425–428.

Boyce, W.T. (1992) The vulnerable child: new evidence, new approaches, *Adv. Pediatr.*, 39:1–33.

Boyles, Jr., J.H. (1977) The validity of using cytotoxic food test in clinical allergy. *Ear Nose Throat J.*, 56(4):168–173.

Bradley, L.A., Alarcon, G.S., Scarinci, I.C., and Haile, J.E. (1992) Pain thresholds and psychological stress among patients with fibromyalgia [abstract], *Arthritis Rheumatol.*, 35:R11.

Braun, B.G. (1984) Towards a theory of multiple personality and other dissociative phenomena, *Psychiatr. Clin. N. Am.*, 7:171–193.

Breier, A., Charney, D.S., and Heninger, G.R. (1985) The diagnostic validity of anxiety disorders and their relationship to depressive illness, *Am. J. Psychiatry*, 142(7):787–797.

Bremner, J.D., Southwick, S.M., Johnson, D.R., Yehuda, R., and Charney, D.S. (1993) Childhood physical abuse and combat-related posttraumatic stress disorder in Vietnam veterans, *Am. J. Psychiatry*, 150(2):235–239.

Bremner, J.D., Randall, P., Scott, T.M., Bronen, R.A., Seibyl, J.P., Southwick, S. M., Delaney, R.C., McCarthy, G., Charney, D.S., and Innis, R.B. (1995a) MRI-based measurement of hippocampal volume in patients with combat-related posttraumatic stress disorder, *Am. J. Psychiatry*, 152(7):973–981.

Bremner, J.D., Randall, P., Scott, T.M., Capelli, S., Delaney, R., McCarthy, G., and Charney, D.S. (1995b) Deficits in short-term memory in adult survivors of childhood abuse, *Psychiatry Res.*, 59:97–107.

Bremner, J.D., Licinio, J., Darnell, A., Krystal, J.H., Owens, M.J., Southwick, S.M., Nemeroff, C.B., and Charney, D.S. (1997a) Elevated CSF corticotropin-releasing factor concentrations in posttraumatic stress disorder, *Am. J. Psychiatry*, 154(5):624–629.

Bremner, J.D., Randall, P., Vermetten, E., Staib, L., Bronen, R.A., Mazure, C., Capelli, S., McCarthy, G., Innis, R.B., and Charney, D.S. (1997b) Magnetic resonance imaging-base measurement of hippocampal volume in posttraumatic stress disorder related to childhood physical and sexual abuse — a preliminary report, *Biol. Psychiatry*, 41:23–32.

Breslau, N., Davis, G.C., Andreski, P., and Peterson, E. (1991) Traumatic events and posttraumatic stress disorder in an urban population of young adults, *Arch. Gen. Psychiatry*, 48:216–222.

Breuer, J. and Freud, S. (1973) *Studies on Hysteria: Physical Mechanisms of Hysteria*, (trans. by J. Strachey), New York: Penguin.

Brewin, C.R. (1987) Cognitive theories of motivation, in Eysenck, H. and Martin, I. (Eds.), *Theoretical Foundations of Behavior Therapy*, New York: Plenum, pp. 277–293.

Brewin, C.R. and Andrews, B. (1992) The role of context and autobiography in cognitive assessment, *Psychol. Inquiry*, 3:229–230.

Brewin, C.R., Andrews, B., and Gotlib, I.H. (1993) Psychopathology and early experience: a reappraisal of retrospective reports, *Psychol. Bull.*, 113(1):82–98.

Breyer, S. (1993) *Breaking the Vicious Cycle: Toward Effective Risk Regation*, Cambridge, MA: Harvard University Press.

Bridges, A.J., Conley, C., Wang, G., Burns, D.E., and Vasey, F.B. (1993) A clinical and immunologic evaluation of women with silicone breast implants and symptoms of rheumatic disease, *Ann. Intern. Med.*, 118:929–936.

Briere, J. (1992a) Medical symptoms, health risk, and history of childhood sexual abuse, *Mayo Clinic Proc.*, 67:527–532.

Briere, J. (1992b) Methodological issues in the study of sexual effects, *J. Consulting Clin. Psychol.*, 60(2):196–203.

Briere, J. and Runtz, M. (1988) Symptomatology associated with childhood sexual victimization in a nonclinical adult sample, *Child Abuse Neglect*, 12, 51–59.

Briere, J. and Conte, J. (1993) Self-reported amnesia for abuse in adults molested as children, *J. Traumatic Stress*, 6:21–31.

Broadbent, D.E. (1971) *Decision and Stress*, New York: Academic Press.

Brodal, A. (1982) *Neurological Anatomy in Relation to Clinical Medicine*, 3rd ed., New York: Oxford University Press.

Brodie, J.D. (1996) Imaging for the clinical psychiatrist: facts, fantasies, and other musings [editorial], *Am. J. Psychiatry*, 153(2):145–149.

Brodsky, C.M. (1983) "Allergic to everything": a medical subculture, *Psychosomatics*, 24(8):731–742.

Brodsky, C.M. (1988) The psychiatric epidemic in the American workplace, *Occup. Med. State Art Rev.*, 3(4):653–662.

Brody, H. (1980) *Placebos and the Philosophy of Medicine: Clinical, Conceptual, and Ethical Issues*, Chicago: University of Chicago Press.

Brown, G.R. and Anderson, B. (1991) Psychiatric morbidity in adult inpatients with childhood histories of sexual and physical abuse, *Am. J. Psychiatry*, 148:55–61.

Brown, T.M., Merritt, W.D., and Evans, D.L. (1988) Psychogenic vocal cord dysfunction masquerading as asthma, *J. Nerv. Ment. Dis.*, 176:308–310.

Browne, A. and Finkelhor, D. (1986) Impact of child sexual abuse: a review of the research, *Psychol. Bull.*, 99:66–77.

Bryer, J.B., Nelson, B.A., Miller, J.B., and Krol, P.A. (1987) Childhood sexual and physical abuse as factors in adult psychiatric illness, *Am. J. Psychiatry*, 144:1426–1430.

Buchwald, D. and Garrity, D. (1994) Comparison of patients with chronic fatigue syndrome, fibromyalgia, and multiple chemical sensitivities, *Arch. Intern. Med.*, 154:2049–2053.

Buchwald, D., Sullivan, J.L., and Komaroff, A.L. (1987) Frequency of "chronic active Epstein-Barr virus infection" in a general medical practice, *JAMA*, 257:2303–2307.

Burg, J.R., Gist, G.L., Allred, S.L., Radtke, T.M., Pallos, L., and Cusack, C.D. (1995) The National Exposure Registry — morbidity analyses of noncancer outcomes from the trichloroethylene subregistry baseline data, *Int. J. Occup. Med. Toxicol.*, 4(2):237–257.

Busch, F.N., Cooper, A.M., Klerman, G.L., Penzer, R.J., Shapiro, T., and Shear, M.K. (1991) Neurophysiological, cognitive-behavioral, and psychoanalytic approaches to panic disorder: toward an integration, *Psychoanalytic Inquiry*, 11:316–332.

Butler, R.W., Braff, D.L., Rausch, J.L., Jenkins, M.A., Sprock, J., and Geyer, M.A. (1990) Physiological evidence of exaggerated startle response in a subgroup of Vietnam veterans with combat-related PTSD, *Am. J. Psychiatry*, 147:1308–1312.

Cahill, L., Prins, B., Weber, M., and McGaugh, J.L. (1994) Beta-adrenergic activation and memory for emotional events, *Nature*, 371:702–704.

Cain, W.S. (1988) Olfaction, in Atkinson, R.C., Herrnstein, J.J., Lindzey, G. et al. (Eds.), *Stevens' Handbook of Experimental Psychology*. Vol. 1. *Perception and Motivation*, 2nd ed., New York: Wiley, pp. 409–459.

California Medical Association, Scientific Board, Task Force on Clinical Ecology (1986) Clinical ecology: a critical appraisal, *West. J. Med.*, 144(2):239–245.

Callender, T.J., Morrow, L., Subramanian, K., Duhon, D., and Ristovv, M. (1993) Three-dimensional brain metabolic imaging in patients with toxic encephalopathy, *Environ. Res.*, 60:296–319.

Cannon, W.B. (1927) The James-Lange theory of emotions: a critical examination and an alternative theory, *Am. J. Psychol.*, 39:106–124.

Cannon, W.B. (1929) *Bodily Changes in Pain, Hunger, Fear and Rage*, 2nd ed., New York: D. Appleton.

Cannon, W.B. (1931) Against the James-Lange and thalamic theories of emotion, *Psych. Rev.*, 38:281–295.

Cannon, W.B. (1957) "Voodoo" death, *Psychosomatic Med.*, 14:182–190.

Cantril, H. and Hunt, W.A. (1932) Emotional effects produced by the injection of adrenalin, *Am. J. Psychol.*, 44:300–307.

Caplan, R.D., Tripathi, R.C., and Naidu, R.K. (1985) Subjective past, present and future fit: effects on anxiety, depression, and other indicators of well-being, *J. Personality Social Psychol.*, 8:180–197.

Carlson, E.T. (1986) The history of dissociation until 1880, in Quen, J.M. (Ed.), *Split Minds/Split Brains: Historical and Current Perspectives*, New York: New York University Press, pp. 7–30.

Carmen E.H., Rieker, P.P., and Mills, T. (1984) Victims of violence and psychiatric illness, *Am. J. Psychiatry*, 141:378–383.

Carus, P. (1974 [1900]) *The History of the Devil and the Idea of Evil*, LaSalle, IL: Open Court Publishing Company.

Chancellor-Freeland, C., Zhu, G.F., Kage, R., Beller, D.I., Leeman, S.E, and Black, P.H. (1995) Substance P and stress-induced changes in macrophages, in Chrousos, G.P. et al. (Eds.), *Stress: Basis Mechanisms and Clinical Implications*, New York: New York Academy of Sciences, pp. 472–484.

Charney, D.S. and Heninger, G.R. (1985) Noradrenergic function and the mechanisms of action of antianxiety treatment. II. The effect of long-term imipramine treatment, *Arch. Gen. Psychiatry*, 42:473–481.

Charney, D.S. and Heninger, G.R. (1986) Abnormal regulation of noradrenergic function in panic disorders. Effects of clonidine in healthy subjects and patients with agoraphobia and panic disorder, *Arch. Gen. Psychiatry*, 43:1042–1054.

Charney, D.S., Heninger, G.R., and Breier, A. (1984) Noradrenergic function in panic anxiety: effects of Yohimbine in healthy subjects and patients with agoraphobia and panic disorder, *Arch. Gen. Psychiatry*, 41:751–763.

Charney, D.S., Deutch, A.Y., Krystal, J.H., Southwick, S.M., and Davis, M. (1993) Psychobiologic mechanisms of posttraumatic stress disorder, *Arch. Gen. Psychiatry*, 50:294–305.

Chavez, L. (1995) The tragic story of Media still lives [editorial], *Denver Post*, p. 4E, December 3.

Chemical Injury Litigation Project (1994) *Notice of Noncompliance with the Americans with Disabilities Act (Title II & II)*, Baltimore, MD: MCS Referral & Resources.

Cherry, N., Hutchins, H., Pace, T., and Waldron, H.A. (1985) Neurobehavioral effects of repeated occupational exposure to toluene and paint solvents, *Br. J. Indust. Med.*, 42:291–300.

Christensen, N.J. and Jensen, E.W. (1995) Sympathoadrenal activity and psychosocial stress, in Chrousos, G.P. et al. (Eds.), *Stress: Basis Mechanisms and Clinical Implications*, New York: New York Academy of Sciences, pp. 640–647.

Christenson, R.M., Walker, J.I., Ross, D.R., and Maltbie, A.A. (1981) Reactivation of traumatic conflicts, *Am. J. Psychiatry*, 138:984–985.

Christianson, S.A. and Loftus, E.F. (1987) Memory for traumatic events, *Appl. Cognitive Psychol.*, 1:225–239.

Christopher, K.I., Wood, R.P., Eckert, R.C., Blager, F.B., Raney, R.A., and Souhrada, J.F. (1983) Vocal-cord dysfunction presenting as asthma, *New Engl. J. Med.*, 308:1566–1570.

Chrousos, G.P. (1995) The hypothalamic-pituitary-adrenal axis and immune-mediated inflammation, *New Engl. J. Med.*, 332(20):1351–1362.

Chrousos, G.P. and Gold, P.W. (1992) The concepts of stress and stress system disorders: overview of physical and behavioral homeostasis, *JAMA*, 267:1244–1252.

Chrousos, G.P., Loriaux, D.L., and Gold, P.W., Eds. (1988) *Mechanisms of Physical and Emotional Stress. Advances in Experimental Medicine and Biology*, Vol. 245, New York: Plenum Press.

Chrousos, G.P. et al., Eds. (1995) *Stress: Basis Mechanisms and Clinical Implications*, New York: New York Academy of Sciences.

Chu, J.A. and Dill, D.L. (1990) Dissociative symptoms in relation to childhood physical and sexual abuse, *Am. J. Psychiatry*, 147:887–892.

Cloninger, C.R., Svrakic, D.M., and Przybeck, T.R. (1993) A psychobiological model of temperament and character, *Arch. Gen. Psychiatry*, 50:975–990.

Claman, H.N. Corticosteroids and lymphoid cells. (1972) Corticosteroids and lymphoid cells, *New Engl. J. Med.*, 287:388–397.

Clark, D.M. (1988) A cognitive model of panic attacks, in Rachman, J. and Mazer J. (Eds.), *Panic: Psychological Perspectives*, Hillsdale, NJ: Lawrence Erlbaum Assoc., pp. 71–89.

Cloninger, C.R., Svrakic, D.M., and Przybeck, T.R. (1993) A psychobiological model of temperament and character, *Arch. Gen. Psychiatry*, 50:975–990.

Cohen, R.A. and O'Donnell, B.F. (1993a) Attentional dysfunction associated with psychiatric illness, in Cohen, R.A. (Ed.), *The Neuropsychology of Attention*, New York: Plenum Press, pp. 275–306.

Cohen, R.A. and O'Donnell, B.F. (1993b) Physiological substrates of attention, in Cohen, R.A. (Ed.), *The Neuropsychology of Attention*, New York: Plenum Press, pp. 115–144.

Cohen, R.E., Creticos, P.S., and Norman, P.S. (1993–94) The effects of guided imagery (GI) on allergic subjects' responses to ragweed-pollen nasal challenge: an exploratory investigation, in Pope, K.S. and Singer, J.L. (Eds.), *Imagination, Cognition, and Personality*, 13(3):259–269.

Cohen, S. (1980) After-effects of stress on human performance and social behaviour. A review of research and theory, *Psychol. Bull.*, 88(1):82–108.

Cohen, S. and Rodriguez, M.S. (1995) Pathways linking affective disturbances and physical disorders, *Health Psychol.*, 14(5), 374–380.

Cohen, S., Evans, G.W., Stokols, D., and Krantz, D. (1986) *Behavior, Health, and Environmental Stress*, New York: Plenum Press.

Cohn, J.R. (1994) Multiple chemical sensitivity or multi-organ dysesthesia, *J. Allergy Clin. Immunol.*, 93:953–954.

Cole, P.M. and Putnam, F.W. (1992) Effect of incest on self and social functioning: a developmental psychopathology perspective, *J. Consult. Clin. Psychol.*, 60:174–184.

Colligan, M.J., Pennebaker, J.W., and Murphy, L.R., Eds. (1982) *Mass Psychogenic Illness: A Social Psychological Analysis*, Hillsdale, NJ: Lawrence Erlbaum Assoc.

Condemi, J.J. (1991) Unproven diagnostic and therapeutic techniques, In Metcalfe, D.D., Sampson, H.A., and Simon, R.A. (Eds.), *Food Allergy: Adverse Reactions to Foods and Food Additives*, Boston: Blackwell Scientific, pp. 392–404.

Connick, N. Hon. (1996) Initial decision in the matter of the disciplinary proceedings regarding the license to practice dentistry in the State of Colorado by Hal A. Huggins, D.D.S., License no. 3057, Case No. DE 95-04, before the State Board of Dental Examiners, State of Colorado.

Consensus Conference (1982) Defined diets and childhood hyperactivity, *JAMA*, 248(3):290–292.

Conway, F. and Siegelman, J. (1978) *Snapping: America's Epidemic of Sudden Personality Change*, New York: Dell.

Coons, P.M., Bowman, E.S., Pellow, T.A., and Schneider, P. (1989) Post-traumatic aspects of the treatment of victims of sexual abuse and incest, *Psychiatric Clin. North Am.*, 12:325–335.

Costa, E. (1985) Benzodiazepine/GABA interactions: a model to investigate the neurobiology of anxiety, in Tuma, A.H. and Maser, J. (Eds.), *Anxiety and the Anxiety Disorders*, Hillsdale, NJ: Lawrence Erlbaum Assoc., pp. 27–52.

Courtois, C.A. (1979) The incest experience and its aftermath, *Victimology*, 4:337–347.

Coyne, J.C. (1976) Depression and the response of others, *J. Abnormal Psychol.*, 85(2):186–193.

Craik, F.I.M. and Lockhart, R.S. (1972) Levels of processing: a framework for memory research, *J. Verbal Learning Verbal Behav.*, 11:671–684.

Crook, W.G. (1975) Food allergy: the great masquerader, *Pediatr. Clin. North Am.*, 22:227–238.

Crook, W.G. (1986) *The Yeast Connection: A Medical Breakthrough*, 3rd ed., Jackson, TN: Professional Books.

Cullen, M.R. (1987) The worker with multiple chemical sensitivities: an overview. *Occup. Med. State Art Rev.*, 2(4):655–662.

Cullen, M.R., Pace, P.E., and Redlich, C.A. (1993) The experience of the Yale Occupational and Environmental Medicine Clinics with multiple chemical sensitivities, 1986–1991, presented at the American Academy of Allergy and Immunology 60th Anniversary Meeting, Multiple Chemical Sensitivities Workshop, March 15, 1993, Chicago, IL, *Toxicol. Indust. Health*, 8(4):15–19.

Cutler, S.E. and Nolen-Hoeksema, S. (1991) Accounting for sex differences in depression through female victimization: childhood sexual abuse, *Sex Roles*, 24:425–438.

Dager, S.R., Cowley, D.S., and Dunner, D.L. (1987a) Biological markers in panic states: lactate-induced panic and mitral valve prolapse, *Biol. Psychiatry*, 22:339–359.

Dager, S.R., Holland, J.P., Cowley, D.S., and Dunner, D.L. (1987b) Panic disorder precipitated by exposure to organic solvents in the work place, *Am. J. Psychiatry*, 144(8), 1056–1058.

Dager, S.R., Marro, K.I., Richards, T.L., and Metzger, G.D. (1994) Preliminary application of magnetic resonance spectroscopy to investigate lactate-induced panic, *Am. J. Psychiatry*, 151(1):57–63.

Dager, S.R., Strauss, W.L., Marro, K.I., Richards, T.L., Metzger, G.D., and Artru, A.A. (1995) Proton magnetic resonance spectroscopy investigation of hyperventilation in subjects with panic disorder and comparison subjects, *Am. J. Psychiatry*, 152(5):666–672.

Daniell, W.E. (1996) Science, integrity and investigators rights: current challenges, presented at Multiple Chemical Sensitivities: State-of-the-Science, October 30–November 1, 1995, Baltimore, MD, *Reg. Toxicol. Pharmacol.*, 24(1):S152–S162.

Daniell, W.E., Stockbridge, H.L., Labbe, R.F. et al. (1997) Environmental chemical exposures and disturbances of heme synthesis, *Environ. Health Perspect.*, 105(Suppl. 1):37–53.

Dantzer, R. and Kelley, K.W. (1989) Stress and immunity: an integrated view of relationships between the brain and the immune system, *Life Sci.*, 44:1995–2008.

Davidoff, L.L. (1992a) Models of multiple chemical sensitivities (MCS) syndrome: using empirical data (especially interview data) to focus investigations, *Toxicol. Indust. Health*, 8(4):229–247.

Davidoff, L.L. (1992b) Letters to the editor, *Arch. Environ. Health*, 47(5):389–390.

Davidoff, L.L. and Fogarty, L. (1994) Psychogenic origins of multiple chemical sensitivities syndrome: a critical review of the research literature, *Arch. Environ. Health*, 49:316–325.

Davidoff, L.L., Callender, T.J., Morrow, L.A., and Ziem, G. (1991) Letter to the editor, *J. Psychosomatic Res.*, 35:621–623.

Davidson, J.R.T., Hughes, D., Blazer, D.G., and George, L.K. (1991) Post-traumatic stress disorder in the community: an epidemiological study, *Psychol. Med.*, 21:713–721.

Davies, R.K. (1979) Incest: some neuropsychiatric findings, *Int. J. Psychiatry Med.*, 9:117–121.

Davis, E.S. (1985) The legal side of ecologic illness, *Ecol. Illness Law Rep.*, 2(6):4–10.

Davis, M. (1992) The role of the amygdala in fear and anxiety, *Ann. Rev. Neurosci.*, 15:353–375.

Deale, A., Chalder, T., Marks, I., and Wessely, S. (1997) Cognitive behavior therapy for chronic fatigue syndrome: a randomized controlled trial, *Am. J. Psychiatry*, 154(3):408–414.

DeBellis, M.D. and Putman, F.W. (1994) The psychobiology of childhood maltreatment, *Child Adolescent Psychiatric Clin. North Am.*, 3(4):663–678.

DeBellis, M.D., Chrousos, G.P., Dorn, L.D., Burke, L., Helmers, K., Kling, M.A., Trickett, P.K., and Putnam, F.W. (1994a) Hypothalamic-pituitary-adrenal axis dysregulation in sexually abused girls, *J. Clin. Endocrinol. Metab.*, 78(2):249–255.

DeBellis, M.D., Lefter, L., Trickett, P.K., and Putnam, F.W. (1994b) Urinary catecholamine excretion in sexually abused girls, *J. Am. Acad. Child Adolescent Psychiatry*, 33:320–326.

DeBellis, M.D., Burke, L., Trickett, P.K., and Putnam, F.W. (1996) Antinuclear antibodies and thyroid function in sexually abused girls, *J. Traumatic Stress*, 9(2):369–378.

Dekker, E. and Groen, J. (1956) Reproducible psychogenic attacks of asthma: a laboratory study, *J. Psychosomatic Res.*, 1:58–67.

Dekker, E., Pelser, H.E., and Groen, J. (1957) Conditioning as a cause of asthmatic attacks: a laboratory study, *J. Psychosomatic Res.*, 2:97–108.

Delgado, J.M.R., Roberts, W.W., and Miller, N.E. (1954) Learning motivated by electrical stimulation of the brain, *Am. J. Physiol.*, 179:587–593.

Demitrack, M.A., Dale, J.K., Straus S.E., Laue, L., Listwak, S.J., Kruesi, M.J.P., Chrousos, G.P., and Gold, P.W. (1991) Evidence for impaired activation of the hypothalamic-pituitary-adrenal axis in patients with chronic fatigue syndrome, *J. Clin. Endocrinol. Metab.*, 73:1224–1234.

Derogatis, L.R. (1983) *SCL-90-R Administration, Scoring and Procedures Manual, II*, Towson, MD: Clinical Psychometric Research.

Deutch, A.Y. and Roth, R.H. (1990) The determinants of stress-induced activation of the prefrontal cortical dopamine system, *Progr. Brain Res.*, 85:367–403.

Deutsch, A. (1980) Tenacity of attachment to a cult leader: a psychiatric perspective, *Am. J. Psychiatry*, 137(12):1569–1573.

Dickey, L.D., Ed. (1976) *Clinical Ecology*, Springfield, IL: Charles C Thomas.

Dirks, J.F., Jones, N.F., and Kinsman, R.A. (1977) Panic-fear: a personality dimension related to intractability in asthma, *Psychosomatic Med.*, 39(2):120–126.

Dirks, J.F., Kinsman, R.A., Staudenmayer, H., and Kleiger, J.H. (1979) Panic-fear in asthma: symptomatology as an index of signal anxiety and personality as an index of ego resources, *J. Nerv. Ment. Dis.*, 167(10):615–619.

Dismukes, W.E., Wade, J.S., Lee, J.Y., Dockery, B.K., and Hain, J.D. (1990) A randomized, double-blind trial of nystatin therapy for the candidiasis hypersensitivity syndrome, *New Engl. J. Med.*, 323(25):1717–1723.

Djuric, V.J. and Bienenstock, J. (1993) Learned sensitivity, *Ann. Allergy*, 71:5–14.

Dohrenwend, B.S., Dohrenwend, B.P., Dodson, M., and Shrout, P.E. (1984) Symptoms, hassles, social support, and life events: problem of confounded measures, *J. Abnormal Psychol.*, 93:222–230.

Dominian, J. (1983) Doctor as prophet, *Br. Med. J.*, 287:1925–1927.

Donnay, A. (1994a) Complaint concerning paper by Drs. Simon, Daniell, Stockbridge, Claypoole, and Rosenstock filed with the Office of Research Integrity, University of Washington, and the Annals of Internal Medicine, Baltimore, MD: MCS Referral & Resources.

Donnay, A. (1994b) Second complaint concerning paper by Drs. Simon, Daniell, Stockbridge, Claypoole, and Rosenstock filed with the Office of Research Integrity, University of Washington, and the Annals of Internal Medicine, Baltimore, MD: MCS Referral & Resources.

Donnay, A. and Ziem, G. (1995) *Protocol for Evaluating Disorders of Porphyrin Metabolism in Chemically Sensitive Patients*, published and distributed by MCS Referral & Resources, 2326 Pickwick Road, Baltimore MD 21207.

Doty, R.L., Deems, D.A., Frye, R.E., Pelberg, R., and Shapiro, A. (1988) Olfactory sensitivity, nasal resistance and autonomic function in patients with multiple chemical sensitivities, *Arch. Otolaryngol. Head Neck Surg.*, 114:1422–1427.

Douglas, M. (1985) *Risk Acceptability According to the Social Sciences*, New York: Russell Sage Foundation.

Downing, C.J. (1988) Expectancy and visual-spatial attention effects on vision, *J. Exper. Psychol.: Human Percept. Performance*, 14:188–197.

Drossman, D.A. (1978) The problem patient: evaluation and care of medical patients with psychosocial disturbances, *Ann. Intern. Med.*, 88:366–372.

Drossman, D.A. (1994) Physical and sexual abuse and gastrointestinal illness: what is the link?, *Am. J. Med.*, 97:105–107.

Drossman, D.A., McKee, D.C., Sandler, R.S. et al. (1988) Psychosocial factors in the irritable bowel syndrome. A multivariate study of patients and nonpatients with irritable bowel syndrome, *Gastroenterology*, 95:701–708.

Drossman, D.A., Leserman, J., Nachman, G., Zhiming, L., Gluck, H., Toomey, T.C., and Mitchell, C.M. (1990) Sexual and physical abuse in women with functional or organic gastrointestinal disorders, *Ann. Intern. Med.*, 113:828–833.

Ducatman, A.M. (1993) MCS: multiple negative findings — implications, history, and predictions, *Occup. Environ. Med. Rep.*, 7(9):73–75.

Dunn, A.J. and Berridge, C.W. (1990) Physiological and behavioral responses to corticotropin-releasing factor administration: is CRF a mediator of anxiety or stress responses? *Brain Res. Rev.*, 15:71–100.

Durkheim, E. (1952 [1930]) *Suicide*, London: Routledge and Kegan Paul.

Dybing, E., Lovik, M., and Smith, E., Eds. (1996) Environmental chemicals and respiratory hypersensitivity: proceedings of the International Programme on Chemical Safety (IPCS) Workshop, November 29–December 3, 1995, Oslo, Norway, IPCS Joint Series 26, *Toxicol. Lett.*, 83(2/3):57–230.

d'Ydewalle, G., Ferson, R., and Swerts, A. (1985) Expectancy, arousal and individual differences in free recall, *J. Memory Language*, 24:519–525.

Dyer, R.S. and Sexton, K. (1996) What can research contribute to regulatory decisions about the health risks of multiple chemical sensitivity?, *Reg. Toxicol. Pharmacol.*, 24(1):S139–S151.

Easterbrook, J.A. (1959) The effect of emotion on cue utilization and the organization of behavior, *Psych. Rev.*, 66(3):183–201.

Eaton, W.W., Kessler, R.C., Wittchen, H.U., and Magee, W.J. (1994) Panic and panic disorder in the United States, *Am. J. Psychiatry*, 151(3):413–420.

Edwards, J.R. and Cooper, C.L. Research in stress, coping, and health: theoretical and methodological issues [editorial], *Psychol. Med.*, 1988, 18:15–20.

Ellefson, R.D. and Ford, R.E. (1996) The polyphyrias: characteristics and laboratory tests, *Reg. Toxicol. Pharmacol.*, 24(1):S119–S125.

Ellenberger, H. (1970) *The Discovery of the Unconscious: The History and Evolution of Dynamic Psychiatry*, New York: Basic Books.

Ellenhorn, M.J. and Barceloux, D.G. (1988) *Medical Toxicology: Diagnosis and Treatment of Human Poisoning*, New York: Elsevier, p. 990.

Ellis, A. and Harper, R.A. (1961) *A Guide to Rational Living*, Hollywood, CA: Wilshire.

Ellis, E. (1986) Clinical ecology: myth and reality, *Buffalo Physician*, 19(5):23–28.

Ellis, H.C., Thomas, R.L., and Rodriguez, I.A. (1984) Emotional mood states and memory: elaborative encoding, semantic processing and cognitive effort, *J. Exp. Psychol.: Learning Memory Cognition*, 10:470–482.

Engdahl, B.E., Speed, N., Eberly, R.E., and Schwartz, J. (1991) Comorbidity of psychiatric disorders and personality profiles of American World War II prisoners of war, *J. Nerv. Ment. Dis.*, 179: 181–187.

Engel, Jr., C.C., Engel, A.L., Campbell, S.J., and McFall, M.E. (1993) Posttraumatic stress disorder symptoms and precombat sexual and physical abuse in Desert Storm veterans, *J. Nerv. Ment. Dis.*, 181(11):683–688.

Engel, G.L. (1978) Psychologic stress, vasodepressor (vasovagal) syncope, and sudden death, *Ann. Intern. Med.*, 89:403–412.

Engen, T., Gilmore, M.M., and Mair, R.G. (1991) Odor memory, in Getchell, T.V. et al. (Eds.), *Smell and Taste in Health and Disease*, New York: Raven Press, pp. 315–327.

Enna, S.J. and Gallagher, J.P. (1983) Biochemical and electrophysiological characteristics of mammalian GABA receptors, *Int. Rev.Neurobiol.*, 24:181–212.

Epstein, S. (1994) Integration of the cognitive and the psychodynamic unconscious, *Am. Psychol.*, 49(8):709–724.

Eriksen, C. and Hoffman, J.E. (1972) Temporal and spatial characteristics of selective encoding from visual displays, *Perception Psychophys.*, 12:201–204.

Etemad, B. (1978) Extrication from cultism, *Curr. Psychiatric Ther.*, 18:217–223.

Evans, G.W., Carrere, S., and Johansson, G. (1989) A multivariate perspective on environmental stress, *Arch. Complex Environ. Stud.*, 1:1–5.

Famularo, R., Kinscherff, R., and Fenton, T. (1992) Psychiatric diagnoses of abusive mothers: a preliminary report, *J. Nerv. Ment. Dis.*, 180:658–661.

Fanselow, M.S. (1986) Conditioned fear-induced opiate analgesia: a competing motivational state theory of stress analgesia, *Ann, N.Y. Acad. Sci.*, 467:40–54.

Farley, M.A. (1993) Legal standards for the admissability of novel scientific evidence, in Saferstein, R. (Ed.), *Forensic Science Handbook*, III, Englewood Cliffs, NJ: Prentice-Hall.

Fauci, A.S. (1978) Mechanisms of the immunosuppressive and anti-inflammatory effects of glucocorticosteroids, *J. Immunopharmacol.*, 9:1–25.

Feingold, B.F. (1975) *Why Your Child Is Hyperactive*, New York: Random House.

Feinstein, H. (1984) *On Becoming William James*, Ithaca: Cornell University Press.

Feldmann, T.B. and Johnson, P.W. (1995) Cult membership as a source of self-cohesion: forensic implications, *Bull. Am. Acad. Psychiatry Law*, 23(2):239–248.

Felitti, V.J. (1991) Long-term medical consequences of incest, rape, and molestation, *South. Med. J.*, 84:328–331.

Felten, D.L., Felten, S.Y., Bellinger, D.L., Carlson, S.L., Ackerman, K.D., Madden, K.S., Olschowki, J.A., and Livnat, S. (1987) Noradrenergic sympathetic neural interactions with the immune system: structure and function, *Immunol. Rev.*, No. 100:225–260.

Felten, S.Y. and Felten, D.L. (1991) Innervation of lymphoid tissue, in Ader, R., Felten, D.L., and Cohen, N. (Eds.), *Psychoneuroimmunology*, 2nd ed., San Diego: Academic Press, pp. 27–71.

Ferguson, A. (1990) Food sensitivity or self-deception? [editorial], *New Engl. J. Med.*, 323(7):476–478.

Feyerabend, P. (1975) *Against Method*, London: Redwood Burn.

Fiedler, N., Maccia, C., and Kipen, H. (1992) Evaluation of chemically sensitive patients, *J. Occup. Med.*, 34(5):529–538.

Fiedler, N., Kipen, H., DeLuca, J., Kelly-McNeil, K., and Natelson, B. (1994) Neuropsychology and psychology of MCS, *Toxicol. Indust. Health*, 10(4/5):545–554.

Figley, C.R., Ed. (1985) *Trauma and Its Wake: The Study and Treatment of Post-Traumatic Stress Disorder*, New York: Brunner/Mazel.

File, S.E., Deakin, J.F.W., Longden, A., and Crow, T.J. (1979) An investigation of the role of the locus coeruleus in anxiety and agonistic behavior, *Brain Res.*, 169:411–420.

Fincher, J. (1984) *The Brain: Mystery of Matter and Mind*, New York: Torstar Books.

Findley, T. (1953) The placebo and the physician, *Med. Clin. North Am.*, 37:1821–1826.

Fink, L.A., Bernstein, D., Handelsman, L., Foote, J., and Lovejoy, M. (1995) Initial reliability and validity of the childhood trauma interview: a new multidimensional measure of childhood interpersonal trauma, *Am. J. Psychiatry*, 152(9):1329–1335.

Fink, M., Taylor, M.A., and Volavka, J. (1969) Anxiety precipitated by lactate [correspondence], *New Engl. J. Med.*, 281:1429.

Finkelhor, D., Hotaling, G., Lewis, I.A., and Smith, C. (1990) Sexual abuse in a national survey of adult men and women: prevalence, characteristics, and risk factors, *Child Abuse Neglect*, 14:19–28.

Fisher, H.K. and Olin, B.M. (1956) The dynamics of placebo therapy: a clinical study, *Am. J. Med. Sci.*, 232:504.

Fisher, S. and Greenberg, R.P. (1989) A second opinion: rethinking the claims of biological psychiatry, in Fisher, S. and Greenberg, R.P. (Eds.), *The Limits of Biological Treatments for Psychological Distress: Comparisons with Psychotherapy and Placebo*, Hillsdale, NJ: Lawrence Erlbaum Assoc., pp. 309–336.

Fitzgerald, F.T. (1983) Science and scam: alternative thought patterns in alternative health care [occasional notes], *New Engl. J. Med.*, 309(17):1066–1067.

Fliess, R. (1973) *Symbol, Dream and Psychosis*, New York: International University Press,.

Folkman, S. and Lazarus, R.S. (1986) Stress processes and depressive symptomatology, *J. Abnormal Psychol.*, 95(2):107–113.

Folkman, S. and Lazarus, R.S. (1988) The relationship between coping and emotion: implications for theory and research, *Soc. Sci. Med.*, 26(3):309–317.

Folkman, S., Lazarus, R.S., Gruen, R.J., and Delongis, A. (1986) Appraisal, coping, health status, and psychological symptoms, *J. Personality Social Psychol.*, 50(3):571–579.

Foote, S.L., Berridge, C.W., Adams, L.M., and Pineda, J.A. (1991) Electrophysiological evidence for the involvement of the locus coeruleus in alerting, orienting, and attending, *Prog. Brain Res.*, 88:521–532.

Forbes, G.B. (1982) Nutrition and hyperactivity [editorial], *JAMA*, 248(3):355–356.

Ford, C.V. (1983) *The Somatizing Disorders: Illness as a Way of Life*, New York: Elsevier.

Ford, C.V. (1986) The somatizing disorders, *Psychosomatics*, 27(5):327–337.

Frank, J.D. and Frank, J.B. (1991) *Persuasion and Healing: A Comparative Study of Psychotherapy*, 3rd ed., Baltimore: Johns Hopkins Press.

Frank, R., O'Neill, J.J., Utell, M.J., Hackney, J.D., Ryzin, J.V., and Brubaker, P.E., Eds. (1985) *Inhalation Toxicology of Air Pollution*, Philadelphia: Agency for Toxic Substances and Disease Registry (ATSDR), Publication #04-872000-17.

Frankenhaeuser, M. (1975) Experimental approaches to the study of catecholamines and emotions, in Levi, L. (Ed.), *Emotions: Their Parameters and Measurement*, New York: Raven Press, pp. 209–234.

Frankenhaeuser, M. and Jarpe, G. (1962) Psychophysiological reactions to infusions of a mixture of adrenalin and noradrenaline, *Scand. J. Psychol.*, 3, 21–29.

Frankenhaeuser, M. and Johansson, G. (1986) Stress at work: psychobiological and psychosocial aspects, *Int. Rev. Appl. Psychol.*, 35:287–299.

Franklin, J. (1994) *Poisons of the Mind*, keynote address, Convention of the Society of Toxicology, Dallas TX, March 16, 1994 (address: School of Journalism and Communication, University of Oregon, Eugene, OR 97403).

Freedman, M.R., Rosenberg, S.J., and Schmaling, K.B. (1991) Childhood sexual abuse in patients with paradoxical vocal cord dysfunction, *J. Nerv. Ment. Dis.*, 179:295–298.

French, Jr., J.R.P., Rogers, W.L., and Cobb, S. (1974) Adjustment as person-environment fit, in Coelho, G., Hamburg, D., and Adams, J. (Eds.), *Coping and Adaptation*, New York: Basic Books, pp. 316–333.

French, T.M. and Alexander, F. (1941) Psychogenic factors in bronchial asthma, *Psychosomatic Med.*, monograph 4.

Freud, A. (1939–1945) Infants without families, *Collected Works*, 3:543–664.

Freud, S. (1894) The justification for detaching from neurasthenia a particular syndrome: the anxiety-neurosis (in Jones, E., Ed. (1959) *The Collected Papers of Sigmund Freud*, Vol. 1, New York: Basic Books, pp. 76–106 [trans. by J. Riviere]).

Freud, S. (1936) *The Problem of Anxiety*, New York: Psychoanalytic Quarterly Press (trans. by H. A. Bunker),

Freud, S. (1952) *Remembering, Repeating, and Working Through, 1914*, standard ed., Vol. 12, London: Hogarth Press, pp.145–150.

Friedman, E.M. and Irwin, M.R. (1995) A role of CRH and the sympathetic nervous system in stress-induced immunosuppression, in Chrousos, G.P. et al. (Eds.), *Stress: Basis Mechanisms and Clinical Implications*, New York: New York Academy of Sciences, pp. 396–411.

Friedman, M.J., Southwick, S.M., and Charney, D.S. (1993) Pharmacotherapy for recently evacuated military casualties, *Military Med.*, 158:493–497.

Friedman, M.J., Schnurr, P.P., and McDonagh-Coyle, A. (1994) Post-traumatic stress disorder in the military veteran, *Psychiatric Clin. North Am.*, 17(2):265–277.

Fukuda, K., Straus, S., Hickie, I., Sharpe,M., Dobbins, J., and Komaroff, A. (1994) The chronic fatigue syndrome: a comprehensive approach to its definition and study, *Ann. Intern. Med.*, 121:953–959.

Funkenstein, D.H. (1955) The physiology of fear and anger, *Sci. Am.*, 192(5):74–80.

Funkenstein, D.H., Greenblatt, M., and Solomon, H.C. (1952) Norepinephrine-like and epinephrine-like substances in psychotic and psychoneurotic patients, *Am. J. Psychiatry*, 108:652–661.

Funkenstein, D.H., King, S.H., and Drolette, M.E. (1957) *Mastery of Stress*, Cambridge: Harvard University Press.

Fuster, J.M. (1989) *The Prefrontal Cortex: Anatomy, Physiology, and Neuropsychology of the Frontal Lobe*, 2nd ed, New York: Raven Press.

Garbarino, J. (1978) The elusive crime of emotional abuse, *Child Abuse Neglect*, 2:89–99.

Gabriel, S.E., O'Fallon, W.M., Kurland, L.T., Woods, J.E., Beard, C.M., and Melton, III, L.J. (1994) Risk of connective-tissue diseases and other disorders after breast implantation, *New Engl. J. Med.*, 330:1690–1702.

Galanter, M. (1982) Charismatic religious sects and psychiatry: an overview, *Am. J. Psychiatry*, 139(12):1539–1548.

Galanter, M. (1990) Cults and zealous self-help movements: a psychiatric perspective, *Am. J. Psychiatry*, 147(5):543–551.

Galapeaux, E.A. (1976) Chemical testing and therapy, in Dickey, L.D. (Ed.), *Clinical Ecology*, Springfield, IL: Charles C Thomas, pp. 402–407.

Galen, R.S. and Gambino, S.R. (1975) *Beyond Normality: The Predictive Value and Efficiency of Medical Diagnosis*, New York: John Wiley & Sons,.

Gallie, W.B. (1952) *Peirce and Pragmatism*, Harmondsworth, UK: Penguin Books.

Gamberale, F. (1985) Use of behavioral performance tests in the assessment of solvent toxicity, *Scand. J. Work Environ. Health*, 11(Suppl. 1):65–74.

Gammage, R.B. and Kaye, S.V., Eds. (1985) *Indoor Air and Human Health*, Chelsea, MI: Lewis Publishers.

Gann, D.S., Dallman, M.F., and Engeland, W.C. (1981) Reflex control and modulation of ACTH and corticosteroids, in McCann, S.M. (Ed.), *Endocrine Physiology*, III, Baltimore, MD: University Park Press.

Garvey, M.J., Cook, B., and Noyes, Jr., R. (1988) The occurrence of a prodrome of generalized anxiety in panic disorder, *Compr. Psychiatry*, 29(5):445–449.

Gelder, M.G. (1986) Panic attacks: new approaches to an old problem, *Br. J. Psychiatry*, 149:346–352.

Gelinas, D.J. (1983) The persisting negative effects of incest, *Psychiatry*, 46:312–332.

Gellhorn, E. (1964) Motion and emotion: the role of proprioception in the physiology and pathology of the emotions, *Psych. Rev.*, 71:457–472.

George, M.S., Ketter, T.A., Parekh, P.I., Horwitz, B., Herscovitch, P., and Post, R.M. (1995) Brain activity during transient sadness and happiness in healthy women, *Am. J. Psychiatry*, 152(3):341–351.

Getchell, T.V., Doty, R.L., Bartoshuk, L.M., and Snow, Jr., J.B., Eds. (1991) *Smell and Taste in Health and Disease*, New York: Raven Press.

Gilbert, C.M. (1988) Psychosomatic symptoms: implications for child sexual abuse, *Issues Ment. Health Nursing*, 9:399–408.

Gilbert, M.E. (1992a) A characterization of chemical kindling with the pesticide endosulfan, *Neurotoxicol. Teratol.*, 14:151–158.

Gilbert, M.E. (1992b) Neurotoxicants and limbic kindling, in Isaacson, R.L. and Jensen, K.F. (Eds.), *The Vulnerable Brain and Environmental Risks*. Vol. 1. *Malnutrition and Hazard Assessment*, New York: Plenum Press, pp. 173–192.

Gilman, S.L. (1982) *Seeing the Insane*, New York: Wiley.

Gloor, P., Olivier, A., Quesney, L.F., Andermann, F., and Horowitz, S. (1982) The role of the limbic system in experiential phenomena of temporal lobe epilepsy, *Ann. Neurol.*, 12:129–144.

Goddard, G.V., McIntyre, D.C., and Leech, C.K. (1969) A permanent change in brain function resulting from daily electrical stimulation, *Exp. Neurol.*, 25:295–330.

Golbert, T.M. (1971) Sublingual desensitization, *JAMA*, 217:1703–1704.

Golbert, T.M. (1975) A review of controversial diagnostic and therapeutic techniques employed in allergy, *J. Allergy Clin. Immunol.*, 56(3):170–190.

Gold, D., Bowden, R., Sixbey, J., Riggs, R., Katon, W.J., Ashley, R., Obrigewitch, R., and Corey, L. (1990) Chronic fatigue: a prospective clinical and virologic study, *JAMA*, 264:48–53.

Gold, P.E. and Zornetzer, S.F. (1983) The mnemon and its juices: neuromodulation of memory processes, *Behav. Neural Biol.*, 38:151–189.

Gold, P.W., Goodwin, F.K., and Chrousos, G.P. (1988a) Clinical and biochemical manifestations of depression: relation to the neurobiology of stress, Part I, *New Engl. J. Med.*, 319:348–353.

Gold, P.W., Goodwin, F.K., and Chrousos, G.P. (1988b) Clinical and biochemical manifestations of depression: relation to the neurobiology of stress, Part II, *New Engl. J. Med.*, 319:413–420.

Goldberg, D. (1996) A dimensional model for common mental disorders, *Br. J. Psychiatry*, 168(Suppl. 30):44–49.

Goldberg, D. and Huxley, P. (1970) *Mental Illness in the Community*, London: Tavistock.

Goldberg, D. and Bridges, K. (1991) Minor psychiatric disorders and neurasthenia in general practice, in Gastpar, M. and Kielholz, P. (Eds.), *Problems of Psychiatry in General Practice*, Lewiston, NY: Hogrefe & Huber, pp. 79–88.

Goldberg, D., Gask, L., and O'Dowd, T. (1989) The treatment of somatization: teaching techniques of re-attribution, *J. Psychosomatic Res.*, 33:689–695.

Golding, J.M., Cooper, M.L., and George, L.K. (1997) Sexual assault history and health perceptions: seven general population studies, *Health Psychol.*, 16(5):417–425.

Goldstein, D. (1995) Clinical assessment of sympathetic response to stress, in Chrousos, G.P. et al. (Eds.), *Stress: Basis Mechanisms and Clinical Implications*, New York: New York Academy of Sciences, pp. 570–593.

Goldstein, M.L. (1968) Physiological theories of emotion: a critical historical review from the standpoint of behavior theory, *Psychol. Bull.*, 69:23–40.

Gomez, R., Schvaneveldt, R.W., and Staudenmayer, H. (1996) Assessing beliefs about "environmental illness/multiple chemical sensitivity", *J. Health Psychol.*, 1(1):107–123.

Gori, G.B. (1996) The role of objective science in policy development: evidence versus conjecture, *Reg. Toxicol. Pharmacol.*, 24(1):S3–S7.

Gothe, C.J., Molin, C.M., and Nilsson, C.G. (1995) The environmental somatization syndrome, *Psychosomatics*, 36(1):1–11.

Gots, R.E. (1993a) *Toxic Risks: Science, Regation, and Perception*, Boca Raton, FL: Lewis Publishers.

Gots, R.E. (1993b) Medical hypothesis and medical practice: autointoxication and multiple chemical sensitivities, *Reg. Toxicol. Pharmacol.*, 18:2–12.

Gots, R.E., Tamar, D.H., Flamm, W.G., and Carr, C.J. (1993) Multiple chemical sensitivities: a symposium on the state of the science, *Reg. Toxicol. Pharmacol.*, 18:61–78.

Graham. D.T., Wolf, S., and Wolff, H.G. (1950) Changes in tissue sensitivity associated with varying life situations and emotions, their relevance to allergy, *J. Allergy*, 21:478–86.

Gray, G.C., Coate, B.D., Anderson, C.M., Kang, H.K., Berg, S.W., Wignall, F.S., Knoke, J.D., and Barrett-Connor, E. (1996) The postwar hospitalization experience of U.S. veterans of the Persian Gulf War, *New Engl. J. Med.*, 335(20):1506–1513.

Gray, H. (1973) *Anatomy of the Human Body*, 29th American ed. (edited by C. M. Goss), Philadelphia: Lea & Febiger.

Gray, J. and Bentovim, A. (1996) Illness induction syndrome: paper I — a series of 41 children from 37 families identified at the Great Ormond Street Hospital for Children NHS trust, *Child Abuse Neglect*, 20(8):655–673.

Gray, T.S. (1991) Amygdala: role in autonomic and neuroendocrine responses to stress, in McCubbin, J.A., Kaufmann, P.G., and Nemeroff, C.B. (Eds.), *Stress, Neuropeptides and Systemic Disease*, San Diego: Academic Press, pp. 37–53.

Green, A.H. (1989) Physical and sexual abuse of children, in Kaplan, H.I. and Sadeck, B.I. (Eds.), *Comprehensive Textbook of Psychiatry*, Vol. 2, 5th ed., Baltimore: Williams & Wilkins, pp. 1962–1970.

Green, A.H., Voeller, K. Gaines, R.W., and Kubie, J. (1981) Neurological impairment in maltreated children, *Child Abuse Neglect*, 5:129–134.

Grinker, R.R. and Spiegel, J.J. (1945) *Men Under Stress*, New York: McGraw-Hill.

Groves, J.E. (1978) Taking care of the hateful patient, *New Engl. J. Med.*, 298(16):883–887.

Guglielmi, R.S., Cox, D.J., and Spyker, D.A. (1994) Behavioral treatment of phobic avoidance in multiple chemical sensitivity, *J. Behav. Ther. Exp. Psychiatry*, 25(3):197–209.

Guidotti, T.L., Alexander, R.W., and Fedoruk, M.J. (1987) Epidemiologic features that may distinguish between building-associated illness outbreaks due to chemical exposure or psychogenic origin, *J. Occup. Med.*, 29:148–150.

Gurvits, T.V., Shenton, M.E., Hokama, H., Ohta, H., Lasko, N.B., Gilbertson, M.W., Orr, S.P., Kikinis, R., Jolesz, F.A., McCarley, R.W., and Pitman, R.K. (1996) Magnetic resonance imaging study of hippocampal volume in chronic, combat-related posttraumatic stress disorder, *Biol. Psychiatry*, 40:1091–1099.

Guthrie, E., Creed, F., Dawson, D., and Tomenson, B. (1991) A controlled trial of psychological treatment for the irritable bowel syndrome, *Gastroenterology*, 100:450–457.

Guyenet, P.G. (1991) Central noradrenergic neurons: the autonomic connection, *Prog. Brain Res.*, 88:365–380.

Gyuk, I., Dietrich, F., and Wisecup, W. (1989) Case study of putative hyper-sensitivity to electric and magnetic fields, in *Contractors Review Book of Abstracts*, Portland, OR, Nov. 13–16, p. A-24,

Halgren, E., Walter, R.D., Cherlow, D.G., and Crandall, P.H. (1978) Mental phenomenon evoked by electrical stimulation of the human hippocampal formation and amygdala, *Brain*, 101:83–117.

Hamilton, V. and Warburton, D.M., Eds. (1979) *Human Stress and Cognition: An Information Processing Approach*, Chichester: Wiley.

Hansel, F.K. (1968) *Allergy and Immunity in Otolaryngology*, 2nd ed., Rochester, MN: American Academy of Ophthalmology and Otolaryngology, pp. 134–135.

Harlow, H.F. (1958) The nature of love, *Am. J. Psychol.*, 13:673–685.

Harlow, H.F. and Zimmerman, R.R. (1959) Affectional responses in the infant monkey, *Science*, 21:421–432.

Haugaard, J.J. and Emery, R.E. Methodological issues in child sexual abuse research, *Child Abuse Neglect*, 1989, 13:89–100.

Havender, W.R. and Flynn, L.T. (1992) *Does Nature Know Best? Natural Carcinogens in American Food*, New York: American Council on Science and Health.

Hedge, A. (1995) Sick Building Syndrome and Mass Psychogenic Illness, presented at the Aspen Environmental Medicine Conference, Aspen, CO, Sept. 9, 1995.

Heilman, K.M. and Watson, R.T. (1977) The neglect syndrome — a unilateral defect of the orienting response, in Harnad, S., Doty, R.W., Jaynes, J., Goldstein, L., and Krauthamer, G. (Eds.), *Lateralization in the Nervous System*, New York: Academic Press, pp.285–302.

Heilman, K.M., Watson, R.T., Valenstein, E., and Goldberg, M.E. (1987) Attention: behavior and neural mechanisms, in Plum, F. (Ed.), *The Handbook of Physiology*. Section 1. *The Nervous System*, Bethesda, MD: American Physiological Society, pp. 461–481.

Hemingway, III, R.B. and Reigle, T.G. (1987) The involvement of endogenous opiate systems in learned helplessness and stress-induced analgesia, *Psychopharmacology*, 93:353–357.

Henle, M. (1974) The cognitive approach: the snail beneath the shell, in Rosner, S. and Abt, L.E. (Eds.), *Essays in Creativity*, Croton-on-Hudson, NY: North River Press, pp. 23–44.

Herman, J.L. (1981) *Father-Daughter Incest*, Cambridge, MA: Harvard University Press.

Herman, J.L. (1992) Complex PTSD: a syndrome of survivors of prolonged and repeated trauma, *J. Traumatic Stress*, 5:377–391.

Herman, J.L. and Schztzow, E. (1987) Recovery and verification of memories of childhood sexual trauma, *Psychoanalytic Psychol.*, 4:1–14.

Herman, J.L., Russell, D., and Trocki, K. (1986) Long-term effects of incestuous abuse in childhood, *Am. J. Psychiatry*, 143:1293–1296.

Herman, J.L., Perry, J.C., and van der Kolk, B.A. (1989) Childhood trauma in borderline personality disorder, *Am. J. Psychiatry*, 146:490–495.

Herrmann, W.M., Ed. (1982) *Electroencephalography in Drug Research*, New York, Stuttgart: Gustav Fischer.

Hess, E.H. and Polt, J.M. (1960) Pupil size as related to interest value of visual stimuli, *Science*, 132:349–350.

Hess, E.H. and Polt, J.M. (1964) Pupil size in relation to mental activity during simple problem solving, *Science*, 143:1190–1192.

Heuser, G., Mena, I., Goldstein, J., Thomas, C., and Alamos, F. (1993) Neurospect findings in patients exposed to neurotoxic chemicals [abstract], *Clin. Nucl. Med.*, 18:923.

HEW (1978) Laetrile: the Commissioner's decision, U.S. Department of Health Education and Welfare, Washington, D.C.: U.S. Government Printing Office, Publication No. 77-3056, p. xii.

Hewer, W. (1983) The relationship between the alternative practioner and his patient, *Psychother. Psychosomatics*, 40:172–180.

Hickie, I. and Wilson, A. (1994) A catecholamine model of fatigue, *Br. J. Psychiatry*, 165:275–276.

Hickie I., Lloyd A., Wakefield D., and Parker, G. (1990) The psychiatric status of patients with chronic fatigue syndrome, *Br. J. Psychiatry*, 156:534–540.

Hickie, I., Hadzi-Pavlovic, D., and Ricci, C. (1997) Reviving the diagnosis of neurasthenia [editorial], *Psychol. Med.*, 37(5):989–994.

Hill, A. B. (1965) The environment and disease: association or causation, *Proc. Roy. Soc. Med.: Sect. Occup. Med.*, 58(5):295–300.

Hines, J.F. (1993) A comparison of clinical diagnoses among male and female soldiers deployed during the Persian Gulf War, *Military Med.*, 158:99–101.

Hobfoll, S.E. (1989) Conservation of resources: a new attempt at conceptualizing stress, *Am. Psychol.*, 44(3):513–524.

Hodgson, M.J., Frohliger, J., Permar, E. et al. (1991) Symptoms and microenvironmental measures in non-problem buildings, *J. Occup. Med.*, 33:527–533.

Hoffer, E. (1989) *The True Believer*, New York: Harper Perennial.

Holmes, G.P., Kaplan, J.E., Stewart, J.A., Hunt, B., Pinsky, P.F., and Schonberger, L.B. (1987) A cluster of patients with a chronic mononucleosis-like syndrome: is Epstein-Barr virus the cause?, *JAMA*, 257:2297–2302.

Holmes, G.P., Kaplan, J.E., Gantz, N.M., Komaroff, A.L., Schonberger, L.B., Straus, S.E., Jones, J.F., Bubois, R.E., Cunningham-Rundles, C., Pahwa, S., Tosato, G., Zegans, L.S., Purtilo, D.T., Brown, N., Schooley, R.T., and Brus, I. (1988) Chronic fatigue syndrome: a working case definition, *Ann. Intern. Med.*, 108(3):387–389.

Holroyd, K.A. (1976) Cognition and desensitization in the group treatment of test anxiety, *J. Consulting Clin. Psychol.*, 44:991–1001.

Honzak, R., Horackova, E., and Culik, A. (1972) Our experience with the effect of placebo in some functional and psychosomatic disorders, *Activitas Nervosa Superior (Prague)*, 14:184–185.

Horney, K. (1937) *The Neurotic Personality of Our Time*, New York: W.W. Norton & Company.

Horowitz, M.J. (1986) *Stress Response Syndromes*, 2nd ed., Northvale, NJ: Jason Aronson.

Horowitz, M.J., Adams, J.E., and Rutkin, B.B. (1968) Visual imagery on brain stimulation, *Arch. Gen. Psychiatry*, 19:469–486.

Horwitz, C.A., Henle W., Henle, G., Rudnick, H., and Latts, E. (1985) Long-term serological follow-up of patients for Epstein-Barr virus after recovery from infectious mononucleosis, *J. Infect. Dis.*, 151:1150–1153.

Howard, L.M. and Wessely, S. (1993) The psychology of multiple allergy, *Br. Med. J.*, 307:747–748.

Howard, P.K. (1994) *The Death of Common Sense: How Law is Suffocating America*, New York: Random House.

Howlett, T.A. and Rees, L.H. (1986) Endogenous opioid peptides and hypothalamo-pituitary function, *Ann. Rev. Physiol.*, 48:527–536.

Hubbard, R.L. (1990) *Clear Body, Clear Mind: The Effective Purification Program*, Los Angeles, CA: Bridge Publications.

Hubel, D.H., Henson, C.O., Rupert, A., and Galambos, R. (1959) "Attention" units in the auditory cortex, *Science*, 129:1279–1280.

Hubert, P. (1991) *Galileo's Revenge: Junk Science in the Court Room*, New York: Basic Books.

Hudson, R.P. (1989) Theory and therapy: ptosis, stasis, and autointoxication, *Bull. History Med.*, 63:392–413.

Hudziak, J.J., Boffeli, T.J., Kriesman, J.J., Battaglia, M.M., Stanger, C., and Guze, S.B. (1996) Clinical study of the relation of borderline personality disorder to Briquet's syndrome (hysteria), somatization disorder, antisocial personality disorder, and substance abuse disorders, *Am. J. Psychiatry*, 153(12):1598–1606.

Hummel, T. (1997) *Odor Perception Studies in Self-Identified "Sensitive" and "Non-Sensitive" Individuals*, paper presented at the ESRI sponsored symposium, Low-Level Environmental Exposures: State-of-the-Science, Seattle, WA, Oct. 21.

Hummel, T., Roscher, S., Jaumann, M.P., and Kobal, G. (1996) Intranasal chemoreception in patients with multiple chemical sensitivity: a double-blind investigation, *Reg. Toxicol. Pharmacol.*, 24:S79–S86.

Hunter, R. S., Kilstrom, N., and Loda, F. (1985) Sexually abused children: identifying masked presentations in a medical setting, *Child Abuse Neglect*, 9:17–25.

Hyams, K.C., Wignall, F.S., and Roswell, R. (1996) War syndromes and their evaluation: from the U.S. Civil War to the Persian Gulf War, *Ann. Intern. Med.*, 125(5):398–405.

Hyman, R. (1993) Occult health practices, in Barrett, S. and Jarvis, W.T. (Eds.), *The Health Robbers: A Close Look at Quackery in America*, Buffalo, NY: Prometheus Books, pp. 55–66.

Hyman, S.E. and Nestler, E.J. (1996) Initiation and adaptation: a paradigm for understanding psychotropic drug action, *Am. J. Psychiatry*, 153(2):151–162.

Insel, T.R., Ninan, P.T., Aloi, J., Jimerson, D.C., Skolnick, P., and Paul, S.M. (1984) A bensodiazepine receptor-mediated model of anxiety: studies in nonhuman primates and clinical implications, *Arch. Gen. Psychiatry*, 41:741–750.

IPCS (International Programme on Chemical Safety): (1996) Conclusions and recommendations of a workshop on multiple chemical sensitivities (MCS), *Reg. Toxicol. Pharmacol.*, 24(1):S188–S189.

Isaacson, R.L. (1982) *The Limbic System*, 2nd ed., New York: Plenum Press.

Isaacson, R.L. and Pribram, K., Eds. (1975) *The Hippocampus*, Vol. I, New York: Plenum Press.

Ishiyama, F.I. (1990) Meaningful life therapy: use of Morita therapy principles in treating patients with cancer and intractable diseases, *Int. Bull. Morita Ther.*, 3:77–88.

Ito, Y., Teicher, M.H., Glod, C.A., Harper, D., Magnus, E., and Gelbard, H.A. (1993) Increased prevalence of electrophysiological abnormalities in children with psychological, physical, and sexual abuse, *J. Neuropsychiatry Clin. Neurosci.*, 5:401–408.

Jacobs, S., Mason, J., Kosten, T., Brown, S., and Ostfeld, A. (1984) Urinary-free cortisol excretion in relation to age in acutely stressed persons with depressive symptoms, *Psychosomatic Med.*, 46:213–221.

Jacobson A. and Richardson, B. (1987) Assault experiences of 100 psychiatric inpatients: evidence of the need for routine therapy, *Am. J. Psychiatry*, 144:908–913.

Jacobson, A. and Herald, C. (1990) The relevance of childhood sexual abuse to adult psychiatric inpatient care, *Hospital Community Psychiatry*, 41:154–158.

Jacoby, L.L., Lindsay, D.S., and Toth, J.P. (1992) Unconscious influences revealed: attention, awareness, and control, *Am. Psychol.*, 47(6):802–809.

James, W. (1884) What is an emotion?, *Mind*, 9:188–205.

James, W. (1890) *Principles of Psychology*, New York: Holt.

Janis, I.L. (1982) Decision making under stress, in Goldberger, L. and Breznitz, S. (Eds.), *Handbook of Stress: Theoretical and Clinical Aspects*, New York: The Free Press, pp. 69–87.

Janis, I.L. and King, B.T. (1954) The influence of role-playing on opinion change, *J. Abnormal Social Psychol.*, 49:211–218.

Jasper, H.H. (1958) Reticular-cortical systems and theories of the integrated action of the brain, in Harlow, H.F. and Woolsey, C.N. (Eds.), *Biological and Biochemical Bases of Behavior*, Madison: University of Wisconsin Press, pp. 37–61.

Jenike, M.A., Baer, L., Summergrad, P., Minichiello, W.E., Holland, A., and Seymour, R. (1990) Sertraline in obsessive-compulsive disorder: a double blind comparison study, *Am. J. Psychiatry*, 147(7):923–928.

Jewett, D.L. (1992a) Diagnosis and treatment of hypersensitivity syndrome: proceedings of the Association of Occupational and Environmental Clinics (AOEC) Workshop on Multiple Chemical Sensitivity, September 20–21, 1991, Washington, D.C., sponsored by Agency for Toxic Substances and Disease Registry (ATSDR), *Toxicol. Indust. Health*, 8(4):111–123.

Jewett, D.L. (1992b) Research strategies for investigating multiple chemical sensitivity, proceedings of the Association of Occupational and Environmental Clinics (AOEC) Workshop on Multiple Chemical Sensitivity, September 20–21, 1991, Washington, D.C., sponsored by Agency for Toxic Substances and Disease Registry (ATSDR), *Toxicol. Indust. Health*, 8(4):175–179.

Jewett, D.L., Fein, G., and Greenberg, M.H. (1990) A double-blind study of symptom provocation to determine food sensitivity, *New Engl. J. Med.*, 323:429–433.

Jones, D.P.H. (1987) The untreatable family, *Child Abuse Neglect*, 11:409–420.

Jones, J.F. (1995) Matter of Fatigue, presented at the Aspen Environmental Medicine Conference, Aspen, CO, September 7–9.

Joseph, R. (1992) The limbic system: emotion, laterality, and unconscious mind, *Psychoanalytic Rev.*, 79(3):405–456.

Jung, C.G. (1967) *Symbols of Transformations*, Bollinger Series, Vol. 5 (trans. by R.F.C. Hull), New York: Princeton University Press.

Jung, C.G. (1971) *Psychological Types*, Bollinger Series, Vol. 6 (trans. by R.F.C. Hull), New York: Princeton University Press.

Kagan, J., Reznick, J.S., and Snidman N. (1988) Biological bases of childhood shyness, *Science*, 240:167–171.

Kahn, E. and Letz, G. (1989) Clinical ecology: environmental medicine or unsubstantiated theory, *Ann. Intern. Med.*, 111(2):104–106.

Kahneman, D. and Tversky, A. (1973) On the psychology of prediction, *Psych. Rev.*, 80:237–251.

Kailin, E.W. and Collier, R. (1971) "Relieving" therapy for antigen exposure, *JAMA*, 217:78.

Kalivas, P.W. and Duffy, P. (1989) Similar effects of daily cocaine and stress on mesocorticolimbic dopamine neurotransmission in the rat, *Biol. Psychiatry*, 25:913–928.

Kalivas, P.W., Richardson-Carlson, R., and Van Orden, G. (1986) Cross-sensitization between foot shock stress and enkephalin-induced motor activity, *Biol. Psychiatry*, 21:939–950.

Kandel, E.R. and Hawkins, R.D. (1992) The biological basis of learning and individuality, *Sci. Am.*, Sept:79–86.

Kang, H.K. and Bullman, T.A. (1996) Mortality among U.S. veterans of the Persian Gulf War, *New Engl. J. Med.*, 335(20):1498–1504.

Kardiner, A. (1941) *The Traumatic Neuroses of War*, New York: Hoeber.

Kasl, S.V. and Cooper, C.L., Eds. (1987) *Stress and Health: Issues in Research Methodology*, Chichester: Wiley.

Katon, W. (1986) Panic disorder: epidemiology, diagnosis and treatment in primary care, *J. Clin. Psychiatry*, 47(10, suppl.):21–27.

Katon, W., Kleinman, A., and Rosen, G. (1982a) Depression and somatization: a review, Part I, *Am. J. Med.*, 72:127–135.

Katon, W., Kleinman, A., and Rosen, G. (1982b) Depression and somatization: a review, Part II, *Am. J. Med.*, 72:241–247.

Katon W., Lin, E., von Korff, M., Russo, J., Lipscomb, P., and Bush, T. (1991) Somatization: a spectrum of severity, *Am. J. Psychiatry*, 148(1):34–40.

Kazdin, A.E. (1976) Statistical analyses for single-case experimental designs, in Hersen, M. and Barlow, D.H. (Eds.), *Single-Case Experimental Designs*, New York: Pergamon Press, pp. 265–316.

Kazdin, A.E. (1977) Research issues in covert conditioning, *Cognitive Ther. Res.*, 1:45–48.

Kazdin, A.E. (1978) Methodological and interpretive problems of single-case experimental designs, *J. Consulting Clin. Psychol.*, 46(4):629–642.

Keane, T.M., Caddell, J.M., and Taylor, K.L. (1988) Mississippi Scale for combat-related posttraumatic stress disorder: three studies in reliability and validity, *J. Consulting Clin. Psychol.*, 56:85–90.

Kehrl, H., Ball, B., Prah, J. et al. (1995) Clinical Evaluation of Persons with Self-Report of Chemical Sensitivity Recruited from the Local Community, presented at the Aspen Environmental Medicine Conference, Aspen, CO, Sept. 7–9.

Kelley, H.H. (1967) Attribution theory in social psychology, in Levine, D. (Ed.), *Nebraska Symposium on Motivation*, Lincoln, NE: University of Nebraska Press, pp. 192–238.

Kellner, R. Somatization — theories and research, *J. Nerv. Ment. Dis.*, 1990, 178:150–160.

Kellner, R. (1991) *Somatization and Hypochondriasis*, New York: Praeger.

Kempe, C.H., Silverman, F.N., Steele, B.F., Droegmueller, W., and Silver, H.K. (1962) The battered child syndrome, *JAMA*, 181(1):105–112.

Kephart, W.M. (1976) *Extraordinary Groups: The Sociology of Unconventional Life-Styles*, New York: St. Marten's Press.

Kimberly, A.L., Vaillant, G.E., Torrey, W.C., and Elder, G.H. (1995) A 50-year prospective study of the psychological sequelae of World War II combat, *Am. J. Psychiatry*, 152(4):516–522.

Kinsbourne, M. (1994) Sugar and the hyperactive child [letter to editor], *New Engl. J. Med.*, 330(5):355–356.

Kinsman, R.A., Dahlem, N.W., Spector, S., and Staudenmayer, H. (1977) Observations on subjective symptomatology, coping behavior and medical decisions in asthma, *Psychosomatic Med.*, 39(3):102–119.

Kipen, H.M., Hallman, W., Kelly-McNeil, K., and Fiedler, N. (1995) Measuring chemical sensitivity prevalence: a questionnaire for population studies, *Am. J. Public Health*, 85(4):574–577.

Kirkpatrick, C.H. (1989) Chronic mucocutaneous candidiasis, *Eur. J. Clin. Microbiol. Infect. Dis.*, 8:448–456.

Kirsch, I. (1978) The placebo effect and the cognitive-behavioral revolution, *Cognitive Ther. Res.*, 2:255–264.

Kirsch, I. and Henry, D. (1977) Extinction versus credibility in the desensitization of speech anxiety, *J. Consulting Clin. Psychol.*, 45:1052–1059.

Kirsch, I., Tennen, H., Wickless, C., Saccone, A.J., and Cody, A. (1983) The role of expectancy in fear reduction, *Behav. Ther.*, 14:520–533.

Kizer, K.W., Joseph, S., Moll, M., and Rankin, J.T. (1995) Unexplained illness among Persian Gulf war veterans in an Air National Guard unit: preliminary report for August 1990–March 1995, *JAMA*, 274(1):16–17.

Klein, D.F. (1981) Anxiety reconceptualized, in Klein, D.F. and Rabkin, J.G. (Eds.), *New Research and Changing Concepts*, New York: Raven Press.

Klein, D.F. (1993) False suffocation alarms, spontaneous panics, and related conditions, *Arch. Gen. Psychiatry*, 50:306–317.

Klein, R. (1989) Introduction to the disorders of the self, in Masterson, J.F. and Klein, R. (Eds.), *Psychotherapy of the Disorders of the Self*, New York: Brunner/Mazel, pp. 30–46.

Kleinman, A. (1982) Neurasthenia and depression: a study of somatization and culture in China, *Culture Med. Psychiatry*, 6:117–190.

Kleinman, A. (1986) *Social Origins of Distress and Disease: Depression, Neurasthenia, and Pain in Modern China*, New Haven: Yale University Press.

Kling, M.A., Roy, A., Doran, A.R., Calabrese, J.R., Rubinow, D.R., Whitfield, Jr., H.J., May, C., Post, R.M., Chrousos, G.P., Gold, P.W. (1991) CSF levels of CRH, ACTH, and SRIF in Cushing's syndrome, major depression, and normal volunteers: physiological and pathophysiological interrelationships, *J. Clin. Endocrinol. Metabol.*, 72:260–271.

Kluft, R.P. (1985) Hypnotherapy of childhood multiple personality disorder, *Am. J. Clin. Hypnosis*, 27:201–210.

Knott, V., Chaudry, R., and Lapierre, Y. D. (1981) Panic induced by sodium lactate: electrophysiological correlates, *Progr. Neuropsychopharmacol.*, 5:511–514.

Knutson, J.F. (1988) Physical and sexual abuse of children, in Rauth, D.K. (Ed.), *Handbook of Pediatric Psychology*, New York: Guilford, pp. 32–70.

Ko, G.N., Elsworth, J.D., Roth, R.H., Rifkin, B.G., Leigh, H., and Redmond, Jr., E. (1983) Panic-induced elevation of plasma MHPG in phobic anxious patients: effects of clonidine and imipramine, *Arch. Gen. Psychiatry*, 40:425–430.

Kojo, I. (1988) The mechanism of the psychophysiological effects of placebo, *Med. Hypotheses*, 27:261–264.

Kohut, H. (1971) *The Analysis of the Self*, New York: International Universities Press.

Kohut, H. (1977) *The Restoration of the Self*, New York: International Universities Press.

Kohut, H. and Wolf, E. (1978) Disorders of the self and their treatment: an outline, *Int. J. Psychoanalysis*, 59:414–425.

Kolb, L.C. (1973) *Modern Clinical Psychiatry*, 8th ed., Philadelphia: W.B. Saunders.

Kolb, L.C. (1987) Neurophysiological hypothesis explaining posttraumatic stress disorder, *Am. J. Psychiatry*, 144:989–995.

Kopin, I.J., Polinsky, R.J, Oliver, J.A., Oddershede, I.R., and Ebert, M.H. (1983) Urinary catecholamine metabolites distinguish different types of sympathetic neuronal dysfunction in patients with orthostatic hypotension, *J. Clin. Endocrinol. Metabol.*, 57:632–637.

Koppel, C. and Fahron, G. (1995) Toxicological and neuropsychological findings in patients presenting to an environmental toxicology service, *Clin. Toxicol.*, 33(6):625–629.

Kosten, T.R., Mason, J.W., Giller, E.L., Ostroff, R.B., and Harkness, L. (1987) Sustained urinary norepinephrine and epinephrine elevation in post-traumatic stress disorder, *Psychoneuroendocrinology*, 12:13–20.

Kostowski, W., Giacolone, E., Garattini, S., and Valzelli, L. (1968) Studies on behavioral and biochemical changes in rats after lesions of the midbrain raphe, *Eur. J. Pharmacol.*, 4:371–376.

Kraepelin, E. (1921) *Manic-Depressive Insanity and Paranoia* (trans. by R.M. Barclay; G.M. Robertson, Ed.), Edinburgh: E. & S. Livingston.

Kreiss, K. (1993) The sick building syndrome in office buildings: a breath of fresh air, *New Engl. J. Med.*, 328:821–827.

Krishnan, K.R.R., Venkataraman, S., Doraiswamy, P.M., Reed, D.A., and Richie, J.C. (1991) Current concepts in hypothalamo-pituitary-adrenal axis regulation, in McCubbin, J.A., Kaufmann, P.G., and Nemeroff, C.B. (Eds.), *Stress, Neuropeptides, and Systemic Disease*, San Diego: Academic Press, pp. 19–35.

Kruesi, M.J.P., Dale, J.K., and Straus, S.E. (1989) Psychiatric diagnoses in patients who have chronic fatigue syndrome, *J. Clin. Psychiatry*, 50:53–56.

Krystal, H. (1978) Trauma and affects, *Psychoanalytic Study Child*, 33:81–116.

Krystal, H., Ed. (1988) *Integration and Self Healing: Affect, Trauma, Alexithymia*, Hillsdale, NJ: Lawrence Erlbaum Assoc.

Krystal, J.H., Bennett, A.L., Bremner, J.D., Southwick, S.M., and Charney, D.S. (1995) Toward a cognitive neuroscience of dissociation and altered mental functions in post-traumatic stress disorder, in Friedman, M.J., Charney, D.S., and Deutch, A.Y. (Eds.), *Neurobiological and Clinical Consequences of Stress: From Normal Adaptation to Post-Traumatic Stress Disorder*, Hagerstown, MD: Lipincott-Raven, pp. 239–269.

Kuhn, T.S. (1970) *The Structure of Scientific Revolutions*, 2nd ed., Chicago: University of Chicago Press.

Kulka, R.A., Schlenger, W.E., Fairbank, J.A., Hough, R.L., Jordan, B.K., Marmar, C.R., and Weiss, D.S. (1990) *Trauma and the Vietnam War Generation: Report of Findings from the National Vietnam Veterans Readjustment Study*, New York: Brunner/Mazel.

Kupfermann, I. (1979) Modulatory actions of neurotransmitters, *Ann. Rev. Neurosci.*, 2:447–465.

Kurt, T.L. (1995a) Sauna-Depuration: Toxicokinetics, presented at 2nd Aspen Environmental Medicine Conference, Aspen, CO, Sept. 7–9.

Kurt, T.L. (1995b) Multiple chemical sensitivities — a syndrome of pseudotoxicity manifest as exposure perceived symptoms, *Clin. Toxicol.*, 33(2):101–105.

Lacey, J.I. and Lacey, B.C. (1958) Verification and extension of the principle of autonomic response stereotypy, *Am. J. Psychol.*, 71:50–73.

Lader, M. (1975) *The Psychophysiology of Mental Illness*, London: Routledge and Kegan Paul.

Lakatos, I. (1970) Falsification and the methodology of scientific research programmes, in Lakatos, I. and Musgrave, A. (Eds.), *Criticism and the Growth of Knowledge: Proceedings of the International Colloquium on the Philosophy of Science, London, 1965*, London: Cambridge University Press, pp. 91–195.

Lamielle, M. (1994) Response panel #1, proceedings of the Conference on Low-Level Exposure to Chemicals and Neurobiologic Sensitivity, April 6–7, 1994, Baltimore, MD, Sponsored by Agency for Toxic Substances and Disease Registry (ATSDR), *Toxicol. Indust. Health*, 10(4/5):327–332.

Landfield, P., Waymire, J., and Lynch, G. (1978) Hippocampal aging and adrenocorticoids: quantitative correlations, *Science*, 202:1098–1101.

Lange, C.G. (1922) The emotions, in Dunlap, K. (Ed.), *The Emotions*, Baltimore, MD: Williams & Wilkins, pp. 33–90.

Lapierre, Y.D., Knott, V.J., and Gray, R. (1984) Psychophysiological correlates of sodium lactate, *Psychopharmacol. Bull.*, 20:50–57.

Lasagna, L., Mosteller, F., von Felsinger, J.M., and Beecher, H.K. (1954) A study of the placebo response, *Am. J. Med.*, 16:770–779.

Lashof, J.C. (1996) *Presidential Advisory Committee on Gulf War Veterans' Illnesses*, Washington, D.C.: U.S. Government Printing Office (ISBN 0-16-048942-3).

Lazarus, R.S. (1966) *Psychological Stress and the Coping Process*, New York: McGraw-Hill.

Lazarus, R.S. and Folkman, S. (1984) *Stress, Appraisal, and Coping*, New York: Springer.

LeDoux, J.E. (1987) Emotion, in Blum, F. (Ed.), *Handbook of Physiology. Section 1. The Nervous System*, Bethesda, MD: American Physiological Society, pp. 419–459.

LeDoux, J.E. (1994) Emotion, memory and the brain, *Sci. Am.*, June:50–57.

LeDoux, J.E. (1995) Emotions: clues from the brain, *Ann. Rev. Psychol.*, 46:209–235.

LeDoux, J.E. (1996) *The Emotional Brain: The Mysterious Underpinnings of Emotional Life*, New York: Simon & Schuster.

Lees-Haley, P.R. and Brown, R.S. (1992) Biases in perception and reporting following a perceived toxic exposure, *Perceptual Motor Skills*, 75:531–544.

Lehman, C.W. (1980) The leukocytic food allergy test: a study of its reliability and reproducibility. Effect of diet and sublingual food drops on this test, *Ann. Allergy*, 45:150–158.

Lehninger, A.L. (1970) *Biochemistry*, New York: Worth Publishers, p. 207.

Leibenluft, E. (1996) Sex is complex [editorial], *Am. J. Psychiatry*, 153(8):969–972.

Lencz, T., McCarthy, G., Bronen, R.A., Scott, T.M., Inserni, J.A., Sass, K.J., Novelly, R.A., Kim, J.H., and Spencer, D.D. (1992) Quantitative magnetic resonance imaging in temporal lobe epilepsy: relationship to neuropathology and neuropsychological function, *Ann. Neurology*, 31:629–637.

LeRoy, J., Davis, T.H., and Jason, L.A. (1996) Treatment efficacy: a survey of 305 MCS patients, *CFIDS Chronicle*, Winter:52–53.

Lesch, K.P., Laux, G., Schulte, H.M., Pfuller, H., and Beckmann, H. (1988) Corticotropin and cortisol response to human CRH as a probe for HPA system integrity in major depressive disorder, *Psychiatry Res.*, 24:25–34.

Lessof, M.H., Wraith, D.G., Merrett, T.G., Merrett, J., and Buisseret, P.D. (1980) Food allergy and intolerance in 100 patients — local and systemic effect, *Q. J. Med.*, 49:259–271.

Leventhal, E.A., Hansell, S., Diefenbach, M., Leventhal, H., and Glass, D.C. (1996) Negative affect and self-report of physical symptoms: two longitudinal studies of older adults, *Health Psychol.*, 15(3):193–199.

Levin, A.S. and Byers, V.S. (1987) Environmental illness: a disorder of immune regulation, *Occup. Med. State Art Rev.*, 2(4):669–681.

Levine, A.G. (1982) *Love Canal: Science, Politics and People*, Lexington, MA: Lexington Books.

Levine, E.S., Litto, W.J., and Jacobs, B.L. (1990) Activity of cat locus coeruleus noradrenergic neurons during the defense reaction, *Brain Res.*, 531:189–195.

Levine, S.E. and Salter, N.E. (1976) Youth and contemporary religious movements: psychological findings, *J. Can. Psychiatry Assoc.*, 21:411–420.

Levinson, D.J. (1978) *The Seasons of a Man's Life*, New York: Ballantine.

Lewis, M. (1995) Self-conscious emotions, *Am. Sci.*, 83:68–78.

Leznoff, A. (1997) Provocative challenges in patients with multiple chemical sensitivity, *J. Allergy Clin. Immunol.*, 99(4):438–442.

Lieberman, P., Crawford, L., Bjelland, J., Connell, B., and Rice, M. (1974) Controlled study of the cytotoxic food test, *JAMA*, 231(7):728–733.

Liebowitz, M.R., Gorman, J.M., Fyer, A.J., Dillon, D.J., and Klein, D.F. (1984) Effects of naloxone on patients with panic attacks, *Am. J. Psychiatry*, 141:995–997.

Lifton, R.J. (1956) "Thought reform" of western civilians in Chinese Communist prisons, *Psychiatry*, 19:173–195.

Limieux, A.M. and Coe, C.L. (1995) Abuse-related posttraumatic stress disorder: evidence for chronic neuroendocrine activation in women, *Psychosomatic Med.*, 57:105–115.

Lin, E.H., Katon, W., Von Korff, M., Bush, T., Lipscomb, P., Russo, J., and Wagner, E. (1991) Frustrating patients: physician and patient perspectives among distressed high users of medical services, *J. Gen. Intern. Med.*, 6:241–246.

Lindemann, E. and Finesinger, J.E. (1938) The effect of adrenalin and mecholyl in states of anxiety in psychoneurotic patients, *Am. J. Psychiatry*, 95:353–370.

Lindsley, D.B. (1951) Emotion, in Stevens, S.S. (Ed.), *Handbook of Experimental Psychology*, New York: Wiley, pp. 473–515.

Lindsley, D.B. (1960) Attention, consciousness, sleep and wakefulness, in Field, J., Magoun, H.W., and Hall, V.E. (Eds.), *Handbook of Physiology*. Section I. *Neurophysiology*, III, Washington, D.C.: American Physiology Society, pp. 1553–1593.

Lingjaerde, O. (1985) Lactate-induced panic attacks: possible involvement of serotonin reuptake stimulation, *Acta Psychiatrica Scandinavica*, 72:206–208.

Lipowski, Z.J. (1987) Somatization: medicine's unsolved problem [editorial], *Psychosomatics*, 28(6):294–297.

Lipowski, Z.J. (1988) Somatization: the concept and its clinical application, *Am. J. Psychiatry*, 145(11):1358–1368.

Lipsitt, D.R. (1970) Medical and psychological characteristics of "crocks", *Psychiatry Med.*, 1:15–25.

Liske, E. and Forster, F.M. (1964) Pseudoseizures: a problem in the diagnosis and management of epileptic patients, *Neurology*, 14:41–49.

Little, G.A., Ballard, R.A., Brooks, J.G. et al. (1987) National Institutes of Health Consensus Development Conference on Infantile Apnea and Home Monitoring, Sept. 1986, *Pediatrics*, 294:1637–1641.

Livingston, K.E. and Hornykiewicz, O., Eds. (1978) *Limbic Mechanism: The Continuing Evolution of the Limbic System Concept*, New York: Plenum Press.

Livingston, R.B. (1978) *Sensory Processing, Perception, and Behavior*, New York: Raven Press.

Livingston, R. (1987) Maternal somatization disorder and Munchausen syndrome by proxy, *Psychosomatics*, 28(4):213–217.

Locke, R.C. (1995) Expert Testimony in the Post-Daubert Era, presented at the Meeting of the California Society of CPAs, Camino Real at Puerto Vallarta, January 12–13, 1995 (address: Landels, Ripley & Diamond, LLP, Hills Plaza, 350 The Embarcadero, San Francisco, CA 94105-1250).

Loewenstein, R.J. (1990) Somatoform disorders in victims of incest and child abuse, in Kluft, R.P. (Ed.), *Incest-Related Syndromes of Adult Psychopathology*, Washington, D.C.: American Psychiatric Press, pp. 75–111.

Loftis, J. and Ross, L. (1973) Retrospective misattribution of a conditioned emotional response, *J. Personality Social Psychol.*, 30(5):683–687.

Lopez-Ibor, Jr., J.J., Soria, J., Canas, F., and Rodriguez-Gamazo, M. (1985) Psychopathological aspects of the toxic oil syndrome catastrophe, *Br. J. Psychiatry*, 147:352–365.

Lorig, T.S. (1989) Human EEG and odor response, *Progr. Neurobiol.*, 33:387–398.

Luborsky, L., Singer, B., and Luborsky, L. (1975) Comparative studies of psychotherapies, *Arch. Gen. Psychiatry*, 32:995–1008.

Lundberg, U. and Frankenhaeuser, M. (1978) Psychophysiological reactions to noise as modified by personal control over noise intensity, *Biol. Psychol.*, 6:51–59.

Luparello, T., Lyons, H.A., Bleecker, E.R., McFadden, Jr., E.R. (1968) Influences of suggestion on airway reactivity in asthmatic subjects, *Psychosomatic Med.*, 30(6):819–825.

Luparello, T.J., Leist, N., Lourie, C.H., and Sweet, P. (1970) The interaction of psychologic stimuli and pharmacologic agents on airway reactivity in asthmatic subjects, *Psychosomatic Med.*, 32(5):509–513.

Lupien, S., Lecours, A.R., Lussier, I., Schwartz, G., Nair, N.P.V., and Meaney, M.J. (1994) Basal cortisol levels and cognitive deficits in human aging, *J. Neurosci.*, 14:2893–2903.

Lykken, D.T. (1968) Neuropsychology and psychophysiology in personality research, in Borgatta, E.F. and Lambert, W.W. (Eds.), *Handbook of Personality Theory and Research*, Chicago: Rand McNally and Company, pp. 413–509.

Lynn, R. (1966) *Attention Arousal and the Orientation Reaction*, Oxford, UK: Pergamon Press.

Maas, J.W., Koslow, S.H., Davis J., Katz, M., Frazer, A., Bowden, C.L., Berman, N., Gibbons, R., Stokes, P., and Landis, H. (1987) Catecholamine metabolism and disposition in healthy and depressed subjects, *Arch. Gen. Psychiatry*, 44:337–344.

Mabray, C.R., Burditt, M.L., Martin, T.L., Jaynes, C.R., and Hayes, J.R. (1982) Treatment of common gynecologic-endocrinologic symptoms by allergy management procedures, *Obstet. Gynecol.*, 59(5):560–564.

Mackenzie, J.N. (1886) The production of the so-called "rose cold" by means of an artificial rose, *Am. J. Med. Sci.*, 91:45–57.

Maclean, P.D. (1949) Psychosomatic disease and the visceral brain. Recent developments bearing on the Papez theory of emotion, *Psychosomatic Med.*, 11:338–353.

Maclean, P.D. (1952) Some psychiatric implications of physiological studies on frontotemporal portion of limbic system (visceral brain), *Electroencephalogr. Clin. Neurophysiol.*, 4:407–418.

MacLean, P.D. (1955) The limbic system ("visceral brain") and emotional behavior, *Am. Med. Assoc. Arch. Neurol. Psychiatry*, 73:130–134.

Maclean, D. and Reichlin, S. (1981) Neuroendocrinology and the immune system, in Ader, R. (Ed.), *Psychoneuroimmunology*, New York: Academic Press, pp. 475–520.

Macmillan, M. (1990) Freud and Janet on organic and hysterical paralyses: a mystery solved?, *Int. Rev. Psychoanalysis*, 17:189–203.

MacQueen, G., Marshall, J. Perdue, M., Siegel, S., and Bienenstock, J. (1989) Pavlovian conditioning of rat mucosal mast cells to secrete rat mast cell protease II. *Science*, 243:83–85.

Maes, M., Bosmans, E., Suy, E., Vandervorst, C., Dejonckheere, C., and Raus, J. (1991) Antiphospholipid, antinuclear, Epstein-Barr and cytomegalovirus antibodies, and soluble interleukin-2 receptors in depressive patients, *J. Affective Disorders*, 21:133–140.

Maes, M. Meltzer, H., Jacobs, J., Suy, E., Calabrese, J., Minner, B., and Raus, J. (1993) Autoimmunity in depression: increased antiphospholipid autoantibodies. *Acta Psychiatrica Scandinavica*, 87:160–166.

Magoun, H.W. (1963) *The Waking Brain*, 2nd ed., Springfield, IL: Charles C Thomas.

Mahl, G.F., Rothenberg, A., Delgado, J.M.R., and Hamlin, H. (1964) Psychological responses in the human to intracerebral electrical stimulation, *Psychosomatic Med.*, 26:337–368.

Maier, S.F. and Seligman, M.E.P. (1976) Learned helplessness: theory and evidence, *J. Exp. Psychol. Gen.*, 105(1):3–46.

Maier, S.F. and Watkins, L. (1998) Cytokines for psychologists: implications of bidirectional immune-to-brain communication for understanding behavior, mood, and cognition, *Psych. Rev.*, 105(1):83–107.

Maier, S.F., Watkins, L.R., and Fleshner, M. (1994) Psychoneuroimmunology: the interface between behavior, brain, and immunity, *Am. Psychol.*, 49(12):1004–1017.

Mandell, A., Chapman, L., Rand, R., and Walter, R. (1963) Plasma corticosteroids: changes in concentration after stimulation of hippocampus and amygdala, *Science*, 139:1212.

Mangoun, G.R. and Hillyard, S.A. (1987) The spatial allocation of attention as indexed by event-related brain potentials, *Human Factors*, 29:195–211.

Mann, C.A., Lubar, J.F., Zimmerman, A.W., Miller, C.A., and Muenchen, R.A. (1991) Quantitative analysis of EEG in boys with attention-deficit-hyperactivity disorder: controlled study with clinical implications, *Pediatr. Neurol.*, 8:30–36.

Manu, P., Matthews, D.A., Lane, T.J., Tennen, H., Hesselbrock, V., Mendola, R., and Affleck, G. (1989) Depression among patients with a chief complaint of chronic fatigue, *J. Affective Dis.*, 17:165–172.

Margraf, J., Ehlers, A., Roth, W.T., Clark, D.B., Sheikh, J., Agras, W.S., and Taylor, C.B. (1991) How "blind" are double-blind studies?, *J. Consulting Clin. Psychol.*, 59(1):184–187.

Mark, A.L. (1990) Regulation of sympathetic nerve activity in mild human hypertension, *J. Hypertension*, 8(Suppl. 7):S67–S75.

Markowitz, A. (1983) The role of family therapy in the treatment of symptoms associated with cult affiliation, in Halperin, D. (Ed.), *Psychodynamic Perspectives on Religion, Sect, and Cult*, Littleton, MA: John Wright, pp. 323–333.

Marshall, E. (1986) Immune system theories on trial, *Science*, 234:1490–1492.

Mason, J.W. (1975) A historical view of the stress field: Part I, *J. Human Stress*, 1:6–12.

Mason, J.W., Giller, E.L., Kosten, T.R., and Harkness, L., (1988) Elevation of urinary norepinephrine/cortisol ratio in posttraumatic stress disorder, *J. Nerv. Ment. Dis.*, 176:498–502.

Mason, J.W., Giller, E.L., Kosten, T.R., and Yehuda, R. (1990) Psychoendocrine approaches to the diagnosis and pathogenesis of post-traumatic stress disorder, in Giller, E. (Ed.), *Biological Assessment and Treatment of PTSD*, Washington, D.C.: American Psychiatric Press, pp. 65–86.

Mason, S.T. and Fibiger, H.C. (1979) Anxiety: the locus coeruleus disconnection, *Life Sci.*, 25:2141–2147.

Masterson, J.F. and Klein, R., Eds. (1989) *Psychotherapy of the Disorders of the Self*, New York: Brunner/Mazel.

Matarazzo, J.D. (1971) The practice of psychotherapy is art and not science, in Mahrer, A.R. and Pearson, L. (Eds.), *Creative Developments in Psychotherapy*, Cleveland, OH: Case Western Reserve University Press, pp. 364–392.

Matarazzo, J.D. (1990) Psychological assessment versus psychological testing: validation from the Binet to the school, clinic and courtroom, *Am. Psychol.*, 45:999–1017.

Mavissakalian, M.R. and Barlow, D.H. Assessment of obsessive-compulsive disorders, in Barlow, D.H. (Ed.), *Behavioral Assessment of Adult Disorders*, New York: Guilford Press, 1981.

May, C.D. (1976) Objective clinical and laboratory studies of immediate hypersensitivity reactions to food in asthmatic children, *J. Allergy Clin. Immunol.*, 58(4):500–515.

May, C.D. (1980) Food allergy — material and ethereal, *New Engl. J. Med.*, 302:1142–1143.

May, R. (1979) *The Meaning of Anxiety*, New York: Pocket Books.

Mayberg, H.S. (1992) Functional brain scans as evidence in criminal court: an argument for caution, *J. Nucl. Med.*, 33(6):18N–25N.

Mayberg, H.S. (1994) Critique: SPECT studies of multiple chemical sensitivity, proceedings of the Conference on Low-Level Exposure to Chemicals and Neurobiologic Sensitivity, April 6–7, 1994, Baltimore, MD, sponsored by Agency for Toxic Substances and Disease Registry (ATSDR), *Toxicol. Indust. Health*, 10(4/5):661–666.

McCann, S.M., Sternberg, E.M., Lipton, J.M., Chrousos, G.P., Gold, P.W., Smith, C.C., Eds. (1998) Neuroimmunomodulation: molecular aspects, integrative systems, and clinical advances. *Ann. N.Y. Acad. Sci.*, vol. 840.

McCourtie, D. (1991) An overview, in *Chronic Diseases in Canada: Environmental Sensitivities Workshop*, Health Protection Branch, Health and Welfare Canada, Ottowa, May 24, 1990.

McEwen, B.S. and Sapolsky, R.M. (1995) Stress and cognitive function, *Curr. Opin. Neurobiol.*, 5:205–216.

McFadden, Jr., E.R. and Zawadski, D.K. (1996) Vocal cord dysfunction masquerading as exercise-induced asthma, *Am. J. Respir. Crit. Care Med.*, 153:942–947.

McFadden, Jr., E.R., Luparello, T., Lyons, H.A., and Bleecker, E. (1969) The mechanism of action of suggestion in the induction of acute asthma attacks, *Psychosomatic Med.*, 31(2):134–143.

McFarlane, A.C. (1988) The aetiology of post-traumatic stress disorders following a natural disaster, *Br. J. Psychiatry*, 152:116–121.

McGaugh, J.L. (1989) Involvement of hormonal and neuromodulary systems in the regulation of memory storage, *Ann. Rev. Neurosci.*, 12:255–287.

McGaugh, J.L. (1992) Affect neuromodulatory systems and memory storage, in Christianson, S.A. (Ed.), *The Handbook of Emotion and Memory: Research and Theory*, Hillsdale, NJ: Lawrence Erlbaum Assoc., pp. 245–268.

McGuire, T.L. and Feldman, M.D. (1989) Psychologic morbidity of children subjected to Munchausen syndrome by proxy, *Pediatrics*, 83:289–292.

McKenna, P.J. (1984) Disorders with overvalued ideas, *Br. J. Psychiatry*, 145:579–585

McNaughton, B.L. (1993) The mechanism of expression of long-term enhancement of hippocampal synapses: current issues and theoretical implications, *Ann. Rev. Physiol.*, 55:375–396.

Meadow, R. (1977) Munchausen syndrome by proxy: the hinterland of child abuse, *Lancet*, 2:343–345.

Meadow, R. (1982) Munchausen syndrome by proxy, *Arch. Dis. Childhood*, 57:92–98.

Meadow, R. (1984) Fictitious epilepsy, *Lancet*, 2(July 7):25–28.

Mearin, F., Cucala, M., Azpiroz, F., and Malagelada, J.R. (1991) The origin of symptoms on the brain-gut axis in functional dyspepsia, *Gastroenterology*, 101:999–1006.

Meggs, W.J., Dunn, K.A., Bloch, R.M., Goodman, P.E., and Davidoff, A.L. (1996) Prevalence and nature of allergy and chemical sensitivity in a general population, *Arch. Environ. Health*, 51(4):275–282.

Meichenbaum, D. (1977) *Cognitive Behavior Modification: An Integrative Approach*, New York: Plenum Press.

Mena, I., Goldstein, J., Jouanne, E., and Lesser, I. (1993) Cerebral hypoperfusion in late life chronic fatigue syndrome (CFS) and late life depression [abstract], *J. Nucl. Med.*, 34:210P.

Mendelson, G. (1987) Measurement of conscious symptom exaggeration by questionnaire: a clinical study, *J. Psychosomatic Res.*, 31(6):703–711.

Menzies, R., Tamblyn, R., Farant J.P., Hanley, J., Nunes, F., and Tamblyn, R. (1993) The effect of varying levels of outdoor-air supply on the symptoms of sick building syndrome, *New Engl. J. Med.*, 328(12):821–827.

Mesulam, M.M. (1990) Schizophrenia and the brain [editorial], *New Engl. J. Med.*, 322(12):842–844.

Michelson, L., Mavissakalian, M., and Marchione, K. (1985) Cognitive and behavioral treatments of agoraphobia: clinical, behavioral and psychophysiological outcomes, *J. Consult. Clin. Psychol.*, 5396):913–925.

Milich, R. and Pelham, W. E. (1986) Effects of sugar ingestion on classroom and playground behavior of attention deficit disordered boys, *J. Consult. Clin. Psychol.*, 54(5):714–718.

Milich, R., Wolraich, M.L., and Lindgren, S.D., (1986) Sugar and hyperactivity: a critical review of empirical findings, *Clin. Psychol. Rev.*, 6:493–513.

Miller, A. (1984) *Thou Shalt Not Be Aware: Society's Betrayal of the Child* (trans. by H. and H. Hannum), New York: Farrar, Straus, Giroux,.

Miller, C.S. (1992) Possible models for multiple chemical sensitivity: conceptual issues and role of the limbic system, *Toxicol. Indust. Health*, 8(4):181–202.

Miller, C.S. (1996) Chemical sensitivity: symptom, syndrome or mechanism for disease?, *Toxicology*, 11:69–86.

Miller, C.S. (1997) Toxicant-induced loss of tolerance — an emerging theory of disease?, *Environ. Health Perspect.*, 105(Suppl. 2):445–453.

Miller, N.E. (1951) Learnable drives and rewards, in Stevens, S.S. (Ed.), *Handbook of Experimental Psychology*, New York: Wiley, pp. 435–472.

Miller, N.E. (1969) Learning of visceral and glandular responses, *Science*, 163:434–445.

Millon, T. and Davis, R.D. (1996) *Disorders of Personality: DSM-IV and Beyond*, 2nd ed., New York: John Wiley & Sons.

Mishkin, M. (1982) A memory system in the monkey, in Broadbent, D.E. and Weiskrantz, L. (Eds.), *The Neuropsychology of Cognitive Function*, London: The Royal Society, pp. 85–95.

Miura, M. and Usa, S. (1970) Psychotherapy of neurosis: Morita therapy, *Psychologia*, 13:18–35.

Moffitt, J. W. and Stagner, R. (1956) Perceptual rigidity and closure as functions of anxiety, *J. Abnorm. Soc. Psychol.*, 52:354–357.

Montanaro, A. and Bardana, Jr., E.J. (1992) The chemically sensitive patient, in Bardana, Jr., E.J., Montanaro, A., and O'Hollaren, M. (Eds.), *Occupational Asthma*, Philadelphia: Hanley and Belfus, pp. 255–266.

Moore, R.Y. and Bloom, F.E. (1978) Central catecholamine neuron systems: anatomy and physiology of the dopamine systems, *Ann. Rev. Neurosci.*, 1:129–169.

Mooser, S.B. (1987) The epidemiology of multiple chemical sensitivities, *Occup. Med. State. Art Rev.*, 2(4):663–668.

Morgan, C.A., Grillon, C., Southwick, S.M., Davis, M., and Charney, D.S. (1996) Exaggerated acoustic startle reflex in Gulf War veterans with posttraumatic stress disorder, *Am. J. Psychiatry*, 153(1):64–68.

Morrison, J. (1989) Childhood sexual histories of women with somatization disorder, *Am. J. Psychiatry*, 146:239–241.

Morrow, L.A., Ryan, C.M., Goldstein, G., and Hodgson, M.J. (1989) A distinct pattern of personality disturbance following exposure to mixtures of organic solvents, *J. Occup. Med.*, 31(9):743–746.

Morse, D.R., Martin, J., and Moshonov, J. (1991) Psychosomatically induced death: Relative to stress, hypnosis, mind control, and voodoo: a review and possible mechanisms, *Stress Med.*, 7:213–232.

Morse, J.C. and Morse, E.L. (1987) Toward a theory of therapy with cultic victims, *Am. J. Psychotherapy*, XLI(4):563–570.

Moruzzi, G. and Magoun, H.W. (1949) Brainstem reticular formation and activation of the EEG, *Electroencephalogr. Clin. Neurophysiol.*, 1:455–473.

Moser, F., Schaeffer, J., Waxman, A.D., Mayberg, H., and Newer, M.R. (1995) New modalities of brain imaging and neurocognitive assessment: emerging areas of interest and controversy related to applications in head injury, neurotoxicology and environmental medicine, *Disability*, 4(2):1–12.

Moses, R. (1978) Adult psychic trauma: the question of early predisposition and some detailed mechanisms, *Int. J. Psychoanalysis*, 54:353–363.

Mumtaz, M.M., Sipes, I.G., Clewell, H.J., and Yang, R.S.H. (1993) Symposium overview: risk assessment of chemical mixtures and biologic and toxicologic issues, *Fundam. Appl. Toxicol.*, 21:258–269.

Munck, A. (1971) Glucocorticoid inhibition of glucose uptake by peripheral tissues: old and new evidence, molecular mechanisms, and physiological significance, *Perspect. Biol. Med.*, 14:265–289.

Murphy, C. and Jinich, S. (1996) Olfactory dysfunction in Down's syndrome, *Neurobiol. Aging*, 17(4):631–637.

Murphy, C., Nordin, S., de Wijk, R.A., Cain, W.S. and Polich, J. (1994) Olfactory-evoked potentials: assessment of young and elderly, and comparison to psychophysical threshold, *Chemical Senses*, 19(1):47–56.

Murthy, R.S. and Isaac, M.K. (1987) Mental health needs of Bhopal disaster victims and training of medical officers in mental health effects, *Indian J. Med. Res.*, 86(Suppl.):51–58.

Naquet, R., Regis, H., Fischer-Williams, M., and Fernandez-Guardiola, A. (1960) Variations in the responses evoked by light along the specific pathways, *Brain*, 83:52–56.

National Research Council (1987) Board of Environmental Studies and Toxicology, Workshop on Health Risks from Exposure to Common Indoor Household Products in Allergic or Chemically Diseased Persons, July 1, 1987.

National Research Council (1992a) *Multiple Chemical Sensitivities: Addendum to Biological Markers in Immunotoxicology*, Washington, D.C.: National Academy Press.

National Research Council (1992b) *Environmental Neurotoxicology*, Washington, D.C.: National Academy Press.

Nauta, W.J.H. and Feirtag, M. (1986) *Fundamental Neuroanatomy*, New York: W.H. Freeman.

Nemiah, J.C. and Uhde, T.W. (1989) Phobic disorders, in Kaplan, H.I. and Sadock, B.J. (Eds.), *Comprehensive Textbook of Psychiatry*, Vol. I, 5th ed., Baltimore, MD: Williams & Wilkins, pp. 972–984.

Nesse, R.M., Cameron, O.G., Curtis, G.C., McCann, D.S., and Huber-Smith, M.J. (1984) Adrenergic function in patients with panic anxiety, *Arch. Gen. Psychiatry*, 41:773–776.

Nethercott, J.R., Davidoff, L.L., Curbow, B., and Abbey, H. (1993) Multiple chemical sensitivities syndrome: toward a working case definition, *Arch. Environ. Health*, 48(1):19–26.

Neveu, P.J. (1988) Cerebral neocortex modulation of immune functions, *Life Sci.*, 42:1917–1923.

Nolen-Hoeksema, S. (1990) *Sex Differences in Depression*, Stanford, CA: Stanford University Press.

Nordin, S., Paulsen, J.S., and Murphy, C. (1995a) Sensory- and memory-mediated olfactory dysfunction in Huntington's disease, *J. Int. Neuropsychol. Soc.*, 1:281–290.

Nordin, S., Monsch, A.U., and Murphy, C. (1995b) Unawareness of smell loss in normal aging and Alzheimer's disease: discrepancy between self-report and diagnosed smell sensitivity, *J. Gerontol. Psychol. Sci.*, 50B(4):P187–P192.

Norris, F.H. (1992) Epidemiology of trauma: frequency and impact of different potentially traumatic events on different demographic groups, *J. Consult. Clin. Psychol.*, 60:409–418.

O'Banion, D.R. (1981) *Ecological and Nutritional Treatment of Health Disorders*. Springfield, IL: Charles C Thomas.

Office of Technology Assessment (1990) *Neurotoxicity: Identifying and Controlling Poisons of the Nervous System*, U.S. Congress, OTA-BA-436, Washington, D.C.: U.S. Government Printing Office, pp.63–77.

Ogata, S.N., Silk, K.R., Goodrich, S., Lohr, N.E., Westen, D., and Hill, E.M. (1990) Childhood sexual and physical abuse in adult patients with borderline personality disorder, *Am. J. Psychiatry*, 147:1008–1013.

Oh, V.M.S. (1994) The placebo effect: can we use it better?, *Br. Med. J.*, 309:69–70.

O'Keefe, J. and Nadel, L. (1978) *The Hippocampus as a Cognitive Map*, New York: Oxford University Press.

Oken, D. (1989) Current theoretical concepts in psychosomatic medicine, in Kaplan, H.I. and Sadock, B.T. (Eds.), *Comprehensive Textbook of Psychiatry*, Vol. II, 5th ed., Baltimore, MD: Williams & Wilkins, pp. 1160–1169.

Oldham, J.M., Skodol, A.E., Kellman, H.D., Hyler, S.E., Doidge, N., and Rosnick, L., Gallaher, P.E. (1995) Comorbidity of axis I and axis II disorders, *Am. J. Psychiatry*, 152(4):571–578.

Oliver, J.E. (1993) Intergenerational transmission of child abuse: rates, research, and clinical implications, *Am. J. Psychiatry*, 150:1315–1324.

Orme, T. (1994) *MCS: Multiple Chemical Sensitivity*, New York: American Council of Science and Health.

Orr, S.P., Lasko, N.B., Shalev, A.Y., and Pitman, R.K. (1995) Physiologic responses to loud tones in Vietnam veterans with posttraumatic stress disorder, *J. Abnormal Psychol.*, 104(1):75–82.

Orr, S.P., Lasko, N.B., Metzger, L.J., Berry, N.J., Ahern, C.E., and Pitman, R.K. (1997) Psychophysiologic assessment of PTSD in adult females sexually abused during childhood, in Yehuda, R. and McFarlane, A.C. (Eds.), *Psychobiology of Posttraumatic Stress Disorder*, New York: New York Academy of Sciences, pp. 491–493.

Oxford English Dictionary (1971) New York: Oxford University Press.

Papez, J.W. (1937) A proposed mechanism of emotion. Archives of Neurological Psychiatry, 38:725–743.

Papp, L.A., Klein, D.F., and Gorman, J.M. (1993) Carbon dioxide hypersensitivity, hyperventilation, and panic disorder, *Am. J. Psychiatry*, 150:1149–1157.

Pavlov, I.P. (1951) *Types of Higher Nervous Activity, Their Relationship to Neuroses and Psychoses, and the Physiological Mechanism of Neurotic and Psychotic Symptoms* (trans. by D. Myshne and S. Belsky), Complete Works, Vol. III, Book 2, Moscow, pp. 344–349 (republished in Popov, Y., Rokhlin, L., and Pavlov, I.P., Eds., *Psychopathology and Psychiatry*, Moscow: Foreign Languages Publishing House, date unspecified, quote from p. 388).

Pearl, R. (1929) *The Rate of Living*, New York: Knopf.

Pearlin, L.I. and Schooler, C. (1978) The structure of coping, *J. Health Soc. Behav.*, 19:2–21.

Pearson, D.J. (1985) Food allergy, hypersensitivity and intolerance, *J. Roy. Coll. Physicians Lond.*, 19(3):154–162.

Pearson, D.J. (1986) Pseudo food allergy, *Br. Med. J.*, 292:221–222.

Pearson, D.J. and Rix, K.J.B. (1987) Psychological effects of food allergy, in Brostoff, J. and Challacombe, S.J. (Eds.), *Food Allergy and Intolerance*, East Sussex, UK: Baillieres Tindall, pp. 688–708.

Pearson, D.J., Rix, K.J.B., and Bentley, S.J. (1983) Food allergy: how much in the mind? A clinical and psychiatric study of suspected food hypersensitivity, *Lancet*, June:1259–1261.

Penfield, W.W. (1956) Functional localization in temporal and deep sylvian areas, *Res. Publ. Assoc. Res. Nerv. Ment. Disorders*, 36:210–226.

Penfield, W.W. and Jasper, H. (1954) *Epilepsy and the Functional Anatomy of the Human Brain*, Boston, MA: Little, Brown.

Penfield, W.W. and Roberts, L. (1959) *Speech and Brain-Mechanisms*, Princeton, NJ: Princeton University Press.

Penfield, W.W. and Perot, P. (1963) The brain's record of auditory and visual experience, *Brain*, 86:595–696.

Pennebaker, J.W. and Watson, D. (1991) The psychology of somatic symptoms, in Kirmayer, L.J. and Robbins, J.M. (Eds.), *Current Concepts of Somatization: Research and Clinical Perspectives*, Washington, D.C.: American Psychiatric Press, pp. 21–35.

Perry, J.C. and Vaillant, G.E. (1989) Personality Disorders, in Kaplan, H.I. and Sadock, B.T. (Eds.), *Comprehensive Textbook of Psychiatry*, 5th ed., Baltimore, MD: Williams & Wilkins, pp. 1352–1386.

Peterson, C. and Seligman, M.E.P. (1984) Causal explanations as a risk factor for depression: theory and evidence, *Psych. Rev.*, 91(3):347–374.

Phillips, K.A., Gunderson, J.G., Hirschfeld, R.M.A., and Smith, L.E. (1990) A review of the depressive personality, *Am. J. Psychiatry*, 147(7):830–837.

Philpott, W.H. and Kalita, D.K. (1980) *Brain Allergies: The Psychonutrient Connection*, New Canaan, CT: Keats Publishing.

Piaget, J. (1970) *Structuralism* (trans. by C. Maschler), New York: Basic Books.

Piaget, J. (1972) *Judgment and Reasoning in the Child* (trans. by M. Warden), Totoway, NJ: Littlefield, Adams, & Company.

Pincus, J.H. and Tucker, G.J. (1985) *Behavioral Neurology*, 3rd ed., New York: Oxford University Press.

Pitman, R.K. (1997) Overview of biological themes in PTSD, in Yehuda, R. and McFarlane, A.C. (Eds.), *Psychobiology of Posttraumatic Stress Disorder*, New York: New York Academy of Sciences, 821:1–9.

Pitman, R.K., Orr, S.P., Forgue, D.F., de Jong, J.B., and Clairborn, J.M. (1987) Psychophysiologic assessment of posttraumatic stress disorder imagery in Vietnam combat veterans, *Arch. Gen. Psychiatry*, 44:970–975.

Pitman, R.K., van der Kolk, B.A., Orr, S.P., and Greenberg, M.S. (1990) Naloxone-reversible analgesic response to combat-related stimuli in posttraumatic stress disorder, *Arch. Gen. Psychiatry*, 47:541–544.

Pitman, R.K., Orr, S.P., and Shalev, A.Y. (1993) Once bitten, twice shy: beyond the conditioning model of PTSD [editorial], *Biol. Psychiatry*, 33:145–146.

Polanyi, M. (1958) *Personal Knowledge*, Chicago: University of Chicago Press.

Pollard, C.A., Pollard, H.J., and Corn, K.J. (1989) Panic onset and mayor life events in the lives of agoraphobics: a test of contiguity. *J. Abnormal Psychol.*, 98(3):318–321.

Pollin, W. and Goldin, S. (1961) The physiological and psychological effects of intravenously administered epinephrine and its metabolism in normal and schizophrenic men: II, *J. Psychiatric Res.*, 1:50–67.

Popper, K.R. (1968) *Conjectures and Refutations: The Growth of Scientific Knowledge*, New York: Harper & Row/Torchbooks.

Posner, M.I. (1988) Structures and functions of selective attention, in Boll, T. and Bryant, B. (Eds.), *Master Lectures in Clinical Neuropsychology*, Washington, D.C.: American Psychological Association, pp. 173–202.

Posner, M.I. and Petersen, S.E. (1990) The attention system of the human brain, *Ann. Rev. Neurosci.*, 13:25–42.

Posner, M.I. and Raichle, M.E. (1994) *Images of Mind*, New York: Freeman.

Post, R.M. (1992) Transduction of psychosocial stress into the neurobiology of recurrent affective disorder, *Am. J. Psychiatry*, 149(8):999–1010.

Post, R.M., Rubinow, D.R., and Ballenger, J.C. (1984) Conditioning, sensitization, and kindling: Implications for the course of affective illness, in Post, R.M. and Ballenger, J.C. (Eds.), *Neurobiology of Mood Disorders*, Baltimore, MD: Williams & Wilkins, pp. 432–466.

Post, R.M., Weiss, S.R.B., and Pert, A. (1987) The role of context in conditioning and behavioral sensitization to cocaine, *Psychopharmacol. Bull.*, 23:425–429.

Post, R.M., Weiss, S.R.B., and Pert, A. (1988a) Cocaine-induced behavioral sensitization and kindling: Implications for the emergence of psychopathology and seizures, *Ann. N.Y. Acad. Sci.*, 537:292–308.

Post, R.M., Weiss, S.R.B., and Pert, A. (1988b) Implications of behavioral sensitization and kindling for stress-induced behavioral change, in Chrousos, G.P., Loriaux, D.L. and Gold, P.W. (Eds.), *Mechanisms of Physical and Emotional Stress*, New York: Plenum, pp. 441–463.

Post, R.M., Weiss, S.R.B., and Pert, A. (1991) Sensitization and kindling effects of chronic cocaine administration, in Lakoski, J.M., Galloway, M.P., and White, F.J. (Eds.), *Cocaine: Pharmacology, Physiology, and Clinical Strategies*, Boca Raton, FL: CRC Press.

Post, R.M., Weiss, S.R.B., Smith, M., Rosen, J., and Frye, M. (1995) Stress, conditioning, and the temporal aspect of affective disorders, in Chrousos, G.P. et al. (Eds.), *Stress: Basis Mechanisms and Clinical Implications*, New York: New York Academy of Sciences, pp. 677–696.

Pribor, E.F. and Dinwiddie, S.H. (1992) Psychiatric correlates of incest in childhood, *Am. J. Psychiatry*, 149:52–56.

Pribor, E.F., Yurzy, S.H., Dean J.T., and Wetzel, R.D. (1993) Briquet's syndrome, dissociation, and abuse, *Am. J. Psychiatry*, 150:1507–1511.

Pribram, K.H. and McGuinness, D. (1975) Arousal, activation, and effort in the control of attention, *Psych. Rev.*, 82(2):116–149.

Putnam, F.W. (1989a) *Diagnosis and Treatment of Multiple Personality Disorder*, New York: Guilford Press.

Putnam, F.W. (1989b) Pierre Janet and modern views of dissociation, *J. Traumatic Stress*, 2:413–429.

Putnam, F.W., Trickett, P.K., Helmers, K., Dorn, L., and Everett, P.B. (1991) *Cortisol Abnormalities in Sexually Abused Girls*, proceedings of the 144th Annual Meeting of the American Psychiatric Association, p. 107.

Quill, T.E. (1985) Somatization disorder: one of medicine's blind spots, *JAMA*, 254(21):3075–3079.

Racine, R. (1978) Kindling: the first decade, *Neurosurgery*, 3(2):234–252.

Rachman, S.J. (1990) *Fear and Courage*, 2nd ed., New York: W.H. Freeman.

Rachman, S.J. and Hodgson, R. (1980) *Obsessions and Compulsions*, Englewood Cliffs, NJ: Prentice-Hall.

Radetsky, P. (1997) *Allergic to the Twentieth Century*, Boston: Little, Brown.

Randolph, T.G. (1945) Fatigue and weakness of allergic origin (allergic toxemia) to be differentiated from "nervous fatigue" or neurasthenia, *Ann. Allergy*, 3:418–430.

Randolph, T.G. (1947) Allergy as a causative factor in fatigue, irritability, and behavior problems of children, *J. Pediatr.*, 31:560–572.

Randolph, T.G. (1955) Depressions caused by home exposures to gas and combustion products of gas, oil and coal, *J. Lab. Clin. Med.*, 46:942.

Randolph, T.G. (1959) Ecologic mental illness — psychiatry exteriorized, *J. Lab. Clin. Med.*, 54:936.

Randolph, T.G. (1962) *Human Ecology and Susceptibility to the Chemical Environment*, Springfield, IL: Charles C Thomas.

Randolph, T.G. and Moss, R.W. (1980) *An Alternative Approach to Allergies*, New York: Lippincott & Crowell.

Rapp, D.J. (1978a) Double-blind confirmation and treatment of milk sensitivity, *Med. J. Aust.*, 1:571–572.

Rapp, D.J. (1978b) Does diet affect hyperactivity?, *J. Learn. Disabil.*, 11(6):383–389.

Rapp, D.J. (1986) Environmental medicine: an expanded approach to allergy, *Buffalo Physician*, 19(5):16–24.

Rauch, S., van der Kolk, B.A., Fisler, R., Alpert, N., Orr, S., Savage, C., Fischman, A., Jenike, M., and Pitman, R. (1996) A symptom provocation study of posttraumatic stress disorder using positron emission tomography and script-driven imagery, *Arch. Gen. Psychiatry*, 53:380–387.

Razran, G. (1971) *Mind in Evolution: East-West Synthesis of Learned Behavior and Cognition*, Boston: Houghton Mifflin.

Rea, W.J. (1977) Environmentally triggered small vessel vasculitis, *Ann. Allergy*, 38:245–251.

Rea, W.J. (1992) *Chemical Sensitivity*. Vol. 1. *Principles and Mechanisms*, Boca Raton, FL: Lewis Publishers.

Rea, W.J. (1994a) *Chemical Sensitivity*. Vol. 2. *Sources of Total Body Load*, Boca Raton, FL: Lewis Publishers.

Rea, W.J. (1994b) Environmental Illness, presentation to the panel on evaluating individuals reporting sensitivities to multiple chemicals, California Department of Health Services, Environmental Health Investigations Branch, May 5, 1994, Berkeley, CA.

Rea, W.J. (1995) *Chemical Sensitivity*. Vol. 3. *Clinical Manifestations of Pollutant Overload*, Boca Raton, FL: Lewis Publishers.

Rea, W.J. (1997) *Chemical Sensitivity*. Vol. 4. *Tools of Diagnosis and Methods of Treatment*, Boca Raton, FL: Lewis Publishers.

Rea, W.J., Bell, I.R., Suits, C.W., and Smiley, R.E. (1978) Food and chemical susceptibility after environmental chemical overexposure: case histories, *Ann. Allergy*, 41:101–110.

Reiman, E.M., Raichle, M.E., Robins, E., Mintun, M.A., Fusselman, M.J., Fox, P. T., Price, J.L., and Hackman, K.A. (1989) Neuroanatomical correlates of lactate-induced anxiety attack, *Arch. Gen. Psychiatry*, 46:493–500.

Report of the Advisory Panel on Environmental Hypersensitivity (1986) Ministry of Health, Province of Ontario, Toronto, Ontario, M7A 2CA.

Rescorla, R.A. (1988) Pavlovian conditioning: it is not what you think it is, *Am. Psychol.*, 43:151–160.

Resnick, H.S., Yehuda, R., Pitman, R.K., and Foy, D.W. (1995) Effect of previous trauma on acute plasma cortisol level following rape, *Am. J. Psychiatry*, 152(11):1675–1677.

Reynolds, D.K. (1976) *Morita Psychotherapy*, Berkeley, CA: University of California Press.

Rinkel, H.J. (1944) Food allergy: the role of food allergy in internal medicine, *Ann. Allergy*, 2:115–124.

Rinkel, H.J., Lee, C.H., Brown, D.W., Willoughby, J.W., and Williams, J.M. (1964) The diagnosis of food allergy, *Arch. Otolaryngol.*, 79:71–79.

Rix, K.J.B., Pearson, D.J., and Bentley, S.J. (1984) A psychiatric study of patients with supposed food allergy, *Br. J. Psychiatry*, 145:121–126.

Roberts, A.H., Kewman, D.G., Mercier, L., and Hovell, M. (1993) The power of nonspecific effects in healing: implications for psychosocial and biological treatments, *Clin. Psychol. Rev.*, 13:375–391.

Roberts, L. (1961) Activation and interference of cortical functions, in Sheer, D.E. (Ed.), *Electrical Stimulation of the Brain*, Austin, TX: University Press, pp. 533–553.

Robertson, W.O. (1994) MCS (multiple chemical sensitivity) by proxy, *Vet. Hum. Toxicol.*, 36:579–580.

Robinson, R.G. (1995) Mapping brain activity associated with emotion [editorial], *Am. J. Psychiatry*, 152(3):327–329.

Rokeach, M. (1968) *Beliefs, Attitudes, and Values*, San Francisco, CA: Jossey-Bass.

Roland, P.E. (1994) *Brain Activation*, New York: Wiley-Liss.

Roland, P.E. and Friberg, L. (1985) Localization of cortical areas activated by thinking, *J. Neurophysiol.*, 53(3):1219–1243.

Rome, H.P. and Braceland, F.J. (1952) The psychological response to ACTH, cortisone, hydrocortisone, and related steroid substances, *Am. J. Psychiatry*, 108:641–650.

Rosch, P.J. (1979) Stress and illness, *JAMA*, 242(5):427–428.

Rose, R.M. (1980) Endocrine responses to stressful psychological events, *Psychiatric Clin. North Am.*, 3:251–276.

Rose, R.M. (1984) Overview of endocrinology of stress, in Brown, G., Koslow, S.H., and Reichlin, S. (Eds.), *Neuroendocrinology and Psychiatric Disorder*, New York: Raven Press, pp. 95–122.

Rosen, E. (1956) Self-appraisal, personal desirability, and perceived social desirability of personality traits, *J. Abnormal Soc. Psychol.*, 52:151–158.

Rosen, J.B., Hitchcock, J.M., Sananes, C.B., Miserendino, M.J.D., and Davis, M. (1991) A direct projection from the central nucleus of the amygdala to the acoustic startle pathway: anterograde and retrograde tracing studies, *Behav. Neurosci.*, 105:817–825.

Rosenberg, C.E. (1962) The place of George M. Beard in nineteenth century psychiatry, *Bull. History Med.*, 36:245–259.

Rosenberg, D.A. (1987)·Web of deceit: a literature review of Munchausen syndrome by proxy, *Child Abuse Neglect*, 11:547–563.

Rosenberg, N.L., Ed. (1995) *Occupational and Environmental Neurology*, Newton, MA: Butterworth-Heineman.

Rosenberg, N.L. (1996a) The neuromythology of silicone breast implants, *Neurology*, 46:308–314.

Rosenberg, N.L. (1996b) Letter to the editor [reply from the author], *Neurology*, 47:1353–1355.

Rosenberg, S.J., Freedman, M.R., Schmaling, K.G., and Rose, C. (1990) Personality styles of patients asserting environmental illness, *J. Occup. Med.*, 32(8):678–681.

Rosenstock, L., Keifer, M., Daniell, W.E., McConnell, R., and Claypoole, K. (1991) Chronic central nervous system effects of acute organophosphate pesticide intoxication, *Lancet*, 338:223–227.

Rosenthal, D. and Frank, J.D. (1956) Psychotherapy and the placebo effect, *Psychol. Bull.*, 53(4):294–302.

Rosenthal, R. (1963) On the social psychology of the psychological experiment: the experimenter's hypothesis as unintended determinant of the experimental results, *Am. Sci.*, 51:268–283.

Rosler, A. (1994) Long-term effects of childhood sexual abuse on the hypothalamic-pituitary-adrenal axis [editorial], *J. Clin. Endocrinol. Metab.*, 78:247–248.

Ross, C.A. (1989) *Multiple Personality Disorder: Diagnosis, Clinical Features, and Treatment*, New York: John Wiley & Sons.

Ross, C.A. (1995) Errors of logic in biological psychiatry, in Ross, C.A. and Pam, A. (Eds.), *Pseudoscience in Biological Psychiatry: Blaming the Body*, New York: John Wiley & Sons, pp. 85–128.

Ross, C.A. and Pam, A. (1995) *Pseudoscience in Biological Psychiatry: Blaming the Body*, New York: John Wiley & Sons.

Ross, C.A., Miller, S.D., Bjornson, L., Reagor, P., Fraser, G.A., and Anderson, G. (1991) Abuse histories in 102 cases of multiple personality disorder, *Can. J. Psychiatry*, 36:97–101.

Ross, M. and Olson, J.M. (1981) An expectancy-attribution model of the effects of placebos, *Psych. Rev.*, 88:408–437.

Rotter, J.B. (1966) Generalized expectancies for internal versus external control of reinforcement, *Psychol. Monogr.*, 80:No. 1.

Roy-Byrne, P.P., Geraci, M., and Uhde, T.W. (1986a) Life events and onset of panic disorder, *Am. J. Psychiatry*, 143(11):1424–1427.

Roy-Byrne, P.P., Uhde, T.W., Post, R.M., Gallucci, W., Chrousos, GP., and Gold, P.W. (1986b) The CRH stimulation test in patients with panic disorder, *Am. J. Psychiatry*, 143:396–399.

Royal College of Physicians, Committee on Clinical Immunology and Allergy (1992) *Allergy: Conventional and Alternative Concepts*, London: Royal College of Physicians.

Russell, D.E.H. (1983) The incidence and prevalence of intrafamilial and extrafamilial sexual abuse of female children, *Child Abuse Neglect*, 7:133–145.

Russell, M., Dark, K.A., Cummins, R.W., Ellman, G., Callaway, E., and Peeke, H.V.S. (1984) Learned histamine release, *Science*, 225:733–734.

Russo, J., Katon, W.J., Sullivan, M. et al. (1995) Severity of somatization and its relationship to psychiatric disorders and personality, *Psychosomatics*, 35:546–556.

Rutter, M. (1984) Psychopathology and development, I: childhood antecedents of adult psychiatric disorder, *Aust. N.Z. J. Psychiatry*, 18:225–234.

Ryan, C.M. and Morrow, L.A. (1992) Dysfunctional buildings or dysfunctional people: an examination of the sick building syndrome and allied disorders, *J. Consult. Clin. Psychol.*, 60(2):220–224.

Sakellaris, P.C. and Vernikos-Danellis, J. (1975) Increased rate of response of the pituitary-adrenal system in rats adapted to chronic stress, *Endocrinology*, 97:597–602.

Salvaggio, J.E. (1996) Understanding clinical immunological testing in alleged chemically induced environmental illness, *Reg. Toxicol. Pharmacol.*, 24(1):S16–27.

Santhuff, P. (1995) Susan Smith: the rest of the story [editorial], *Denver Post*, p. 7B, January 2.

Sapolsky, R.M. (1985) A mechanism for glucocorticoid toxicity in the hippocampus: increased neuronal vulnerability to metabolic insults, *J. Neurosci.*, 5:1228–1232.

Sapolsky, R.M. (1992) *Stress, the Aging Brain, and the Mechanisms of Neuron Death*, Cambridge, MA: MIT Press.

Sapolsky, R.M., Krey, L.C., and McEwen, B.S. (1984) Stress down-regulates corticosterone receptors in a site-specific manner in the brain, *Endocrinology*, 114:287–292.

Sapolsky, R.M., Krey, L.C., and McEwen, B.S. (1985) Prolonged glucocorticoid exposure reduces hippocampal neuron number: implications for aging, *J. Neurosci.*, 5:1222–1227.

Sapolsky, R.M., Krey, L.C., and McEwen, B.S. (1986) The neuroendocrinology of stress and aging: the glucocorticoid cascade hypothesis, *Endocrine Rev.*, 7:284–301.

Saporta, Jr., J.A. and Gans, J.S. (1995) Taking a history of childhood trauma in psychotherapy, *J. Psychother. Practice Res.*, 4(3):194–204.

Sarason, I.G. (1975) Test anxiety, attention, and the general problem of anxiety, in Spielberger, C.D. and Sarason, I.G. (Eds.), *Stress and Anxiety*, Vol. 1, Washington, D.C.: Hemisphere, pp. 165–198.

Sarason, I.G. (1985) Cognitive processes, anxiety and the treatment of anxiety disorders, in Tuma, A.H. and Maser, J. (Eds.), *Anxiety and the Anxiety Disorders*, Hillsdale, NJ: Lawrence Erlbaum Assoc., pp. 87–108.

Satel, S.L., Southwick, S.M., and Gawin, F.M. (1991) Clinical features of cocaine-induced paranoia, *Am. J. Psychiatry*, 148:495–499.

Scarinci, I.C., McDonald-Haile, J., Bradley, L.A., and Richter, J.E. (1994) Altered pain perception and psychosocial features among women with gastrointestinal disorders and history of abuse: a preliminary model, *Am. J. Med.*, 97:108–118.

Schachter, S. (1964) The interaction of cognitive and physiological determinants of emotional state, in Berkowitz, L. (Ed.), *Advances in Experimental Social Psychology*, Vol. I, New York: Academic Press, pp. 49–80.

Schachter, S. and Singer, J.E. (1962) Cognitive, social, and physiological determinants of emotional state, *Psych. Rev.*, 69:379–399.

Schaumburg, H.H. and Spencer, P.S. (1987) Recognizing neurotoxic disease, *Neurology*, 37:276–278.

Schein, E.H. (1956) The Chinese indoctrination program for prisoners of war: a study of attempted "brainwashing", *Psychiatry*, 19:149–172.

Schneider, F., Gur, R.E., Alavi, A., Seligman, M.E.P., Mozley, L.H., Smith, R.J., Mozley, P.D., and Gur, R.C. (1996) Cerebral blood flow changes in limbic regions induced by unsolvable anagram tasks, *Am. J. Psychiatry*, 153(2):206–212.

Schnurr, P.P., Friedman, M.J., and Rosenberg, S.D. (1993) Preliminary MMPI scores as predictors of combat-related PTSD symptoms, *Am. J. Psychiatry*, 150(3):479–483.

Schottenfeld, R.S. (1987) Workers with multiple chemical sensitivities: a psychiatric approach to diagnosis and treatment, *Occup. Med. State Art Rev.*, 2:739–753.

Schottenfeld, R.S. and Cullen, M.R. (1986) Recognition of occupation-induced posttraumatic stress disorders, *J. Occup. Med.*, 28(5), 365–369.

Schvaneveldt, R.W. and Meyer, D.E. (1973) Retrieval and comparison processes in semantic memory, in Kornblum, S. (Ed.), *Attention and Performance*, IV, New York: Academic Press, pp. 395–409.

Schwartz, G.E., Bell, I.R., Dikman, Z.V., Fernandez, M., Kline, J.P., Peterson, J.M., and Wright, K.P. (1994) EEG responses to low-level chemicals in normals and cacosmics, *Toxicol. Ind. Health*, 10(4/5):633–643.

Schwartz, S.P., White, P.E., and Hughes, R.G. (1985) Environmental threats, communities, and hysteria, *J. Public Health Policy*, 6(1):58–77.

Schwarz, E.D. and Perry, B.D. (1994) The post-traumatic response in children and adolescents, *Psychiatric Clin. North Am.*, 17(2):311–326.

Seeman, M.V. (1997) Psychopathology in women and men: focus on female hormones, *Am. J. Psychiatry*, 154(12):1641–1647.

Seligman, M.E.P. (1975) *Helplessness: On Depression, Development, and Death*, San Francisco, CA: Freedman.

Seligman, M.E.P. (1995) The effectiveness of psychotherapy: the Consumer Reports study, *Am. Psychol.*, 50(12):965–974.

Seligman, M.E.P. and Nolen-Hoeksema, S. (1987) Explanatory style and depression, in Magnusson, D. and Ohman, A. (Eds.), *Psychopathology*, New York: Academic Press.

Sells, S.B. (1970) On the nature of stress, in McGrath, J.E. (Ed.), *Social and Psychological Factors in Stress*, New York: Holt.

Selner, J.C. (1989) Workup of the chemically sensitive patient, *Masters Allergy*, 1:8–16.

Selner, J.C. (1991) Book review of "Chemical Exposures, Low Levels and High Stakes" by N. Ashford and C.S. Miller, *Ann. Allergy*, 67:456.

Selner, J.C. (1996) Chamber challenges: the necessity of objective observation, *Reg. Toxicol. Pharmacol.*, 24(1):S87–S95.

Selner, J.C. and Staudenmayer, H. (1985) The practical approach to the evaluation of suspected environmental exposures: chemical intolerance, *Ann. Allergy*, 55(5):665–673.

Selner, J.C. and Staudenmayer, H. (1986) The relationship of the environment and food to allergic and psychiatric illness, in Young, S.H., Rubin, J.M., and Daman, H.R. (Eds.), *Psychophysiological Aspects of Allergic Disorders*, New York: Praeger, pp. 102–146.

Selner, J.C. and Condemi, J.J. (1988) Unproven diagnostic and treatment methods, in Middleton, Jr., E., Reed, C.E., Ellis, E. F., Adkinson, Jr., N.F., and Yunginger, J.W. (Eds.), *Allergy: Principles and Practice*, St. Louis, MO: Mosby, pp. 1571–1597.

Selner, J.C. and Staudenmayer, H. (1991) Food allergy: psychological considerations, in Metcalfe, D.D., Sampson, H.A., and Simon, R.A. (Eds.), *Food Allergy: Adverse Reactions to Foods and Food Additives*, Boston: Blackwell Scientific, pp. 370–381.

Selner, J.C. and Staudenmayer, H. (1992a) Psychological factors complicating the diagnosis of work-related chemical illness, *Immunol. Allergy Clin. North Am.*, 12(4):909–919.

Selner, J.C. and Staudenmayer, H. (1992b) Neuropsychophysiologic observations in patients presenting with environmental illness, *Toxicol. Indust. Health*, 8(4):145–155.

Selner, J.C., Staudenmayer, H., Koepke, J.W., Harvey, R., and Christopher, K. (1987) Vocal cord dysfunction: the importance of psychologic factors and provocation challenge testing, *J. Allergy Clin. Immunol.*, 79(5):726-733.

Selye, H. (1936) A syndrome produced by diverse nocuous agents, *Nature*, 138:32.

Selye, H. (1946) The general adaptation syndrome and the diseases of adaptation, *J. Allergy*, 17:231–247, 289–323, 358–398.

Selye, H. (1982) History and present status of the stress concept, in Goldberger, L. and Breznitz, S. (Eds.), *Handbook of Stress: Theoretical and Clinical Aspects*, New York: Free Press/Macmillan, pp. 7–17.

Shalev, A.Y., Orr, S.P., Peri, T., Schreiber, S., and Pitman, R.K. (1992) Physiologic responses to loud tones in Israeli patients with posttraumatic stress disorder, *Arch. Gen. Psychiatry*, 49:870–875.

Shapiro, A.K. (1964) Factors contributing to the placebo effect: their implications for psychotherapy, *Am. J. Psychother.*, 18(Suppl.):73–88.

Shapiro, Jr., D.H., Schwartz, C.E., and Astin, J.A. (1996) Controlling ourselves, controlling our world: psychology's role in understanding positive and negative consequences of seeking and gaining control, *Am. Psychol.*, 51(12):1213–1230.

Shapiro, T. (1993) On reminiscences, *J. Am. Psychoanalyt. Assoc.*, 41:395–421.

Sharpe, M., Peveler, R., and Mayou, R. (1992) The psychological treatment of patients with functional somatic symptoms: a practical guide, *J. Psychosomatic Res.*, 36(6):515–529.

Sharpe, R. (1995) U.S. agency issues set of definitions on the disabled, *Wall Street Journal*, p. B11, March 16.

Shear, M.K. (1988) Cognitive and biological models of panic: toward an integration, in Rachman, S. and Maser, J.D. (Eds.), *Panic: Psychological Perspectives*, Hillsdale, NJ: Lawrence Erlbaum Assoc., pp. 51–70.

Shear, M.K. (1996) Factors in the etiology and pathogenesis of panic disorder: revisiting the attachment-separation paradigm, *Am. J. Psychiatry*, 153(7):125–136.

Shear, M.K., Kligfield, P., Harschfield, G., Devereux, R.B., Polan, J.J., Mann, J.J., Pickering, T., and Frances, A.J. (1987) Cardiac rate and rhythm in panic patients, *Am. J. Psychiatry*, 144(5):633–637.

Shear, M.K., Cooper, A.M., Klerman, G.L., Busch, F.N., and Shapiro, T.A (1993) psychodynamic model of panic disorder, *Am. J. Psychiatry*, 150(6):859–866.

Shedler, J., Mayman, M., and Manis, M. (1993) The illusion of mental health, *Am. Psychol.*, 48(11):1117–1131.

Shengold, L. (1989) *Soul Murder: The Effects of Childhood Abuse and Deprivation*, New York: Ballantine Books.

Shorter, E. (1992) *From Paralysis to Fatigue: A History of Psychosomatic Illness in the Modern Era*, New York: Free Press/Macmillan.

Shusterman, D. (1992) Critical review: the health significance of environmental odor pollution, *Arch. Environ. Health*, 47(1):76–87.

Shusterman, D. and Dager, S. (1991) Prevention of psychological disability after occupational respiratory exposures, *Occup. Med. State Art Rev.*, 6, 11–27.

Shusterman, D., Balmes, J., and Cone, J. (1988) Behavioral sensitization to irritants/odorants after acute over exposures, *J. Occup. Med.*, 30(7):565–567.

Shusterman, D., Lipscomb, J., Neutra, R., and Satin, K. (1991) Symptom prevalence and odor-worry interaction near hazardous waste sites, *Environ. Health Perspect.*, 94:25–30.

Siegel, J.M., Sorenson, S.B., Golding, J.M., Burnam, M.A., and Stein, J.A. (1987) The prevalence of childhood sexual assault: the Los Angeles epidemiologic catchment area project, *Am. J. Epidemiol.*, 126:1141–1153.

Sierra Club (1981) *Training Materials on Toxic Substances* and *Tools for Effective Action*, San Francisco, CA: Sierra Club.

Silk, K.R., Ed. (1994) *Biological and Neurobehavioral Studies of Borderline Personality Disorder*, Washington, D.C.: American Psychiatric Press.

Silk, K.R., Lee, S., Hill, E.M., and Lohr, N.E. (1995) Borderline personality disorder symptoms and severity of sexual abuse, *Am. J. Psychiatry*, 152(7):1059–1064.

Simon, B. (1992a) "Incest — see under Oedipus complex": the history of an error in psychoanalysis, *J. Am. Psychoanalyt. Assoc.*, 40:955–988.

Simon, G.E. (1992b) Psychiatric treatments in multiple chemical sensitivity, proceedings of the Association of Occupational and Environmental Clinics (AOEC) Workshop on Multiple Chemical Sensitivity, September 20–21, 1991, Washington, D.C., sponsored by Agency for Toxic Substances and Disease Registry (ATSDR), *Toxicol. Indust. Health*, 8(4):67–72.

Simon, G.E., Katon, W.J., and Sparks, P.J. (1990) Allergic to life: psychological factors in environmental illness, *Am. J. Psychiatry*, 147:901–906.

Simon, G.E., Daniell, W., Stockbridge, H., Claypoole, K., and Rosenstock, L. (1993) Immunologic, psychological and neuropsychological factors in multiple chemical sensitivity, *Ann. Intern. Med.*, 19(2):97–103.

Simon, T.R. (1992c) Breast implants and organic solvent exposure can be associated with abnormal cerebral SPECT studies in clinically impaired patients [abstract], *Radiology*, 185:234.

Singer, J.E. (1982) Yes, Virginia, there really is a mass psychogenic illness, in Colligan, M.J., Pennebaker, J.W., and Murphy, L.R. (Eds.), *Mass Psychogenic Illness: A Social Psychological Analysis*, Hillsdale, NJ: Lawrence Erlbaum Assoc., pp. 127–135.

Smith, Jr., G.R. and McDaniel, S.M. (1983) Psychologically mediated effect on the delayed hypersensitivity reaction to tuberculin in humans, *Psychosomatic Med.*, 45:65–70.

Smith, Jr., G.R., Monson, R.A., and Ray, D.C. (1986) Psychiatric consultation in somatization disorder, *New Engl. J. Med.*, 314(22):1407–1413.

Smith, S. (1983) Stigma: accept it and win, *Human Ecologist* (newsletter for the Human Ecology Action League), 22:4–8.

Snyder, H.S. and Weiss, E. (1989) Hysterical stridor: a benign cause of upper airway obstruction, *Ann. Emerg. Med.*, 18(9):991–994.

Society of Nuclear Brain Imaging Council (1996) Ethical clinical practice of fuctional brain imaging, *J. Nucl. Med.*, 37(7):1256–1259.

Sokolov, Y.N. (1963) *Perception and the Conditioned Reflex*, Oxford, UK: Pergamon Press.

Solnit, A. and Kris, M. (1967) Trauma and infantile experiences: a longitudinal perspective, in Furst, S. (Ed.), *Psychic Trauma*, New York: Basic Books.

Solomon, G.F. (1995) *Immune and Nervous System Interactions: An Analytic Bibliography Supporting Key Postulates on Communication Links, Similarities, and Implications*, Malibu, CA: The Fund for Psychoneuroimmunology.

Southall, D.P., Stebbens, V.A., Rees, S.V., Lang, M.H., Warner, J.O., and Shinebourne, E.A. (1987) Apnoeic episodes induced by smothering: two cases identified by covert video surveillance, *Br. Med. J.*, 294(June 27):1637–1641.

Southall, D.P., Plunkett, M.C.B., Banks, M.W., Falkov, A.F., and Samuels, M.P. (1997) Covert video recordings of life-threatening child abuse: lessons for child protection, *Pediatrics*, 100(5):735–760.

Southwick, S.M., Krystal, J., Johnson, D., and Charney, D.S. (1992) *Neurobiology of PTSD: Annual Review of Psychiatry*, Washington, D.C.: American Psychiatric Association Press.

Southwick, S.M., Bremner, D., Krystal, J. H., and Charney, D.S. (1994) Psychobiologic research in post-traumatic stress disorder, *Psychiatric Clin. North Am.*, 17(2):251–264.

Southwick, S.M., Morgan, C.A., Darnell, A., Bremner, D., Nicolaou, A.L., Nagy, L.M., and Charney, D.S. (1995a) Trauma-related symptoms in veterans of Operation Desert Storm: a 2-year follow-up, *Am. J. Psychiatry*, 152(8):1150–1155.

Southwick, S.M., Yehuda, R., and Morgan, III, C.A. (1995b) Clinical studies of neurotransmitter alterations in post-traumatic stress disorder, in Friedman, M.J., Charney, D.S., and Deutch, A.Y. (Eds.), *Neurobiological and Clinical Consequences of Stress: From Normal Adaptation to PTSD*, Philadelphia: Lippincott-Raven, pp. 335–349.

Southwick, S.M., Morgan, C.A., Bremner, A.D., Grillon, C.G., Krystal, J.H., Nagy, L.M., and Charney, D.S. (1997a) Noradrenergic alterations in posttraumatic stress disorder, in Yehuda, R. and McFarlane, A.C. (Eds.), *Psychobiology of Posttraumatic Stress Disorder*, New York: New York Academy of Sciences, pp. 125–141.

Southwick, S.M., Morgan, C.A., Nicolaou, A.L., and Charney, D.S. (1997b) Consistency of memory for combat-related traumatic events in veterans of operation Desert Storm, *Am. J. Psychiatry*, 154(2):173–177.

Spacapan, S. and Cohen, S. (1983) Effects and aftereffects of stressor expectations, *J. Personality Soc. Psychol.*, 45:1243–1254.

Sparks, P.J., Daniell, W., Black, D.W., Kipen, H.M., Altman, L.C., Simon, G.E., and Terr, A.I. (1994) Multiple chemical sensitivity syndrome: a clinical perspective. I. Case definition, theories of pathogenesis, and research needs, *J. Occup. Med.*, 36(7):718–730.

Spencer, P.S. and Schaumburg, H.H., Eds. (1980) *Experimental and Clinical Neurotoxicology*, Baltimore, MD: Williams & Wilkins.

Spencer, P.S. and Schaumburg, H.H. (1985) Organic solvent neurotoxicity: facts and research needs, *Scand. J. Work Environ. Health*, 11(Suppl. 1):53–60.

Sperry, R.W. (1988) Psychology's mentalist paradigm and the religion/science tension, *Am. Psychol.*, 43(8):607–613.

Sperry, R.W. (1993) The impact and promise of the cognitive revolution, *Am. Psychol.*, 48(8):878–885.

Spielberger, C.D., Ed. (1972) *Anxiety: Current Trends in Theory and Research, Vols. 1 and 2*, New York: Academic Press.

Spitz, R.A. (1945) Hospitalism: an inquiry into the genesis of psychiatric conditions in early childhood, *Psychoanalyt. Study Child*, 1:53–74.

Spitz, R.A. (1946) Hospitalism: a follow-up report on investigation described in Volume I, 1945, *Psychoanaly. Study Child*, 2:113–117.

Spitzer, M., Kwon, K.K., Kennedy, W., Rosen, B.R., and Belliveau, J.W. (1995) Category-specific brain activation in fMRI during picture naming, *Neural Rep.*, 6:2109–2112.

Springs, F.E. and Friedrich, W.N. (1992) Health risk behaviors and medical sequelae of childhood sexual abuse, *Mayo Clin. Proc.*, 67:527–533.

Spyker, D.A. (1995) Multiple chemical sensitivities — syndrome and solution, *Clin. Toxicol.*, 33(2):95–99.

Squire, L.R. (1986) Mechanisms of memory, *Science*, 232:1612–1619.

Squire, L.R. (1987) Memory: neural organization and behavior, in Blum, F. (Ed.), *Handbook of Physiology. Section 1. The Nervous System*, Bethesda, MD: American Physiological Society, pp. 295–371.

Squire, L.R. (1992) Memory and the hippocampus: a synthesis from findings with rats, monkeys, and humans, *Psych. Rev.*, 99:195–231.

Squire, L.R., Knowlton, B., and Musen, G. (1993) The structure and organization of memory, *Ann. Rev. Psychol.*, 44:453–495.

Stanford, S.C. and Salmon, P. (1993) *Stress: From Synapse to Syndrome*, London: Academic Press.

Starkman, M.N., Gebarski, S.S., Berent, S., and Schteingart, D.E. (1992) Hippocampal formation volume, memory dysfunction, and cortical levels in patients with Cushing's syndrome, *Biol. Psychiatry*, 32:756–765.

Staudenmayer, H. (1975) Understanding conditional reasoning with meaningful propositions, in Falmange, R.J. (Ed.), *Reasoning: Representation and Process*, Hillsdale, NJ: Lawrence Erlbaum Assoc., pp. 55–79.

Staudenmayer, H. (1996) Clinical consequences of the EI/MCS "diagnosis": two paths, *Reg. Toxicol. Pharmacol.*, 24:S96–S110.

Staudenmayer, H. (1997) Multiple chemical sensitivities or idiopathic environmental intolerances (EI): psychophysiological foundation of knowledge for a psychogenic explanation [editorial], *J. Allergy Clin. Immunol.*, 99(4):434–437.

Staudenmayer, H. and Selner, J.C. (1987) Post-traumatic stress syndrome (PTSS): escape in the environment, *J. Clin. Psychol.*, 43(1):156–157.

Staudenmayer, H. and Camazine, M. (1989) Sensing type personality, projection and universal "allergic" reactivity, *J. Psychol. Type*, 18:59–62.

Staudenmayer, H. and Kramer, R.E. (in preparation) Psychogenic chemical sensitivity: psychogenic pseudoseizures elicited by provocation challenges with fragrances.

Staudenmayer, H. and Selner, J.C. (1990) Neuropsychophysiology during relaxation in generalized, universal "allergic" reactivity to the environment: a comparison study, *J. Psychosomatic Res.*, 34(3):259–270.

Staudenmayer, H. and Selner, J.C. (1995) Failure to assess psychopathology in patients presenting with chemical sensitivities, *J. Occup. Med.*, 37(6):704–709.

Staudenmayer, H., Kinsman, R.A., and Jones, N.F. (1978) Attitudes toward respiratory illness and hospitalization in asthma: Relationships with personality, symptomatology and treatment responses, *J. Nerv. Ment. Dis.*, 166:624–634.

Staudenmayer, H., Selner, J.C., and Buhr, M. (1993a) Double-blind provocation challenges in 20 patients presenting with "multiple chemical sensitivity", *Reg. Pharmacol. Toxicol.*, 18:44–53.

Staudenmayer, H., Selner, M.E., and Selner, J.C. (1993b) Adult sequelae of childhood abuse presenting as environmental illness, *Ann. Allergy*, 71(6):538–546.

Steele, B. (1970) Parental abuse of infants and small children, in Anthony, J. and Benedek, T. (Eds.), *Parenthood*, New York: Little, Brown, pp. 449–477.

Steele, B. (1976) Violence within the family, in Helfer, R. and Kempe, C. (Eds.), *Child Abuse and Neglect*, Cambridge, MA: Ballinger.

Stein, M.B., Koverola, C., Hanna, C., Torchia, M.G., and McClarty, B. (1997) Hippocampal volume in women victimized by childhood sexual abuse, *Psychol. Med.*, 27:951–959.

Stein, M.B., Walker, J.R., Anderson, G., Hazen, A.L., Ross, C.A., Eldridge, G., and Forde, D.R. (1996) Childhood physical and sexual abuse in patients with anxiety disorders and in a community sample, *Am. J. Psychiatry*, 153(2):275–277.

Steinberg, A.B. (1992) Systemic lupus erythematosus, in Wyngaarden, J.B., Smith, L.H., and Bennett, J.C. (Eds.), *Cecil Textbook of Medicine*, Philadelphia: W.B. Saunders, pp. 1522–1527.

Sternbach, R.A. Psychological dimensions and perceptual analyses, including pathologies of pain, in Carterett, E.E. and Friedman, M.D. (Eds.), *Handbook of Perception*, Vol. 6B, New York: Academic Press, 1978, pp. 231–261.

Stewart, D.E. (1990) The changing faces of somatization, *Psychosomatics*, 31(2):153–158.

Stewart, D.E. and Raskin, J. (1985) Psychiatric assessment of patients with "20th century disease" ("total allergy syndrome"), *Can. Med. Assoc. J.*, 133:1001–1006.

Still, G.F. (1902) Some abnormal psychical conditions in children (the Goulstonian lectures), *Lancet*, 1:1008–1012, 1077–1082, 1163–1168.

Stone, M.H. (1989) Individual psychotherapy with victims of incest, *Psychiatric Clin. North Am.*, 12(2):237–255.

Straight, J.M. and Vogt, Jr., R.F. (1997) The use of biologic markers in the clinical and epidemiologic evaluation of exposures and health effects related to indoor pollution, in Bardana, Jr., E.J. and Montanaro, A. (Eds.), *Indoor Air Pollution and Health*, New York: Marcel Dekker, pp. 305–349.

Straus, S.E., Dale, J.K., Tobi, M., Lawley, T., Preble, O., Blaese, R., Hallahan, C., and Henle, W. (1988) Acyclovir treatment of the chronic fatigue syndrome: lack of efficacy in a placebo-controlled trial, *New Engl. J. Med.*, 319:1692–1698.

Strickland, B.R. (1978) Internal-external expectancies and health-related behaviors, *J. Consult. Clin. Psychol.*, 46(6):1192–1211.

Sutker, P.B., Winstead, D.K., Galina, Z.H., and Allain, Jr., A.N. (1991) Cognitive deficits and psychopathology among former prisoners of war and combat veterans of the Korean conflict, *Am. J. Psychiatry*, 148:67–72.

Sutker, P.B., Davis, J.M., Uddo, M., and Ditta, S.R. (1995a) Assessment of psychological distress in Persian Gulf troops: ethnicity and gender comparisons, *J. Personality Assess.*, 64(3):415–427.

Sutker, P.B., Vasterling, J.J., Brailey, K., and Allain, Jr., A.N. (1995b) Memory, attention, and executive deficits in POW survivors: contributing biological and psychological factors, *Neuropsychology*, 9:118–125.

Sutton, R.E., Koob, G.F., Le Moal, M., Rivier, J., and Vale, W. (1982) Corticotropin releasing factor produces behavioral activation in rats, *Nature*, 297:331–333.

Svrakic, D.M., Whitehead, C., Przybeck, T.R., and Cloninger, C.R. (1993) Differential diagnosis of personality disorders by the seven-factor model of temperament and character, *Arch. Gen. Psychiatry*, 50:991–999.

Swanson, L.W. (1983) The hippocampus and the concept of the limbic system, in Seifert, W. (Ed.), *Neurobiology of the Hippocampus*, London: Academic Press, pp. 3–19.

Sykes, C.J. (1992) *A Nation of Victims: The Decay of the American Character*, New York: St. Marten's Press.

Tabershaw, I.R. and Cooper, W.C. (1966) Sequelae of acute organic phosphate poisoning, *J. Occup. Med.*, 8:5–20.

Taerk, G.S., Toner, B.B., Salit, I.E., Garfinkel, P.E., and Ozersky, S. (1987) Depression in patients with neuromyasthenia (benign myalgic encephalomyelitis), *Int. J. Psychiatry Med.*, 17:49–56.

Talley, N.J., Fung, L.H., Gilligan, I.J., McNeil, D., and Piper, D.W. (1986) Association of anxiety, neuroticism, and depression with dyspepsia of unknown cause, *Gastroenterology*, 90:886–892.

Teicher, M.H. (1988) Biology of anxiety, *Med. Clin. North Am.*, 72:791–814.

Teicher, M.H., Glod, C.A., Surrey, J., and Swett, C. (1993) Early childhood abuse and limbic system ratings in adult psychiatric outpatients, *J. Neuropsychiatry*, 5:301–306.

Teicher, M.H., Ito, Y., Glod, C.A., Schiffer, F., and Gelbard, H.A. (1994) Early abuse, limbic system dysfunction, and borderline personality disorder, in Silk, K.R. (Ed.), *Biological and Neurobehavioral Studies of Borderline Personality Disorder*, Washington, D.C.: American Psychiatric Press, pp. 177–207.

Teicher, M.H., Ito, Y., Glod, C.A., Andersen, S.L., Dumont, N., and Acerman, E. (1997) Preliminary evidence for abnormal cortical development in physically and sexually abused children using EEG coherence and MRI, in Yehuda, R. and McFarlane, A.C. (Eds.), *Psychobiology of Posttraumatic Stress Disorder*, New York: New York Academy of Sciences, pp. 160–175.

Terr, A.I. (1986a) Environmental illness: a clinical review of 50 cases, *Arch. Intern. Med.*, 146:145–149.

Terr, A.I. (1986b) Food allergy: a manifestation of eating disorder?, *Int. J. Eating Disorders*, 5(3):575–579.

Terr, A.I. (1987a) "Multiple chemical sensitivities": immunologic critique of ecology theories and practice, *Occup. Med. State Art Rev.*, 2:683–694.

Terr, A.I. (1987b) Clinical ecology [editorial], *J. Allergy Clin. Immunol.*, 79:423–426.

Terr, A.I. (1989) Clinical ecology in the workplace, *J. Occup. Med.*, 31:257–261.

Terr, A.I. (1992) Multiple chemical sensitivities syndrome, *Immunol. Allergy Clin. North Am.*, 12(4):897–908.

Terr, A.I. (1993a) Immunologic issues in "multiple chemical sensitivities", *Reg. Toxicol. Pharmacol.*, 18:54–60.

Terr, A.I. (1993b) Multiple chemical sensitivities [editorial], *Ann. Intern. Med.*, 119(2):163–164.

Terr, A.I. (1993c) Unconventional theories and unproven methods in allergy, in Middleton, Jr., E., Reed, C.E., Ellis, E.F., Adkinson, Jr., N.F., Yunginger, J.W., and Busse, W.W. (Eds.), *Allergy: Principles and Practice*, Vol II, 4th ed., St. Louis, MO: Mosby, pp. 1767–1793.

Third Task Force for Research Planning in Environmental Health Science (1984) *Human Health and the Environment: Some Research Needs*, Bethesda, MD: National Institutes of Health, publication no. 86-1277.

Thoits, P.A. (1983) Life events and psychological distress, in Kaplan, H.B. (Ed.), *Psychological Stress: Trends in Theory and Research*, New York: Academic Press, pp. 33–103.

Thomas, L. (1980) Report of the Governor's Panel To Review Scientific Studies and the Development of Public Policy on Problems Resulting from Hazardous Wastes, transmitted to Governor Hugh Carey and Members of the New York Legislature, October 8.

Thompson, G.M., Hon., et al. (1985) Report of the Ad Hoc Committee on Environmental Hypersensitivity Disorders, Ministry of Health, Province of Ontario, Toronto, Canada.

Tollefson, L. (1993) Multiple chemical sensitivity: controlled scientific studies as proof of causation, *Reg. Toxicol. Pharmacol.*, 18(1):32–43.

Truss, C.O. (1986) *The Missing Diagnosis*, 2nd ed., Birmingham, AL: Missing Diagnosis.

Tsien, A. and Spector, S.L. (1997) Role of provocation challenge in assessing indoor air pollution, in Bardana, Jr., E.J. and Montanaro, A. (Eds.), *Indoor Air Pollution and Health*, New York: Marcel Dekker, pp. 387–420.

Tversky, A. and Kahneman, D. (1971) Belief in the law of small numbers, *Psychol. Bull.*, 76:105–110.

Tyrer, P.J. (1985) Neurosis divisible?, *Lancet*, i, 685–688.

Tyrer, P.J. (1990) The division of neurosis: a failed classification, *J. Roy. Soc. Med.*, 83:614–616.

Updegraff, T.R. (1977) Food allergy and cytotoxic test, *Ear Nose Throat J.*, 56:450–459.

Ur, E., White, P.D., and Grossman, A. (1992) Hypothesis: cytokines may be activated to cause depressive illness and chronic fatigue syndrome, *Eur. Arch. Psychiatry Clin. Neurosci.*, 241:371–322.

Ursin, H., Baade, E., and Levine, S., Eds. (1978) *Psychobiology of Stress: A Study of Coping Men*, New York: Academic Press.

U.S. Congress, Office of Technology Assessment (1992) *Neurotoxicity: Identifying and Controlling Poisons of the Nervous System*, New York: Van Nostrand Reinhold.

U.S. House Subcommittee on Oversight and Investigation (1980) *Hazardous Waste: Memphis, Tennessee, Area*, serial no. 96-144, Washington, D.C.: U.S. Government Printing Office, April 2.

Valzelli, L. (1982) Serotonergic inhibitory control of experimental aggression, *Psychopharmacol. Res. Commun.*, 12:1–13.

Vance, M.L. (1994) Medical progress: hypopituitarism, *New Engl. J. Med.*, 330(23):1651–1662.

van der Hart, O. and Friedman, B. (1989) A reader's guide to Pierre Janet on dissociation: a neglected intellectual heritage, *Dissociation*, 2:3–16.

van der Kolk, B.A. (1987) *Psychological Trauma*, Washington, D.C.: American Psychiatric Press.

van der Kolk, B.A. (1988) The trauma spectrum: the interaction of biological and social events in genesis of the trauma response, *J. Traumatic Stress*, 1(3):273–290.

van der Kolk, B.A. (1994) The body keeps the score: memory and the evolving psychobiology of posttraumatic stress, *Harvard Rev. Psychiatry*, 1(5):253–265.

van der Kolk, B.A. and Greenberg, M.S. (1987) The psychobiology of the trauma response: hyperarousal, constriction, and addiction to traumatic reexposure., in van der Kolk, B.A. (Ed.), *Psychological Trauma*, Washington, D.C.: American Psychiatric Press, pp. 63–87.

van der Kolk, B. A. and van der Hart, O. (1989) Pierre Janet and the breakdown of adaptation in psychological trauma, *Am. J. Psychiatry*, 146(12):1530–1540.

van der Kolk, B. A. and van der Hart, O. (1991) The intrusive past: the flexibility of memory and the engraving of trauma, *Am. Imago*, 48(4):425–454.

van der Kolk, B.A. and Saporta, J. (1991) The biological response to psychic trauma: mechanisms and treatment of intrusion and numbing, *Anxiety Res.*, 4:199–212.

van der Kolk, B. A., Greenberg, M. S., Boyd, H., and Crystal, J. (1985) Inescapable shock, neurotransmitters, and addiction to trauma: toward a psychobiology of post-traumatic stress, *Biol. Psychiatry*, 20:314–325.

van der Kolk, B.A., Greenberg, M.S., Orr, S.P., and Pitman, R.K. (1989) Endogenous opioids, stress induced analgesia, and in posttraumatic stress disorder, *Psychopharmacol. Bull.*, 25:417–421.

van der Kolk, B.A., Roth, S., and Pelcovitz, D. (1992) *Field Trials for DSM-IV, Post Traumatic Stress Disorder II: Disorders of Extreme Stress*, Washington, D.C.: American Psychiatric Press.

van der Kolk, B.A., Pelcovitz, D., Roth, S., Mandel, F.S., McFarlane, A., and Herman, J.L. (1996) Dissociation, somatization, and affect dysregulation: the complexity of adaptation to trauma, *Am. J. Psychiatry*, 153(7):83–93.

Vasterling, J.J., Brailey, K., Constans, J.I., and Sutker, P.B. (1998) Attention and memory dysfunction in posttraumatic stress disorder, *Neuropsychology*, 12(1):125–133.

Venables, P.H. and Christie, M.J. (1980) Electrodermal activity, in Martin, I. and Venables, P.H. (Eds.), *Techniques in Psychophysiology*, New York: John Wiley & Sons, pp. 3–67.

Vercoulen, J.H.M.M., Swanink, C.M.A., Zitman, F.G., Vreden, S.G.S., Hoofs, M.P.E., Fennis, J.F.M., Galama, J.M.D., van der Meer, J.W.M., and Bleijenberg, G. (1996) Randomized, double-blind, placebo-controlled study of fluoxetine in chronic fatigue syndrome, *Lancet*, 347:858–861.

Vickery, D. (1992) Therapist as victim: a preliminary discussion, *Dissociation*, 5(2):155–158.

Vineis, P., Schulte, P.A., and Vogt, Jr., R.F. (1993) Technical variability in laboratory data, in Schulte, P.A. and Perrera, F. (Eds.), *Molecular Epidemiology: Principles and Practice*, New York: Academic Press, pp. 109–135.

Vlachakis, N.D., DeGuia, D., Mendlowitz, M., Antram, S., and Wolf, R.L. (1974) Hypertension and anxiety. A trial with epinephrine and norepinephrine infusion, *Mt. Sinai J. Med.*, 41:615–625.

Volow, M.R. (1986) Pseudoseizures: an overview, *South. Med. J.*, 79:600–607.

Wachtel, P.L., Ed. (1982) *Resistance: Psychodynamic and Behavioral Approaches*, New York: Plenum Press.

Wada, J.A. (1990) *Kindling 4*, New York: Plenum Press.

Wada, J.A., Sato, M., and Corcoran, M.E. (1974) Persistent seizure susceptibility and recurrent spontaneous seizures in kindled cats, *Epilepsia*, 15:465–478.

Waddell, W.J. (1993) The science of toxicology and its relevance to MCS, *Reg. Toxicol. Pharmacol.*, 18,13–22.

Walker, E.F. and DiForio, D. (1997) Schizophrenia: a neural diathesis-stress model, *Psych. Rev.*, 104(04):667–685.

Walker, E.F., Katon, W., Harrop-Griffiths, J., Holm, L., Russo, J., and Hickok, L.R. (1988) Relationship of chronic pelvic pain to psychiatric diagnoses and childhood sexual abuse, *Am. J. Psychiatry*, 145:75–80.

Walker, E.F., Katon, W.J., Roy-Byrne, P.P., Jemelka, R.P., and Russo, J. (1993) Histories of sexual victimization in patients with irritable bowel syndrome or inflammatory bowel disease, *Am. J. Psychiatry*, 150:1502–1511.

Walker, E.F., Katon, W.J., Hanson, J., Harrop-Griffiths, J., Holm, L., Jones, M.L., Hickok, L.R., and Russo, J. (1995) Psychiatric diagnoses and sexual victimization in women with chronic pelvic pain, *Psychosomatics*, 36(6):531–540.

Wallace, L.A., Pellizzari, E.D., Hartwell, T.D., Sparacino, C., Whitmore, R., Sheldon, L., Zelon, H., and Perritt, R. (1987) The TEAM study: personal exposures to toxic substances in air, drinking water, and breath of 400 residents of New Jersey, North Carolina, and North Dakota, *Environ. Res.*, 43:290–307.

Warburton, D.M. (1979) Physiological aspects of information processing and stress, in Hamilton, V. and Warburton, D.M. (Eds.), *Human Stress and Cognition: An Information Processing Approach*, New York: John Wiley & Sons, pp. 33–65.

Warwick, H.M.C. and Marks, I.M. (1988) Behavioral treatment of illness phobia and hypochondriasis, *Br. J. Psychiatry*, 152:239–241.

Weaver, V.M. (1996) Medical management of the multiple chemical sensitivity patient, *Reg. Toxicol. Pharmacol.*, 24(1):S111–S115.

Weber, R.W., Hoffman, M., Raine, Jr., D.A., and Nelson, H.S. (1979) Incidence of bronchoconstriction due to aspirin, azo dyes, non-azo dyes, and preservatives in a population of perennial asthmatics, *J. Allergy Clin. Immunol.*, 64(1):32–37.

Weil-Malherbe, H., Axelrod, J., and Tomchick, R. (1959) Blood-brain barrier for adrenalin, *Science*, 129:1226–1227.

Weiner, B. (1985) An attributional theory of achievement motivation and emotion, *Psych. Rev.*, 92(4):548–573.

Weiner, H. (1977) *Psychobiology and Human Disease*, New York: Elsevier.

Weiner, H. (1985) The concept of stress in the light of studies on disasters, unemployment and loss: a critical analysis, in Zales, M.R. (Ed.), *Stress in Health and Disease*, New York: Brunner/Mazel, pp. 24–94.

Weiss, S.R.B., Murman, D., Post, R.M., and Pert, A. (1986) Conditioning in Cocaine-Induced Behavioral Sensitization, 16th Annual Meeting, Society for Neuroscience Abstracts, Washington, D.C., Nov. 9–14, 12:914 (abstract 249.13).

Werry, J.S. (1976) Food additives and hyperactivity [editorial], *Med. J. Aust.*, 2(8):281–282.

Wessely, S. (1990) Old wine in new bottles: neurasthenia and "ME", *Psychol. Med.*, 20:35–53.

Wessely, S. and Wardle, C. The current literature: "mass sociogenic illness", *Br. J. Psychiatry* 1990, 157:421–424.

Wessely, S., Chalder, T., Hirsch, S., Wallace, P., and Wright, D. (1996) Psychological symptoms, somatic symptoms, and psychiatric disorder in chronic fatigue and chronic fatigue Syndrome: a prospective study in the primary care setting, *Am. J. Psychiatry*, 153(8):1050–1059.

Whelan, E.M. (1985) *Toxic Terror*, Ottowa, IL: Jameson Books.

Whelan, E.M. (1993) The food-fear epidemic, in Barrett S. and Jarvis, W.T. (Eds.), *The Health Robbers*, Buffalo, NY: Prometheus Books, pp. 67–82.

White, R.F. and Proctor, S.P. (1992) Research and clinical criteria for development of neurobehavioral test batteries, *J. Occup. Med.*, 34(2):140–148.

White, R.F. and Proctor, S.P. (1997) Solvents and neurotoxicity, *Lancet*, 349:1239–1243.

Whitehorn, J.C. (1959) Goals of psychotherapy, in Rubinstein, E.A. and Parloff, M.B. (Eds.), *Research in Psychotherapy*, Washington, D.C.: American Psychological Association, pp. 1–9.

Witorsch, P. and Schwartz, S.L. (1994) Conditions with an uncertain relationship to air pollution: Sick building syndrome, multiple chemical sensitivities, and chronic fatigue syndrome, in Witorsch, P. and Spagnolo, S.V. (Eds.), *Air Pollution and Lung Disease in Adults*, Boca Raton, FL: CRC Press, pp. 285–300.

Witorsch, P., Ayesu, K., Balter, N.J., and Schwartz, S.L. (1995) Multiple Chemical Sensitivity: Clinical Features and Causal Analysis in 61 Cases, presented at the North American Congress of Clinical Toxicology Annual Meeting, Rochester, NY, September 17.

Wolf, S. (1959) The pharmacology of placebos, *Pharmacol. Rev.*, 11:689–704.

Wolfe, F., Cathey, M.A., and Hawley, D.J. (1994) A double-blind placebo-controlled trial of fluoxetine in fibromyalgia, *Scand. J. Rheumatol.*, 23:255–259.

Wolfe, J. and Charney, D.S. (1991) Use of neuropsychological assessment in posttraumatic stress disorder, *Psychol. Assess.*, 3(4):573–580.

Wolfe, J., Brown, P.J., and Bucsela, M.L. (1992) Symptom responses of female Vietnam veterans to Operation Desert Storm, *Am. J. Psychiatry*, 149(5):676–679.

Wolff, H.G. (1950) Life stress and bodily disease — a formulation: the nature of stress in man, in *Proceedings of the Association for Research in Nervous and Mental Disease*, Baltimore, MD: Williams & Wilkins, pp. 1059–1094; also in Wolff, H.G., Wolf, Jr., S., and Hare, C.E., Eds. (1950) *Life Stress and Bodily Disease*, Baltimore, MD: Williams & Wilkins.

Wolpe, J. (1958) *Psychotherapy by Reciprocal Inhibition*, Stanford, CA: Stanford University Press.

Wolpe, J. and Rowan, V.C. (1988) Panic disorder: a product of classical conditioning, *Behav. Res. Ther.*, 26(6):441–450.

Wolraich, M., Milich, R., Stumbo, P., and Schultz, F. (1985) Effects of sucrose ingestion on the behavior of hyperactive boys, *J. Pediatr.*, 106:675–682.

Wolraich, M.L., Lindgren, S.D., Stumbo, P.J., Stegink, L.D., Appelbaum, M.I., and Kiritsy, M.C. (1994) Effects of diets high in sucrose or aspartame on the behavior and cognitive performance of children, *New Engl. J. Med.*, 330(5):301–307.

Wong, N. (1989) Theories of personality and psychopathology: classical psychoanalysis, in Kaplan, H.I. and Sadock, B.J. (Eds.), *Comprehensive Textbook of Psychiatry*, Vol. I, 5th ed., Baltimore, MD: Williams & Wilkins, pp. 356–403.

World Health Organization (1992) The Tenth Revision of the International Classification of Diseases and Related Health Problems (ICD-10), Geneva: WHO.

Wurtman, R.J. (1983) Neurochemical changes following high-dose aspartame with dietary carbohydrates [correspondence], *New Engl. J. Med.*, 309:429–430.

Wyatt, G.E. and Peters, S.D. (1986) Issues in the definition of child abuse in prevalence research, *Child Abuse Neglect*, 10:231–40.

Yehuda, R. and Antelman, S.M. (1993) Criteria for rationally evaluating animal models of posttraumatic stress disorder, *Biol. Psychiatry*, 33:479–486.

Yehuda, R., Southwick, S.M., Nussbaum, G., Wahby, V., Giller, E.L., and Mason, J.W. (1990) Low urinary cortisol excretion in patients with posttraumatic stress disorder, *J. Nerv. Ment. Dis.*, 178:366–369.

Yehuda, R., Lowry, M.T., Southwick, M.D., Shaffer, D., and Giller, Jr., E.L. (1991) Lymphocyte glucocorticoid receptor number in posttraumatic stress disorder, *Am. J. Psychiatry*, 148:499–504.

Yehuda, R., Southwick, S., Giller, E.L., Ma, X., and Mason, J.W. (1992) Urinary catecholamine excretion and severity of PTSD symptoms in Vietnam combat veterans, *J. Nerv. Ment. Dis.*, 180:321–325.

Yehuda, R., Resnick, H., Kahana, B., and Giller, Jr., E.L. (1993) Long-lasting hormonal alterations in extreme stress in humans: normative or maladaptive?, *Psychosomatic Med.*, 55:287–297.

Yehuda, R., Boisoneau, D., Lowy, M.T., and Giller, Jr., E.L. (1995a) Dose-response changes in plasma cortisol and lymphocyte glucocorticoid receptors following dexamethasone administration in combat veterans with and without posttraumatic stress disorder, *Arch. Gen. Psychiatry*, 52:583–593.

Yehuda, R., Kahana, B., Binder-Brynes, K., Southwick, S.M., Mason, J.W., and Giller, E.L. (1995b) Low urinary cortisol excretion in holocaust survivors with posttraumatic stress disorder, *Am. J. Psychiatry*, 152(7):982–986.

Yehuda, R., Keefe, R.S.E., Harvey, P.D., Levengood, R.A., Gerber, D.K., Geni, J., and Siever, L.J. (1995c) Learning and memory in combat veterans with posttraumatic stress disorder, *Am. J. Psychiatry*, 152:137–139.

Yerkes, R.M. and Dodson, J.D. (1908) The relation of strength of stimulus to rapidity of habit formation, *J. Comp. Neurol. Psychol.*, 18:459–482.

Young, E., Patel, S., Stoneham, M., Rona, R., and Wilkinson, J.D. (1987) The prevalence of reaction to food additives in a survey population, *J. Roy. Coll. Physicians Lond.*, 21(4):241–247.

Zelazo, P.D., Carter, A., Reznick, J.S., and Frye, D. (1997) Early development of executive function, *Rev. Gen. Psychol.*, 1(2):198–226.

Ziem, G.E. and Davidoff, L.L. (1992) Illness from chemical "odors": is the health significance understood? [editorial], *Arch. Environ. Health*, 47(1):88–91.

Zola, I.K. (1966) Culture and symptoms — an analysis of patients' presenting complaints, *Am. Sociol. Rev.*, 31:615–630.

Zola-Morgan, S.M. and Squire, L.R. (1990) The primate hippocampal formation: evidence for a time-limited role in memory storage, *Science*, 250:288–290.

Zornetzer, S.F. (1978) Neurotransmitter modulation and memory: a new neuropharmacological phrenology, in Lipton, M.S., DiMascio, A., and Killiam, K.F. (Eds.), *Psychopharmacology: A Generation of Progress*, New York: Raven Press, pp. 637–649.

Zornetzer, S.F. and Gold, M. (1976) The locus coeruleus: its possible role in memory consolidation, *Physiol. Behav.*, 16:331–336.

Index